Transport Science and Technology

Transport Science and Technology

Edited by

KONSTADINOS G. GOULIAS
Department of Geography
University of California
Santa Barbara,
United States

ELSEVIER

Amsterdam • Boston • Heidelberg • London • New York • Oxford
Paris • San Diego • San Francisco • Singapore • Sydney • Tokyo

Elsevier
The Boulevard, Langford Lane, Kidlington, Oxford OX5 1GB, UK
Radarweg 29, PO Box 211, 1000 AE Amsterdam, The Netherlands

First edition 2007

British Library Cataloguing in Publication Data
A catalogue record for this book is available from the British Library

Library of Congress Cataloging-in-Publication Data
A catalog record for this book is available from the Library of Congress

ISBN-13: 978-0-08-044707-0
ISBN-10: 0-08-044707-4

For information on all Elsevier publications
visit our website at books.elsevier.com

Printed and bound in The Netherlands

07 08 09 10 11   10 9 8 7 6 5 4 3 2 1

CONTENTS

Logistics, Supply Chains, and Intermodal Systems

PREFACE

Transportation evolved in the past 30 years as a self standing field of research and education. This book and the conference from which we extracted a sample of presented papers showcase this unique nature of transportation practice and research. The conference Transportation Science and Technology Congress (TRANSTEC) ATHENS 2004 was held in September 1-5, 2004, at the landmark Athens Hilton Hotel, following the Athens 2004 Olympic Games. Dedicated to the truly Olympic achievements of our transportation profession, this book illustrates creativity and innovation in science and technology.

TRANSTEC's objectives were to assemble a wide range of case studies, motivate collaborations among professionals that do not usually meet in other venues, identify themes and methods that are shared by different specialties, and gather specialists to celebrate the science and technology excellence creating a unique forum for exchange of ideas across the entire spectrum of the transportation industry. There were 85 presentations and workshops from 24 countries and 150 attendees.

There are five groups of chapters in this book that start with a selection of review contributions describing the state of the art in simulation, capacity and traffic operations, soft computing for modeling and simulation, and innovations in transportation materials research. This is followed by a section dedicated to the host country illustrating the context within which the Olympic Games were planned and delivered, the solutions to transportation problems and impressive technologies employed leading to the most successful Olympic Games. The remaining three sections take us to an exciting trip around the world showcasing first the importance of human-centered designs in the section on "systemic and systematic approaches to human performance and behavior." Then, as a reflection of today's information era a group of chapters shows the pioneering science and technology role of transport systems in the section on "information systems, communication, management, and control." The final section offers a rich set of complex solutions on "logistics, supply chains, and intermodal systems" and includes contributions from maritime transportation.

There is no doubt the TRANSTEC success is due to the persons that worked hard to make it a reality. First of all, the impeccable conference organization characterized by a uniquely Hellenic hospitality and warmth is due to the *ALVIA DMC* and the creative genius of Nassos Stevenson. Inspirational guidance and support also came from Konstantinos Zekkos of *"DROMOS" Consulting Ltd* who also ensured the attendance of many of our local colleagues. The warm welcome and insights about the transportation and tourism relationship of the Deputy Minister of Tourism Anastasios Liaskos set the stage for a memorable experience.

As always a successful meeting is due to its scientific committees. TRANSTEC was fortunate to have Chandra Bhat (University of Texas), Lily Elefteriadou (University of Florida), George A. Giannopoulos (Aristotle University of Thessaloniki), Dimitrios G. Goulias (University of Maryland), Loukas Kalisperis (The Pennsylvania State

University), Assad Khattak (University of North Carolina), Ryuichi Kitamura (Kyoto University), Hani S. Mahmassani (University of Maryland), John M. Mason, Jr. (The Pennsylvania State University), Mirko Novak (Czech Technical University of Prague), Ram M. Pendyala (Arizona State University), Konstantinos M. Zekkos ("DROMOS" Consulting Ltd), and Athanassios K. Ziliakopoulos (University of Thessaly).

As always the ELSEVIER staff excellence in producing high quality publications is clearly demonstrated in this document.

Many thanks to all new and old friends that made TRANSTEC and this book possible.

Konstadinos G. Goulias
University of California Santa Barbara

The conference participants at the Athens Hilton conference site overlooking the Acropolis

Transport Science and Technology
K.G. Goulias, editor

CHAPTER 1

Introduction to science, technology, and transport systems

Konstadinos G. Goulias, University of California Santa Barbara

INTRODUCTION

The need to provide for safe, reliable, and efficient movement of people and goods is held as the core mission of transportation systems and services. In pursuing this mission, one can identify at least five influential themes that are embedded in the creation and operation of transportation systems. These themes are: a) behaviour, b) design, c) performance, d) technology, and e) chance.

Behaviour is of paramount importance from two perspectives: the travellers and the operators. It includes our everyday behaviour but it is also based on our values, perceptions, intentions, attitudes, and the exchange of material and nonmaterial entities. In-depth understanding of this human nature is essential to the planning, design, and operational analysis of transportation systems. In fact, transportation specialties are interested in these aspects and research on this is extremely active, accelerating in pace during the past few years, and aiming not only to understand but to also predict human behavior. The second theme, *design*, contains the traditional component of engineering system design that is now at a very mature state. There is also another component that is emerging as an exciting new field, which is a human centered design of systems and services and incorporates ideas from field studies and disciplines that are not traditionally associated with typical transportation studies. *Performance*, the third theme, is of paramount importance again from two perspectives: the human performance, particularly in human machine interaction, and the system and its components' performance. This area spans a wide spectrum from the materials used to build the systems to the operations of an entire system itself. Unavoidably *technology* provides the tools used to make the systems and services work with prominent place taken today by information and telecommunication technologies. This is particularly present in this book because of the important role technology played in the recent years but also because this is the area where we see galloping advances. *Chance* is a theme that is becoming increasingly prominent in the science and engineering of transportation systems and services and it is increasingly used to account for our inability to control all the factors impacting our systems. We also realize that we do not have sufficient knowledge about the world surrounding us and therefore our inventions need to take this into account. Most important, however, chance is also used to account for the rich nature and unpredictability of human behavior in the interactions with other humans, the anthropogenic world, and nature itself.

All five themes interact and influence each other very often in ways that we need to understand and they are all reflected in each chapter of this book. As the chapters in this book show, interaction among these themes is a process that spans the wide spectrum of transport science and technology. The process itself is unique to transportation and in addition to transferring methods from other disciplines to engineering system design (interdisciplinary approach), it is also emerging as a scientific and technological field with its own principles, methods, and techniques much like medicine. This book is a small contribution to this emerging transdisciplinary nature of transport science and technology.

BOOK ORGANIZATION

Instead of offering a comprehensive review of transport science and technology we selected a sample of a few interesting aspects that demonstrate the synergy of the five themes discussed above. Selection of the chapters was from the more than 80 presentations given at the Transport Science and Technology Congress (TRANSTEC) in Athens in September, 2004. Emphasis is given to a balanced representation of the five themes above but also representation from the different schools of thought around the world and the variety of specific missions in transportation research and practice.

This book is divided into five sections. We start with a state-of-the-art section in which there are four overview chapters on selected aspects. The second chapter by Lily Elefteriadou on "Highway capacity analysis in the U.S: state of the art and future directions" offers an informative review of the most popular manual/handbook for traffic operators and the determination of highway capacity. Then, Aruna Sivakumar and Chandra Bhat in the third chapter give us another state of the art review of emerging simulation methods that are increasingly employed by many transportation planning models and they are very useful for other applications as well. The fourth chapter by Ondrej Pribyl is a summary of a training workshop at TRANSTEC and shows how techniques of soft computing developed in other fields found their home in traffic operations and transportation planning. This section closes with a chapter by Dimitrios Goulias illustrating technological and methodological advances in materials research. One common thread among all chapters is the extensive use of probability and statistics that by now is no longer an innovation but a tool of the transportation trade.

In the second section of the book and reflecting the TRANSTEC venue and the excitement of the Athens Olympics we dedicate the chapters to the Hellenic transportation systems. The first chapter by George Giannopoulos offers a comprehensive review of transportation in Greece within the broader context of the European Union. In the second chapter we find the blueprint of planning for the Athens 2004 Olympic games written by John Frantzeskakis. Undoubtedly the phenomenal success and exemplary organization of these Olympics is partially to be attributed to blueprints of this type. Liza Panagiotopoulou in Chapter 8 describes a typical example of advanced technology at the service of transportation management and operations, which is the "eye in sky" employed and tested during the Athens Olympics. This section concludes with another major technology application in the Northern Greek provinces along the ancient Egnatia Odos that is today a high speed freeway. Konstantinos Koutsoukos and Lefteris Koutras provide a complete description of the technical and institutional issues and the solutions and technologies employed.

Human nature is examined in depth within the third section of the book on human performance and behaviour. The section starts with Mirko Novak's overview on attention decreases setting the stage for potential solutions to one of the most severe problems of transportation today (i.e., the large amount of fatalities and injuries travellers suffer every year worldwide). Then, in Chapter 11 Tomas Brandesky reviews some key ideas of creative human reasoning and offers a proposal for microsopic simulation to mimic human reasoning. Chapter 12 is the third contribution of the Czech Technical University in this section in which Zdeněk Votruba, Mirko Novák, and Jaroslav Veselý argue convincingly that uncertainty should be taken into account in the design of man-machine interfaces to study reliability. The other chapters in this section switch gear to the study of behaviour. Chapter 13 is a unique contribution in which Goran Vuk and Tine Lund Jensen demonstrate differences between model predictions and observed changes using a before-after methodology for the newly completed Copenhagen Metro. In the same spirit of developing new methods Pat Burnett in Chapter 14 argues that qualitative research methods have a place in the toolbox of planners and engineers and should be given a more careful consideration. In Chapter 15 Tor Vorraa shows how one commercial software represents toll systems and how the software can be used to produce impact scenarios. This section ends with a chapter written by Natalija Jolić and Zvonko Kavran on simulation modeling for transportation, land use and development decisions in which decision making and behaviour are integrated to form a comprehensive planning process.

In the fourth section of this book nine chapters are dedicated to information systems, communication, management, and control. First, we find three papers that illustrate the complex relationship among time use, technology ownership, telecommunications, and travel behaviour. The first paper by Kuniaki Sasaki, Kazuo Nishii, Ryuichi Kitamura, and Katsunao Kondo, offers new evidence on the intra-household relationship between telecommunication, activity participation, and travel using survey data. Kim and Goulias in the following two papers analyze the determinants of telecommunication technology ownership in Chapter 18 and then in Chapter 19 show that mode choice is influenced by telecommunications technologies in complex ways. Studying information technology and its impact on transportation systems also requires examining the behaviour of other agents such as commercial operators. In Chapter 20 Vladimir Momčilović, Vladimir Papić, Olivera Medar and Aleksandar Manojlović demonstrate how a decision support system can change the behavior of commercial operators to achieve lower energy consumption and fuel emissions. In the second portion of this section a group of chapters is dedicated to information systems in the context of traffic operations and control. Hartwig Hetzheim and Wolfram Tuchscheerer describe applications of mathematical methods to video camera image processing for traffic operations. Then, Mashrur Chowdhury and K.-C. Wang, discuss the design and simulation of a new intelligent traffic sensor network that provides real time information for management and control. The wide availability of traffic management center data motivates the Discrete Time Markov Chain approach to estimate Origin-Destination-specific travel times developed by Jiyoun Yeon and Lily Elefteriadou and described in Chapter 23. Turning to intersection traffic control and the effectiveness of countdown signals, Scott Washburn, Deborah Leistner, and Byungkon Ko, in Chapter 24 find them having a positive influence on pedestrian crossing behaviour. This section closes with another chapter in the area of operations and safety. In Chapter 25 Konstantina Gkritza, John Collura, Samuel C. Tignor, and Dusan Teodorovic, provide a comprehensive report about safety and operations of signal preemption for emergency vehicles.

The final section of this book is dedicated to the evaluation of logistics, supply chains, and intermodal systems. Chapter 26, the opening chapter of this section, written by Evelyn Thomchick, provides a discussion of transportation policy perceptions by government agencies, transportation service providers, and the users of transportation in the context of supply chain management. Then, Katherine Jefferson describes a case study on how performance measurement was used for staff classification and compensation levels, alternative work schedules, equipment procurement proposals and work practices. This is followed by a chapter written by Nikolina Brnjac, Dragan Badanjak, and Vinko Jenić, in which they examine hackepack technology as a rational solution for optimally distributing the movement of goods among modes. Along similar optimality objectives Gaetano Fusco and Maria Pia Valentini, in Chapter 29 describe a procedure for real-time management of an urban logistic centre using a dynamic vehicle routing formulation. This is followed in Chapter 30 by a hierarchical distribution system to provide logistic services design by Tomasz Ambroziak, Marianna Jacyna, Mariusz Wasiak. In the next chapter Marianna Jacyna illustrates the use of multiobjective optimisation to perform multicriteria evaluation of a network.

The last chapters of this book are dedicated to an area that Greece plays a worldwide leadership role - maritime transportation. Chapter 32, by Orestis Schinas and Harilaos Psaraftis, continues along the thinking of multicriteria evaluations to offer an overall evaluation of the Greek coastal shipping companies with emphasis on the needs of lenders and investors. This is followed by another chapter, Chapter 33, by the same two authors that describes a new evaluation tool for port management. In the following chapter, Dimitris Lyridis, Pavlos Zacharioudakis, and Dimitris Chatzovoulos describe a new index of the tanker market that is more comprehensive and reliable than existing and widely used indices. Chapter 35, by Maria Boile, Sotiris Theofanis, and Lazar Spasovic, shows how the port authority investment problem can be mathematically formulated and solved while considering interactions among all players, the direct impact of ports, and the inland transportation system. The book concludes with Chapter 36 that illustrates through a case study the substantial infrastructure development required to use maritime facilities as floating accommodations during special events. The case study is the Port of Piraeus and its modification for the Athens 2004 Olympics offering many lessons that can be used in other contexts and circumstances.

Transport Science and Technology
K.G. Goulias, editor

CHAPTER 2

HIGHWAY CAPACITY ANALYSIS IN THE U.S: STATE OF THE ART AND FUTURE DIRECTIONS

Lily Elefteriadou, Ph.D., University of Florida

INTRODUCTION

The U.S. Highway Capacity Manual (HCM) has been providing techniques for estimating capacity and assessing the traffic operational quality of transportation facilities for over 50 years. The first HCM was published in 1950, and it was the first document to quantify the concept of capacity for transportation facilities. The second edition of the HCM appeared in 1965; it defined the concept of level of service (LOS) and also distinguished between operational, design and planning analyses. Subsequent editions and updates appeared in 1985, 1994, and 1997, while the most recent edition was published in 2000.

The development of the HCM is guided by the Transportation Research Board's (TRB) committee on Highway Capacity and Quality of Service (HCQS), which is responsible for approving its contents. The HCQS committee currently consists of 33 members, and includes researchers, government agency representatives, and private consultants. It is organized into several subcommittees, each one responsible for designated sections of the HCM. Research pertaining to highway capacity analysis is funded by various agencies (such as the National Cooperative Highway Research Program – NCHRP, the Federal Highway Administration – FHWA, State Departments of Transportation, and others), and is regularly reviewed by the appropriate subcommittee(s). These later deliberate and vote on whether research results should be incorporated into the HCM. If the response is positive, the research and recommended changes to the HCM, are submitted to the HCQS committee for its consideration and vote. The HCQS committee meets twice a year: in January during the Annual TRB meeting in Washington DC, and sometime during the summer. Information on the membership, meetings and activities of the committee can be obtained at: www.ahb40.org.

The most recent edition of the HCM (2000) includes some significant changes and additions from previous versions. New methodologies have been developed for various facility types, and new substantive material has been added to the HCM. The objectives of this paper are to:

- Highlight changes and additions in the latest edition of the HCM;
- Identify shortcomings of the HCM 2000, particularly with respect to freeway analyses;
- Discuss topics under consideration by the HCQS committee for future HCM editions; and
- Propose recommended research directions for highway capacity and quality of service analyses.

The next part of the paper discusses the HCM 2000 in general, and outlines its contents, while the third part presents some of its shortcomings. The fourth part focuses on freeway analysis methods, including changes in the HCM 2000, and shortcomings of the existing methods. The fifth part provides recommendations for future research on highway capacity and quality of service issues, while the last part contains a summary of the paper along with final conclusions and recommendations.

THE HCM 2000 CONTENTS

The HCM 2000 has been significantly expanded compared to its previous editions, and is now organized into the following five parts:
Part I - Overview (Chapters 1 - 6)
Part II - Concepts (Chapters 7-14)
Part III - Methodologies (Chapters 15-27)
Part IV - Corridor and Areawide Analyses (Chapters 28-30)
Part V - Simulation and Other Models (Chapter 31)

Both metric and English unit versions have been published. The HCM 2000 is also available on CD-ROM, with the option to be able to link directly to software that implements the HCM procedures. Several vendors have developed computer programs, however the HCQS committee does not evaluate or ensure the quality of any software packages. Highlights for each of the five parts are provided in the following paragraphs.

Part I – Overview

The first part of the manual includes an overview of the HCM (Chapter 1), presents the basic concepts of capacity and quality of service (Chapter 2), discusses the HCM applications (Chapter 3) and use of its results (Chapter 4), and includes a significantly expanded glossary of terms (Chapter 5) and list of symbols used throughout the manual (Chapter 6).

Part II – Concepts

The second part introduces the basic concepts for the facility types defined in the HCM. Chapter 7 contains information on traffic flow parameters, and Chapter 8 discusses traffic characteristics. The information in both these chapters is similar to the content of the introductory chapters in the previous edition of the HCM (1997). An overview of the analytical procedures provided in the manual, and guidance on their application is provided in Chapter 9. A new section on accuracy and precision is also included in this chapter, alerting the user on the limitations of the accuracy and precision of the methods included in the manual, however no specific estimates or confidence intervals are given for any of the HCM models.

Chapters 10 through 14 include general concepts for the following five facility types: urban streets, pedestrian and bicycle facilities, highways, freeways, and transit. These chapters also include recommended reasonable approximations for input parameters to each of the respective methodologies, intended for use in planning applications, when the user has very little information regarding the facility being analyzed.

Part III – Methodologies

Chapters 15 through 27 in Part III contain the methodologies for analyzing 13 facility types: urban streets (previously titled arterial streets), signalized intersections, unsignalized intersections, pedestrians, bicycles, two-lane highways, multilane highways, freeway facilities, basic freeway segments, freeway weaving, ramps and ramp-junctions, interchange ramp terminals, and transit.

New methodologies are presented for bicycle facilities, two-lane highways, freeway facilities, and transit. In the signalized intersections chapter, a new methodology is added to estimate the back-of-queue, and new saturation adjustment factors are provided for pedestrian and bicycle effects. For unsignalized intersections, a new 95th percentile queue estimation method is included. The chapter on interchange ramp terminals is new, however it is only conceptual and does not provide an analytical methodology. There is a recently completed project (NCHRP 3-60) for updating this chapter, and the committee has voted to publish a circular with the new procedures, which is expected to be released in 2006. The chapter on pedestrians includes the consideration of additional pedestrian facilities, previously not addressed.

For the freeway and multilane highways chapters, new passenger car equivalency (PCE) values are provided. The freeway weaving chapter includes capacity values for a variety of weaving segments, based on weaving segment type, free flow speed, length of weaving segment, and weaving ratio. In the ramps and ramp-junctions chapter, a new set of speed prediction models are provided, for considering the traffic operational quality of the entire cross section, rather than only the two rightmost lanes.

Part IV – Corridor and Area-wide Analyses

Part IV is new in this edition of the manual and includes methods for conducting corridor and area-wide analyses. Chapter 28 presents the analysis framework, and discusses system performance measures, and measurements of traveller perceptions. Chapter 29 provides information on combining individual facility, segment and point analyses into estimates of overall corridor performance measures, particularly for use in planning studies. Chapter 30 contains guidance on how to use the HCM analysis procedures for area-wide analyses, particularly those involving regional travel demand forecasting models and long-range transportation plans.

Part V – Simulation and Other Models

The material included in Part V (Chapter 31) of the manual is also new in this edition, and it contains information on simulation and other models. It presents information on a variety of alternative models that can be used for traffic operational analysis, and it gives guidance on the selection of such models and on procedures for their application. The material in this chapter was based on a paper by Elefteriadou et al. (1999).

SHORTCOMINGS OF THE HCM 2000

The following is a (subjective) list of shortcomings of the HCM in general:

- Measures of effectiveness and performance measures should be emphasized, rather than the qualitative LOS designations A through F.
- The existing methodologies should be extensively validated with field data from around the US.
- Greater emphasis should be placed on at-capacity and oversaturated conditions.
- Uncertainty and variability in the HCM models and analysis procedures should be considered.
- The HCM should provide more detailed guidance on the use of simulation and other models for highway capacity analysis.
- User perceptions should be considered in traffic operational analyses.
- Traffic analysis from a systems perspective should be facilitated.

The first item is still being debated within the HCQS committee, and it is not certain yet what direction the next HCM version may take. The second item is now being considered by the HCQS committee, and several approaches to solving this problem are being discussed. The last five items are discussed in greater length in the fifh part of the paper.

HIGHLIGHTS AND SHORTCOMINGS OF FREEWAY ANALYSIS METHODS

Freeway analysis includes one freeway systems chapter (freeway facilities), and three freeway segment chapters (basic freeway segments, freeway weaving, and ramp junctions). The freeway facilities methodology is new, and it relies on the results of segment analyses. In addition, there are new passenger car equivalency (PCE) values that apply to all freeway-related chapters, which were developed using simulation, and are based on equivalent density. These new PCE values are in general lower than the old ones, reflecting the improvements in heavy vehicle performance over the past several years. The driver population factor is still included in the methodology. The user is prompted to obtain and use local data, however no guidance is given on what data to collect, when and how. The following paragraphs discuss the changes, additions and shortcomings in each of the four freeway-related chapters.

Highlights and Shortcomings - Freeway Facilities

A freeway facility is defined as being composed of three types of segments: basic freeway segments, weaving segments and ramp junctions. The methodology integrates the methods of analysis for these three types of segments, and provides performance measures in space and time. The methodology can consider freeway facilities of up to 12 miles, and can analyze oversaturated conditions, provided that the first and the last analysis intervals are undersaturated. The methodology is based on shockwave analysis to handle queue backup. When demand exceeds capacity during a given interval, the excess demand is transferred to the next interval. A four step-process is used for each bottleneck encountered, following a pre-specified sequence for analyzing cells. The user can obtain estimates of speed, travel time, density, flow rate, v/c ratio, and congestion status for each cell in the time-space domain. These can be aggregated as the user desires, over the length of the facility, the analysis time, and also the entire time-space domain. Guidance is provided in the chapter on adjustments to segment capacity due to construction, weather, and accidents.

The methodology does not consider multiple overlapping bottlenecks, and the user is referred to Part V – Simulation, for addressing such cases. Another issue that should be addressed in future editions of the HCM is that there are no clear rules to differentiate between some types of segments. For example, the HCM does not provide any guidance on what should the maximum length of a weaving segment be. When a weaving segment is very long it may be more appropriate to consider it as a lane addition followed by a lane drop.

Highlights and Shortcomings - Freeway Weaving

The analysis of freeway weaving segments is still based on the old methodology (since the 1970s) and data. The most important changes to this chapter include the calculation of density for weaving segments, and the provision of capacity estimates. These capacity estimates are based on the assumption that capacity is reached when density is 27 pc/km/ln

(43 pc/mi/ln). Capacities are provided as a function of weaving type (A, B or C), free flow speed, volume ratio, length of the weaving segment, and number of lanes.

For this chapter, a rigorous methodology is needed, to at least include validation and adjustment of methods with more recent data and/or simulation, and a capacity estimation method. This later should consider weaving demands in terms of origin-destinations, rather than weaving ratios. The next edition of the HCM should expand the methodology to address all possible weaving configurations (i.e., weaving on urban streets, major weave sections, Type C weaves with left side off-ramps). A new NCHRP project is currently under way to address some of these issues, and it is expected to be completed by 2008.

Highlights and Shortcomings - Freeway Ramp Junctions

This chapter provides the methodology for analyzing merge and diverge areas, and has not changed significantly from the previous HCM edition. The most significant change is the addition of speed prediction models for the lanes outside the influence area (which only includes the two right-most lanes). In addition, discussion on the capacity of merge and diverge areas is now included in the chapter.

The next edition of the HCM should consider a new capacity estimation procedure for ramp junctions considering the probability of breakdown. Also, rigorous analysis methods for oversaturated conditions should be provided. Analysis procedures for ramp metering considering its impacts on capacity and quality of service should be added. Lastly, improved analysis of multi-lane on- and off- ramps should be developed and added in the chapter.

FUTURE DIRECTIONS IN HIGHWAY CAPACITY ANALYSIS

Five general issues have been identified as research topics that should be given high priority in next editions and updates of the HCM:

- Definition of capacity that considers research on probabilistic breakdown occurrence;
- Guidance on uncertainty and variability in the HCM models;
- Guidance on simulation model usage for highway capacity analysis;
- Level of Service (LOS) designations based on user perception;
- Highway capacity analysis tools that consider a systems perspective.

Each of these five issues is discussed in some detail in the following paragraphs.

Definition Of Capacity

The term "capacity" has been used to quantify the traffic-carrying ability of transportation facilities. The definition of capacity and its numerical value have evolved over time. The current published version of the *Highway Capacity Manual* (2000) defines the capacity of a facility as "...the maximum hourly rate at which persons or vehicles reasonably can be expected to traverse a point or a uniform section of a lane or roadway during a given time period, under prevailing roadway, traffic and control conditions (HCM, p. 2-2)." For freeway facilities, capacity values are given as 2,250 pcphpl, and up to 2,400 pcphpl, as a function of free flow speed.

Implied in the current definition and understanding of freeway capacity is the notion that the facility will "break down" (transition from an uncongested state to a congested state) when demand exceeds a specified capacity value. Elefteriadou et al (1995) showed however that breakdown does not necessarily occur always at the same demand levels, but can occur when flows are lower or higher than those traditionally accepted as capacity. Observation of field data showed that, at ramp merge junctions, breakdown may occur at flows lower than the maximum observed, or capacity flows. Furthermore it was observed that, at the same site and for the same ramp and freeway flows, breakdown may or may not occur. In a subsequent paper, Lorenz and Elefteriadou (2001) conducted an extensive analysis of speed and flow data collected at two freeway-bottleneck-locations in Toronto, Canada. The authors confirmed that the existing freeway capacity definition does not accurately reflect the relationship between breakdown and flow rate, and indicated that freeway capacity may be more adequately described by incorporating a probability of breakdown component in the definition. Given these findings, it is important for the HCM to reconsider the existing capacity definition. Also, from the point of view of the operator, if the sources of variability are known, throughput can be maximized by minimizing the probability of breakdown. Even postponing breakdown using ramp-metering and other similar techniques until later in the peak hour, would have beneficial results in the traffic operational quality.

Guidance on Uncertainty and Variability

The HCM 2000 briefly discusses accuracy and precision, but the methods presented do not consider the uncertainty associated with inputs, and models, nor their associated impacts. In general, uncertainty may occur because of incomplete information, because of simplifications and approximations, and regarding the form and structure of models (Morgan and Henrion, 1990). Sources of uncertainty include random error and statistical variation, systematic error and subjective judgment, linguistic imprecision, variability over time and space, randomness and unpredictability and approximations.

In highway capacity analysis, researchers have recently begun exploring this topic. Roess (2001) reported that for uninterrupted flow facilities a 10% error in inputs and model parameters yields:

For Basic Freeway Sections:	> 15% error in density
For Weaving Sections:	> 33% error in density
For Ramp Junctions:	> 29%-43% error in density

Kyte and Zahib (2001) found that uncertainty is highest at high v/c when the need for precision is highest. They also stated that the largest source of uncertainty is demand volume, and concluded that users need to understand uncertainty in results, and the error propagation process. They stated that field measurements have uncertainty, and recommended that there is a need to eliminate factors with little effect on final results, and also to validate methods with field data.

As shown, the implications of ignoring uncertainty are enormous, resulting in great errors when planning, designing and operating transportation facilities. Consideration of uncertainty in traffic operational analysis is essential for the following reasons:

- To help in deciding whether to expend resources to acquire additional information;
- To guide the design and refinement of a model and help select the appropriate level of detail for each component; and
- To help when combining uncertain information from different sources, by using appropriate weighting.

Guidance on Simulation Model Usage

Simulation is often used to address issues that cannot be effectively resolved using the HCM. There is an abundance of simulation programs for studying different highway facilities, and these programs differ both in scope and capabilities. The HCM 2000 includes some discussion on the use of simulation and other models for traffic operational analysis and provides brief examples for their application. It doesn't however go as far as naming specific models, making the use of examples and illustrations difficult. The main reason that specific models are not named, is that any mention of a specific commercial model may be perceived as an "endorsement" from the HCQS committee and TRB. In addition, given the frequent changes and upgrades in commercially available software, it is difficult for the HCQS committee to either include a comprehensive list of models, or select and discuss specific models according to given criteria.

The HCM however should provide some guidance: a) on the situations not effectively addressed by the HCM methodologies, and the types of models that would be appropriate for each of these, and b) on how to use and compare performance measures from alternate models, considering their respective definitions in the HCM and in other models. This information should be provided along with each HCM methodology, to assist users and facilitate access to this information.

Level of Service (LOS) Designations Based on User Perception

The issue of whether to include LOS designations within the HCM continues to be discussed within the HCQS committee. The letter designations indicating quality of service have been used widely in a variety of settings (including state and local legislation, and by non-transportation professionals), making it very difficult to eliminate them. On the other hand, the implications of using qualitative LOS are that there are discontinuities at the boundaries, and that transportation professionals in general don't refer nor use the numerical values of the respective measures of performance. Thus, there are two questions that should be considered regarding LOS:

- Is there a rational way to use qualitative LOS designations, so that the arbitrariness of the groupings and the boundaries can be considered?
- Should the user perception of quality of service be considered, and if yes, how?

Regarding the first question, one study has looked at user perception of the quality of service and of delay at a signalized intersection (Pecheaux et al, 2000) to examine the relationship between user perception and LOS. A second study has looked at the possibility of using fuzzy clustering and user perception to obtain LOS designations (Fang et al 2001). Regarding the second question, the existing measures of performance and LOS designations do not take into account user perception of the quality of service. Travellers and users of the transportation system probably perceive their travel time and speed (rather than density or delay), as well as the presence of congestion, and based on these make travelling decisions (time and mode of departure, route, etc.). Thus it would be desirable for traffic operational analyses to be able to predict and provide such measures, which could assist in modelling transportation systems, and eventually in planning and designing transportation facilities and networks.

Highway Capacity Analysis Tools for a Systems Perspective

It is important to be able to consider and conduct highway capacity analysis for highway facility systems, because there are substantive interactions between both sequential and parallel facilities. For sequential facilities, a bottleneck affects adjoining facilities, either by "starving" the demand downstream, or by building up a queue upstream. In the case of parallel facilities, congestion on one facility would change the demand patterns on the other. An important performance measure, travel time, can best be obtained and predicted over the highway system, rather than a point or a segment. Furthermore, given the discussion above on user perception from a transportation network perspective, it is important to be able to assess and analyze the interactions between the quality of service and travel decisions. Understanding this relationship would more closely approximate human behaviour, and would allow transportation professionals to plan, design and operate transportation facilities more effectively.

SUMMARY AND CONCLUSIONS

This paper briefly presented highlights for the HCM 2000 and discussed some of its shortcomings. It also presented recommendations on research that is required to strengthen the HCM. In summary, these recommendations include:

- Clearly define capacity considering breakdown as a probabilistic event.
- Consider uncertainty in the inputs, modeling and outputs of the HCM procedures.
- Provide additional guidance on the use of simulation and other models, including the use of measures of effectiveness as provided by these models.

- Re-evaluate the use of LOS designations, and consider the importance of user perception.
- Emphasize the importance of traffic analysis from a systems perspective, and provide comprehensive procedures that can be effectively used by planners and designers.

REFERENCES

Elefteriadou, L., G. List, J. Leonard, H. Lieu, M. Thomas, R. Giguere, R. Brewish, G. Johnson, *"Beyond the Highway Capacity Manual: A Framework for Selecting Simulation Models in Traffic Operational Analyses"*, Transportation Research Record 1678, National Academy Press, 1999, pp. 96-106.

Elefteriadou, L., R.P. Roess, W.R. McShane, *"The Probabilistic Nature of Breakdown at Freeway - Merge Junctions"*, Transportation Research Record 1484, National Academy Press, 1995, pp. 80-89.

Fang, F.C., L. Elefteriadou, K. Pechaux, M. Pietrucha, *"Using Fuzzy Clustering of User estimated delay to Define Level of Service at Signalized Intersections"*, submitted to the Transportation Research Board, Washington DC, 2002

Highway Capacity Manual 2000, Transportation Research Board, National Research Council, Washington DC., 2000

Kyte and Zahib (2001), *"Uncertainty and Precision for Intersection Analysis"*, presented at the Transportation Research Board Meeting- Session 104, Washington D.C., January 2001

Lorenz, M., L. Elefteriadou, *"Defining Highway Capacity as a Function of the Breakdown Probability"*, forthcoming, Transportation Research Record, and presented at the Transportation Research Board Meeting, Washington DC, January 2001.

May, A.D., *"Traffic Flow Fundamentals"*, Prentice-Hall, 1994

Morgan, M.G. and M. Henrion, *"Uncertainty - A Guide to Dealing with Uncertainty in Quantitative Risk and Policy Analysis"*, Cambridge University Press, New York, 1990

Pecheux, K.K., M.T. Pietrucha, P.P. Jovanis, *"Evaluation of Average Delay as a Measure of Effectiveness for Signalized Intersections"*, *80th TRB Annual Meeting*, Jan. 2001, Washington, D.C.

Roess R. and E. Prassas, *"Accuracy and Precision in Uninterrupted Flow Analysis"*, presented at the Transportation Research Board Meeting- Session 104, Washington D.C., January 2001

Transport Science and Technology
K.G. Goulias, editor

CHAPTER 3

EMERGING SIMULATION-BASED METHODS

Aruna Sivakumar

RAND Europe (Cambridge) Ltd., Westbrook Centre, Milton Road, Cambridge CB4 1YG, United Kingdom, Tel: +44 1223 353 329, Fax: +44 1223 358 845, Email: arunas@rand.org

Chandra R. Bhat

The University of Texas at Austin, Dept. of Civil, Arch. & Environmental Engineering, 1 University Station C1761, Austin TX 78712-0278, USA. Tel: +1 512 232 6272, Fax: +1 512 475 8744, Email: bhat@mail.utexas.edu

INTRODUCTION

The incorporation of behavioral realism in econometric models helps establish the credibility of the models outside the modeling community, and can also lead to superior predictive and policy analysis capabilities. Behavioral realism is incorporated in econometric models of choice through the relaxation of restrictions that impose restrictive behavioral assumptions regarding the underlying choice process. For example, the extensively used multinomial logit (MNL) model has a simple form that is achieved by the imposition of the restrictive assumption of independent and identically distributed error structures (IID). But this assumption also leads to the not-so-intuitive property of independence from irrelevant alternatives (IIA).

The relaxation of behavioral restrictions on choice model structures, in many cases, leads to analytically intractable choice probability expressions, which necessitate the use of numerical integration techniques to evaluate the multidimensional integrals in the probability expressions. The numerical evaluation of such integrals has been the focus of extensive research dating back to the late 1800s, when multidimensional polynomial-based cubature methods were developed as an extension of the one-dimensional numerical quadrature rules (see Press et al., 1992 for a discussion). These quadrature-based methods, however, suffered from the "curse of dimensionality"; and so pseudo-Monte Carlo (PMC) *simulation methods* were proposed in the 1940s to overcome this problem. The PMC simulation approach has an expected integration error of $N^{-0.5}$, which is independent of the number of dimensions 's' and thus provides a great improvement over the quadrature-based methods. Several variance reduction techniques (example, Latin Hypercube Sampling or LHS) have since been developed for the PMC methods, which potentially lead to even more accurate integral evaluation with fewer draws. Despite the improvements achieved by these variance reduction techniques, the convergence rate of PMC methods is generally slow for simulated likelihood estimation of choice models.

Extensive number theory research in the last few decades has led to the development of a more efficient simulation method, the quasi-Monte Carlo (QMC) method. This method uses the basic principle of the PMC method in that it evaluates a multi-dimensional integral by replacing it with an average of the values of the integrand computed at N discrete points. However, rather than using random sequences, QMC methods use low discrepancy, deterministic, quasi-Monte Carlo (or QMC) sequences that are designed to achieve a more even distribution of points in the integration space than the PMC sequences.

Quasi-Monte Carlo Simulation

Over the years, several different quasi-random sequences have been proposed for QMC simulation. Many of these low-discrepancy sequences are linked to the van der Corput sequence, which was originally introduced for dimension s = 1 and base b = 2. Sequences based on the van der Corput sequence are also referred to as the reverse radix-based sequences (such as the Halton sequence). To find the nth term, x_n, of a van der Corput sequence, we first write the unique digit expansion of n in base b as:

$$n = \sum_{j=0}^{\infty} a_j(n) b,^j \text{ where } 0 \le a_j(n) \le b-1 \text{ and } b^J \le n \le b^{J+1} \quad (1)$$

This is a unique expansion of n that has only finitely many non-zero coefficients $a_j(n)$. The next step is to evaluate the radical inverse function in base b, which is defined as

$$\phi_b(n) = \sum_{j=0}^{\infty} a_j(n) b^{-j-1} \quad (2)$$

The van der Corput sequence in base b is then given by $x_n = \phi_b(n)$, for all $n \ge 0$. This idea, that the coefficients of the digit expansion of an increasing integer n in base b can be used to define a one-dimensional low-discrepancy sequence, inspired Halton to create an s-dimensional low-discrepancy Halton sequence by using s van der Corput sequences with relatively prime bases for the different dimensions (Halton, 1970).

An alternative approach to the generation of low-discrepancy sequences is to start with points placed into certain equally sized volumes of the unit cube. These fixed length sequences are referred to as (t,m,s)-nets, and related sequences of indefinite lengths are called (t,s)-sequences. Sobol suggested a multi-dimensional (t,s)-sequence using base 2, which was further developed by Faure who suggested alternate multidimensional (t,s)-sequences with base $b \geq s$. For a detailed description of the (t,s)-sequences, see Niederreiter (1992).

Irrespective of the method of generation, the even distribution of points provided by the low-discrepancy QMC sequences leads to efficient convergence for the QMC method, generally at rates that are higher than the PMC method. In particular, the theoretical upper bound for the integration error in the QMC method is of the order of N-1.

Despite these obvious advantages, the QMC method has two major limitations. First, the deterministic nature of the quasi-random sequences makes it difficult to estimate the error in the QMC simulation procedure (while there are theoretical results to estimate integration error with the QMC sequence, these are much too difficult to compute and are very conservative upper bounds). Second, a common problem with many low-discrepancy sequences is that they exhibit poor properties in higher dimensions. The Halton sequence, for example, suffers from significant correlations between the radical inverse functions for different dimensions, particularly in the larger dimensions. A growing field of research in QMC methods has resulted in the development, and continuous evolution, of efficient randomization strategies (to estimate the error in integral evaluation) and scrambling techniques (to break correlations in higher dimensions).

Research Context

Research on the generation and application of randomized and scrambled QMC sequences clearly indicates the superior accuracy of QMC methods over PMC methods in the evaluation of multidimensional integrals (see Morokoff and Caflisch, 1994, 1995). In particular, the advantages of using QMC simulation for such applications in econometrics as simulated maximum likelihood inference, where parameter estimation entails the approximation of several multidimensional integrals at each iteration of the optimization procedure, should be obvious. However, the first introduction of the QMC method for the simulated maximum likelihood inference of econometric choice models occurred only in 1999, when Bhat proposed and tested Halton sequences for mixed logit estimation and found their use to be vastly superior to random draws. Since Bhat's initial effort, there have been several successful applications of QMC methods for the simulation estimation of flexible discrete choice models, though most of these applications have been based on the Halton sequence (see, for example, Revelt and Train, 2000; Bhat, 2001; Park et al., 2003; Bhat and Gossen, 2004). Number theory, however, abounds in many other kinds of low-discrepancy sequences that have been proven to have better theoretical and empirical convergence properties than the Halton sequence in the estimation of a single multidimensional integral. For instance, Bratley and Fox (1988) show that the Faure and Sobol sequences are superior to the Halton sequence in terms of accuracy and efficiency. There have also been several numerical studies on the simulation estimation of a single multidimensional integral that present significant improvements in the performance of QMC sequences through the use of scrambling techniques (Kocis and Whiten, 1997; Wang and Hickernell, 2000). It is, therefore, of interest to examine the performances of the different QMC sequences and their scrambled versions in the simulation estimation of flexible discrete choice models.

In the following sections, we present the results of a study undertaken to numerically compare the performance of different kinds of low-discrepancy sequences, and their scrambled and randomized versions, in the simulated maximum likelihood estimation of the mixed logit class of discrete choice models. Specifically, we examine the extensively used Halton sequence and a special case of (t.m.s)-nets known as the Faure sequence. The choice of the Faure sequence was motivated by two reasons. First, the generation of the Faure sequence is a fairly straightforward and non-time consuming procedure. Second, it has been proved that the Faure sequence performs better than the more commonly used Halton sequence in the evaluation of a single multi-dimensional integral (Kocis and Whiten, 1997).

The primary objective of the study was to compare the performance of the Halton and Faure sequences against the performance of a stratified random sampling PMC sequence (the LHS sequence) by constructing numerical experiments within a simulated maximum likelihood inference framework[1]. The numerical experiments also included a comparison of scrambled versions of the QMC sequences against their standard versions to examine potential improvements in performance through scrambling. Further, the total number of draws of a QMC sequence required for the estimation of a Mixed Multinomial Logit (MMNL) model can be generated either with or without scrambling across observations (a description of these methods is provided in the following section), and both these approaches were also compared in the study. Another important point to note is that the standard and scrambled versions of the QMC and the LHS sequences are all generated as uniformly-distributed sequences of points. In this study, we also tested and compared the Box-Müller and the Inverse Normal transformation procedures to convert the uniformly-distributed sequences to normally-distributed sequences that are required for the estimation of the random coefficients MMNL model.

The following sections present a brief background on the alternative sequences and methodologies, describe the evaluation framework and experimental design employed in the study, and discuss the computational results. The performances of the various non-scrambled and scrambled QMC sequences in the different test scenarios are evaluated based on their ability to efficiently and accurately recover the true model parameters.

[1] Sandor and Train (2004) perform a comparison of four different kinds of (t,m,s)-nets, the standard Halton, and random-start Halton sequences against simple random draws. They estimate a 5-dimensional mixed logit model using 64 QMC draws per observation, and compare the bias, standard deviation and RMSE associated with the estimated parameters. In the study presented here, we have conducted numerical experiments both in 5 and 10 dimensions in order that the comparisons may capture the effects of dimensionality. For the 5-dimensional mixed logit estimation problem, we also examined the impact of varying number of draws (25, 125 and 625). Finally, we examined the performance of the Faure sequence and the LHS method, along with the Halton sequence, and considered different scrambling variants of these sequences.

BACKGROUND FOR GENERATION OF ALTERNATIVE SEQUENCES

In this section we describe the various procedures used to generate PMC and QMC sequences for the numerical experiments. Specifically, we present the methods for the generation of PMC sequences using the LHS procedure, and the generation of the QMC sequences proposed by Halton and Faure; the scrambling strategies and randomization techniques applied in the study; the generation of sequences with and without scrambling across observations; and basic descriptions of the Box-Müller and Inverse Normal transforms.

PMC Sequences

A typical PMC simulation uses a simple random sampling (SRS) procedure to generate a uniformly-distributed PMC sequence over the integration space. An alternate approach known as Latin Hypercube sampling (LHS), that yields asymptotically lower variance than SRS, is described in the following sub-section.

Latin Hypercube Sampling (LHS)

The LHS method was first proposed as a variance reduction technique (McKay et al., 1979) within the context of PMC sequence-based simulation. The basis of LHS is a full stratification of the integration space, with a random selection inside each stratum. This method of stratified random sampling in multiple dimensions can be easily applied to generate a well-distributed sequence. The LHS technique involves drawing a sample of size N from multiple dimensions such that for each individual dimension the sample is maximally stratified. A sample is said to be maximally stratified when the number of strata equals the sample size N, and when the probability of falling in each of the strata equals N-1.

An LHS sequence of size N in K dimensions is given by

$$\psi_{lhs}^{(N)} = ((p - \xi)/N), \qquad (3)$$

where, $\psi_{lhs}^{(N)}$ is an NxK matrix consisting of N draws of a K-dimensional LHS sequence; p is an NxK matrix consisting of K different random permutations of the numbers 1,...,N; ξ is an NxK matrix of uniformly-distributed random numbers between 0 and 1; and the K permutations in p and the NK uniform variates ξ_{ij} are mutually independent.

In essence, the LHS sequence is obtained by slightly shifting the elements of an SRS sequence, while preserving the ranks (and rank correlations) of these elements, to achieve maximal stratification. For instance, in a 2-dimensional LHS sequence of 6 (N) points, each of the six equal strata in either dimension will contain exactly one point (see Figure 1).

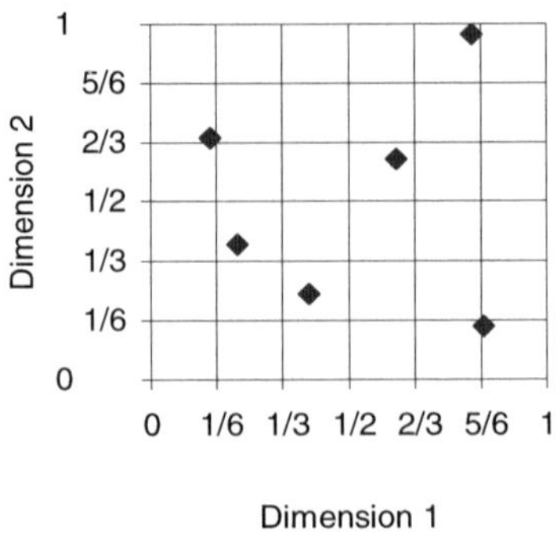

Figure 1. Uniformly distributed LHS sequence in 2 dimensions (N = 6)

QMC Sequences

QMC sequences are essentially deterministic sets of low-discrepancy points that are generated to be evenly distributed over the integration space. The following sub-sections describe the procedures used in the study to generate the standard Halton and Faure sequences.

Halton Sequence

The standard Halton sequence in s dimensions is obtained by pairing s one-dimensional van der Corput sequences based on s pairwise relatively prime integers, $b_1, b_2, ..., b_S$ (usually the first s primes) as discussed earlier. The Halton sequence is based on prime numbers, since the sequence based on a non-prime number will partition the unit space in the same way as each of the primes that contribute to the non-prime number. Thus, the nth multidimensional point of the sequence is as follows:

$$\phi(n) = (\phi_{b_1}(n), \phi_{b_2}(n), ..., \phi_{b_s}(n)) . \quad (4)$$

The standard Halton sequence of length N is finally obtained as

$$\psi_h^{(N)} = [\phi(1)', \phi(2)', ..., \phi(n)']' . \quad (5)$$

The Halton sequence is generated number-theoretically as described above rather than randomly and so successive points of the sequence "know" how to fill in the gaps left by earlier points (see Figure 2), leading to a more even distribution within the domain of integration than the randomly generated LHS sequence.

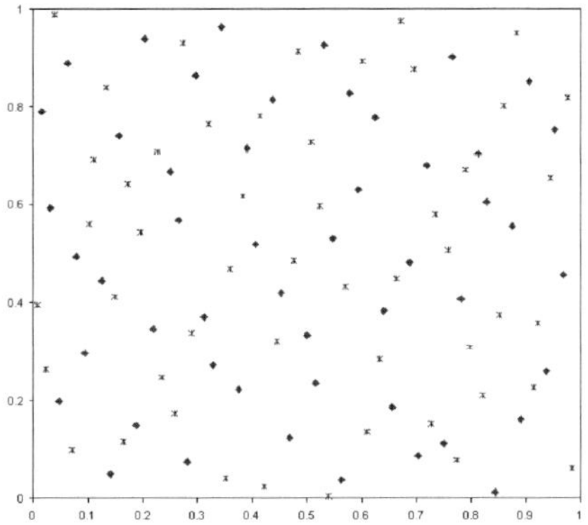

Figure 2. First 100 points of a 2-dimensional Halton sequence

Faure Sequence

The standard Faure sequence is a (t,s)-sequence designed to span the domain of the s-dimensional cube uniformly and efficiently. In one dimension, the generation of the Faure sequence is identical to that of the Halton sequence. In s dimensions, while the Halton sequence simply pairs s one-dimensional sequences generated by the first s primes, the higher dimensions of the Faure sequence are generated recursively from the elements of the lower dimensions. So if b is the smallest prime number such that $b \geq s$ and $b \geq 2$, then the first dimension of the s-dimensional Faure sequence corresponding to n can be obtained by taking the radical inverse of n to the base b:

$$\phi_b^1(n) = \sum_{j=0}^{J} a_j^1(n) b^{-j-1} \tag{6}$$

The remaining dimensions are found recursively. Assuming we know the coefficients $a_j(n)$ corresponding to the first (k-1) dimensions, the coefficients for the kth dimension are generated as follows:

$$a_j^k(n) = \sum_{i \geq j}^{J} {}^iC_j a_i^{k-1}(n) \bmod b, \tag{7}$$

where ${}^iC_j = i!/\, j!(i-j)!$ is the combinatorial function. Thus the next level of coefficients required for the kth element in the s-dimensional sequence is obtained by multiplying the coefficients of the (k-1)th element by an upper triangular matrix C with the following elements.

$$C = \begin{bmatrix} {}^0C_0 & {}^1C_0 & {}^2C_0 & {}^3C_0 & \cdots \\ 0 & {}^1C_1 & {}^2C_1 & {}^3C_1 & \\ 0 & 0 & {}^2C_2 & {}^3C_2 & \\ 0 & 0 & 0 & {}^3C_3 & \\ \vdots & & & & \end{bmatrix}$$

These new coefficients $a_j^k(n)$ are then reflected about the decimal point to obtain the kth element as follows:

$$\phi_b^k(n) = \sum_{j=0}^{J} a_j^k(n) b^{-j-1}, \; 2 \le k \le s \tag{8}$$

This recursive procedure generates the s points corresponding to the integer n in the Faure sequence based on b $(\geq s)$. Thus the nth multidimensional point in the sequence is

$$\phi(n) = (\phi_b^1(n), \phi_b^2(n), \ldots, \phi_b^S(n))$$

The standard Faure sequence of length N is then obtained in the same manner as the standard Halton sequence:

$$\psi_f^{(N)} = [\phi(1)', \phi(2)', \ldots, \phi(n)']' \tag{9}$$

Faure sequences are essentially (t,m,s)-nets in any prime b with $b \geq s$ and $t = 0$. A Faure sequence of b_m points is generated to be evenly distributed over the integration space, such that if we plot the sequence in the integration space together with the elementary intervals of area b_{-m}, exactly one point will fall in each elementary interval (illustration in Figure 3 adapted from Ökten and Eastman, 2004).

Earlier studies have shown that for higher dimensions, the properties of the Faure sequence are poor for small values of n in equation (9) (Fox, 1986). We overcome this in our study by dropping the first 100000 multidimensional points for all the standard and scrambled Faure sequences generated.

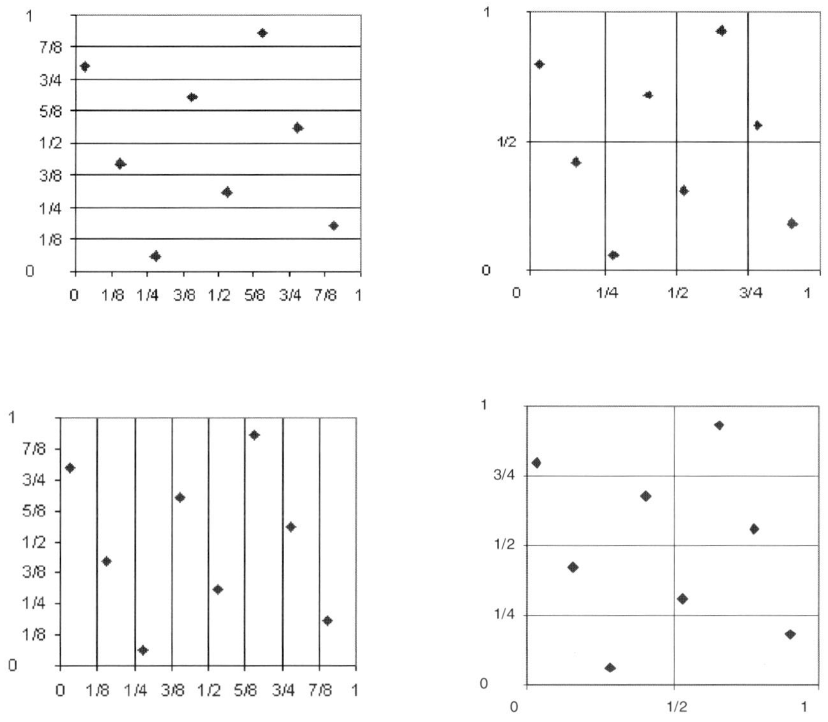

Figure 3. (0,3,2)-net in base 2 with elementary intervals of area 1/8 (Modified from Ökten and Eastman, 2004)

Scrambling Techniques used with QMC Sequences

Research has shown that many QMC sequences have poor properties in higher dimensions, which can be alleviated using suitable scrambling techniques. The standard Halton sequence, for instance, suffers from significant correlations between the radical inverse functions at higher dimensions. For example, the fourteenth dimension (corresponding to the prime number 43) and the fifteenth dimension (corresponding to the prime number 47) consist of 43 and 47 increasing numbers, respectively. This generates a correlation between the fourteenth and fifteenth coordinates of the Halton sequence as illustrated in Figure 4a. The standard Faure sequence, on the other hand, forms distinct patterns in higher dimensions that also cover the unit integration space in diagonal strips, thus showing significantly higher discrepancies in the higher dimensions. Figure 4b illustrates this in a plot of the fifteenth and sixteenth coordinates of the Faure sequence.

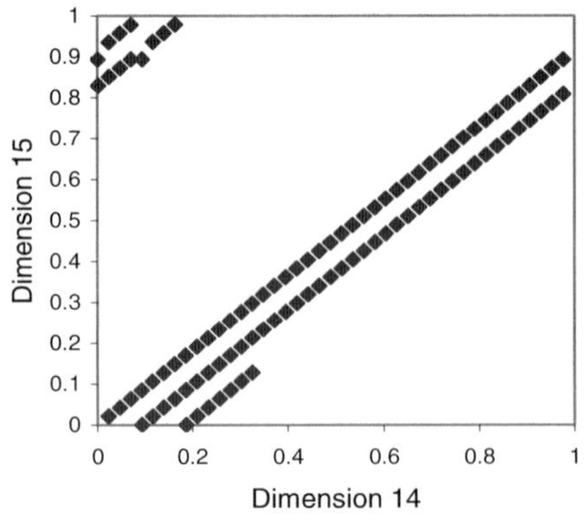

Figure 4a. Standard Halton sequence (Source: Bhat, 2003)

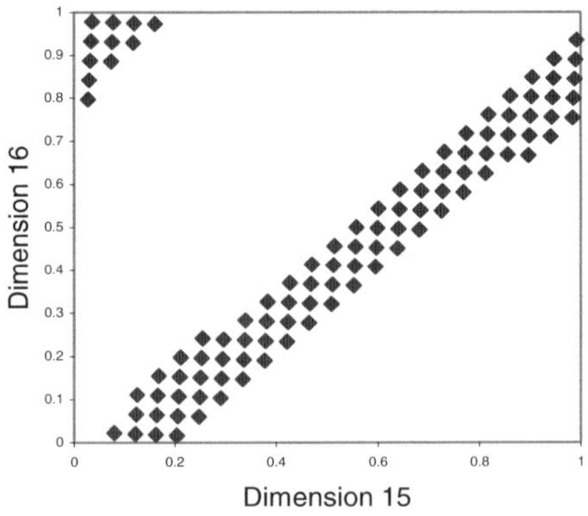

Figure 4b. Standard Faure sequence

Several scrambling techniques have been suggested to redistribute the points and thus improve the uniformity of the QMC sequences in higher dimensions. In this study, we have implemented the Braaten-Weller scrambling for Halton sequences, and the Random Digit and Random Linear scrambling for Faure sequences. Each of these methods is described in greater detail in the following sections.

Braaten-Weller Scrambling

Braaten and Weller (1979), in their paper, describe a permutation of the coefficients $a_j(n)$ in equation (1) that minimizes the discrepancy of the resulting scrambled Halton sequence. Their method suggests different permutations for different prime numbers, thus effectively breaking the correlation across dimensions. Braaten and Weller have also proved that their scrambled sequence retains the theoretically appealing N-1 order of integration error of the standard Halton sequence.

Figure 5a presents the Braaten-Weller scrambled Halton sequence in the fourteenth and fifteenth dimensions. The effectiveness of this scrambling technique in breaking correlations is evident from a comparison of Figures 4a and 5a.

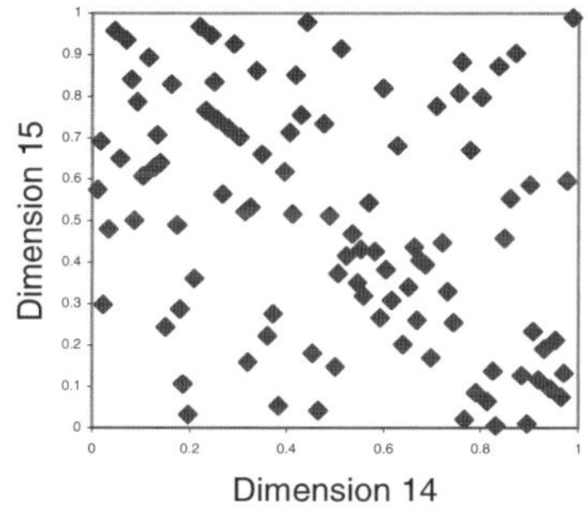

Figure 5a. Braaten-Weller Scrambled Halton Sequence

To illustrate the Braaten-Weller scrambling procedure, take the 5th number in base 3 of the Halton sequence, which in the digitized form is 0.21 ($or\ \frac{7}{9}$). The suggested permutation for the coefficients $(0, 2, 1)$ for the prime 3 is $(0, 1, 2)$, which when expanded in base 3 translates to $1\times3^{-1} + 2\times3^{-2} = \frac{5}{9}$. The first 8 numbers in the standard Halton sequence corresponding to base 3 are $(\frac{1}{3}, \frac{2}{3}, \frac{1}{9}, \frac{4}{9}, \frac{7}{9}, \frac{2}{9}, \frac{5}{9}, \frac{8}{9})$. The Braaten-Weller scrambling procedure yields the following scrambled sequence: $(\frac{2}{3}, \frac{1}{3}, \frac{2}{9}, \frac{8}{9}, \frac{5}{9}, \frac{1}{9}, \frac{7}{9}, \frac{4}{9})$.

Random Digit Scrambling

The Random Digit scrambling approach for Faure sequences is conceptually similar to the Braaten-Weller method, and suggests random permutations of the coefficients $a_j^k(n)$ to scramble the standard Faure sequence (see Matoušek, 1998, for a description).

Figure 5b presents the Random Digit scrambled Faure sequence in the fifteenth and sixteenth dimensions. A comparison of Figures 4b and 5b indicates that the Random Digit scrambling technique is very effective in breaking the patterns in higher dimensions and generating a more even distribution of points.

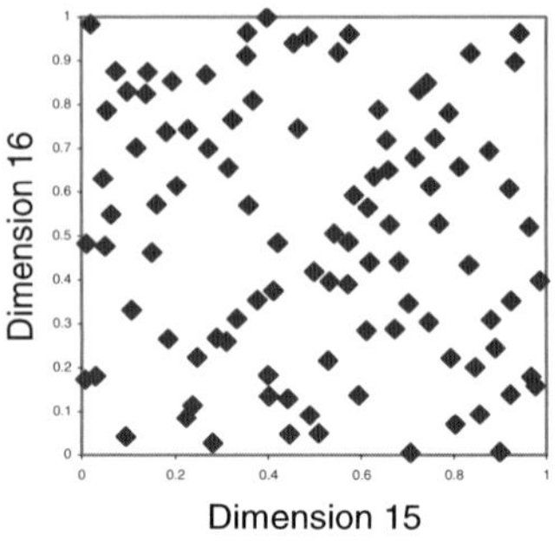

Figure 5b. Random Digit Scrambled Faure Sequence

The Random Digit scrambling technique uses independent random permutations for each coefficient in each dimension of the sequence. For example, consider the following 5-dimensional Faure sequence,

$\{\{(2,1,0),(2,3,1),(2,4,2),(4,2,3),(1,0,4)\},\{(1,0,0),(3,2,1),(0,2,4),(0,4,4),(4,4,0)\}\}$.

In each of the 5 dimensions, the vector's base 5 expansion has 3 digits, which implies that we need 15 independent random permutations $\pi = (\pi_1,......,\pi_{15})$. π_1, for example, could be the following permutation

$\pi_1(0)=4; \pi_1(2)=0; \pi_1(3)=1; \pi_1(4)=3.$

So when all 15 permutations are applied to the sequence, we obtain the scrambled Faure sequence as follows

$$\{\{(\pi_1(2),\pi_2(1),\pi_3(0)),(\pi_4(2),\pi_5(3),\pi_6(1)),(\pi_7(2),\pi_8(4),\pi_9(2)),$$
$$(\pi_{10}(4),\pi_{11}(2),\pi_{12}(3)),(\pi_{13}(1),\pi_{14}(0),\pi_{15}(4))\},$$
$$\{(\pi_1(1),\pi_2(0),\pi_3(0)),(\pi_4(3),\pi_5(2),\pi_6(1)),(\pi_7(0),\pi_8(2),\pi_9(4)),$$
$$(\pi_{10}(0),\pi_{11}(4),\pi_{12}(4)),(\pi_{13}(4),\pi_{14}(4),\pi_{15}(0))\}\}.$$

Random Linear Scrambling

The Random Linear Scrambling technique for Faure sequences proposed by Matoušek (1998) is based on the concept of cleverly introducing randomness in the recursive procedure of generating the coefficients for each successive dimension.

Figure 5c presents the Random Linear scrambled Faure sequence in the fifteenth and sixteenth dimensions. A comparison of Figures 4b and 5c indicates that the Random Linear scrambling method results in a much more even distribution of points in the fifteenth and sixteenth coordinates than the Random Digit scrambling method (Figure 5b)[2].

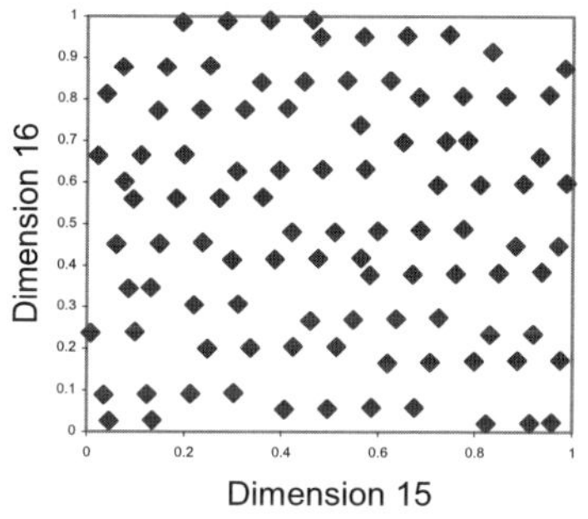

Figure 5c. Random Linear Scrambled Faure Sequence

The Random Linear scrambling approach of Matoušek is easily implemented by modifying the upper triangular combinatorial matrix C used in generating Faure sequences. A linear combination AC+B is used in the place of the matrix C, where A is a randomly generated matrix and B is a random vector, both consisting of uniform random variates U[0, b-1].

Randomization of QMC Sequences

The Halton and Faure sequences described in the preceding sub-sections, and their scrambled versions discussed above, are fundamentally deterministic and do not permit the practical estimation of integration error. Since a comparison of the performance of these sequences necessitates the computation of simulation variances and errors, it is necessary to randomize these QMC sequences. Randomization of QMC sequences is a technique that introduces randomness into a deterministic QMC sequence while preserving the equidistribution property of the sequence (see Shaw, 1988; Tuffin, 1996).

The numerical experiments in this study used Tuffin's randomization procedure (see Bhat, 2003, for a detailed explanation of the randomization procedure) to perform 20 estimation runs for each test scenario. The results of these 20 estimation runs were then used to compute the relevant statistical measures.

---

[2] The behavior of the Random Linear scrambling technique seemed to not always be predictable in terms of uniformity of coverage. In particular, the results of the Random Linear scrambling method for the nineteenth and twentieth dimensions of the Faure sequence were observed to be rather poor as the redistribution of points occurs in a fixed pattern.

Generation of Draws with and without Scrambling across Observations

The simulated maximum likelihood estimation of an MMNL with a K-dimensional mixing distribution involves generating a K-dimensional PMC or QMC sequence for a specified number of draws 'N' for each individual in the dataset. Therefore estimating an MMNL model on a dataset with Q individuals will require an N×Q K-dimensional PMC or QMC sequence, where each set of N K-dimensional points computes the contribution of an individual to the log-likelihood function. A PMC or QMC sequence of length N×Q can be generated either as one continuous sequence of length N×Q or as Q independent sets of length N each. In the case of PMC sequences, both these approaches amount to the same since a PMC sequence is identical to a random sequence with each point of the sequence being independent of all the previous points. In the case of QMC sequences, Q independent sets of length N can be generated by first constructing a sequence of length N and then scrambling it Q times, which is known as generation with scrambling across observations. The other alternative of generating a continuous QMC sequence of length N×Q is known as generation without scrambling across observations. QMC sequences generated with and without scrambling across observations exhibit different properties (see Train, 1999; Bhat, 2003; Sivakumar et al., 2005).

The study presented here examines the performance of the various scrambled and standard QMC sequences generated both with and without scrambling across observations.

Box-Müller and Inverse Normal Transforms

The standard and scrambled versions of the Halton and Faure sequences, and the LHS sequence are generated to be uniformly distributed over the multidimensional unit cube. Simulation applications, however, may require these sequences to take on other distributional forms. For example, the estimation of the MMNL model described in the following section requires normally-distributed multivariate sequences that span the multidimensional domain of integration. The transformation of the uniformly-distributed LHS and QMC sequences to normally-distributed sequences can be achieved using either the inverse standard normal distribution or one of the many approximation procedures discussed in the literature, such as the Box-Müller Transform, Moro's method and Ramberg and Schmeiser approximation. The study presented here compares the performances of the inverse normal and the Box-Müller transforms.

If Y is a K-dimensional matrix of length N*Q containing the uniformly-distributed LHS or QMC sequence, the inverse normal transformation yields $X = \Phi^{-1}(Y)$, where X is a normally-distributed sequence of points in K-dimensions. The Box-Müller method approximates this transformation as follows. The uniformly-distributed sequence of points in Y are transformed to the normally-distributed sequence X using the equations

$$X_{ij} = \cos(2\pi Y_{i(j+1)})\sqrt{-2\log Y_{ij}} \text{ and } X_{i(j+1)} = \sin(2\pi Y_{i(j+1)})\sqrt{-2\log Y_{ij}}, \qquad (10)$$

for all i = 1, 2, ... N*Q, and j = 1, 3, 5, ... K-1, assuming that K is even. If K is odd, then we simply generate an extra column of the sequence and perform the Box-Müller transform with the K+1 even columns. The $(K+1)^{th}$ column of the transformed matrix X can then be dropped.

EVALUATION FRAMEWORK

We evaluate the performance of the sequences discussed above in the context of the simulated maximum likelihood estimation of an MMNL model using simulated datasets. This section describes in detail the evaluation framework used in the numerical experiments in the study. All the numerical experiments were implemented using the GAUSS matrix programming language.

Simulated Maximum Likelihood Estimation of the MMNL Model

In the numerical experiments in this study, we used a random-coefficients interpretation of the MMNL model structure. However, the results from these experiments can be generalized to any model structure with a mixed logit form. The random-coefficients structure essentially allows heterogeneity in the sensitivity of individuals to exogenous attributes. The utility that an individual q associates with alternative i is written as:

$$U_{qi} = \beta_q' x_{qi} + \varepsilon_{qi} \tag{11}$$

where, x_{qi} is a vector of exogenous attributes, β_q is a vector of coefficients that varies across individuals with density $f(\beta)$, and ε_{qi} is assumed to be an independently and identically distributed (across alternatives) type I extreme value error term. With this specification, the unconditional choice probability of alternative i for individual q, P_{qi}, is given by the following mixed logit formula:

$$P_{qi}(\theta) = \int_{-\infty}^{\infty} L_{qi}(\beta) f(\beta \mid \theta)\, d(\beta), \quad L_{qi}(\beta) = \frac{e^{\beta' x_{qi}}}{\sum_j e^{\beta' x_{qj}}}, \tag{12}$$

where, β represents parameters which are random realizations from a density function f(.) called the mixing distribution, and θ is a vector of underlying moment parameters characterizing f(.). While several density functions may be used for f(.), the most commonly used is the normal distribution with θ representing the mean and variance.

The objective of simulated maximum likelihood inference is to estimate the parameters 'θ' by numerical evaluation of the choice probabilities for all the individuals using simulation. Using 'N' draws from the mixing distribution f(.), each labelled β_n, n = 1,...,N, the simulated probability for an individual can be calculated as

$$SP_{qi}(\theta) = (1/N) \sum_{n=1,\dots,N} L_{qi}(\beta^n). \tag{13}$$

$SP_{qi}(\theta)$ has been proved to be an unbiased estimate of $P_{qi}(\theta)$ whose variance decreases as the number of draws 'N' increases. The simulated log-likelihood function is then computed as

$$SLL(\theta) = \sum_{q=1,\ldots,Q} \ln(SP_{qi}(\theta)), \qquad (14)$$

where i is the chosen alternative for individual q. The parameters 'θ' that maximize the simulated log-likelihood function are then calculated. Properties of this estimator have been studied, among others, by Lee (1992) and Hajivassiliou and Ruud (1994).

Experimental Design

The data for the numerical experiments conducted in this study were generated by simulation. Two sample data sets were generated containing 2000 observations (or individuals q in equation 11) and four alternatives per observation. The first data set was generated with 5 independent variables to test the performance of the sequences in 5 dimensions. The values for each of the 5 independent variables for the first two alternatives were drawn from a univariate normal distribution with mean 1 and standard deviation of 1, while the corresponding values for each independent variable for the third and fourth alternatives were drawn from a univariate normal distribution with mean 0.5 and standard deviation of 1. The coefficient to be applied to each independent variable for each observation was also drawn from a univariate normal distribution with mean 1 and standard deviation of 1 ($i.e., \beta_{qi} \sim N(1,1), q = 1,2,\ldots,2000$ and $i = 1,\ldots,4$). The values of the error term, ε_{qi}, in equation (11) were generated from a type I extreme value (or Gumbel) distribution, and the utility of each alternative was computed. The alternative with the highest utility for each observation was then identified as the chosen alternative. The second data set was generated similarly but with 10 independent variables to test the performance of the sequences in 10 dimensions.

Test Scenarios

The study uses the simulated datasets described above to numerically evaluate the performance of the LHS sequence, and the standard and scrambled versions of the Halton and Faure sequences within the MMNL framework. Random-coefficients mixed logit models, in 5 and 10 dimensions, were first estimated using a simulated estimation procedure with 20,000 random draws (N = 20,000 in equation 13). The resulting estimates were declared to be the "true" parameter values. The various sequences were then evaluated by computing their abilities to recover the "true" model parameters. This technique has been used in several simulation-related studies in the past (see Bhat, 2001; Hajivassiliou et al., 1996).

We tested the performance of the standard Halton, Braaten-Weller scrambled Halton, standard Faure, Random Digit Scrambled Faure, Random Linear Scrambled Faure, and LHS sequences. For each of these six sequences we tested cases with 25, 125 and 625 draws (N in equation 13) for 5 dimensions and with 100 draws for 10 dimensions.

COMPUTATIONAL RESULTS

The estimation of the 'true' parameter values served as the benchmark to compare the performances of the different sequences. The performance evaluation of the various sequences was based on their ability to recover the true model parameters accurately. Specifically, the evaluation of the proximity of estimated and true values was based on two performance measures: (a) root mean square error (RMSE), and (b) mean absolute percentage error (MAPE). Further, two properties were computed for each performance measure: (a) bias, or the difference between the mean of the relevant values across the 20 runs and the true values, and (b) total error, or the difference between the estimated and true values across all runs[3].

One general note before we proceed to present and discuss the results. The Box-Müller transform method to translate uniformly-distributed sequences to normally-distributed sequences performed worse than the inverse normal transform method almost universally for all the scenarios we tested (this is consistent with the finding of Tan and Boyle, 2000). We therefore present only the results of the inverse transform procedure here (the Box-Müller results are available from the authors).

The computational results are divided into four tables (Tables 1a-1d), one each for 25, 125, 625 (5 dimensions) and 100 draws (10 dimensions). In each table, the first column specifies the type of sequence used. The second column indicates whether the sequence is generated with or without scrambling across observations. The remaining columns list the RMSE and MAPE performance measures for the estimators in each case.

Table 1a. Evaluation of ability to recover model parameters: 5 dimensions, 25 draws

Sequence Type	Scrambling across observations	RMSE		MAPE	
		Bias	Total error	Bias	Total error
Standard Halton	No Scrambling	0.2987	0.3275	30.6976	30.6976
	Scrambling	0.2817	0.2997	29.7409	29.7409
Braaten-Weller Scram. Halton	No Scrambling	0.3157	0.3515	32.5745	32.5745
	Scrambling	0.2948	0.3259	30.4528	30.4544
Standard Faure	No Scrambling	0.2586	0.2869	27.2551	27.2551
	Scrambling	0.2374	0.2887	24.0570	24.0937
Random Digit Scram. Faure	No Scrambling	0.2955	0.3332	28.8420	28.8420
	Scrambling	0.2947	0.3541	29.8144	29.8144
Random Linear Scram Faure	No Scrambling	0.2677	0.2978	27.9082	27.9082
	Scrambling	0.2848	0.3209	29.4035	29.4035
LHS	N/A	0.2650	0.3059	27.7668	27.7668

[3] We also computed the simulation variance, i.e.; the variance in relevant values across the 20 runs and the true values. However, we chose not to discuss the results of those computations here in order to simplify presentation and also because the total error captures simulation variance.

Table 1b. Evaluation of ability to recover model parameters: 5 dimensions, 125 draws

Sequence Type	Scrambling across observations	RMSE		MAPE	
		Bias	Total error	Bias	Total error
Standard Halton	No Scrambling	0.0538	0.0672	5.6565	6.0881
	Scrambling	0.0560	0.0627	5.9892	6.0709
Braaten-Weller Scram. Halton	No Scrambling	0.0383	0.0560	4.0664	5.1062
	Scrambling	0.0445	0.0646	4.7313	5.9334
Standard Faure	No Scrambling	0.0393	0.0553	4.1668	4.5773
	Scrambling	0.0455	0.0630	4.8227	5.3210
Random Digit Scram. Faure	No Scrambling	0.0298	0.0489	3.1551	4.2517
	Scrambling	0.0432	0.0563	4.5803	5.0752
Random Linear Scram Faure	No Scrambling	0.0364	0.0534	3.9041	4.4663
	Scrambling	0.0310	0.0450	3.2947	4.1762
LHS	N/A	0.0715	0.0789	7.5294	7.6367

Table 1c. Evaluation of ability to recover model parameters: 5 dimensions, 625 draws

Sequence Type	Scrambling across observations	RMSE		MAPE	
		Bias	Total error	Bias	Total error
Standard Halton	No Scrambling	0.0088	0.0189	0.8701	1.6096
	Scrambling	0.0065	0.0161	0.6021	1.3830
Braaten-Weller Scram. Halton	No Scrambling	0.0069	0.0177	0.7053	1.5221
	Scrambling	0.0060	0.0170	0.6013	1.4086
Standard Faure	No Scrambling	0.0070	0.0131	0.7148	1.1309
	Scrambling	0.0047	0.0129	0.3596	1.0538
Random Digit Scram. Faure	No Scrambling	0.0025	0.0138	0.2354	1.1987
	Scrambling	0.0059	0.0174	0.5914	1.4629
Random Linear Scram Faure	No Scrambling	0.0049	0.0161	0.4702	1.4698
	Scrambling	0.0035	0.0152	0.3423	1.2542
LHS	N/A	0.0152	0.0311	1.5890	2.7455

Table 1d. Evaluation of ability to recover model parameters: 10 dimensions, 100 draws

Sequence Type	Scrambling across observations	RMSE		MAPE	
		Bias	Total error	Bias	Total error
Standard Halton	No Scrambling	0.2224	0.2692	26.6145	26.8211
	Scrambling	0.1953	0.2489	23.5067	23.9490
Braaten-Weller Scram. Halton	No Scrambling	0.1681	0.2500	19.8661	21.4625
	Scrambling	0.3297	0.3666	30.2559	30.5939
Standard Faure	No Scrambling	0.1969	0.3114	22.1754	26.5580
	Scrambling	0.2337	0.3068	27.7484	29.8256
Random Digit Scram. Faure	No Scrambling	0.1844	0.2577	21.8181	22.4525
	Scrambling	0.1998	0.2585	24.5396	24.7051
Random Linear Scram Faure	No Scrambling	0.1740	0.2266	20.9043	21.2949
	Scrambling	0.1802	0.2679	20.7861	22.5148
LHS	N/A	0.2213	0.3013	25.6583	26.5579

The different test scenarios of the QMC sequences in 5 dimensions clearly indicate that a larger number of draws results in lower bias, and total error. However, the margin of improvement decreases as the number of draws increases. The following are other key observations from our analysis.

1. At very low draws, the standard versions of the Halton and Faure sequences perform better than the scrambled versions. However, the bias and total error of the estimates is very high and we strongly recommend against the use of 25 or less draws in simulation estimation.

2. The scrambled versions of both the Halton and Faure sequences perform better than their standard versions at 125 draws (for 5 dimensions) and 100 draws (for 10 dimensions). At 625 draws for 5 dimensions, the standard versions of both the Halton and Faure sequences perform marginally better than their scrambled versions in terms of total error but yield much higher bias. Overall, using about 100-125 draws with scrambled versions of QMC sequences seems appropriate (though one would always gain by using a higher number of draws at the expense of more computational cost).

3. The Faure sequence generally performs better than the Halton sequence across both 5 and 10 dimensions. The only exception is the case of 100 draws for 10 dimensions, which indicates that, in terms of bias, the Braaten-Weller scrambled Halton sequence without scrambling across observations performs slightly better than the corresponding Random Linear scrambled Faure. However, this difference is marginal and the Random Linear scrambled Faure clearly yields the lowest total error.

4. Among the Faure sequences, the Random Linear and Random Digit scrambled Faure sequences perform better than the standard Faure (except the case with 25 draws for 5 dimensions, which we anyway do not recommend; see point 1 above). However, between the two scrambled Faure versions there is no clear winner.

5. The Random Linear scrambled Faure with scrambling across observations performs better than without scrambling across observations for 5 dimensions (for 125 and 625 draws). For 10 dimensions, the Random Linear scrambled Faure with scrambling across observations performs slightly less well than without scrambling across observations. However, this difference is rather marginal.

6. The Random Digit scrambled Faure performs better when generated without scrambling across observations in all the cases.

7. Overall, the results of the study indicate that the Random Linear and Random Digit scrambled Faure sequences are among the most effective QMC sequences for simulated maximum likelihood estimation of the MMNL model. While both the scrambled versions of the Faure sequence perform well in 5 dimensions, the Random Digit scrambled Faure without scrambling across observations performs marginally better. In 10 dimensions, on the other hand, the Random Linear scrambled Faure without scrambling across observations yields the best performance both in terms of bias and total error.

8. Our study also strongly recommends the use of the inverse transform to convert uniform QMC sequences to normally-distributed sequences.

CONCLUSIONS

Simulation techniques have evolved over the years, and the use of quasi-Monte Carlo (QMC) sequences for simulation is slowly beginning to replace pseudo-Monte Carlo (PMC) methods, as the efficiency and faster convergence rates of the low-discrepancy QMC sequences makes them more desirable. There have been several studies comparing the performance of different QMC sequences in the evaluation of a single multidimensional integral, and recommending them over the traditional PMC sequence. The use of QMC sequences in the simulated maximum likelihood estimation of flexible discrete choice models, which entails the estimation of parameters by the approximation of several multidimensional integrals at each iteration of the optimization procedure, is, however, relatively recent.

In this chapter, we present the results of a study undertaken to experimentally compare the overall performance of the Halton and Faure sequences and their scrambled versions, against each other and against the LHS sequence in the context of the simulated likelihood estimation of an MMNL model of choice. Brief, yet complete, methodologies for the generation of alternative QMC sequences are also presented here. The results of our analysis indicate that the Faure sequence consistently outperforms the Halton sequence, and that both the Faure and Halton sequences provide significant advantages over traditional PMC simulation methods. The Random Linear and Random Digit scrambled Faure sequences, in particular, are among the most effective QMC sequences for simulated maximum likelihood estimation of the MMNL model.

REFERENCES

Bhat, C.R. (2001). Quasi-random maximum simulated likelihood estimation of the mixed multinomial logit model. *Transportation Research Part B*, **35(7)**, 677-693.

Bhat, C.R. (2003). Simulation estimation of mixed discrete choice models using randomized and scrambled halton sequences. *Transportation Research Part B*, **37(9)**, 837-855.

Bhat, C.R. and R. Gossen (2004). A mixed multinomial logit model analysis of weekend recreational episode type choice. *Transportation Research Part B*, **38(9)**, 767-787.

Braaten, E. and G. Weller (1979). An improved low-discrepancy sequence for multidimensional quasi-monte carlo integration. *Journal of Computational Physics*, **33**, 249-258.

Bratley, P. and B.L. Fox (1988). Implementing sobol's quasi-random sequence generator. *ACM Transactions on Mathematical Software*, **14**, 88-100.

Fox, B.L. (1986). Implementation and relative efficiency of quasi-random sequence generators. *ACM Transactions on Mathematical Software*, **12(4)**, 362-376.

Hajivassiliou, V.A. and P.A. Ruud (1994). Classical estimation methods for LDV models using simulation. In: *Handbook of Econometrics* (Engle, R.F. and D.L. McFadden, eds.), Vol. IV, pp. 2383-2441. Elsevier, New York.

Hajivassiliou, V.A., McFadden, D.L. and P.A. Ruud (1996). Simulation of multivariate normal rectangle probabilities and their derivatives: theoretical and computational results. *Journal of Econometrics*, **72**, 85-134.

Halton, J.H. (1970). A retrospective and prospective survey of the monte carlo method. *SIAM Review*, **12(1)**, 1-63.

Kocis, L. and W.J. Whiten (1997). Computational investigations of low-discrepancy sequences. *ACM Transactions on Mathematical Software*, **23(2)**, 266-294.

Lee, L-F. (1992). On efficiency of methods of simulated moments and maximum simulated likelihood estimation of discrete choice models. *Econometric Theory*, **8**, 518-552.

Matoušek, J. (1998). On the L2-discrepancy for anchored boxes. *Journal of Complexity*, **14**, 527-556.

McKay, M.D., Conover, W.J. and R.J Beckman (1979). A comparison of three methods for selecting values of input variables in the analysis of output from a computer code. *Technometrics*, **21**, 239-245.

Morokoff, W.J. and R.E. Caflisch (1994). Quasi-random sequences and their discrepancies. *SIAM Journal of Scientific Computation*, **15(6)**, 1251-1279.

Morokoff, W.J. and R.E. Caflisch (1995). Quasi-monte carlo integration. *Journal of Computational Physics*, **122**, 218-230.

Niederreiter, H. (1992). *Random Number Generation and Quasi-Monte Carlo Methods.* SIAM, Philadelphia.

Ökten, G. and W. Eastman (2004). Randomized quasi-monte carlo methods in pricing securities. *Journal of Economic Dynamics & Control*, **28(12)**, 2399-2426.

Park, Y.H., Rhee, S.B. and E.T. Bradlow (2003). An integrated model for who, when, and how much in internet auctions. Working Paper, Department of Marketing, Wharton.

Press, W.H., Teukolsky, S.A., Vetterling, W.T., and B.P. Flannery (1992). *Numerical Recipes in C: The Art of Scientific Computing*. Second Edition, Cambridge University Press, Massachusetts.
Revelt, D. and K. Train (2000). Customer-specific taste parameters and mixed logit: household's choice of electricity supplier. Economics Working Papers E00-274, Department of Economics, University of California, Berkeley.
Sandor, Z. and K. Train (2004). Quasi-random simulation of discrete choice models. *Transportation Research Part B*, **38(4)**, 313-327.
Sivakumar, A., Bhat, C.R. and G. Ökten (2005). Simulation estimation of mixed discrete choice models with the use of randomized quasi-monte carlo sequences: a comparative study. *Transportation Research Record*, **1921**, 112-122.
Shaw, J.E.H. (1988). A quasirandom approach to integration in bayesian statistics. *The Annals of Statistics*, **16(2)**, 895-914.
Tan, K.S. and P.P Boyle (2000). Applications of randomized low discrepancy sequences to the valuation of complex securities. *Journal of Economic Dynamics & Control*, **24**, 1747-1782.
Train, K. (1999). Halton sequences for mixed logit. Working Paper No. E00-278, Department of Economics, University of California, Berkeley.
Tuffin, B. (1996). On the use of low-discrepancy sequences in monte carlo methods. *Monte Carlo Methods and Applications*, **2**, 295-320.
Wang, X. and F.J. Hickernell (2000). Randomized halton sequences. *Mathematical and Computer Modelling*, **32**, 887-899.

Transport Science and Technology
K.G. Goulias, editor

CHAPTER 4

COMPUTATIONAL INTELLIGENCE IN TRANSPORTATION: SHORT USER-ORIENTED GUIDE

Ing. Ondřej Přibyl, Ph.D.
Czech Technical University in Prague
Faculty of Transportation Sciences
Na Florenci 25, Praha 1, 110 00
Czech Republic

Introduction

This paper provides a very brief introduction to computational intelligence and its application to transportation. The paper complements the many interesting reviews of existing literature and applications, such as in Teodorovic and Vukadinovic, 1998; and Avineri, 2005. This paper has different objectives. It provides guidance for transportation practitioners who are facing real word problems and are interested in applying nowadays popular methods from the field of artificial intelligence or soft computing. It provides a short introduction to this field and basic overview of the theory behind these methods. The major focus of the presentation is on showing the strong and weak points of each of these methods and a proper application field.

In the past, often practitioners tend to disapprove of computational intelligence based on their attempts to use or reviews of methods. Unfortunately, in most cases negative experience with these methods is due to inappropriate applications of a model. Most models have strengths in specific contents and only if they are applied correctly. This paper provides guidance in this direction and emphasizes the important features of the methods.

Key areas, terminology and brief history

Probably the most known term in the general public is the so called ***artificial intelligence*** (AI). It is very popular nowadays even though it is a rather old discipline. The AI started in the 1940s when Norbert Wiener published a book called *Cybernetics or Control and Communication in the Animal and the Machine* (Wiener, 1948). The AI does not have a unified definition at the moment. For example, the following views on AI are rather common:

Thinking Humanly `"The automation of activities that we associate with human thinking, activities such as decision-making, problem solving, learning..."	Thinking Rationally "The study of mental faculties through the use of computational models."
Acting Humanly "The study of how to make computers do things at which, at the moment, people are better."	Acting Rationally "The branch of computer science that is concerned with the automation of intelligent behavior."

Fig. 1: Different approaches to the meaning of AI (Adopted from: Russell and Norvig, 2003)

In general artificial intelligence can be understood as a subject dealing with computational models that use strong symbolic manipulation. The artificial intelligence is a fast developing subject, however, the following areas are understood as its core topics (Bonissone, 2000):

- Natural language processing
- Computer vision
- Robotics
- Problem solving and planning
- Learning
- Expert systems

In this paper we focus on a related area, **soft computing.** It is originally a branch of artificial intelligence, however it follows its own path. Soft computing differs from conventional (hard) computing in that, unlike hard computing, it is tolerant of imprecision, uncertainty, partial truth, and approximation. The methods belonging to this field are motivated by human mind and human reasoning. Because soft computing does not focus on symbolic manipulation and rather uses extensive numerical computation, it is also known as **computational intelligence.**

Soft computing consists of the following major areas:

- Fuzzy systems (FS),
- Artificial neural networks (ANN),
- Evolutionary or genetic algorithms (GA), and
- Probabilistic Reasoning (PR) (belief networks, chaos theory and learning theory).

The last issue will not be covered in this chapter. We focus on the more traditional methods in the rapidly evolving area.

The following features characterize soft computing (adopted from Jang, 1997):

- It is meant to describe and understand **human expertise**
- Consists of **biologically inspired computing models** and **new optimization techniques**
- It uses **strong numerical computation** and so called **model-free learning**
- It is meant to be used in **new application domains** from **real- world**
- The methods aim to be **robust and fault tolerant and** are meant to be used **for solving of non-linear tasks**

If we look at the features described above, it is clear that transportation and traffic related problems are really suitable for using soft computing. The problems we are facing are from the real world, they are usually strongly non-linear, often described in linguistic terms, and the output should be understood by humans. In many cases, standard methods do not bring satisfactory results and we need to look for some alternative and more advanced methods. Methods from the field of soft computing can be in some cases the solution.

Before we proceed with particular methods and describe their main features, we bring to your attention the following table. It describes the major milestones that changed the world of artificial intelligence and soft computing.

Tab. 1: Major milestones in the history of AI and soft computing (adopted from Jang, 1997)

	Conventional AI	Neural Networks	Fuzzy Systems	Other
1940s	Cybernetics (Norbert Wiener)	McCulloch - Pitts neuron model		
1950s	Artificial Intelligence	Perceptron		
1960s	Lisp Language	Adaline, Madaline	Fuzzy sets (L.Zadeh)	
1970s	Knowledge engineering (Expert systems)	Back-propagation alg., Cognitron	Fuzzy controller	Genetic algorithm
1980s		Self-organizing map Hopfield Net. Boltzman machine	Fuzzy modelling (TSK model)	Artificial life Immune modelling
1990s			Neuro-Fuzzy modelling ANFIS	Genetic programing

Overview of methods

Artificial Neural Networks

An Artificial Neural Network is an information processing paradigm that is inspired by the way biological nervous systems, such as the brain, process information. It is composed of a large number of simple processing elements (neurones). A biological neuron and its artificial model are depicted in Fig. 2.

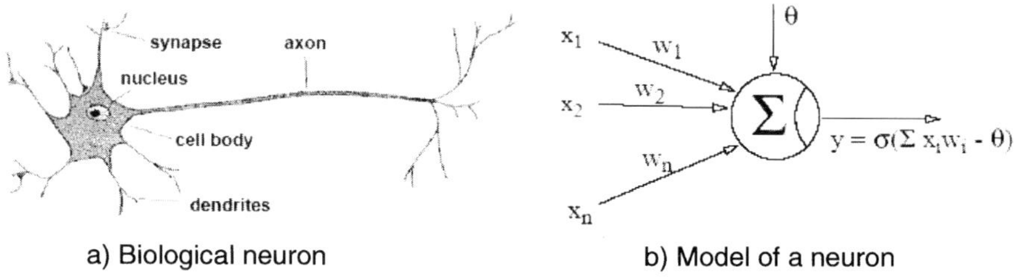

Fig. 2: Scheme of a) a biological neuron and its b) mathematical model

The neuron processes weighted inputs and in case their sum exceeds given threshold, Θ, the signal activates the output. The strength of an ANN is not in the number of neurons, but in their high interconnectivity. The neurons work in unison on solving a given problem in a network. According to the type of interconnection, two basic types of ANN are distinguished: multilayer feedforward neural network, and recurrent neural networks. Here we focus on the first type only. An example of a multilayer feedforward network is in Fig. 3. In this example there is one input layer, one hidden layer and one output layer. The interconnections among neurons are characterized by their weights.

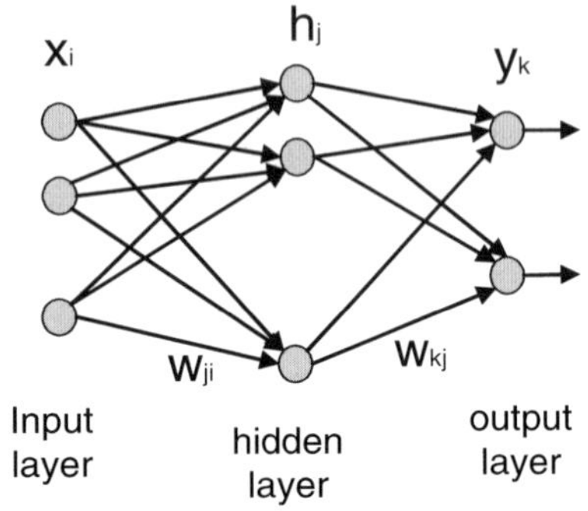

Fig. 3: Example of a multilayer neural network

Learning in neural nets is primarily a process of adjusting these weights. The most popular method of learning is called *back-propagation*. At the beginning of the process, the network is initialized by setting these weights to small random numbers. The network is then presented with some input data and the desired output value(s). Based on the input values and the initial weights the ANN provides an output. The net's output is compared to the desired output and using partial derivations of the difference, the weights are modified so that the network's output matches the desired value. This is propagated backwards through the network and this is why it got the name backpropagation. In this matter all input-output data pairs are presented to the network. This whole process is repeated until some stopping criterion is met (for example decrease of the mean square error between the network's output and the desired value below some predefined value). For more information about learning of ANN and the mathematics behind see for example Jang (1997).

Features of artificial neural networks

Advantages

Adaptive learning – ANN have the ability to learn how to solve problems based on the data given for training. ANN are also able to adapt to new situations.

Parallelism – ANN are inherently parallel and naturally amenable to expression in a parallel notation and implementation on parallel hardware.

Distributed memory - in ANN, memory is distributed over many units giving resistance to noise. It ensures strong fault tolerance and robustness. If some neurons are destroyed, or their connections altered, the behavior of the network as a whole is only slightly degraded (fault tolerance via redundant information coding).

Real-time operation - computations in ANN may be carried out in parallel. Special hardware devices are being designed and manufactured which take advantage of this capability. On the other hand training can take longer time when performed on a non-parallel computer.

Universal aproximator – multilayered feedforward neural network having at least one hidden layer was proved to be a universal function approximator (Hornik et al., 1989).

Drawbacks

Black box – ANN are often understood as black boxes. Apart from defining the general architecture of a network, the user has no other means to effect the processing than to feed it input and watch it train and await the output. Neural networks cannot explain the results they obtain; their rules of operation are completely unknown.

No a priori information – This is a similar property to the previous one. No a priori information can be put in the network. It learns from scratch just based on the data provided. In systems where human experience exists it is real drawback.

Data dependency - the performance of a network is sensitive to the quality and type of preprocessing of the input data. Also it is known to be a data hungry algorithm.

ANN in transportation applications

The strength of ANN is in function approximations and predictions in an environment where their high tolerance to error can be used. We should consider using ANN if the target function is unknown and we expect it to be non-linear. On the other hand, we should ensure enough of input data. ANN are useful for direct processing of raw detector data.

It is clear that there are many areas in traffic and transport system that meets these concerns. A nice overview of transport related problems approached by ANN is in Himanen et al. (1998). Here we name only the major fields in which ANN have been successfully used:

- Driver Behavior
    - Modeling drivers behavior in signalized urban Intersection
    - Driver decision making model
- Traffic Flow
    - Intersection control
    - Estimation of Speed-Flow relationship
- Transportation planning and management
    - Trip generation model
    - Urban public transport equilibrium
    - Incident detection
    - Prediction parking characteristics
    - Travel time prediction

Fuzzy Systems

A classical set is a set with crisp boundaries, for example a set A of students in a class. For each student we can decide whether each student belongs to the set A or not. However, the crisp sets often do not reflect the nature of human thoughts, which tend to be abstract and imprecise. We can look at the following example. For many control tasks, we need to quantify intensity of vehicles into several categories. Here we can describe a high intensity of vehicles as the case when there are more than 1200 vehicles per hour. The problem is that according to the common Aristotelian's logic, a flow of 1199 vehicles per hour belongs to medium intensity and a flow of 1201 vehicles per hour belongs to high intensity. The difference of only two vehicles leads to the usage of a different set. The example is demonstrated in Fig. 4 a.

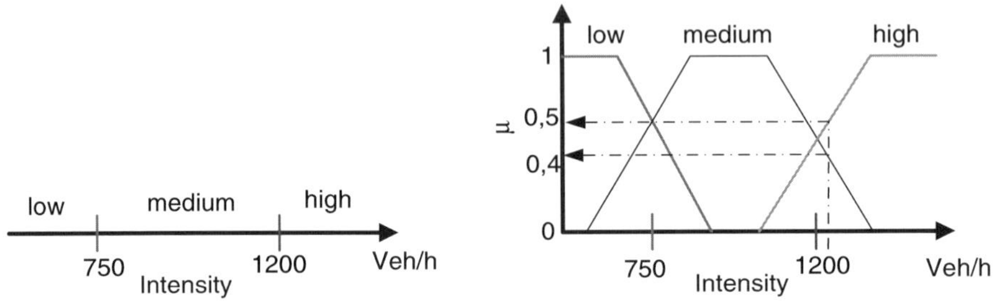

a) Example of crisp sets b) Example of fuzzy sets

Fig. 4: Principle of a) crisp and b) fuzzy sets

In contrast to classical set, a fuzzy set is a set without crisp boundary. The transition between "to belong to a set" and "not to belong to a set" is gradual. It is expressed by, so called, *degree of membership*, μ, that takes values between 0 and 1. Here it is important to stress that the degree of membership does not correspond to probability. Even though it has similar meaning, the mathematical probabilistic properties are not met (for example P(A)+P(~A) = 1).
The function that expresses the degree of membership is called *membership function (MF)*. They can have different shape (Konar, 1999), but the most common are *triangular*, *Gaussian*, *sigmoidal*, or *trapezoidal* MF.

The previous example of traffic flow in fuzzy sets is depicted in Fig. 4 b. In this figure we can see the simple trapezoid membership functions for low, medium and high intensity. In this example, the intensity of 1201 vehicles per hour belongs to the set high intensity with the degree of membership $\mu = 0{,}5$, to the set medium intensity with degree $\mu = 0{,}4$, and finally to the set low intensity with degree $\mu = 0$. The difference of two vehicles leads in this case only to small difference in the degrees of membership.

The concept of fuzzy sets is used in a popular computing framework known as *fuzzy system*, *fuzzy inference system* (FIS), or *fuzzy model*. The basic structure of a FIS consists of three components: a *rule base,* which contains a selection of fuzzy rules; *a database* (or dictionary), which defines the membership functions used in the fuzzy rules; and a *reasoning mechanism*, which performs the inference procedure upon the rules and given facts to derive a reasonable output. The principle of using FIS for engineering applications is depicted in Fig. 5.

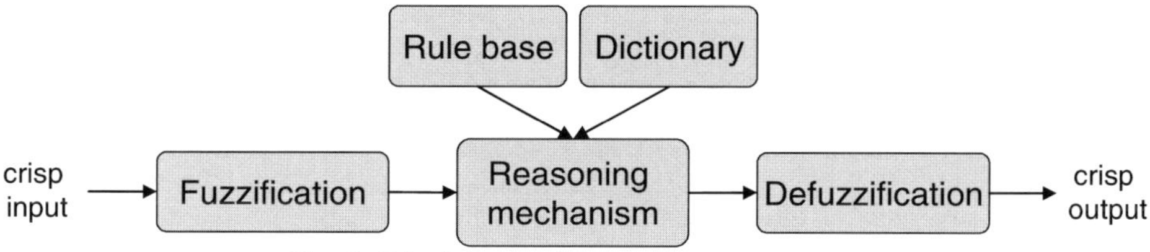

Fig. 5: Principle of a fuzzy inference system

There are two most common FISs: Mamdani FIS and Takagi-Sugeno FIS. The differences between them are in the consequent of their fuzzy rules.

The *Mamdani fuzzy inference system* was proposed as the first attempt to control a steam engine and boiler combination by a set of linguistic control rules obtained from experienced human operators. The fuzzy rules for Mamdani FIS have fuzzy sets on output. A rule can be expressed by following:

$$IF\ x\ is\ A\ AND\ y\ is\ B\ THEN\ z\ is\ C, \quad (13)$$

where A,B and C are fuzzy sets.

The *Takagi-Sugeno fuzzy model* (TSK model) was proposed by Takagi, Sugeno and Kang in an effort to develop a systematic approach to generate fuzzy rules from a given input-output data set. A typical fuzzy rule in a Sugeno fuzzy model has the form

$$IF\ x\ is\ A\ AND\ y\ is\ B\ THEN\ z = f(x,y), \quad (14)$$

where A *and B* are fuzzy sets in the antecedent, while $z = f(x,y)$ is a crisp function in the consequent. Usually $f(x,y)$ is a polynomial in the input variables x and y.

The principle of a TSK model (which is similar to other types of FIS) can be understood from the following example depicted in Fig. 6. Here we consider a simple system with two input variables, *Intensity (x)* and *Occupancy (y)*, and two rules only. The input domain of variable x has two membership functions I1 and I2; the input domain for variable y has also two membership functions O1 and O2. The rules can be written down as follows:

1. IF x is I1 AND y is O1 THEN $f_1 = p_1x + q_1y + r_1$
2. IF x is I2 AND y is O2 THEN $f_2 = p_2x + q_2y + r_2$

where p_i, q_i and r_i are parameters. According to the actual inputs, the degree of membership is computed. In the example, the function *minimum* expresses the term AND in the rule. This minimum is denoted w_i and is understood as weight of given rule. The output of the overall system, z, is computed as a weighted sum of all rules.

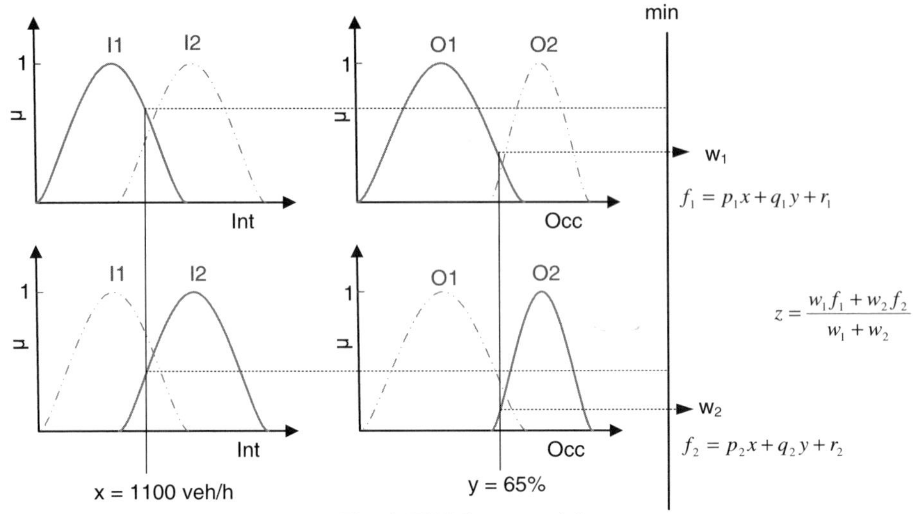

Fig. 6: TSK fuzzy model

One of the major complications in using FIS is its design. It requires a lot of experience and knowledge since no single universal algorithm exists. The following techniques are the most common ways to design a FIS (Babuska, 1998). Either we can use some well structured and complex expert knowledge or we need to use a data driven approach:

1. Expert knowledge
2. Data driven approaches
    a. Grid partitioning
    b. Cluster analysis
    c. Least square identification
    d. Decision tree technique

Features of fuzzy systems

Advantages

Comprehensive knowledge representation – in fuzzy systems, knowledge is represented in the form of comprehensive linguistic rules. It implies that the resulting systems and models are transparent and understandable to human experts.

Dealing with uncertainty – the concept of fuzzy sets enables using of imprecise information such as high intensity. It does not require exact input data. It is extremely useful in any system that uses human input or needs to produce human understandable output.

Robustness – the ability to deal with uncertainty has another important feature. The FS is very robust.

Uses expert knowledge – the knowledge of experts can be easily used in FS to define its rules and control mechanism.

Transparent models – the relationships between inputs and outputs are explicitly defined using the if-then rules.

Using of prior knowledge – contrary to many other systems, FS enable using of prior knowledge.

Drawbacks

Complicated design – there is no simple and standardized way on how to transform knowledge from experts to FS. Even when human operators exist, their knowledge is often incomplete and episodic, rather then systematic.

No general calibrating procedure exists – a formal set of procedures to calibrate if-then rules do not exist.

Curse of dimensionality – the complexity of a FS grows rapidly as the number of input variables and the if-then rules increases. This is true especially when using data driven identification of the system.

FS in transportation applications

The fuzzy systems are suitable mainly for tasks that deal with the following problems:

- When human reasoning and decision-making are involved
    - Supervising, planning, scheduling
- Various types of information are involved
    - Measurements and linguistic information
- Problems using natural language
- Very complex systems
- When there is some prior heuristic knowledge

A detailed overview of transportation related problems is provided in Teodorovic and Vukadinovic (1998) and includes fuzzy traffic control systems, solving route choice problems, controlling a fleet in river traffic, fuzzy decision making, fuzzy scheduling, multi-objective decision making and many others.

Genetic Algorithms

A genetic algorithm is a stochastic process that mimics natural process of biological evolution. It follows the basic principles stated by Darwin, such as "survival of the fittest" in the process of natural selection of species (Konar, 1999). GAs have been successfully used in the field of optimization, machine learning, scheduling, planning and others (Wang and Xue, 2002).

In general, genetic algorithms are used in optimization tasks to find extremes of a function. They perform a kind of *parallel stochastic search*. Contrary to other optimization methods, they are based on a population of solutions, which explains the term parallel. The population covers a range of possible outcomes. Every candidate solution (not only those optimal) is usually represented in a form of a binary string known as *chromosome*. Contrary to hill climbing methods in which a derivation of the fitness function is computed, in GA solutions are identified purely on a fitness level, and therefore local optima are not distinguished from other equally fit individuals. Those solutions closer to the global optimum will thus have higher fitness values. Successive generations improve the fitness of individuals in the population until the optimization convergence criterion is met. Due to this probabilistic nature GA tends to the global optimum, however for the same reasons GA models cannot guarantee finding the optimal solution.

The principle of hill climbing methods is depicted in Fig. 7a. The fitness function in this case is rather complex, so the found solution depends strongly on the starting point (initialization). In this figure there are three starting points, only one of which finds the global extreme. The range of starting points that will lead to finding the global extreme for this function is rather narrow in this case (in Fig. 7b are the borders depicted by red dotted lines). If the starting point is outside these only a local extreme will be found.

In Fig. 7 b an example of an initial population in a genetic algorithm is depicted. The population is usually randomly distributed across the search space. The algorithm identifies the individuals with the optimizing fitness values, and those with lower fitness will naturally get discarded from the population. The solutions with lower probability to appear in the next generation (with higher values of fitness function) are depicted by dotted line in this figure.

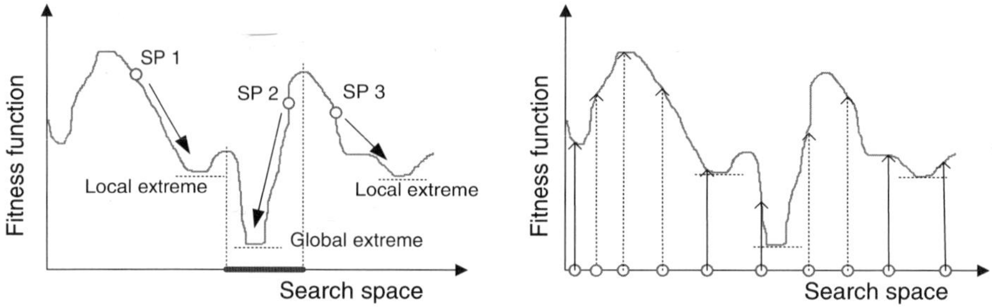

a) gradient descent methods b) genetic algorithms

Fig. 7: Principle of a) gradient descent method and b) simple genetic algorithm.
Legend: SP...starting point

The basic steps of a simple GA are depicted in Fig. 8.

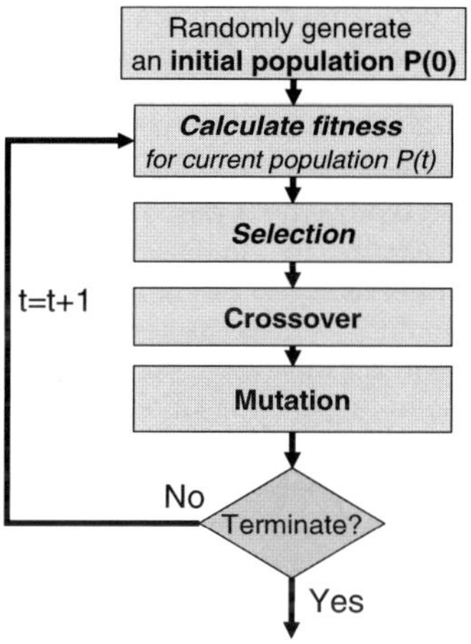

Fig. 8: The principle and basic operators of genetic algorithms

First, a whole population of these chromosomes is generated. This process is usually random. The number of chromosomes in each population is one of the parameters that must be determined by the model developer, however typically it is in the range of tens or hundreds. Using operators known as *selection*, *mutation*, and *crossover* (also called *recombination*) a new population is generated.

Selection chooses those chromosomes in the population that have a good potential for further reproducing. This is the survival-of-the-fittest mechanism as adopted from genetics. Selection is determined based on a *fitness function*, which describes their degree of correctness. However, the fitness function states only the probability of selection. A random mechanism is applied to the whole population so even some chromosomes with a lower value of fitness function can be selected (even though with lower probability). The most common types of a selection operator are *roulette wheel selection* or *tournament selection* (Konar, 1999). The roulette wheel selection chooses individuals with a probability proportional to their relative fitness. Each individual has a section of a roulette wheel according to its relative fitness. Another common selection operator is tournament selection. The individuals in this selection compete directly with each other in a group of *k* elements (*k* is the size of tournament). The best individual(s) is(are) chosen to remain in the new population. The most widely spread tournament has size $k=2$.

The function of the remaining operators is to create new chromosomes. Crossover exchanges subparts of two chromosomes and generates their offsprings. In most applications, a simple *one-point uniform crossover*, or its modifications are used. First, the point of crossover, *cp*, will be randomly determined so that $1 \leq cp \leq q-1$. The second parts of both parent chromosomes will be exchanged and two new offsprings generated.

The operator *mutation* randomly changes the value of some bit in a chromosome. It can be viewed as local improvement method. For each bit in the chromosome, a randomly generated number is compared to a given threshold, i.e., probability of mutation *Pm*. If it exceeds this value, the value of the given bit is swapped (from zero changes to one and on the contrary).

This iterative process continues until one of the termination criteria is met: the maximum number of iteration allowed has been reached, there is no improvement for a predefined number of iterations, or a known acceptable solution has been reached.

Even though GA perform stochastic search it is in no sense random. Briefly stated, genetic algorithms can be viewed as search procedures based on the mechanics of natural selection and genetics. Goldberg (1998) described the fundamental principles of particular operators. *Selection together with the mutation operator* can be viewed as a form of continual improvement, similar to hill climbing methods. Mutation creates variants in the neighborhood of current individuals and selection accepts those individuals with high probability. *Selection together with the crossover operator* equals to an intelligent innovation. Sets of good solution features are combined together with the potential for large improvements in the solution. This operator "jumps" into new areas of the function we want to optimize.

The key problem in setting any genetic algorithm is in the balance between these two features. If the continual improvement is stronger than innovation, the solution will most likely converge to a local extreme. This case is called *premature convergence*. On the other hand if the innovation operator is stronger, the algorithm works as a random search since no local improvement is provided. However, there are ways how to overcome this limitation. An example of a genetic algorithm that is less sensitive to the setting of its parameters was, for example, introduced in Pribyl (2005).

Features of genetic algorithms

Many of the properties of GA have been already discussed above, but this section aims to summarize their major features one more time.

Advantages

Independency on function property – contrary to hill climbing optimization methods, GA perform well in problems for which the fitness landscape is complex - where the fitness function is discontinuous, noisy, changes over time, or has many local optima.

GA are intrinsically parallel - due to their parallelism genetic algorithms are particularly well-suited to solving problems where the space of all potential solutions is truly large - they are able to explore the solution space in multiple directions at once.

Does not distinguish between local and global extremes - solutions are identified purely on a fitness level, and therefore local optima are not distinguished from other equally fit individuals.

Multi-criterial tasks - GA are suitable for multi-criterial tasks in which so called Pareto optimal or also non-dominated solutions are found. This is due to the parallelism, the ability to search large spaces and also its performance on a fitness level.

Suitable for NP hard problems – since GA is a heuristics, it is suitable for solving NP-hard (Non-deterministic Polynomial-time hard) problems for which no algorithm exists.

Drawbacks

Complicated setting of the parameters – there are many parameters in GA that needs to be set in the design procedure (fitness function, size of population, selection of right operators and their parameters, stopping criterion, and many others).

Premature convergence – improper setting of the parameters leads often to a premature convergence, finding of a local optimum instead of the global optimum.

GA in transportation applications

Genetic algorithms make it possible to explore a far greater rage of potential solutions than do other optimization techniques. They are especially well suited for solving of NP-hard problems (which means problems that cannot be solved by an algorithm in a polynomial time). They do not ensure finding of the extreme, but they will converge in a relatively short time to a solution that is close to optimal. In transportation there are many problems that are NP-hard or need to search large spaces and GAs have been successfully used to find a solution. Some examples follow (Sayers and Anderson, 1999; Sadek et al., 1997; Tsai et al., 2004; Liepins et al., 1990; or Anderson et al. 1998)

- NP-hard problems
  - Traveling Salesman Problem
  - Dynamic traffic assignment
  - Vehicle Routing Problems
  - Shortest Route Problem
  - Vehicle Scheduling Problem
  - Vehicle Fleet Size Problem
- Multi-criteria transportation problems
- Search and optimization
- Traffic signal optimization
- Synthetic schedule simulation in planning models
- Tuning of parameters of fuzzy systems or neural networks

Hybrid Systems

In the previous parts, we focused on the core methods of soft computing. The conclusion was that each of these methods has its advantages and drawbacks as well as application fields. We can stress, for example, the following issues (Abraham, 2002):

☺ Neural networks are able to learn
☹ A neural network acts as a black box
☺ Fuzzy systems enable using of human knowledge in simple linguistic terms and rules
☹ It is difficult to build and set parameters of a fuzzy model
☺ Genetic algorithms are suitable for optimization tasks in large spaces

For this reason, researchers started to look for alternative algorithms that would combine the strong properties of particular methods and omit their limitations. The evolution of the four major hybrid systems is depicted in Fig. 9.

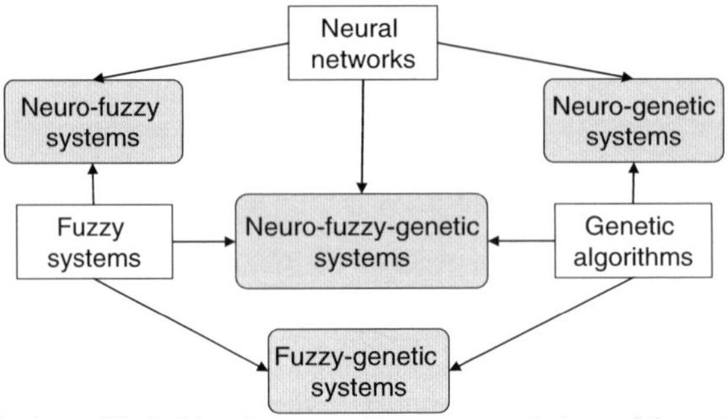

Fig. 9: The evolution of hybrid soft computing systems (adopted from Abraham, 2002)

In the following paragraphs only the major features will be described. More details can be found in Abraham (2002), or in Bonissone (2000).

Neuro-Fuzzy systems
There are many different approaches to combine fuzzy systems and neural networks. Their objective is to have a robust non-linear system that can incorporate expert knowledge, is easily understandable, has defined structure and is able to learn (Konar,1999).
An example that is rather widespread is called ANFIS. The acronym ANFIS derives its name from *Adaptive Neuro-Fuzzy Inference System.* The ANFIS was created by Jyh-Shing R. Jang (Jang, 1997) in order to combine the advantages of both fuzzy systems and artificial neural networks.

ANFIS is a class of adaptive networks that are functionally equivalent to fuzzy inference systems. Contrary to common ANN, it has a fixed number of layers (five) and the neurons in each layer have a specific function. For example neurons in the first layer assign the membership number for each fuzzy set, and neurons in the second layer perform the operation AND. Given an input-output dataset, the parameters of membership functions (MFs) are modified (this process is termed *learning*) using a well-known back-propagation algorithm or hybrid algorithm based on a combination of back-propagation and least squares estimator. After learning we have a structure that corresponds with reality and whose structure can be easily interpreted. By using ANN for solving our problem, though, we also face some complications connected with the *data driven* nature of this approach such as sensitivity to the quality of training data set. An example in trip generation is Pribyl and Goulias (2003).

Neuro-Genetic systems
The success of neural networks largely depends on their architecture, their training algorithm, and the choice of features used in training. Unfortunately, determining the architecture of a neural network is a trial-and-error process; the learning algorithms must be carefully tuned to the data; and the relevance of features to the classification problem may not be known a priori. Genetic algorithms have been successfully used in the past to select the architecture of the neural networks, select relevant features and train the networks (Yao, 1999).

Fuzzy-Genetic systems

Two major ways to combine fuzzy logic and genetic algorithms exists. Fuzzy logic can be used to improve the behavior of genetic algorithms, or genetic algorithms can be used to help set up parameters of fuzzy logic (Cordón et al., 2001). Here we focus only on the later application since it is more popular tool. Even though fuzzy systems are very popular nowadays and can be used for solving many problems, one major complication appears in most cases – it is not an easy task to design the fuzzy system and properly tune its parameters. An alternative approach to an ANFIS described in the previous section is so-called fuzzy genetic algorithms. In these systems, genetics algorithms are used to properly set many of the parameters of a fuzzy system: membership functions, rule base, fuzzy operators and others.

Neuro-Fuzzy-Genetic systems

Even when neural networks are integrated with fuzzy systems it is not ensured that the learning algorithm within the neural part will converge and the tuning of the FS will succeed. The performance can be then further improved by combining this system with genetic algorithms (Bonissone, 2000).

Summary

This paper aimed to provide an insight into the field of soft computing and its application. Any analyst interested in applying such model must be aware of its major features. The previous sections aimed to provide an overview of these features. To summarize, the following table compares the methods discussed using a set of important criteria, which are crucial in the phase of model selection. Basically, for each of the methods described, the table answers the following questions:

- Is the method based on some mathematical model?
- Is it robust with respect to outliers in the data?
- Does it explain its output in comprehensive terms?
- Can it handle small data sets for calibration or training?
- Can it resist uncertainty in data?
- Can it resist partial damage within the model's structure?
- Is it able to adapt to new situation or learn from examples?
- Is it suitable for nonlinear tasks?
- Does it incorporate expert knowledge and/or a priori information?
- Is the reasoning process visible?

Not always, the questions can be answered by simple *yes* or *no*. Very often the models fulfill the criteria only partially. The discussion was provided in the sections above. In this table, for simplicity, only three levels of the dependency are provided:

☺ Yes, the model can handle this feature and it is suitable for such applications

😐 The use of the model is limited for such feature

☹ No, the model cannot handle this feature

Answering these questions during the model selection phase can support the decision and lead to selection of a proper tool needed for solving a problem.

	Mathematical model	Can resist outliers	Explains output	Suits small data sets	Uncertainty tolerance	Fault tolerance	Learning ability	Nonlinearity	Expert knowledge	Reasoning process is visible
FIS	☺	😐	☺	☺	☺	☺	☹	☺	☺	☺
ANN	☹	☹	☹	☹	☺	☺	☺	☺	☹	☹
GA	☹	😐	☹	☺	☺	☺	😐	☺	☹	☹
Symbolic AI	😐	☹	☺	😐	☹	☹	☹	😐	☺	☺
Regression	☺	☹	😐	☹	☹	☹	😐	😐	😐	☺

Fig. 10: Key properties of particular models and systems in soft computing, AI and their comparison to statistical regression.
Legend: ☺...Yes; 😐... Somewhat; ☹... No.

Soft computing methods and algorithms have a tremendous potential in solving science and engineering problems that are typically present in transportation applications. They are suitable for transport and traffic related applications since they can address all major complications we are facing, such as nonlinearity, considerable amounts of input data, substantial uncertainty and vagueness in data, variability of human behavior, and the inherent dynamics within these systems.

Soft computing is not a solver for all problems though. It does not outperform standard methods in all cases. Each particular problem has to be carefully revised and the appropriate tool must be selected. Hopefully this chapter provides the base information required in choosing the proper method. Unfortunately, a complete overview of applications of soft computing to transport related tasks could not be provided within such limited space. However, several interesting references to surveys of existing literature have also been provided. The readers are encouraged to check these references to gain more understanding of the progress and latest development within this field.

References:

Abraham, A. (2002) "Intelligent Systems: Architectures and Perspectives" In Recent Advances in Intelligent Paradigms and Applications, Abraham A., Jain L. and Kacprzyk J. (Eds.), Studies in Fuzziness and Soft Computing, Springer Verlag Germany, ISBN 3790815381, Chapter 1, pp. 1-35, 2002.

Anderson, J.M., T.M. Sayers, and M.G.H.Bell. Optimization of a Fuzzy Logic Traffic Signal Controller by a Multiobjective Genetic Algorithm, In *Proceedings of the Ninth International Conference on Road Transport Information and Control*, pages 186-190, London, April 1998.

Avineri, E. (2005), Soft Computing Applications in Traffic and Transport Systems: A Review. In: Hoffmann, F., Koppen, M., Klawonn, F. and Roy, R. (Eds.), Soft Computing: Methodologies and Applications, Springer Series on Studies in Fuzziness and Soft Computing, Springer-Verlag, Germany, 17-25. ISBN: 3 540 25726 8.

Babuska, R. *Fuzzy Modeling for Control*. Boston, USA: Kluwer Academic Publishers, 1998.

Bonissone, P.P. „Hybrid Soft Computing: Where are we Going?" In ECAI 2000, Proceedings of the 14th European Conference on Artificial Intelligence, Berlin, Germany, August 20-25, 2000, IOS Press. pp: 739-746.

Cordón, O., F. Herrera, F. Gomide, F. Hoffmann and L. Magdalena, ``Ten years of genetic-fuzzy systems: a current framework and new trends, *Proceedings of Joint 9th IFSA World Congress and 20th NAFIPS Internation Conference, pp. 1241-1246, Vancouver - Canada, 2001.*

Goldberg, D.E. "The Race, the Hurdle, and the Sweetspot: Lessons From Genetic Algorithms for the Automation of Design Innovation and Creativity.," IlliGAL Report No. 98007. April 1998.

Himanen, V., P.Nijkamp, A. Reggiani, and J. Raitio (Editors). Neural Networks in Transport Applications. Ashgate Publishing Ltd., England. ISBN: 1 84014 808X, 1998.

Hornik, K., M. Stinchcombe, and H. White. *Multilayer feedforward networks are universal approximators*. Neural Networks, 2:359--366, 1989.

Jang, J.-S. R., C.-T. Sun, and E. Mizutani. Neuro-Fuzzy and Soft Computing. Prentice Hall:1997.

Konar, A. Artificial Intelligence and Soft Computing: Behavioral and Cognitive Modeling of the Human Brain. CRC Press, 1999.

Liepins, G. E., M. R. Hilliard, J. Richrdson, and M. Palmer, „*Genetic algorithm applications to set covering and traveling salesman problems*," in Operations Research and Arti cial Intelligence: the Integration of Problem Solving Strategies, Brown and White, eds., pp. 29-57, Kluwer Academic Press, 1990.

Přibyl, O., "Clustering of Activity Patterns Using Genetic Algorithms", In *Soft Computing: Methodologies and Applications. Series: Advances in Soft Computing*. Edited by Hoffmann, F.; Köppen, M.; Klawonn, F.; Roy, R.., Springer. 2005. p. 37-52.

Přibyl O. and K. G. Goulias (2003) Application of Adaptive Neuro-Fuzzy Inference System to Analysis of Travel Behavior. Transportation Research Record, Journal of the Transportation Research Board, 1854, pp. 180-188.

Russell S. and Norvig, P.``*Artificial Intelligence: A Modern Approach.*" 2nd Ed., Prentice Hall, 2003.

Sadek, A. W., B. L. Smith, and M. J. Demetsky. (1997). Dynamic Traffic Assignment: A Genetic Algorithms Approach. In *Transportation Research Record 1588.*

Sayers, T. M. and J. M. Anderson. "The multi-objective optimisation of a traffic control system", In Proceedings of 14th International Symposium on Transportation and Traffic Theory, pages 153-176, Avi Ceder (editor). Haifa, Israel, July 1999. Technion-Israel, Institute of Technology, Transportation Research Institute.

Teodorovic, D. & Vukadinovic, K., Traffic Control and Transport Planning: A Fuzzy Sets and Neural Networks Approach, Kluwer Academic Publishers, Boston/ Dordrecht/London, (1998).

Tsai, B.-W., V. N. Kannekanti, and J.T. Harvey. „Application of genetic algorithm in asphalt pavement design". In Transportation Research Record. Issue Number: 1891, Publisher: Transportation Research Board. ISSN: 0361-1981. (2004). pp: 112-120.

Wang, H., and D. Xue. "An Intelligent Zone-Based Delivery Scheduling Approach." *Computers in Industry* 48 (2002): 109-125.

Wiener, N. (1948) *Cybernetics: or Control and Communication in the Animal and the Machine*, Cambridge, MA: The MIT Press.

Yao, X. (1999). Evolving artificial neural networks. *Proceedings of the IEEE*, 87 (9), pp. 1423-1447.

K.G. Goulias, editor

CHAPTER 5

DEVELOPMENT OF HIGH PERFORMANCE AND INNOVATIVE INFRASTRUCTURE MATERIALS

Dimitrios Goulias, Associate Professor, Department of Civil and Environmental Engineering, University of Maryland, College Park. dgoulias@umd.edu

INTRODUCTION

Highway agencies are spending every year billions of dollars in infrastructure expenditures. A big portion of their budget is often absorbed in maintenance and construction of the infrastructure so as to maintain the conditions of the facilities at an acceptable level. Year after year the required budget to achieve so is increasing in an exponential manner due to the ever growing use of the infrastructure, its aging and fast decay, and the limited increase in available funding. These along with the "environmental pressure" on reducing the amount of materials and wastes in land fills has generated a significant impetus on the improvement of existing construction materials and the production of long lasting and high performance materials for transportation facilities. This paper describes some of the efforts in this area along with specific case studies and challenges from a variety of different materials used in infrastructure construction.

IMPETUS FOR INNOVATIVE AND HIGH PERFORMANCE MATERIALS

Several reasons have generated the need for improved, long lasting and high performance materials. These include among other:

- **INCREASING LOADING CONDITIONS**
- **CHANGING ENVIRONMENTS**
- **POOR PERFORMANCE OF CURRENT MATERIALS**
- **NEED FOR ENHANCED BEHAVIOR OF EXISTING MATERIALS**
- **LIFE - CYCLE DESIGN**
- **INNOVATIVE CONSTRUCTION TECHNIQUES**
- **NEW MIXTURE DESIGN & PRODUCTION METHODS**
- **NEW APPLICATIONS**
- **WASTE PRODUCTS AND RECYCLING MANDATES**
- **DEVELOPMENT OF PERFORMANCE SPECS**
- **INNOVATION**

Over the years loading conditions have been steadily increasing due to the need to provide a more efficient transportation of people and goods. While weight control has been relatively successful on the interstate highway network due to the presence of the weight stations that is not the case for local and rural roads. Furthermore, in some states triple trailers are permitted so as to carry goods over long distances most efficiently. In port facilities the use of the container dollies often requires stronger and better materials able to resist highly concentrated loads, extreme shearing and abrasive forces, and repetitive / impact loads. Often the environmental conditions in such cases combined with the extreme loadings pose additional requirements for better design and performance of materials. There are several examples in both literature and construction practice of poor performance and early deterioration of facilities due to inappropriate selection and design of materials. In addition, increased and unexpected loadings and high traffic, extreme environments, and in many cases inappropriate adaptation of materials and mixtures with successful performance in other regions, do not perform well without appropriate evaluation of the adoption of such materials and techniques in the new environment and the construction practices of the region of interest. This is typically the result of an "indiscriminate" transferability of materials, mixtures and construction techniques without an extensive evaluation of their adaptation and performance to local materials and construction practice, and local loading and environmental conditions. An example of such a case has been the experience with the asphalt rubber binder and mixtures described later in the paper, due to a "rushed" need to address the 1991 ISTEA requirements and environmental mandates on the use of recycled and waste materials in federally funded projects. It is now the focus of several agencies to reduce waste. For example, NYC has a target to achieve and eventually exceed 40% recycling of waste materials, such as plastics, glass, metals and ash (Goulias et. al. 1998a). Such emphasis has been often translated to the development of innovative highway and building materials, such as rubber modified asphalt, glass filled concrete, composites for utility poles and marine piles. Some of these cases are presented next along with the primary challenges for their development. The reclaiming and recycle-ability of infrastructure materials has lead to concepts such as "waste as resource," "life –cycle design", "multi life-cycle and zero emission.," proposed by state DOTs, UNDP, other local, state and federal agencies, and the infrastructure industry.

Overall modified materials, such as polymer modified asphalt or concrete mixtures have been shown to provide enhanced behaviour and improved performance. Furthermore, they provide specific properties for addressing particular requirements needed for specific applications (for example the case of polymer modified concrete for bridge decks). Clearly, in each case such mixtures and technology has to be adapted to the local conditions and materials, as well as construction practices to avoid premature failures and undesirable results. Innovation in construction technology is often associated with the development of improved materials. This has been the case for example of the roller compacted concrete that now can be used in new applications.

The development of new mix design methods, such as SUPERPAVE has improved significantly both the selection and design of asphalt binders and mixture. The overall trend of the infrastructure industry is to move towards a performance based approach for specifying and designing materials, and away from the "method" and eventually "end-results" requirements and procedures. Such transition typically requires lengthy experimentation and performance data for i) identifying the performance related parameters, ii) identifying easy and quick methods for their evaluation, iii) collecting performance data over a broad spectrum of input material parameters, loading and environmental conditions, and finally developing the required performance models. In this process economics become part of the process since i) in material selection and mixture design the lower cost for the best or an acceptable performance is searched, while ii) in the case of material specifications the economic implications need to be quantifying for defining the pay schedules for superior and/or inferior quality materials. As in the case of asphalt mixtures, the concrete industry is now aggressively investigating the possibility of replacing ACI 211 with a performance based mix design methodology.

Finally innovation and the implementation of nanotechnology in materials are generating a new field for the development of improved and long lasting materials. The research community and the industry has been introducing concepts such as "zero maintenance" pavements, "fast truck" construction and materials, smart and self healing materials, self consolidated concrete, and other. Several of these concepts require the development of new and/or improved materials.

CHALLENGIES OF INNOVATIVE AND HIGH PERFORMANCE MATERIALS

In most cases the successful development of innovative and high performance materials requires the coordination and joint involvement of the scientific community (universities and national laboratories), the manufacturing and construction industry, and the infrastructure agencies. Following are some examples of such successful approach that lead to the development, and in many cases the successful implementation, of innovative and long lasting materials along with the specific challenges.

Behaviour and Fatigue Modelling of High Performance Concrete for Pavements

The mid 1990s "European tours" on concrete pavements organized by AASHTO and the concrete industry lead to the identification of the need to adopt, develop and examine new materials, and design and construction methods for concrete pavements. To address this need several innovative projects were undertaken in the US identified as Technical Evaluation 30 projects (TE- 30) on High Performance Concrete Pavements. The Maryland project had as objective the development of fiber reinforced (FRC) and low shrinkage (LS) concrete mixtures to be used in highway pavement operations (Goulias 2002a). The success of this project was the result of a joint effort and involvement between the scientific community, the concrete industry and material suppliers, instrumentation experts, and infrastructure agencies (FHWA and MSHA). While FRC and LS concrete mixtures have been used in other areas of civil engineering, in pavements the expected benefits range from: improved flexural fatigue resistance, reduction in crack development, reduction in slab warping, increased pavement slab longevity. In this project a polypropylene fibrillated fiber was used, along with water reducer and air-entrained admixtures so as to achieve an acceptable level of workability and air content since they are highly affected by the introduction of fibers. The compressive, Figure1, and flexural test results indicated that a more ductile failure was observed in which the concrete was able to withstand large deformation without a destructive failure associated typically with a very brittle conventional paving mix. To examine the effects on the post-peak plastic behaviour of the FRC mixtures extensive toughness analysis were conducted. For this investigation the stress-strain diagrams revealed that as the fiber content increased the amount of energy absorbed by concrete in the plastic region increased. Furthermore, for the same FRC mix the amount of energy absorbed at later stages of the plastic region increased as well. Thus, while stiffness of the mixture decreased with fiber content, FRC was able to absorb higher amounts of plastic energy and deformation. This is particularly important for pavements where with repeated load applications there is a progressive nucleus of cracking damage, permitting thus a longer life expectancy of the pavement even though eventually larger deflections are experienced. To study the impact on fatigue damage, flexural fatigue analysis was conducted. Fatigue testing typically is associated with high variability in testing. However, as shown in Figure 2, specific FRC mixtures may provide better fatigue performance since at the same stress level have resisted higher number of load application to failure (Nf) that conventional concrete. From this analysis fatigue models, like the one shown next, were developed for each one of the mixtures so as to consider such enhancement into pavement design:

- Fatigue model for 0.2% FRC ($R^2 = 0.94$)

$$\text{Log}\,(Nf) = 1083 - 36\,fc \qquad (1)$$

where, *fc* is the flexural fatigue stress equal to the stress level multiplied by the modulus of rupture(MOR), and *Nf* is the number of cycles to failure. Furthermore, models relating mixture properties to fatigue were investigated so that can be used for future analysis where any of the mixture composition or property is changed without having to actually conduct fatigue testing. Clearly, the assumption of this premise is that such models can be used for mixtures similar to the one that the models were developed with. An example of such a model with an $R^2 = 0.76$ is provided next:

$$y = 130.29 - 25.85x_1 - 0.0115x_2 - 0.57x_3 - 0.27x_4 - 3.847x_5 \qquad (2)$$

where y is the LOG(Nf), x_1 is LOG(MOR), x_2 is the inverted slump x_3 is the air content, x_4 is the unit weight, and x_5 is the applied stress level.

In order to evaluate the LS mixture shrinkage testing and analysis were conducted as well. Two methods of reducing mixture shrinkage were considered. In the first case the water to cement ratio was reduced since ACI studies and modeling have related the degree of shrinkage to the amount of the cementitious paste in the mix. The second case considered using a larger aggregate size since such aggregate constrains the volume change of the mix as it hardens. The lab testing indicated that the two LS mixtures provided similar shrinkage response and lower values than the FRC mixtures.

The results of the laboratory testing were used in order to identify the mixtures to be used in the field testing. For this reason it was decided to build experimental sites on an ongoing concrete pavement construction site, US-50, for both initial field testing and long term in-service load testing. Three experimental sites were thus built with the control mix, the 0.1% FRC and the LS mixture with the large aggregate size. Instrumentation was inserted into the experimental sites including surface strain gages, LVDTs, thermocouples, vibrating wire gages, and clip gages. The test sites were loaded initially with a single and a tandem axle truck for evaluating the "as built" material properties through back-calculation techniques, and using appropriate finite element grids, Figure 3, for evaluating the relative behavior between the three mixtures and built a base model for future behavior and condition monitoring of the three mixtures and pavement sections. The results from the initial field load test were coupled with atmospheric data since the pavement response throughout the day is affected from such local conditions. The in situ material properties were complying with the laboratory testing results. The within and between section variability were acceptable small, indicating a good construction uniformity. Overall the pavement sections with the three mixtures behave in a predictable and similar manner and their long term performance is currently being evaluating for meaningful evaluation of their relative behavior, performance, and longevity. Finally, the LS mixture provided lower field shrinkage measurements during the initial stages.

Figure 1. Fiber Reinforced Concrete Failure Mode

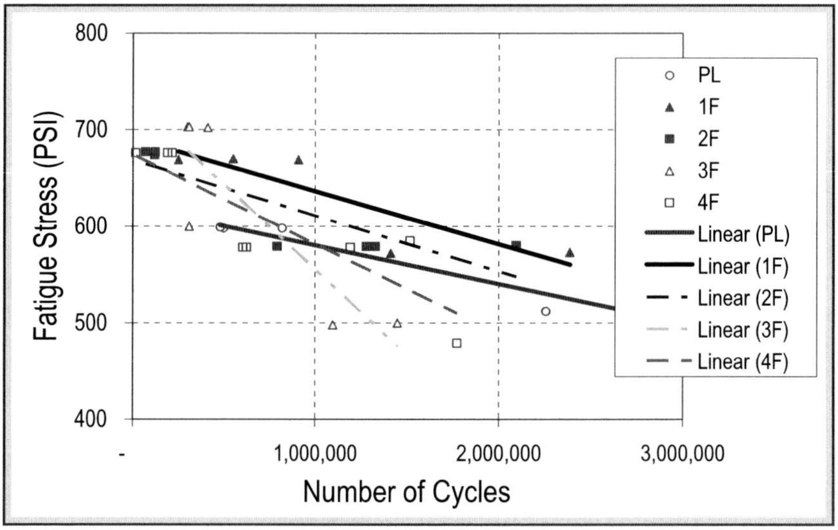

Figure 2. Fatigue Analysis

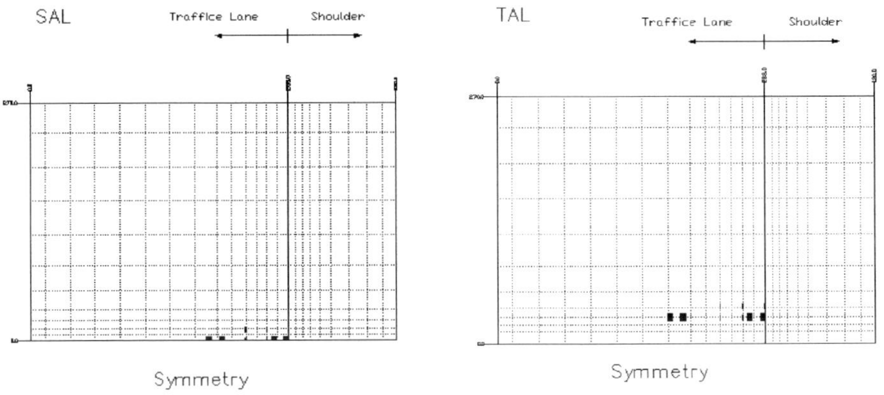

Figure 3. Finite Element Mesh and Loadings

Note: SAL = single axle load, TAL = Tandem Axle Load

Ductile Behaviour and Micro-porosity Modelling of Rubber Filler Concrete

The use of rubber particles in modifying the brittle behaviour of concrete has been examined to a lesser degree. Such type of investigation has been often promoted by the desire to address the environmental need on the use of waste products in infrastructure materials and the desire to improve concrete brittle destructive failure and enhance its plastic behaviour as well as its' durability. To evaluate such effects experimentation was carried out involving the use of commercially available crumb rubber as a replacement of fine aggregate, since substituting coarse aggregate has been shown to significantly reduce concrete strength and stiffness. As in the case of FRC, rubber modified concrete produces a more ductile behaviour in concrete and increases the amount of energy absorption (Goulias et. al. 1998b). Rubber modified concrete (RMC) has also been shown to improve concrete durability. For replacement within 10% by volume of fine aggregate it has been shown that reduction in concrete strength and stiffness properties are within acceptable limits while concrete durability has been improved since a lower reduction in strength and stiffness properties is observed with increasing freeze-thaw cycles (Goulias et. al. 1998b). Macro-porosity modelling was used to better understand the behaviour and performance of RMC and explain some of these effects (Eldin 1993). The general form of the macro-porosity model that has been proposed to relate the effects of voids and rubber aggregate on concrete properties is:

$$f_{rel} = [1 - a/a_{cr})]1.0 \exp [- \delta(a/a_{cr})] \qquad (3)$$

Where f_{rel} is the value of strength as a fraction of the strength of pore free concrete, a is the relative volume of macro-porosity in the concrete mass equal to the volume of rubber in relation to the volume of concrete, a_{cr} is the critical volume of macro-porosity (i.e., macro-porosity at which the strength becomes zero), and δ is the experimental parameter which depends on the stress level in the specimen.

The analysis on RMC have shown a very good agreement between experimental results and the macro-porosity model, Figure 4, with an $R^2 = 0.99$. While the elastic rubber may be considered as additional voids within the concrete mixture, the space it occupies is not accessible to moisture and eventually prompted to freeze thawing damage. Furthermore, failure analysis has shown that elastic rubber tends to arrest crack propagation within the mass of concrete, improving thus its durability as well.

Asphalt Rubber Binders and Mixtures and Development of Performance Based - Integrated Mix Design Methodology

The asphalt industry has a long history of designing and experimenting with modified binders and mixtures. Since the 1991 ISTEA requirements and local recycling mandates on the use of waste materials, the investigation of rubber in asphalt mixtures has been intensified. In many cases the inception and initial experimentation with rubber modified binders and mixtures was initiated in the US back in the 60's, but both technological and legal issues have limited the interest on these. However, "progressive" states that had the ability to further evaluate their performance eventually incorporated these materials in their routine paving operations. One of the main challenges in these modified mixtures is to design the rubber modified binder. The polymerization process is affected from the rubber physical and chemical properties, the properties of the asphalt cement and the conditions of blending. Transferability from region to region has been the case since the design reaction time depends on all of the above parameters. Another obstacle has been the limited response, at the time, of SUPERPAVE mix design methodology and its performance models on modified binders. Thus, several of the experimentations have been based on Marshall design criteria, and often complemented with enhanced mixture property and performance requirements. Such complementary performance requirements often included stiffness, toughness and resilient modulus evaluation, creep/rutting (permanent deformation), fatigue and low temperature cracking testing. A project in this area, undertaken by universities, the asphalt industry and materials suppliers, and DOTs' (US DOT, NJDOT) lead to the design and performance evaluation of asphalt rubber binders and mixtures. These mixtures were then used in the construction of experimental test sites on I-95 and several US routes. One of the major outcomes of that study included the development of an integrated mix design methodology (Goulias 2002b). The methodology considers asphalt mix design and pavement design criteria for establishing the requirements for asphalt mixture composition. Figure 5 presents the steps of the integrated mix design method that was developed in this investigation. As can be seen from Figure 5, pavement design analysis is used for conducting an analytical evaluation of the pavement structure to be built, so as to determine the stress and strains that eventually the asphalt mixture will be experiencing in the field. These parameters are used in identifying the range of values that the asphalt mixtures will be experiencing, and thus consider them in the laboratory performance evaluation of the mixtures. The performance requirements for the mixtures are based on both pavement design and mix design criteria and include requirements related to excessive subgrade deformation, fatigue cracking, rutting analysis, low temperature cracking models, and moisture damage analysis. In this process mixture performance models have to be developed and used (Goulias 2002b). The versatility of such method is that can be easily adapted to incorporate any models related to the specific characteristics of the mixtures of interest (conventional or modified, SMAs. etc).

Once the performance models are developed and/or calibrated the laboratory testing is carried out to identify the materials (binder and aggregate characteristics) and the desired mixture proportioning that provides an acceptable performance. This analysis can be used in conjunction with the SUPERPAVE criteria and volumetric analysis for arriving to the final asphalt mix design. As can be seen from this figure several iterations might be needed to fine tune the asphalt mix design. Also, in situations where limited alternative materials exist, fine tuning recommendations for the pavement structural design may be suggested.

Thus overall, the advantages of the proposed integrated method are that: i) it can include any prediction models that better represent a specific type of asphalt mixtures, ii) can be calibrated with local materials and local conditions, iii) can consider the external stress levels that the mixtures will be exposed to (from the pavement design analysis); iv) it is performance based, and can incorporate the SUPERPAVE binder and aggregate material selection requirements, as well as the volumetric mix design analysis.

Creep Evaluation and Modelling of Innovative Polymer based Reinforced Composites

With the increasing demand of traditional materials and the need for longer lasting and better performance materials the composites industry has risen to the challenge to often retrofit structural elements with composites. The attractiveness and challenges of polymer based composites for use in civil engineering applications, like the example of highway utility poles and marine piles, include:

- properties comparable or better than traditional materials, in this case wood;
- increased life cycle and reduced maintenance;
- light weight properties with savings in material and installation cost;
- potential use of recycled plastics;
- stiffness comparable or better (when reinforced and/or new resins are used) than traditional materials.

- creep behavior of polymer based composites.
- lack of performance data.
- adaptation of appropriate safety factor for design and certification.

To examine and quantify some of these an extensive experimental study was undertaken using recycled plastic resins for developing composite light weight highway poles and marine piles. The initial steps of the analysis included: i) an assessment of existing composites in carrying out the design and in-service loads; and ii) an investigation on the potential adaptation of design standards and safety factors applied to traditional materials through a preliminary feasibility analysis. This preliminary analysis (Goulias et. al., 2001) indicated the need for developing new and/or reinforcing existing polymer resins so as to address structural and design requirements, and enhance creep performance of the composites. In this study the existing HDPE recycled based resins were improved by increasing fiberglass fiber reinforcement content from 20% to 40%, reducing contaminant content from the recycling process by further washing and separation stages in resin reclamation, and increasing the content of chemical additives. Such improvements have lead to an increase in modulus up to 140% and compressive and flexural strength up to 230% and 240% respectively. Furthermore the impact on creep, one of the major weaknesses of polymer based composites, is shown in Figure 6 for one of these reinforced polymer based resins. As can be seen from the figure the majority of the axial compressive strain takes place during the first 150 to 200 hours of loading. At the long term the axial compressive creep tends to level asymptotically according to the following equations that take into account the effects of two different stress levels:

$$Y = 0.0007\,\mathrm{Ln}(X) - 0.0001 \quad \text{with } R^2 = 0.98 \qquad (4)$$

$$Y = 0.0005\,\mathrm{Ln}(X) - 0.0005 \quad \text{with } R^2 = 0.97 \qquad (5)$$

where y is the axial compressive strain and x is the duration of loading. These relationships are particularly valuable in estimating long term creep to be used i) in pole design, and ii) for in-service loading conditions so as to develop prototypes for further testing. The results of the creep analysis were used in conjunction with the ANSI design standards for the development of alternative utility poles. A prototype composite pole is currently being evaluated for long term load and performance testing in field conditions. Further modelling of the effects of mixture composition and resin weight on strength and creep properties is examined. Such relationships are particularly helpful for predicting the impact of changing resin composition on properties without having to repeat extensive experimental testing.

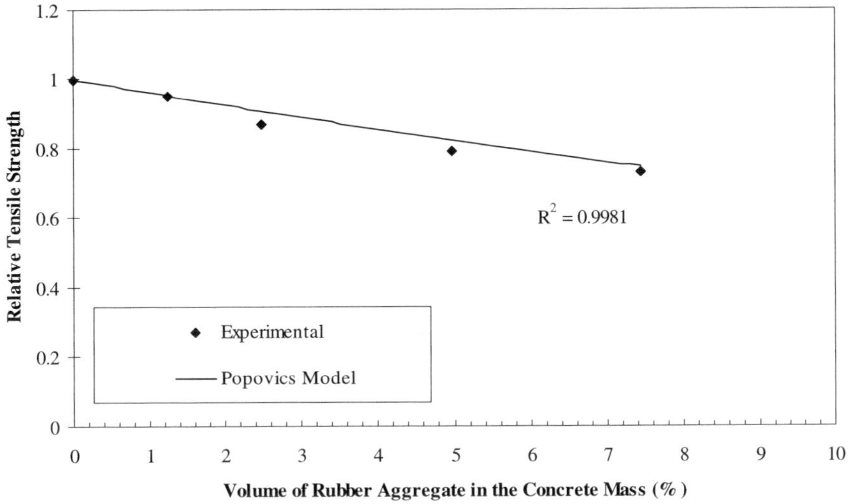

Figure 4. Micro-porosity modelling of Rubber Modified Concrete

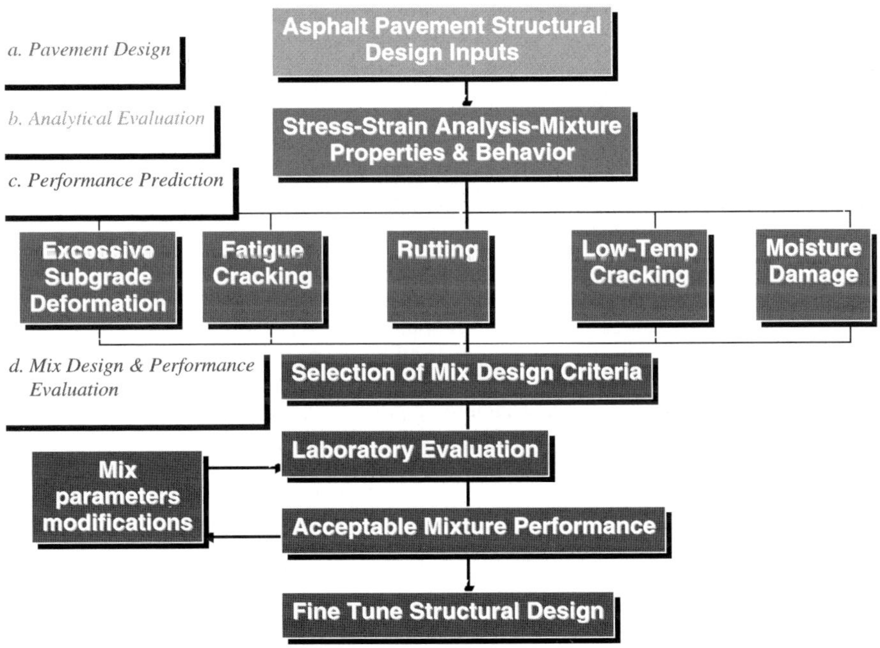

Figure 5. Integrated Mix Design for Rubber Modified Mixtures

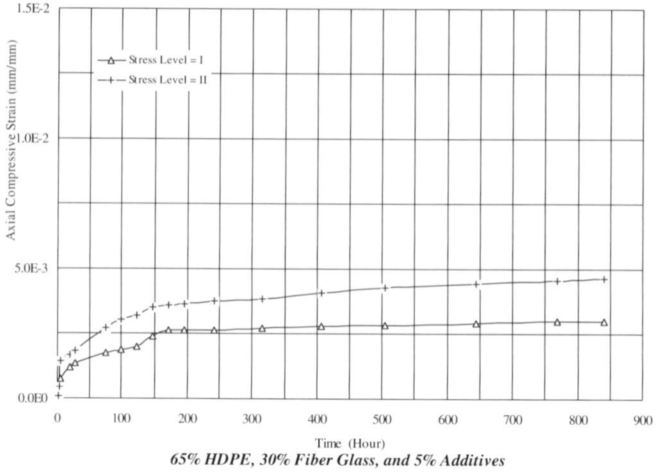

Figure 6. Creep Evaluating and Modelling of Reinforced Polymer based Composite Resin

CURRENT AND FUTURE TRENDS

The development of innovative and high performance materials in these areas is ongoing so as to address some of the critical issues identified in this paper, as well as fine tuning the design of these materials. Sensor technology as well as nano-structure analysis is playing a significant role in the enhancement of existing materials and the development of innovative and improved materials. As an example, sensor technology has made a significant advancement in the development of infrastructure monitoring and the development of smart materials and structures particularly important into seismic regions. Nanotechnology is currently being used for the investigation of "multi-scale" analysis and modelling and the development of innovative infrastructure materials. Such approach is expected to relate macro (mechanical properties, performance and visual scale analysis – 1x to10x magnification) to meso, micro, and nano scale analysis (molecular scale - 1,000,000x magnification) for a better understating, modelling and design of infrastructure materials. This approach is expected to eventually provide the opportunity to "manipulate" at the molecular level materials for producing innovative, longer lasting and improved performance materials. Examples of the current and future work undertaken in this area by the scientific community and the construction industry include:

- analysis through scan tomography, fluid flow modeling, & computational fluid multi-scale chemo-mechanical models;
- multi-scale kinetics-based mechanistic models;
- porosity dynamics;
- durability models based on frost mechanics and transport properties;
- fracture evolution using imaging techniques fracture mechanics models;
- modeling explosive spalling due to elevated temperatures with heat and mass transfer;
- constitutive models based on damage mechanics and FEM-based analysis;
- developing multi-scale performance based mix design methodologies.

REFERENCES

Eldin, N., A. Senouci. (1993). Rubber-Tire Particles as Concrete Aggregate. *Materials in Civil Engineering. J.*, 478 –496.

Goulias D.G. and I. Juran (1998a). Use of Recycled Plastic Resins in Infrastructure Construction Materials, *Journal of Solid Waste Technology and Management. J.*, 25, Vol 2, 105-111.

Goulias, D. G., A.H. Ali. (1998b) Evaluation of Rubber Filled Concrete and Correlation Between Destructive and Non-Destructive Testing Results. *Cement, Concrete and Aggregate. J.*, 1,, Vol 20, 140-144.

Goulias D. G., and A. Ali. (2001). Reinforced Recycled Polymer Based Composites for Highway Poles. *Solid Waste Technology and Management. J.*, 2, Vol 27, 62-68.

Goulias, D.G., (2002). Characterization and Performance of High Performance Concrete for Pavements, *High Performance Structures and Composites*, WIT Press, 329-335.

Goulias D.G., (2002b). Design, Behaviour, and Benefits of Highway Materials Using Recycled Tire Rubber," *Beneficial Use of Recycled Materials in Transportation Applications*, Recycled Materials Resource Center. 545- 552

K.G. Goulias, editor

CHAPTER 6

TRANSPORT POLICY AND RESEARCH ISSUES IN GREECE AND THE EU: Current facts, prospects, and priorities

George A. Giannopoulos, Aristotle University of Thessaloniki & Hellenic Institute of Transport

This paper presents the situation as regards Transport research in Greece and the EU in general. It gives first the basic facts as regards research funding and organization in Greece, and then it goes on to give in more detail data on statistics and indicators about the funding of such research in Greece and the EU. It also presents data from the so called innovation scoreboard of the EU in which one can see the progress made by each country on the application of new technologies, innovation, and research.

In the second part, the paper presents a concise view of the Transportation policy of the EU, which was introduced in 2001 with the well-known White Paper on Transport policy. It also presents the relevant policies followed by the Greek governments and it concludes that these policies should focus more on the interests of Greece to take advantage of the emerging opportunities from the opening of the Eastern European countries (and for some of them joining the EU), the increased European profile and aspirations of Turkey, and the prospects of a pacified Arab world. The paper also includes a section describing the Hellenic Institute of Transport as the sole Organization of the country specifically devoted to Transport research.

INTRODUCTION AND BASIC FACTS

Research is an indispensable parameter for the development progress and prospects of a country. Transport research, addressing one of the most fundamental activities in our societies and economies, is therefore of outmost importance. As such, it has mostly been related (in both Greece and the EU) to Transport policy objectives. Thus it is very appropriate in this paper while reviewing the current state of Transport research in Greece and the EU, to also see the main points of Greek and EU Transport policy which, after all, will set the priorities and main funding of Transport research in the future in this part of the world.

We begin with some introductory facts about the organization and funding instruments for Transport research in Greece and the EU.

The main body in Greece for establishing the priorities, scheduling activities and financing research, is the *General Secretariat of Research and Technological Development (GSRT),* under the *Ministry of Development Research and Technology (MR-T).* The Ministry of Transport, Ministry of Environment – Planning – public works, and the Ministry of Merchant Marine (the three Ministries responsible for various aspects of the Transport system) do not finance Transport research programmes although they may occasionally co-finance EU funded projects or studies, as decided on project by project basis.

MR-T was introduced in 1982 (Law 1266/82) while the legal framework under which research is conducted was established in 1985 with law 1514 /85. The same year with law 1558/85 the GSRT was established. In the same year the first research and development programmes were established. These were the Industrial Research and Development Program (abbreviated in Greek as *PAVE*), evolving to the Research Workforce Enforcement program (known as *PENED*). In 1990 the first operational program for research and technology (known as EPET I) was approved and financed under the framework of the 1st Community Support Framework. Similar objectives with EPET I were addressed by another program, STRIDE HELLAS, in 1992. In 1994, the realization of the second operational program for research and technology, EPET II, was initiated, covering the technological evolution in the country for the period from 1994 to 2000.

Exploiting the experience stemming from the above, the GSRT is currently financing an extensive Operational Program on Competitiveness (EPAN), which aims at the upgrading of the technological competence of the Greek industries, the development of the business market competition, fostering new employment opportunities, and contributing to the general improvement of the quality of life.

In almost all of these programmes Transport subjects are directly or indirectly involved. One call within EPET II was specifically dedicated to Transport. The subject matter for this call was determined by a special study on the future technological needs of Greece in the field of Transport and (GSRT, 1999) which built upon the material of a previous similar study (GSRT, 1994).

The General Secretariat of Research and Technology (GSRT), as the beneficiary of the CSF, announces and finances research projects, which are undertaken in the country by consortia comprising of Universities, research centers, industries, or individual researchers, as well as research activities within the framework of international agreements with other countries.

A second major source of Transport research funding in Greece comes from the various research programmes of the EU. The overall EU research budget of the current Research Framework Programme (the 6th FP) i.e. for the period 2002 – 2006, totals approximately 20 billion EUROS. Transport research activities represent more than 7% of this budget i.e. approximately 1.5 Billion EUROS (1.2 Billion US Dollars). These funds are allocated via competitive research bidding and are channelled through various programmes of research which fall, broadly speaking, under three groups:

- The research Framework Programme of the EU, which finances very many research activities in all fields including Transport. This programme is constructed in 4-year periods (the current one is the 6th so we are in the 6th Framework Programme – 6th FP). The current FP co-finances (up to 50%) research projects covering all modes and fields of Transport as well as ICT (Information Communication Technologies) applications in the field of Transport. These research projects are assigned to Trans-European consortia through a competitive bidding process based on work-plans published before each call. There are approximately 150 research projects assigned so far concerning Transport, and these are supervised (according to their specific subject matter) by 3 relevant Directorate Generals (DG' s) i.e.: the DG on Research and

Technology (RTD), the DG on Transport and Energy (DG TREN), and the DG Information Society (DG INFSO).

- Various research programmes issued by responsible DG s in specific fields e.g.: for the Trans-European Networks in Transport (TEN-Ts), or the development of the so-called Sea-Motorways, and various other specific issues of EU' s Transport Policy.
- Other EU related programmes and funding agencies most notably the Regional Development Fund (ERDF) which finances the INTERREG programmes, the Marie Curie programme and other training and educational activities programmes, etc.

In the European R&TD programs, the Greek research teams cooperate with other relevant "actors" and industries from other member-states of EU. These programs are mainly the Framework Programs and are financed by the structural programs of the European Commission.

BASIC STATISTICS ON RESEARCH AND INNOVATION IN GREECE

The main data concerning the number and characteristics of the various entities that are involved in Transport research in Greece come from the data gathered by the National Institute for Transport Research and an earlier survey contacted by the General Secretariat of Research and Technology of the Ministry of Development (GSRT, 1999). From this survey we can mention some interesting facts, which may still be used to get a representative picture of Transport research in Greece.

The total number of (private and public) entities that were included in the survey was 350. Of these regularly involved in Transport research projects were found to be 60 while 100 more were involved occasionally in transport research projects.

From that survey the following results can be reported.

- The percentage of the total turnover devoted to research activities is shown in Table 1. Approximately one third of the companies spend less than 5% of their turnover for research, one fifth (primarily the dedicated research Institutes and Universities) spends more than 50%, while another 15% spends between 20 and 50%.
- The great majority (approx 65%) of the (60) organizations regularly involved in transport research projects, deals with general applied research subjects and does not specialize or focus in a specific field. The great majority of these subjects refer to ICT applications in Transport, efficiency improvements; organizational managerial improvements etc (see Table 2).
- Of the rest, one 10 % is focused in operation – exploitation of networks, 6% with fuels and energy savings, 10% with the vehicles, and a 5% with the infrastructure (materials, and method of construction).
- There is good availability of high standard educated personnel to do the research (researchers).

Table 1: Percentage split of organizations in terms of research spending as a % of turnover (1999 data)

% of entities	research spending as % of turnover

18%	Less than 1%
13%	between 1 and 5%
23%	between 5 and 20%
13%	between 20 and 50%
18%	More than 50%
15%	No reply

Table 2: Percentage split of research subjects (1999)

Principal field of research activity	% of organizations
Various (not focused)	37%
ICT applications in Transport	27%
Operation – exploitation of Networks	9%
Transport systems	6%
Fuels-energy savings	6%
Materials and method of construction of transport infrastructure	5%
Vehicles – rolling stock	5%
Methods of vehicle construction	4%
New concepts for transport systems	1%

In their totality the private organizations involved in transport research have done so through their participation in EU or Greek government funded research projects as part of a consortium. This has happened exclusively since the mid 1990's.

As regards the "weaknesses" that are realized within the Greek system of Transport research the following points have been noted:

- Difficulty to find the necessary "matching funds" in order to participate in EU funded research projects.
- Non-existence of databases and data in general on which to base new research.
- Not adequate support for transport research work on the part of the Ministries and other governmental Organizations.
- The general innovation and research infrastructure of the country is not satisfactory and is in effect a disincentive for undertaking the expense and complexity of advanced research projects.

The general conclusion of that 1999 survey was that while Greece has very competent and of high standard human resources for research work, it is left behind the other EU member countries as far as adequate soft and hard infrastructures for such research.

THE POSITION OF GREECE IN THE EUROPEAN RESEARCH AND INNOVATION SCOREBOARD

Greek participations in the various EU funded Transport research programmes is, relative to the country' s contribution and general position within the EU, very high. In the 4th Framework Programme (1995 – 98) Greek participations in the various Transport research projects accounted for 7% of the total budget (as compared to 1.5% contribution of the country to the funding of these programmes). In the 5th FP (1999 – 2002) this percentage was 5.6 %, and in the current 6th FP (2002 – 2006) it seems to have dropped even further. Although the figures of Greek participation in EU funded Transport research, in all FP's so far, are quite high relative to the country's financial contribution to these programmes, their downward trend must be attributed to the increasing beauraucracy and financial complexities of EU research funding as well as to the fact that there is an increasing difficulty to secure the remaining 50% of funds.

Some specific indicators of research activity can show expenditure and status of research spending in Greece as compared to that of other countries. Some results from the European innovation scoreboard (European Commission 2001A) are very interesting.

One such indicator is the public or private sector spending for research as a percentage of GDP. For private spending Figure 1 shows the relevant data for all countries of the EU plus US and Japan. Public sector in Greece contributes more than 70% of the R&TD expenditures,

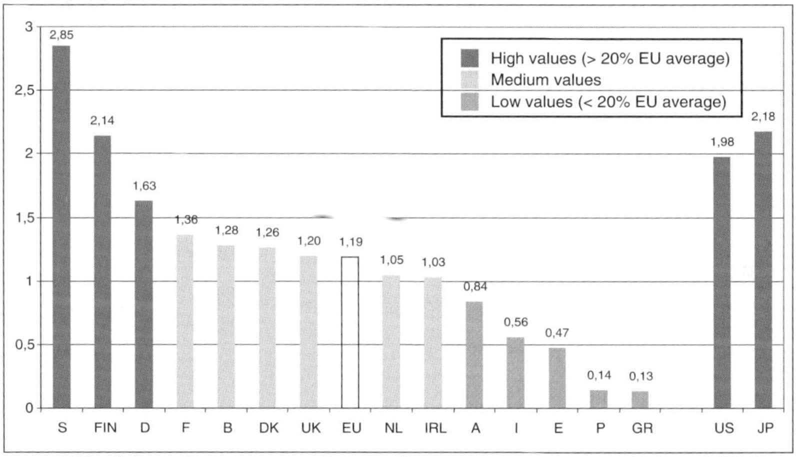

versus 25% of the private sector, whereas European averages are 44.2 and 53.9, respectively.

Figure 1: Private sector spending for research as a percentage of GDP in 15 EU countries (2001)

As shown in Figure 1, the private sector spending for research as a percentage of GDP was in 2001 for Greece 0.13 %, i.e. the lowest in all 15 EU member countries. This compares with an average 1.19% for the whole of the EU, 1.98% for the US and 2.18% for Japan.

As concerns public spending (Figure 2) the Greek percentage is one before the end with 0.38% as compared to 0.66% for the average EU figure, and 0.56 for the US and 0.70% for Japan (all figures for 2001).

The expenditure of private companies in new technologies and new products (innovations) as a percentage to their turnover is for Greece 1.6% as compared to 3.7% of the EU average (Figure 3).

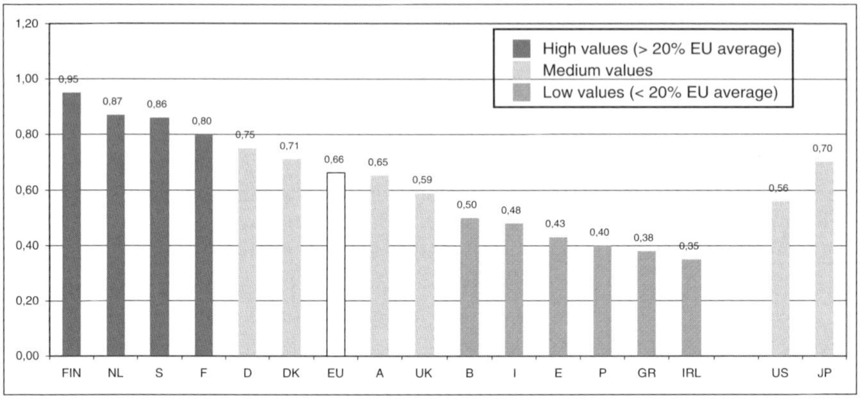

Figure 2: Public R&D expenditure as % of GDP (Government and Higher Education Institutions)

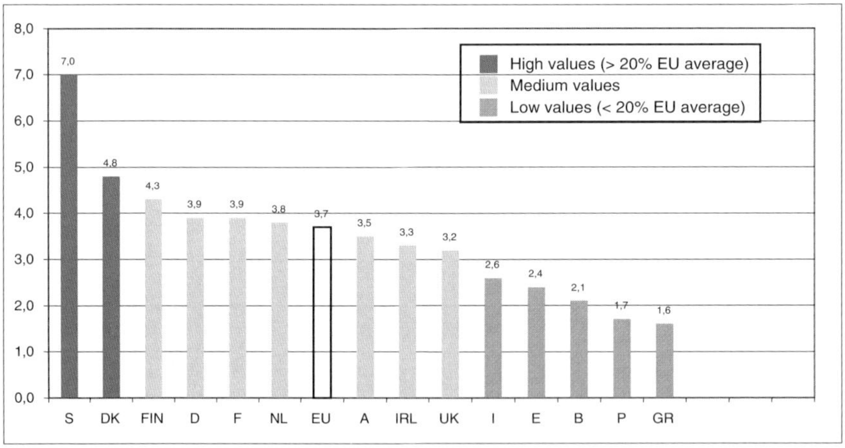

Figure 3: Innovation Expenditure (% of all turnovers in Manufacturing)

Similar results we find as concerns the indicator for the venture capital investments as a percentage of GDP (Figure 4). Greece turns out among the last countries again, with 0.04% as compared to 0.108% of the EU average.

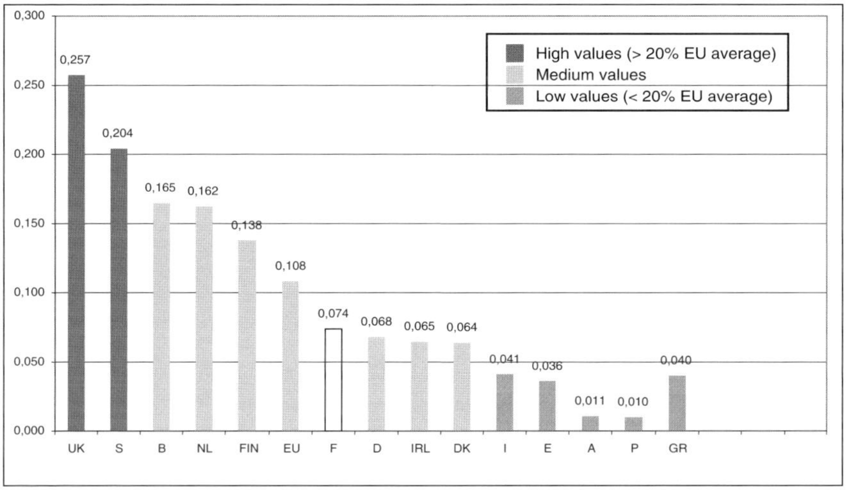

Figure 4: Innovation Finance output and markets (High Tech venture capital as % of GDP)

Of the 18 indicators shown in the scoreboard (European Commission 2001A) Greece as compared to the other 14 EU member countries, scores last in 8 of them, one before last in 3 of them, has a better position (towards the middle) in 4 of them, while no data are shown for the rest 3.

The indicators in which Greece scores better are:

- Percentage of population in highest education (16.9% as compared to 21.2% average EU).
- New capital for ICT companies, and
- Percentage of GDP that is represented by ICT products and services.

However, it is worth mentioning that in all indicators showing the rates of changes (i.e. improvement) over the last 5-year period, Greece comes first among all 15 EU member countries.

THE ORGANISATION OF RESEARCH FUNDING IN GREECE

The organization of research funding in Greece gives also an indication of the research supervision and monitoring structure in the country. The interested reader can find more details in terms of a number of indicators, shown below, in a recent publication of the Hellenic Institute of Transport (HIT 2004). The indicators are the following [1]:

- **GBAORD** – Government Budget Appropriations or Outlays for R&D

[1] These are the indicators for scientific and technological research, used by EUROSTAT and based on the methodology introduced by the well-known Frascati manual.

- o **GERD** – Gross Domestic Expenditures on R&D, which comprises also of the following more analytical indicators:
- o **BERD** – Expenditure on R&D in the Business Enterprise Sector
- o **GOVERD** – Government Intramural Expenditure on R&D
- o **HERD** – Expenditure on R&D in the Higher Education Sector
- o **PNP** – Expenditures on R&D in the Private Non Profit Institutions Sector
- o **R&D Personnel**, which is sub-divided in the following indicators:
  1. Business Enterprise Sector R&D Personnel
  2. Government Institutions R&D Personnel
  3. Higher Education Institutions R&D Personnel
  4. PNP Institutions R&D Personnel

Some basic facts are the following.

Public funds consist of funds from the Tactical Budget and the Public Investment Program, as well as of programs, which are co-financed by the Structural Funds, and the Community programs for R&D.

Private funds are those attributed to research co-financing of the private sector for "public" funded programs, or totally financed projects by the private sector.

As concerns the public funds these come from the following sources:

- o Overall Tactical Budget (Ministry of Education, Ministry of Development, Ministry of Agriculture, etc): Allocated to Universities, (University) and various research Institutes under the jurisdiction of each Ministry. A graphical representation of the distribution of these funds (indicator **GBAORD** above) is shown in Figures 5 and 6 below while Table 4 shows the percentage split of these funds among the various receiving Ministries and Organizations.
- o Public Investment Program: Financing of Scientific Technological Parks, National networks of R&T, etc (via the Ministry of Development).
- o GSRT' s Tactical Budget: Greek contributions to International research Organizations such as the European Science Foundation, International Foundation of Astronomy, European Space Organization, EUREKA, etc. Also financing of Research Centers and Institutes that fall under its jurisdiction.
- o GSRT' s programmes that are financed by the (EU' s) Structural Funds. Theses are placed under the so-called Community Support Frameworks (CSF) and are supervised by the Ministry of Economy and Finance.
- o VAT of technological equipment of GSRT: Financing educational institutes, research center and institutes.
- o Tactical budget and public investment program: Assignment studies, programs to public or private non-profit organizations under GSRT' s jurisdiction.
- o 1% of budget of the National Defense investment Programme allocated to defense research and technology.

The various outlays via which the government funds for R & D are distributed are shown in Figure 5.

The agencies involved in the distribution of Government budget appropriations for R&D are shown in Figure 6 while Table 3 shows the percentage split of these funds between the various agencies for 4 years in the recent past.

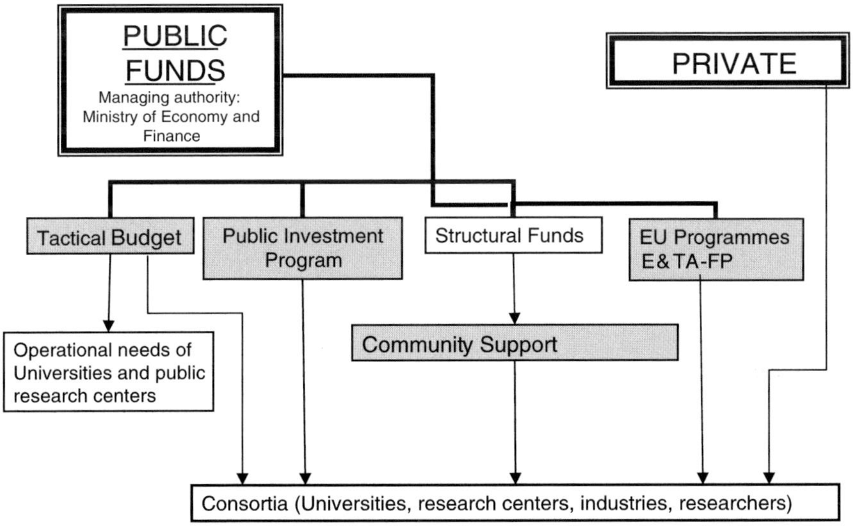

Figure 5: Source and outlays of funds for research and technological innovation development in Greece.

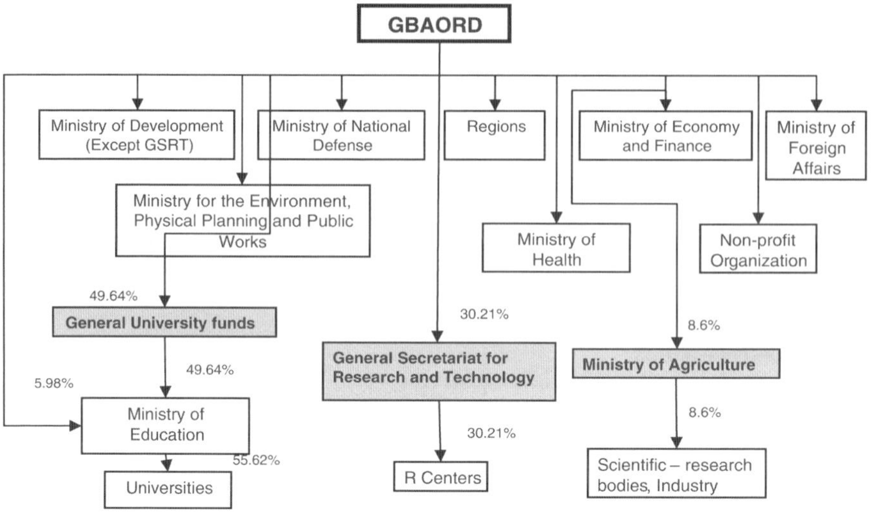

Figure 6: Agencies involved in the distribution of Government Budget Appropriations or Outlays for R&D in Greece (GBAORD).

We will close this section by referring again to the most well known indicator of all i.e. the proportion of the total Gross Expenditure for R&D (GERD) over the Gross Domestic Product (GDP) for Greece and the EU, for a number of years to show the trends. Figure 7 below, shows the relevant figures. It must be noted that the figures are for total (i.e. public and private) expenditure. Although this proportion has been increasing steadily over the past 20 years, it is for Greece very low as compared to the average for the EU. In 1999 it was 0.68 as compared to 1.93 for the EU for the same year (in red). This percentage dropped to 0.51 (for Greece) in 2001 and it rose again in 2003 when the relevant figures were 0.8 for Greece and 1.98 for the EU.

Greece will therefore have to spend much more and much faster on R&TD than the rest of the EU, especially now that the EU heads of state have recently pledged to increase the EU spending on R&TD to 3% of its GDP by 2010.

Table 3: Percentage spread of GBAORD in Greece, among the various Ministries and research Organizations (1998-2001).

Research funding Organizations	1998	1999	2000	2001
GSRT	34,79	33,56	31,91	30,21
Ministry of development (except GSRT)	1,64	2,72	2,00	2,11
Ministry of Agriculture	11,01	9,08	8,21	8,61
Ministry of National Defense	1,25	1,24	1,21	1,26
Ministry of Economy & Finance	0,69	0,95	0,91	0,92
Ministry of Education	49,57	51,34	54,37	55,62
Ministry of Foreign Affairs	0,08	0,04	0,03	0,04
Ministry for the Environment, Physical Planning and Public Works	0,28	0,16	0,21	0,06
Ministry of Culture	0,69	0,74	0,88	0,93
Ministry of Health & Welfare	0,00	0,17	0,10	0,00
Ministry of Labor & Social Affairs	0,00	0,00	0,00	0,00
Regions	0,00	0,00	0,11	0,23
Non profit organizations	0,00	0,00	0,06	0,01
Total	100	100	100	100

Source: GSRT

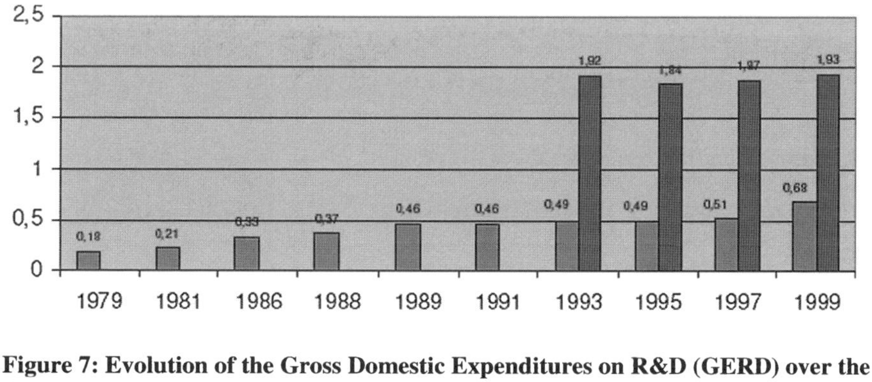

Figure 7: Evolution of the Gross Domestic Expenditures on R&D (GERD) over the Gross Domestic Product in Greece and the EU (in red).

TRANSPORT POLICIES IN GREECE AND THE EU

A. Basic principles and provisions of the EU Transport policy

In September 2001 the EU published the long awaited white paper detailing its Transport policy for the next decade. The title of this white paper speaks for itself: ***"European Transport policy for 2010: A time to decide"***[2]. In it the Commission states the prime goals of its Transport policy for the decade, its priorities in fulfilling these goals, and the policy measures to achieve them.

There are four prime goals set out as follows:

1. *Shifting the balance between the modes of Transport*
2. *Eliminating Bottlenecks in traffic flows in congested networks (all modes)*
3. *Placing the Users at the heart of Transport Policy measures, and*
4. *Managing the Globalization of Transport.*

Within each one of them, the white paper describes the priorities and the specific measures and actions that it will take to fulfill them. In all, some 60 specific policy measures are stated in the paper, which will be taken at EU level under this policy until 2010. At the same time specific milestones are set along the way, notably for monitoring exercises and a mid-term review of the policies followed (i.e. in 2005) in order to check whether the precise targets are being attained and what are the adjustments that are necessary. A most notable feature of the white paper is the specific mention (and hopefully commitment) that the Transport Policy is to be made consistent and adjusted continuously with regard to other relevant Commission policies namely: the *economic*, the *urban and land-use*, the *social and education*, the *urban transport*, the *budget and fiscal*, the *competition*, and the *transport research* policies.

By implementing these 60 policy measures the Commission expects that there will be, by 2010, a *marked break* in the link between Transport growth and Economic Growth, without there being any need to restrict the mobility of people and goods. For example they expect that between 1998 and 2010 there will be an increase of 38% in road haulage instead of 50% if current trends prevail, and in passenger transport by car an increase of 21% against a rise of 43% in GDP.

A brief reference to the main areas and priorities where these 60 policy measures refer (as well as the most notable of these measures) is given below.

1. ***Revitalizing the railways***: Here the priority is to open up rail markets with further improvements in interoperability and safety, not only in international services (as decided already in December 2000), but also in the national ones i.e. lifting of the cabotage principle. Also the commitment is made that a network of railway lines must be dedicated exclusively to goods services.
2. ***Improving quality in the road transport sector***: The measures foreseen under this policy area include: modernization of the way in which road transport services are

[2] Report no. COM (2001) 370, DG TREN, published 12th September 2001.

operated (while complying with the social regulations and workers rights), and tightening inspection procedures in order to put an end to practices preventing fair competition. Also to put up legislation to protect carriers from consignors and enable them to revise their tariffs in the event of rises in fuel prices.

3. ***Promoting transport by sea and inland waterways***. Short Sea shipping is seen as a desirable alternative to build "*veritable sea motorways*" within the framework of the Trans-European Networks. Also tougher rules on maritime safety, a genuinely European maritime traffic management system, as well as a tonnage based taxation system are to be set in place. For the inland waterways their position as "*intermodal waterway branches*" is foreseen, and modern transshipment facilities as well as revised inland waterway vessel characteristics are to be promoted.
4. ***Striking a balance between air transport growth and the environment***: Here a reorganization of European air transport is foreseen to create the "*European single sky*" as concerns air traffic management. Also to expand airport capacity while at the same time introducing new regulations to reduce noise and pollution caused by aircraft.
5. ***Turning "intermodality" into reality:*** This area is aimed at technical harmonization and interoperability between systems particularly for containers, and to promote "*sea motorways*" by targeting innovative appropriate initiatives. This, last, policy is to be effected through a new Community support programme called "***Marco Polo***".
6. ***Building the Trans-European Transport Networks:*** Based on the experience gained so far from the development of the Trans-European Networks (especially the development of the 14 priority projects adopted by the Essen European Council and the application of the guidelines adopted in the 1996 European Parliament and Council decision, the white paper states that the Commission will concentrate on the revision of the current Community guidelines for the development of the Trans-European Networks. This revision will aim at removing the bottlenecks in the railway network, and completing the routes that are identified as priorities for absorbing traffic flows generated by the enlargement, particularly in frontier regions. The new revision will also be aimed particularly at introducing the concept of "*sea motorways*", developing airport capacity, linking the outlying regions on the European continent more effectively, and connecting the networks of the candidate countries to the networks of the EU countries.
7. ***Improving Road safety:*** A number of actions are planned at improving road safety from exchanging good practices, to harmonizing signs (especially for dangerous black spots) and rules for checks and penalties for international commercial transport.
8. ***Adopting a common policy for charging for transport:*** The general principle is the equal treatment for operators and between the modes of transport as regards the price for using infrastructure. Two basic guidelines are adopted in that respect:
   - Harmonization of fuel taxation for commercial users, particularly in road transport, and
   - Alignment of the principles for charging for infrastructure use (integrating the external costs[3]).
9. ***Recognizing the rights and obligations of the users:*** Perhaps the most significant feature of this new Transport policy of the Commission is the recognition of the rights and obligations of the **users**. In this respect Community legislation is to be put in place for helping transport users to understand, and exercise their rights. For example, air

[3] As described in the so-called Costa report no. A5-034/2000 of the European Parliament.

passenger's rights to information, and compensation for denied boarding due to overbooking and compensation in the event of an accident.

10. ***Developing high quality urban transport:*** In this respect the Commission places emphasis on exchanging of good practices aiming at making better use of public transport and existing infrastructure. All measures to improve the quality of urban transport must be compatible with the requirements for sustainable environment and the Kyoto treaty provisions.
11. ***Putting research and technology at the service of clean, efficient transport:*** Under this area specific actions are promised for cleaner, and safer road and maritime transport and on integrating intelligent systems in all modes to make for efficient infrastructure management. Specific mention is made to the expectations from the new 6th Framework Programme (6th FP) for Research and Development and the new policy for creating the integrated "*European Research Area – ERA*", and the *e-Europe* action plan. Also in line with the policy priorities under the previous areas some specific foci for the research, are mentioned for: safety standards in tunnels, harmonization of the means of payment for certain infrastructure (particularly motorway tolls), and improving the environmental impacts of air transport (noise, safety, and fuel consumption).
12. ***Managing the effects of Globalization:*** This area calls for actions that will strengthen the Commission's position and presence in international Organizations concerning Transport, such as the International Maritime Organization (IMO), the International Civil Aviation Organization (ICAO), and the Danube Commission.
13. ***Developing medium and long-term environmental objectives for a sustainable transport system:*** This area aims at creating a package of proposals for measures that if implemented by 2010 will re-direct the common transport policy towards meeting the need for *sustainable development*. The priority is set to proceed to the adoption of pro-active measures (some of them admittedly difficult to accept by the public) for the implementation of new forms of regulation in order to channel future travel demand for mobility and to ensure that the whole of Europe's economy develops in a sustainable fashion.

B. The Greek Transport Policies

For Greece the question of setting and implementing a coherent Transport Policy has been a difficult one for many years. All Greek governments of the last 20 years have almost unanimously followed the policies of the EU, implementing them with a time delay of several years.

It is now of imminent importance, and has been suggested by this author repeatedly in the past (see for example Giannopoulos, 2004A and 2004B), to change this state of affairs. The following thoughts may offer a good indication as to the axes along which a new Greek Transport policy should develop.

For many decades the main preoccupation of Greek Transport policy was to keep open and relatively inexpensive the transport routes connecting Greece with the western European countries of the EU. The same policy was followed even after Greece's accession in the EU, in 1981. This position was influenced by the notion that Greece was a so-called "peripheral" country i.e. one that needed the use the infrastructures of other countries in order to reach its vital markets. Today, after the geopolitical changes that occurred in this part of the world after 1990, things have changed radically. Greece is now at the crossroads of 4 major political and economic regions that create substantial transport flows, which could usefully use the Greek transport infrastructures as their major transit routes to and from western and central Europe. These regions are (see also Figure 8).

- To the West is the bulk of the EU "old" member states to which Greece is a member since 1981.
- To the North are the countries of the old Eastern block many of which have recently (as of 1st May 2004) become full members of the EU, and the rest are (or soon are going to be) candidate countries.
- To the East is the fast developing region of Turkey, which may soon become a candidate country to join the EU, and in any case is associated with it.
- Finally to the South there is Israel as well as the various Arab countries with predominant one Egypt and Saudi Arabia. This region after many years of war and conflict is expected to calm down and prosper hopefully within this decade.

Compared to all these regions, with the exception of the first one, Greece has the highest level of development and political and social stability. It is therefore fair to say that she can play a more central role and it is no longer a "peripheral" country. On the contrary it is one that could play the role of an important transport node offering high level transport and logistics services and making an effective bridge linking the developing countries of the region to the developed ones of the EU and western and Northern Europe.

Correspondingly the new Greek Transport policy, while of course following the basic lines of the EU Transport policy, must focus on the development of its infrastructures and services to become this "central South Eastern European Transport node" discussed earlier. This new course should take into account the 5 major transportation axes that exist in the area. These are shown in Figure 9 below and are the following:

1. The Noth – North Eastern axis via Bulgaria, Rumania and Hungary.
2. The North one, via the new countries of the ex Yugoslavia (axis no 10A).
3. The Western axis of the Adriatic Sea linking Greece and Italy.
4. The Southern one from Suez, which could usefully be diverted to a major Transportation hub that is recommended to be developed in Northern Crete (see Figure 9).
5. The East – Western axis via Ismir in Turkey to the port of Volos in Greece and then linking to the Adriatic axis (see Figure 9).

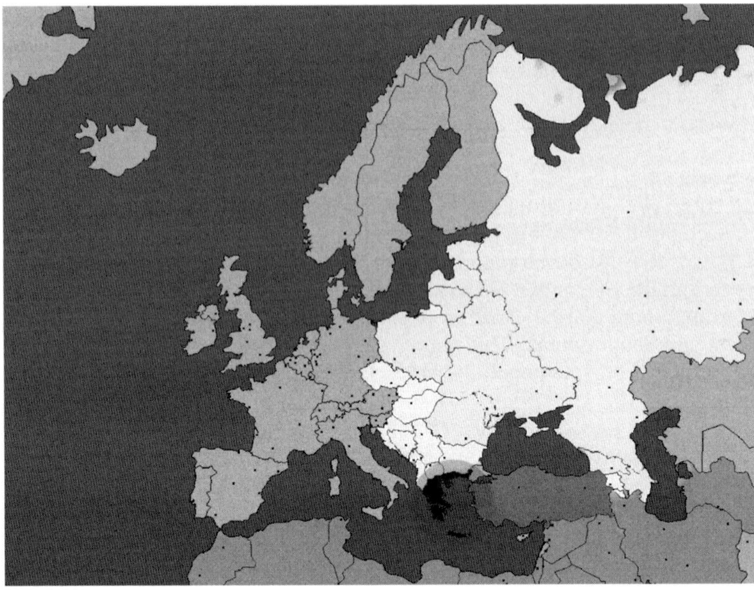

Figure 8: The "central" position of Greece within the 4 sociopolitical regions of South Eastern Europe.

The view of this paper is that the development and proper functioning of all these 5 major axes and their related infrastructures, that would provide full and technologically advanced transport and logistics services, should be the focus of the Greek Transport policy for the next decade.

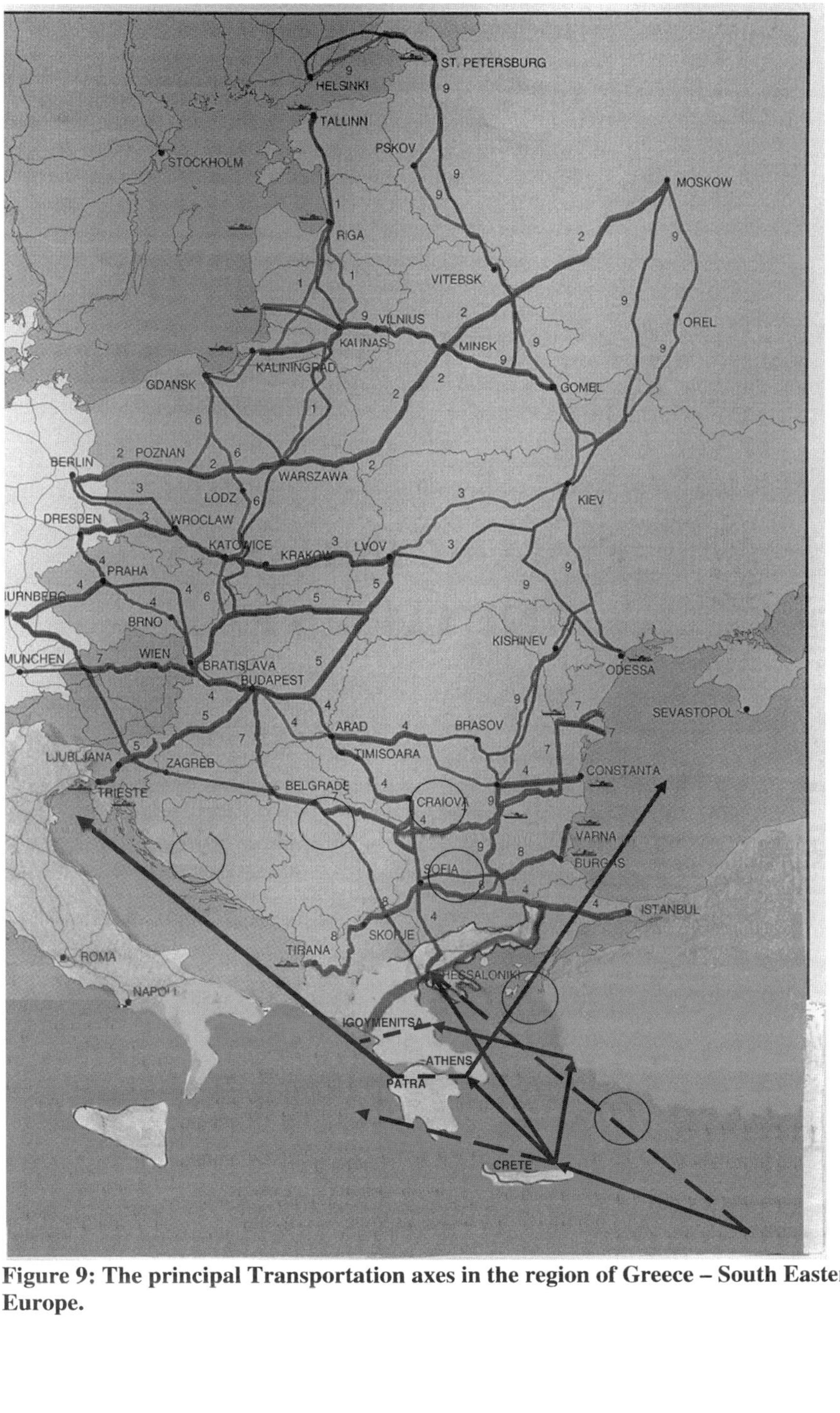

Figure 9: The principal Transportation axes in the region of Greece – South Eastern Europe.

We can also see the following as further challenges for Greek Transport Policy in the enlarged European Union and the new socio-political realities:

1. An overall task is to harmonise transport policies in the region with the EU-Strategies on the one hand and the needs for sustainable development on the other. This means to "decouple" transport growth and GDP growth by the shift from road to rail, water and public passenger transport. Today, in the region of South East Europe 79% of passenger transport and 44% of goods transport is done by road. If nothing is done, CO2 emissions from all transport should increase by 40% by 2010. Enlargement will contribute to the growth of transport needs in the next decades. The growing need for individual mobility and the transition to an global economy will lead to 38% more goods transport and 24% more passenger transport in Europe as a whole, with much higher rates for South Eastern Europe. If the current development continues, transport of heavy vehicles will increase by 50% by 2010. In view of the already existing problems of congestion in the area, this would hardly be tolerable.
2. Together with enlargement, there is an obvious need to make other modes of transport (other than road) more attractive. The European Council of Göteborg has called for a shift of balance between transport modes, basically from road to rail, inland waterways and short sea shipping. The objective of the Commission is to stabilise in 2010 the repartition of transport modes at the level of 1998. This implies to reduce the growth of road transport, whereas rail and inland waterways should triple its growth figures. In our view, this is a realistic and an ambitious objective at the same time. The example of Japan (passenger) and USA (freight transport) shows that railways can successfully operate in developed societies. This shift towards rail must be pursued by the Transport policies of the Greek and the other governments in the area for the next decade and beyond. Enlargement is an opportunity to for railways: the importance of railways in the Eastern part of Europe is higher than in Western Europe (40% vis-à-vis 8%). A feasible goal here should be to maintain the modal share of rail freight transport in the area in 2010 at 35%. Railways could play an essential role in solving transport problems: less pollution, less congestion and fewer accidents. Thus, the revival of railways must be one of the priorities of the transport policy in the area. There is a need therefore for putting in place urgently a "railway package". The key contents of this new railway package should be:
   - The opening of the railway market to competition, including the separation of the infrastructure developer and manager, responsible for the network infrastructure, and the railway undertakings (operators). The basic legislation has been put together by the Commission in Brussels, but the various countries must adopt and specify the details in their National legislation. This has not been done by any of the governments in the area yet.
   - Facilitating full access to the (rail) network of the area as well as of the rest of the European Union's by defining clear rules for the allocation of capacity and improving the infrastructure to key end users.
   - Improving the railway infrastructure at some carefully selected key sections so as to make the maximum of improvement in the operational parameters of the network while spending reasonable amounts of money i.e. within the financing capabilities of the governments in the area.
   - Promoting combined transports (and primarily road / rail) in the area while creating a network of functional Transportation nodes integrated and working together as a system.

3. Improvement of the existing rail network infrastructure is another challenge. Infrastructure has to be improved and bottlenecks have to be eliminated. Insufficient networks in the Candidate Countries and congestion all over the EU may seriously affect our chances of pursuing the sound policies outlined above. The Trans European Networks in the area should be adapted to the needs of the new Member States and to the increasing transport needs from East to West and from South-eastern to Central Europe.
4. The maritime transport network provides a number of challenges too. After the accession of Malta and Cyprus to the EU, the EU merchant fleet will almost double. Romania and Turkey have also important fleets. In the accession negotiations, the EU has emphasised the need to enhance safety in maritime transport in Europe and to promote short-sea shipping. The European Union has adopted an important package of new legislation on maritime safety – the so-called "Erika package". Since EU Member States and Candidate Countries like Romania and Bulgaria will have to implement these measures, maritime safety in the Black Sea and in the Eastern Mediterranean should be seen as a major challenge in the Transport policies in the area too. The EU must continue to assist the new Member States to build up the administrative capacity in the maritime transport sector notably by training inspectors and administrative staff responsible for enforcing its maritime transport legislation.

THE HELLENIC INSTITUTE OF TRANSPORT

The *Hellenic Institute of Transport (HIT)* is the National Organization devoted to the promotion and execution of Transport research in Greece. It was established in March 2000, by Presidential Decree 77/2000 as part of the National, *Center for Research and Technology Development Hellas (CERTH)*. It is a "private status" legal entity under the supervision of the General Secretariat for Research and Technology of the Ministry of Industry, Research and Technology. It is based in Thessaloniki, Greece.

Its Director, substituted by a deputy Director, manages the Institute. The Institute's policy is formulated in consultation with a five member Scientific Council of the Institute (SCI), which includes senior members of the Institute's personnel. The ultimate deciding body is the Governing Board of the CERTH.

The basic scope of services of HIT is to provide a center of excellence in the field of Transport with highly specialized research services offered to government and third party organizations and bodies, and providing support for the conduct of Transport research in Greece. It is also devoted to providing support for the formulation of Transport policy by the relevant Ministries and other government bodies.

The scope of services covers all areas of Transport and in particular the organization, operation, planning, construction of infrastructure, standardization, economic analysis, management, vehicle technology, and impact assessment, of land, maritime, air, and multimodal transport services. The HIT co-operates and interacts with similar organizations and Institutes of the EU and other countries, and represents Greece in the relevant international fora.

The specific areas of HIT's priority activities, can be described as follows :

- Scientific and research support for Transport Policy formulation, to Ministries and other Organizations involved in Transport Policy and control in Greece.
- Specialized research in the field of Transport.
- Organization and operation of a documentation center in the field of Transport.
- Development and maintenance of Data Bases covering important areas of Transport Operation in Greece.
- Transport Research evaluation and appraisal.
- Support of Standardization work in the field of Transport, and issuance of handbooks, rules and guidelines concerning the operation of the Transport system.
- Representation of Greece, in Transport Research and other relevant scientific fora abroad.
- Investigation of user requirements, and adaptation of (transport) research results to industry and users needs.
- Transfer and integration actions of Transport research with the activities and needs of the "Transport Industry" and the Transport Users.
- Organization of Training and professional education Seminars and Programmes.
- Contribution to quality control in the field of Transport.
- Publication and dissemination activities (including Conferences and regular publications).
- Promotion of bilateral as well as multilateral co-operation between Greece and other countries in the field of Transport, with emphasis in the countries of South East Europe.
- Organization of exchanges and placement of young scientists to relevant organizations and companies for practical experience.

Although all areas of Transport research are "covered" by the Institute's activity and scope, it is to be noted that particular emphasis and priority is given to research with a view to developing knowledge and expertise on the particular Greek conditions and requirements that influence the operation of the Transport system and the development of its infrastructure.

The HIT operates in an independent fashion within its scope and objectives. Its internal organization reflects the priorities and scope of its services providing at the same time the necessary flexibility to pursue new directions of research to suit the changing requirements of Transport users in the country. It has permanent personnel (Researchers of A, B, C, and D category), and personnel under research contracts of specific or unspecified time duration. It also employs outside experts and counselors, and has permanent co-operation with the country's major University research centers and Laboratories. It benefits from the administration services of CERTH and it is subject to an annual audit by registered auditors.

Besides its permanent staff, many outside scientists, engineers, planners, and economists specializing in the field of Transport, pull their resources together to offer services to the Institute.

Agreements of co-operation exist with many National Institutes and Universities in other countries. Since February 2003, the Director of HIT (Prof Giannopoulos) is chairman of the

European Conference of Transport Research Institutes (ECTRI) which is the European body encompassing all the major Transport research Institutes of most European countries.

CONCLUSIONS

Transport research in Greece although one of the most developed sectors of research in general, is still very much below the desired levels of funding at least when compared to the rest of the EU countries. The main financing source of the research in the country is the government via a number of channels and funding lines most of which have directly or indirectly their origin in EU' s funds. The GSRT remains the largest source of funding for this research and it is no coincidence that the Hellenic Institute of Transport (National Centre for Research and Technological Development) was created by GSRT, which supervises it and partly finances it.

Transportation research in both Greece and the EU is seen as contributing to the achievement of the overall Transportation policies and is orientated accordingly. Thus, the main challenges for transport policies must be seen as challenges for the research work too. These (transport policy) challenges for the Greek government should be adapted to the particular needs and objectives of a "Greek policy" which should be:

- Strengthening of the new "central" position role of the country to become a nodal point in the international transports of the area.
- Linking effectively with the transportation networks of the neighbouring countries and of the Union as a whole.
- Harmonising the various National legislation provisions to those of the EU and among themselves.
- Improving safety and efficiency of the networks in all modes of Transport (by a combination. Of improvements in infrastructure, and ICT applications).

For all the governments in the area of South Eastern Europe - Eastern Mediterranean the objectives should be:

- To strengthen the position of railways by priority in the freight market by restructuring and modernising railway companies and opening the market. Measures in this respect would be to open to competition the railway market, facilitate access to the network, and promote combined (road / rail) transport.
- To improve infrastructures at key sections of the network, in order to reduce infrastructure bottlenecks and to successfully integrate the international rail and road network of the area into the overall Trans European Networks. Private capital will have to play an important role there.
- To implement and enforce EU standards in the maritime transport concerning safety and efficiency (Erika and other packages).

REFERENCES

European Commission (2001A), "European Innovation Scoreboard 2001", Document SEC(2001)1414, published in *CORDIS Focus* Supplement, CORDIS issue no 18, September 2001.

European Commission (2001B), "White Paper: European Transport Policy for 2010: A time to decide", Document COM(2001)370, Brussels 12/9/01.

GSRT (General Secretariat of Research and Technology) (1994), "Study for the forecasting of impacts of new technologies in the field of Transport in Greece", TRADEMCO Consultants Athens, March 1994.

GSRT (General Secretariat of Research and Technology) (1999), "Technological priorities and needs in the field of Transport", TRUTh SA Consultants Athens, November 1999.

HIT (2004), "Transport research in Greece (Basic statistical data, Organizational and financial issues, related to demand and supply of research)", Hellenic Institute of Transport, National Centre for Research and Technology, ECTRI report, September 2004.

Giannopoulos G.A (2004A), " Towards a new Greek Transport Policy in the framework of the country' s new geopolitical position", Paper presented at the 2nd International Conference on Greek Transport research, Hellenic Institute of Transport and the Hellenic Institute of Transportation Engineers, Athens February 2004.

Giannopoulos, G.A. (2004B), "Transport policy issues for South East Europe", paper for World Conference on Transport Research, Istanbul 2004.

K.G. Goulias, editor

CHAPTER 7

PLANNING ATHENS TRANSPORTATION FOR THE OLYMPIC GAMES AND A FIRST EVALUATION OF RESULTS

J.M. Frantzeskakis, Professor Emeritus, National Technical University of Athens

ABSTRACT

The chapter summarizes the huge effort made in the transportation sector and the difficulties encountered as well as the large possibilities to exploit, after the games, the new and improved infrastructure and the outcome of the systematic effort to implement Transportation System and Demand Management and the related Information Campaigns and Police Enforcement. After the assignment of the games to Athens, following a period of reduced activities, a Transportation Division of the Organizing Committee for the Athens 2004 Olympic Games (ATHOC) was created. With the aid of Consultants, they completed in March 2001 an Olympic Transport Strategic Plan and in March 2002 an Operational Plan and Programme based on the proposals made in the Chapter on Transportation of the Candidacy File. While following the progress and coordinating the efforts made by the numerous other agencies involved, the Transportation Division and their Consultants prepared estimates of Olympic and normal traffic movements through appropriate established models (EMME2, SATURN etc.) and models prepared especially for Athens (DEMAND). Specific studies and designs followed on the access to the venues, estimates of needs for Olympic vehicles and related fleet management etc. The transportation projects carried out by various Ministries and Agencies are summarized. Conclusions and a post-games appraisal are given.

INTRODUCTION

Transportation Planning and Implementation for the Olympic Games is a very complex task which becomes more and more difficult as more athletic and other events are added and the size of the Olympic Family as well as the number of spectators increase. In the case of

Athens, the extra safety measures taken because of the increased worldwide threat for terrorism has created additional difficulties.

Although the number, characteristics, origin/destinations and routes of the trips of the Olympic Family and Spectators (visitors or inhabitants of the city) to and from the athletic and other programmed events (e.g. opening and closing ceremonies) are more or less known and therefore can easily by simulated through computer programmes, the normal trips of the city as well as all other trips of the visitors cannot be easily predicted. Furthermore, the continuous variations of flows due to the changing origin and destination of the trips (various athletic and non athletic venues, varying arrival and departures times and number of spectators etc.) as well as the mixing of Olympic and normal city trips, increase the difficulty to predict the size and location of peaks in traffic flows.

The experience from previous Olympic Games is extremely useful in predicting flows and planning/implementing the proper traffic system and demand management. Unfortunately, although a large amount of information, data, analysis and evaluation of results exist for every city where Olympic Games took place in the past, no systematic comparative studies are available to be used as starting point and as guidelines for planning/implementing transportation in the future. It is suggested that I0C should consider the assignment of such a comparative study and preparation of guidelines to an international group of experts involved in the transportation planning and implementation of the most recent Olympic Games.

Furthermore, I0C should also consider minimizing the continuous increase of the athletic activities removing games and not adding new ones, also allowing for a better distribution of activities in more than one cities of the host country. In Greece no new games were added and it was allowed to carry out certain soccer games in four large cities (Thessaloniki, Patras, Volos and Iraklion) and the shot put in the Stadium of Olympia, in an imposing environment where the ancient Olympic Games were taking place. On the contrary it was not allowed to carry out rowing in the excellent lake of Ioannina (380kms from Athens), the established location of such games in Greece. As a result very expensive installations were made in Shinias, in spite to the strong opposition of Environmental Organizations.

This chapter summarizes the huge effort made in the transportation sector and the difficulties encountered as well as the large possibilities to exploit, after the games, the new and improved infrastructure and the outcome of the systematic effort to implement Transportation System and Demand Management and the related Information Campaigns and Police Enforcement. Conclusions and a post-game appraisal are given.

CANDIDACY FILES

In the Transportation sector, the effort started at the late 80's with the preparation of the Candidacy File for the 1996 Olympic Games. It was imperative to convince the International Olympic Committee that our small country could ensure the proper conditions in a metropolis experiencing the worst traffic problem in Europe, where inhabitants have a reduced traffic consciousness, are not always properly informed and where police enforcement and traffic management are not systematically implemented. For this reason, in the Candidacy Files for the 1996 and 2004 Olympics, a large number of construction projects to expand or improve the transportation facilities as well as a systematic traffic management effort were programmed. Thus, the drawbacks of the Athens traffic situation at that time were converted into advantages because of:

1. The additional capacity provided through the numerous projects under Construction or Planning/Design to be completed in time for the games

2. The large margins for improvement through Systematic Traffic Management and Police Enforcement and the use of new Technologies

3. The favorable locations of all major trip generators along or near an "Olympic Ring" which offer hourly capacities ranging from 6.000 to 12.000 vehicles anded alternative routings. Olympic lanes for the exclusive use of the Olympic Family and special express buses for spectators secured unhindered movement for these critical categories

4. The fact that the completion of the above improvements just before the Games will leave no time margins for saturation

5. The existing experience in facing special traffic problems by complete or partial (even and odd numbers) prohibition of private cars circulation in two predetermined areas, the 13sq.kms "inner ring" covering the central area and the 140sq.kms "outer ring".

Thus, in the report of the Evaluation Committee, no negative comments were made for the traffic in Athens, as made for the other major Candidate City, Rome.

OLYMPIC TRANSPORT STRATEGIC AND OPERATION PLANS

Following a considerable starting delay, the Transport Division of the Organizing Committee for the Athens 2004 Olympic Games (ATHOC) was organized. With the assistance of Consultants and in close cooperation with the competent services of the Ministries involved (mainly Environment, Physical Planning and Public Works-Transport and Communications-Public Order) they prepared an Olympic Transport Strategic Plan (ATHOC 2001) and subsequently an Operation Plan (ATHOC 2002) which became the basis for programming and implementing all related studies and designs.

Exploiting the experience of the more recent Olympic Games, especially those of Sydney attended by staff and consultants of the Transport Division of the ATHOC, some of whom have worked on various posts, the Olympic Transportation System proposed in the Candidacy File was checked and finalized. A *Strategic Plan,* showing the venues and the transportation network (Fig. 1) and describing the objectives, basic strategic directions and programmes was prepared, considering the special conditions of Athens.

It should be stressed that the Road Network shown in the Strategic Plan has been totally implemented, except for two limited sections i.e., the extension of Kimis Avenue from Attiki Odos to National Road No. 1 and the section of the Attiki Odos-Rafina Freeway near Rafina. Both the above sections were not proposed in the Candidacy file.

In the *Operation Plan*, a first elaboration of the Strategic Plan was carried out concerning Policy and Priority Measures, Transportation System and Demand Management, Communication Policies and Public Information, Test Events, Special Needs etc.

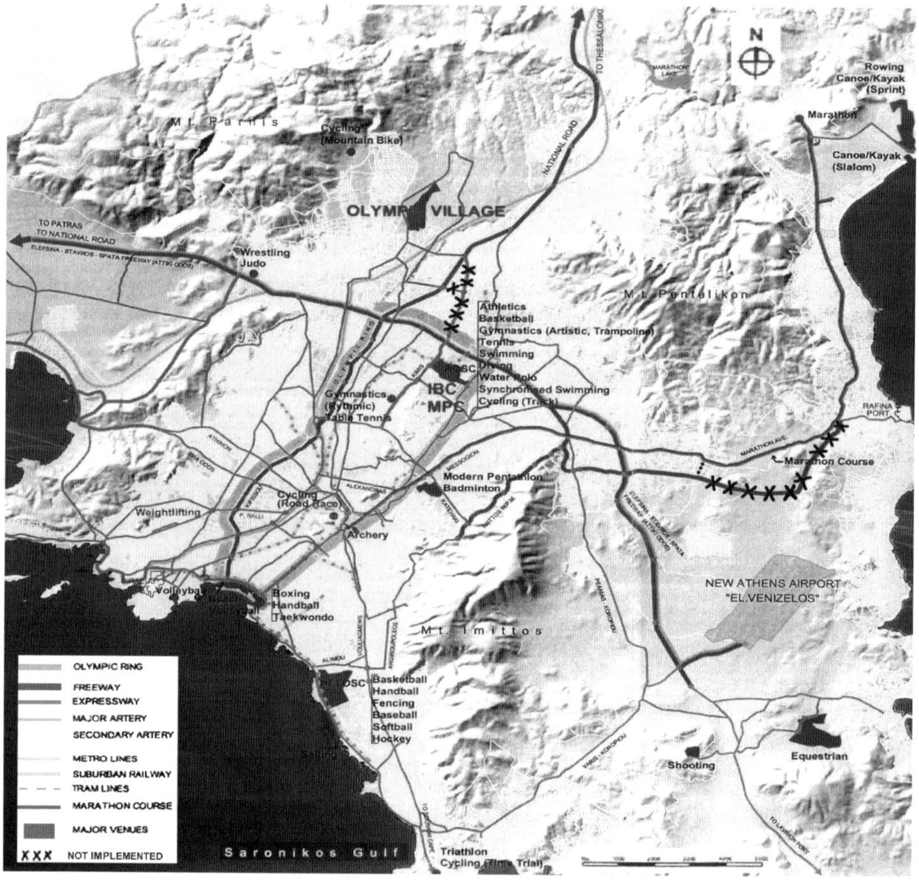

Figure 1. Olympic Transport Strategic Plan

COMPUTATIONS, FUNCTIONAL DESIGN OF ACCESSES AND TRAFFIC MANAGEMENT

Expected Olympic and normal city movements were estimated for the whole city and for the access to the various Olympic venues at critical periods through *EMME 2* Model. *SATURN* Model was used to make more detailed estimates for the access to all athletic and the major non-athletic venues. Programme *DEMAND* was prepared for the Athens Olympic Games to estimate trip demand in half hour periods for each venue, category of users (Olympic Family, Spectators), mode of transport and games programme (DENCO et al 2003). All the above estimates were updated when necessary (e.g. adjustments in Games programme). The SATURN model output, given in Fig. 2, illustrates Olympic family vehicles and spectator buses traffic flows estimated for the morning peak of August 20 on all Olympic Avenues.

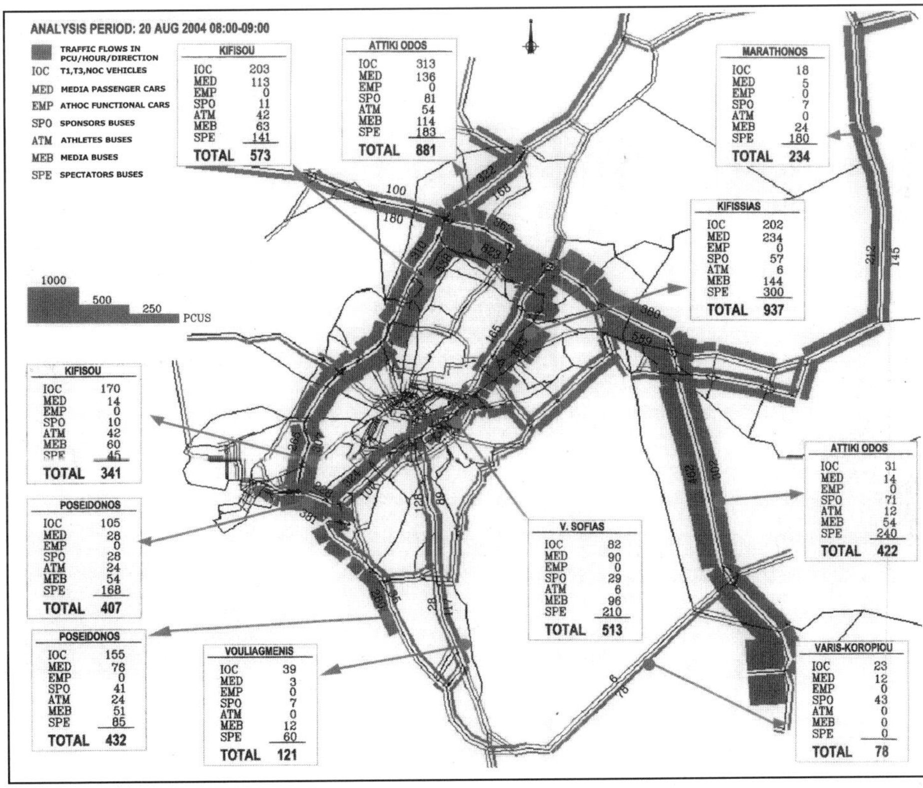

Figure 2. Predicted Traffic Volumes of Olympic Family Vehicles and Spectators Buses

On the basis of the traffic flows estimated through the above models, functional designs were carried out for all critical road sections and intersections, Park and Ride areas, Olympic Lanes, Express bus lines to venues, etc. Furthermore, the need for special management measures were studied. Parking Control Zones (ZES) were defined around all athletic venues to discourage the use of private cars. In these zones parking was allowed only for residents and employees who where supplied with special cards. Zones of Controlled Entrance and Traffic (ZEEK) were established at critical locations near the venues, where traffic was allowed only to accredited vehicles. Fig. 3 illustrates the access road network in the area of the Athens Olympic Sports Complex (AOSC) and the ZES and ZEEK areas around the Centre, while Fig.4 illustrates estimated peak hour Olympic traffic volumes on all access routes to the AOSC.

Olympic Lanes for the exclusive use of the Olympic Family, the express spectators buses to the venues and emergency vehicles were provided along the Olympic Avenues. Restrictions, such as banning of trucks at certain areas and time periods, prohibition of turning movements, enforcement of parking prohibition at critical locations etc. were established according to the estimated traffic flows. Furthermore, additional restrictions were studied to be applied at certain locations and periods in case they will be justified by actual traffic flows (even-odd number circulation, complete banning of private cars etc.).

ATHENS TRAFFIC MANAGEMENT STUDIES

Parallel to the above Traffic Management Studies made by ATHOC 2004 especially for the period of the Olympic Games, the Ministry of Environment Physical Planning and Public Works has assigned 5 Traffic Management Studies for 5 sectors covering the whole of Attica in order to increase the capacity and level of service of the main road network through local low cost improvements such as road section and intersection improvements, prohibition of left turns, systematic enforcement of parking prohibition at critical locations, etc. Except for limited cases implemented only for the duration of the Games (e.g. prohibition of left turns where they hinder the movements on the special Olympic lanes), these improvements aim to exploit the maximum possible capacity of the main road system of the city after the Games. Unfortunately, due to starting delays, some of these improvements, especially those for which the Municipalities had initial objections, were not implemented before the Olympic Games.

OLYMPIC ROAD PROJECTS

Beyond the Road Projects, programmed for the Athens Area before the assignment of the Olympic Games, which were completed in time (Attiki Odos Freeway, National Highways etc.), a special programme was prepared for the immediate implementation of a large number of additional road projects ("Olympic Highway Projects") (Table 1). All these projects, included in the Candidacy File, were also completed in time before the Games.

Figure 3. Athens Olympic Sports Complex. Access Road Network and Control Zones

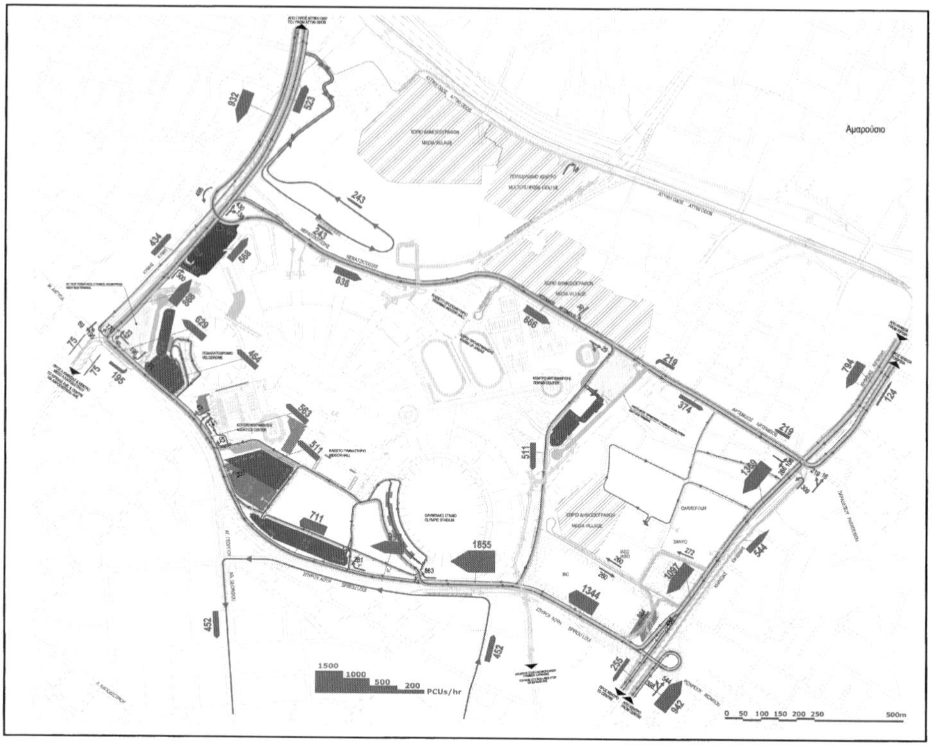

Figure 4. Predicted Olympic Traffic Volumes at the Athens Olympic Sports Complex 27 August 2004, 08.00-09.00

Table 1. OLYMPIC HIGHWAY PROJECTS

		Cost m €
1.	Interchange Kifissou Ave. -Posseidonos Ave.	46
2.	Extension Kifissou Ave., 3.6 km	247
3.	Poseidonos Coastal Ave., Interchange Alimou	16
4.	Poseidonos Coastal Ave., Ag. Kosmas- Elliniko	12
5.	Kiffisias Ave., Interchange Farou Psihikou	15
6.	Traffic Improvements in the Area of the AOSC	66
7.	Kimis Ave. from E.O.1 to Olympic Village	46
8.	K. Souliou Ave. and Sxoinia Ave.	22
9.	Marathon Route, 26 km	53
10.	Staurou-Rafinas Ave.	41
11.	Access to Equestrian Centre	12
12.	Connection to Shooting Centre	3
13.	Varis-Koropiou and Koropiou By-pass, 11 km	58
	Total:	**637**

PUBLIC TRANSPORT

Special attention was given to improve Public Transport, considering the fact that all spectators movements were programmed to be served exclusively by Mass Transportation. Besides the measures to discourage the use of private cars such as the Parking Control Zones mentioned above, special express bus lines to the venues were introduced and the frequency of selected existing bus lines was increased. Furthermore, two new fixed rail systems were introduced: a tramway connecting the western coast to the centre and a suburban railroad to the airport. The existing new metro lines were extended and the old metro line, servicing the Piraeus Port and Centre, the large sports complexes, the Athens City Centre, and the airport, was considerably improved (capacity, stations etc.) (Fig. 5).

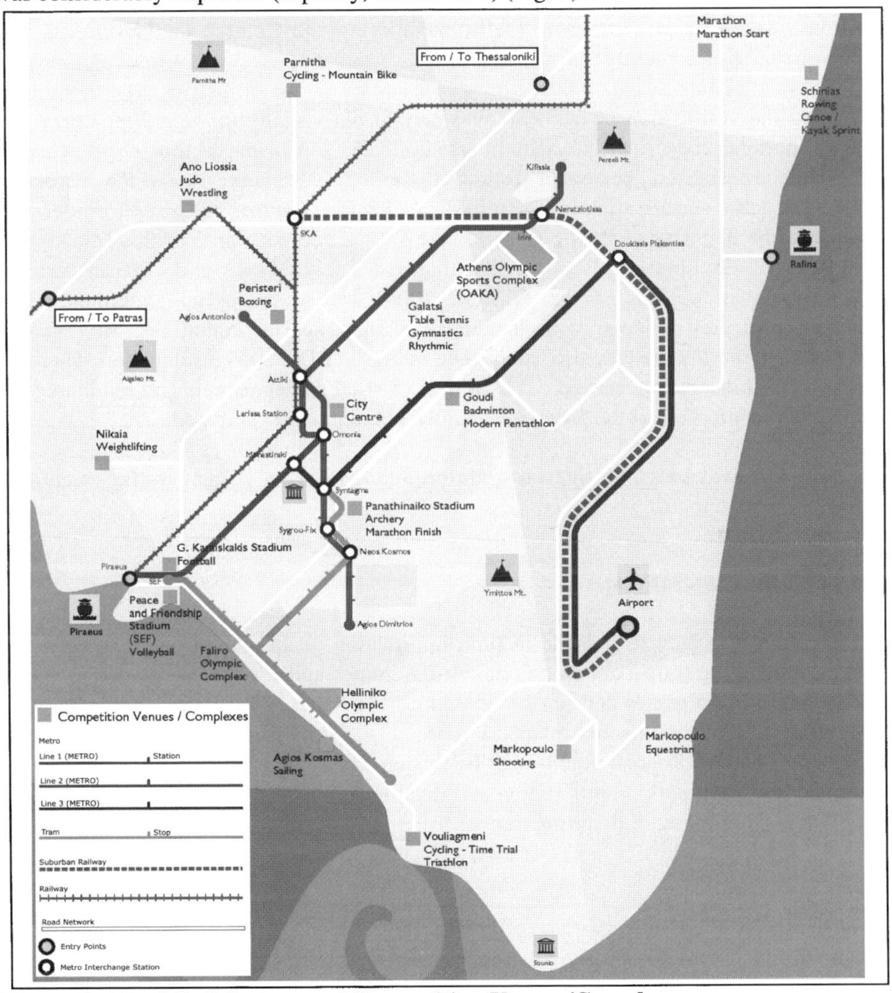

Figure 5. Fixed Rail System and Competition Venues/Complexes

SIGNING

Special Olympic information signs were provided on the accesses and within the venues. Special horizontal and vertical signs were also provided along the Olympic Lanes. Furthermore, the inadequate signing system of the city was improved.

SIGNALIZATION AND TRAFFIC CONTROL CENTRES

The existing signalization system in the Region of Attica was upgraded and extended by replacing the existing installations and incorporating in the Central System 150 isolated intersections. An advanced system of Traffic Monitoring and Control was developed. The collection of traffic data (number of vehicles, speeds occupation etc.) was carried out through 75 machine vision cameras and 2000 detectors located on all major arteries. The monitoring of traffic and the verification of incidents was carried out visually through 208 supervision cameras (Chouliara, 2004). All information was collected and managed in a *Traffic Control Centre* where specialized personnel followed the operation and made the necessary adjustments to the signalization programmes. A *Traffic Control Operation Room* was established in the Ministry of Public Order where specialized personnel of this Ministry, of the Ministries of Environment Physical Planning and Public Works and of Transport and Communications as well as of the ATHOC 2004, were deciding on additional traffic management measures to cope with incidents or other special conditions. An *Olympic Transport Operation System* was also established in the ATHOC 2004 headquarters where the Olympic Family movements, carried out by a fleet of 4000 passenger cars and minibuses and 1800 buses was controlled on the basis of a detailed operation plan prepared.

Twenty four Variable Message signs were informing drivers on current traffic conditions, indicating alternative routes when necessary.

PUBLIC AWARENESS

Special attention was paid to provide in time information to the public on the special and continuously changing traffic conditions during the games and for the use of proper means and routes both for the games and for the normal city activities. A detailed spectators guide was prepared while a large amount of special maps (Fig. 5) and pamphlets were available for additional information. Regular Radio and Television spots were broadcasted regarding the use of public transport, parking prohibition at critical locations, etc. as well as information on special cases of prohibiting or diverting traffic.

WEAKNESSES AND RELATED PROBLEMS DURING PREPARATION

Before testing the results of the whole effort in real conditions and prepare final appraisals and conclusions one could mention the following weaknesses which, although finally overcomed, created various problems during the preparation period.

1. Large starting delays after the assignment of the Games. This is the main weakness which contributed in most of the problems encountered during the preparation period.

2. Difficulties in coordinating the large number of Services and Consultants involved.

3. Delays in the Implementation of Completed Designs (e.g. Athens Traffic Management Studies – 5 sectors).

4. Delays in Studying and Implementing planned measures and actions (e.g. ZEEK, ZES, Olympic Lanes) and related public information campaigns.

5. Lack of proper surveys to improve the estimates made of the Athens population to stay in the City and the Greek visitors from Greece and abroad.

6. Test events in certain critical, from the point of view of transportation, venues under conditions as close as possible to those expected during the Games.

EXPLOITATION OF OLYMPIC LEGACY

One must also stress the legacy to the city after the games i.e. the new and improved infrastructure and the outcome of the systematic effort to implement Transportation System and Demand Management. Following the Games and the long construction period before the Games where Athenians experienced serious problems in their movements, they now have the opportunity to use a non-congested major road network and a high quality public transport including two new rail modes: the tramway and the suburban rail.

To avoid the future effects of the fast increase of car ownership (150 cars/1000inh. in 1980, more than double at present), Athenians should try to make the best use of this new transportation system by reducing the use of passenger cars. The number of trips per year by public transport which declined from a maximum of 480 in 1965 to a low of 140 around 1990, has increased since then to more than 200. The Government is helping this effort by continuing the improvement of Public Transport, providing Park & Ride areas, systematic enforcement of illegal parking prohibition and related transportation campaigns.

AN EVALUATION OF RESULTS

The Success

After the completion of the Athens Olympic Games, their success was universally recognized. Specially, the Athens Transportation System, although badly congested before the Olympic Games, presented no serious problems during the Games in spite to the several initial delays in the extension or improvement of the infrastructure and in traffic management.

Main factors contributing to success

A first appraisal of Transportation during the games shows that the main factors contributing to the successful service provided to the Olympic Family, to the Spectators as well as to the normal traffic of the City were:

1. The comprehensive initial planning (Candidacy File, Strategic and Operational Plans) of the Transportation Facilities in relation to the location of the major areas of activities (venues, Olympic and Press Villages, IBC, MPC etc.)

2. The completion of all Transportation projects planned and programmed in the Candidacy File.

3. The successful application of Traffic Management (Olympic Lanes, prohibition of left turns, parking prohibition at critical locations, odd-even circulation numbers within inner ring etc.).

4. Respect of users to prohibitions, although complete enforcement was impossible due to the magnitude of the task.

5. Substantial increase in the use of Public Transport due to:

    - Major extensions and improvement in Level of Service of Existing Public Transport

    - Introduction of two new rail Mass Transport Means (Suburban RR, Tram)

    - Special Express Bus Lines

- Extension of Service after Midnight

- No charge for Olympic Games ticket holders

- Prohibition of spectators parking within and in the viscinity of Venues

- Fear for Traffic Congestion

- Extensive campaigns

6. Reduced Demand

   - Reduced number of foreigner visitors due to unjustified phobia for terrorism

   - Athenians on August vacations

REFERENCES

ATHOC 2004, Transport Division (2001). *Olympic Transport Strategic Plan*

ATHOC 2004, Transport Division (2002). *Olympic Transport Operation Plan*

Chouliara T. (2004). The Traffic Management System in the Attica Region *Bulletin of the Hellenic Institute of Transportation Engineers*, **141**

DENCO, DROMOS, Brown & Root, Plannet/Ernst & Young (2003). *Use of SATURN model to Analyse Olympic Traffic. 3rd Implementation Phase*, ATHOC, Transport Division

Transport Science and Technology
K.G. Goulias, editor

CHAPTER 8

"EYE IN THE SKY PROJECT": INTELLIGENT TRANSPORT INFRASTRUCTURE FOR SUPPORTING TRAFFIC MONITORING AND MOBILITY INFORMATION

Liza Panagiotopoulou, GEOTOPOS S.A. Athens, Greece

INTRODUCTION

The "Eye in the Sky" project developed an Intelligent Transport Infrastructure based on the synergy of earth observation, mobile communications and digital mapping technologies. The project's overall objective was to provide commercially viable integrated solutions addressing issues of traffic monitoring, fleet management, customized mobility information and emergency services support. The test area of the proposed services was the sky and city of Athens, which hosted the 2004 Olympic Games.

The project promotes scientific and technological innovation by utilising existing state-of-the-art technologies in novel applications and integrating diverse disciplines and data to successfully handle the addressed issues.

The basic structure of the project relied on the use of Floating Car Data (FCD) technology and Low Altitude Platforms (helicopter). A fleet of vehicles "float" throughout the road network measuring speed and travel profiles in addition to the positioning information recorded from a GPS receiver. The data is transmitted via an existing terrestrial GSM network to the Centre. The data is processed in real time, using algorithms specifically designed for urban road networks, and traffic load (traffic flow) for the entire network is calculated. High-resolution digital imagery of the urban area is provided by the camera on-board the helicopter and is processed in real time for providing traffic measurements (density on links). Fusion of the traffic information derived from optical and FCD data provides high quality, reliable and up to date traffic information.

Eye in the Sky project was partially funded by the European Commission DG INFSO under the IST Programme. Its total cost rose up to 4,135,593 € while Commission's funding was 2,029,454 €.

Project Innovation

"Eye in the Sky" introduced technological innovations that provided unique solutions to the issues addressed. FCD algorithms especially designed for urban areas simulated with very good accuracy the traffic situation on the entire network, which is impossible to achieve with existing methods such as induction loops, infra-red sensors and CCTV surveillance. The optical data that complements the FCD approach provided traffic information either for the areas not covered by the FCD or for improving the quality of FCD data. In addition, the system was independent from terrestrial infrastructure. There was no need for extensive terrestrial infrastructure such as cable networks, power stations, CCTV systems. The fleet of FCD vehicles were equipped with GPS receivers and GSM compatible devices for localization and communication purposes, whereas the helicopter had autonomous positioning and communication equipment that enabled it to operate independent of any terrestrial means.

OBJECTIVES

The project's overall objective was to provide integrated solutions addressing issues of Traffic Monitoring, Fleet Management and Customized Mobility Information. "Eye in the Sky" proposed to adapt earth observation technology and terrestrial mobile communication networks to traffic monitoring and management requirements using new specialized software and methodologies designed for the urban environment. The goal was to establish the foundations for the development of healthy commercially-driven services that respond to needs and expectations of society within the European community. The cost-effectiveness of the applications proposed, derived from the fact that the technologies used were proven, technical know-how was abundant hence minimizing research, development and training costs. Successful application of the proposed services in Athens could initiate the development of modern market activities in other peer cities inside the European Union.

GENERAL TECHNICAL APPROACH

The "Eye in the Sky" project aimed to develop and validate two new sets of services based on different state-of-the-art technologies. However, in this paper only the first set of services which is related to intelligent transport infrastructure will be presented, as it is extensively related to Intelligent Transportation Systems Technologies and Uses. More specifically, the service addressed Traffic Monitoring, Fleet Management and the provision of Customised Mobility Information. The second set of services involved Emergency Support for Crises which exceeds the scope of this paper and will therefore not be discussed any further. Detailed information about both services as well as all other issues concerning the project can also be found on the internet in the "Eye in the Sky" website (www.isky.gr).

In the case of the first set of services, the basic structure of the project relied on Floating Car Data (FCD) technology and Low-Altitude Platforms (helicopter). A fleet of vehicles "float" throughout the road network measuring speed and travel profiles in addition to the positioning information recorded from a GPS receiver. The data was transmitted via an existing terrestrial GSM network to a Control Centre and processed in real time, using algorithms specifically designed for urban road networks, and traffic load for the entire network was calculated. The Control Centre has the ability to communicate with the FCD fleet, providing information and instructions for their movements.

This methodology provided dynamic management of a fleet of vehicles in addition to the traffic diagnosis performed centrally. In parallel, the helicopter which flies above the urban area transmitted high resolution digital imagery (acquired by a camera on board) to the Control Centre which was integrated with the FCD traffic data to improve their quality.

The digital imagery was also used to provide traffic information in areas not covered by the FCD fleet. The final product was high quality, up to date, dynamic traffic information which provided the base for Fleet Management services, Traffic Monitoring and Management, and provision of Customized Mobility Information.

This traffic information was used for providing "dynamic" guidance and customized mobility information (e.g. travel times) to private, public and commercial vehicles by using various mobile devices (cellular phone, PDA etc.). This information can facilitate the "floating" fleet of vehicles by allowing them to adapt their travel paths to the conditions of the network.

The proposed integrated solutions offered dependable, cost effective and user-friendly services to respond to essential needs and expectations of the society. The services offered were considered in the context of any-where/any-time access and can ultimately be tailored to individual needs.

ANALYTICAL TECHNICAL APPROACH

More analytically, the traffic information was acquired by the combined use of Floating Car Data transmitted via terrestrial mobile communication networks and aerial images captured by a camera on board a helicopter. A general overview of the System is presented in Figure 1. This information is used for Traffic Monitoring, Dynamic Fleet Management and Customized Mobility information.

Traffic monitoring was achieved by the combined use of the terrestrial and the airborne part.

The Terrestrial part included the "floating" vehicle fleet which drove throughout the road network measuring speed and travel profiles in addition to the positioning information recorded from the GPS receivers. Furthermore, it consisted of the Centre were the data was transmitted via the GSM network and where it was processed in real time. Using algorithms specifically designed for urban road networks, traffic load for the entire network was calculated.

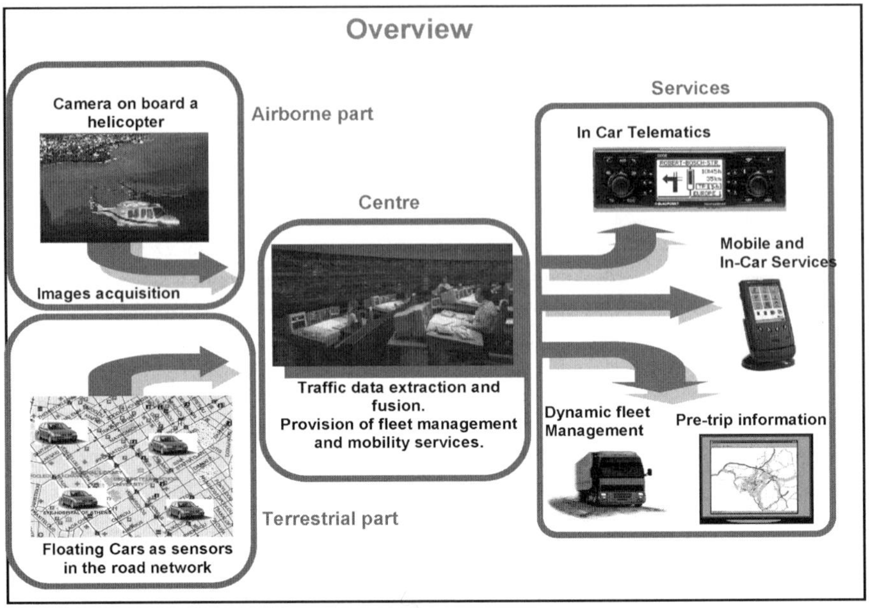

Figure 1. Overview of "Eye in the Sky" Services.

On the other hand the Airborne part involved the helicopter which acquired the digital images of the urban area with the camera on board. The images were transmitted via a Video Transmission System to the Centre where they were processed in real time providing traffic measurements for the road network.

Another important operation of the system was the Traffic Data Fusion. This involved fusion of the traffic data acquired by the aerial images and the FCD in order to provide high quality, reliable and up to date traffic information. An overview of the Traffic Monitoring based on Data Fusion is presented below in Figure 2.

Dynamic Fleet Management

Traffic information acquired with the combined use of airborne and terrestrial part is used for managing fleets of vehicles. This traffic information is the basis for providing route planning and dynamic route guidance according to the actual traffic status and a given routing priority (fastest, shortest) (See also Figure 3).

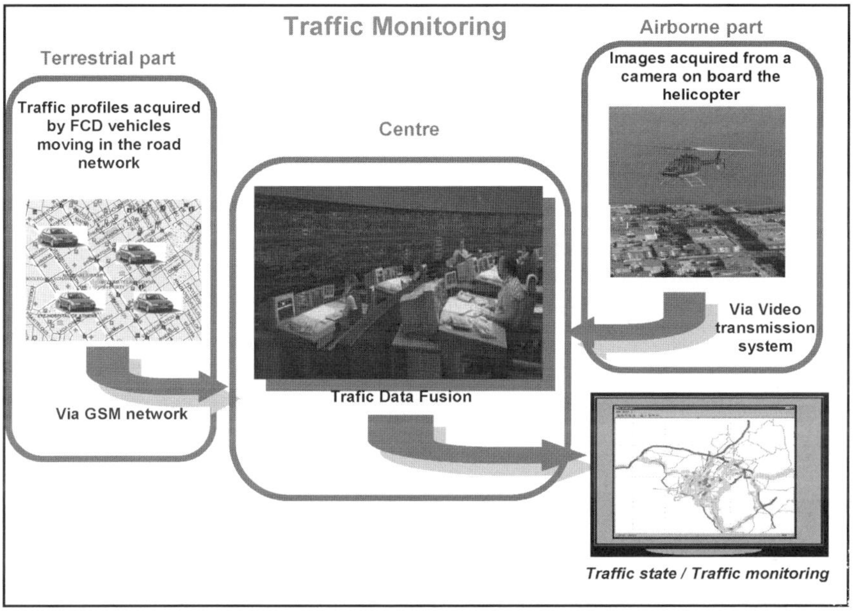

Figure 2. Traffic Monitoring.

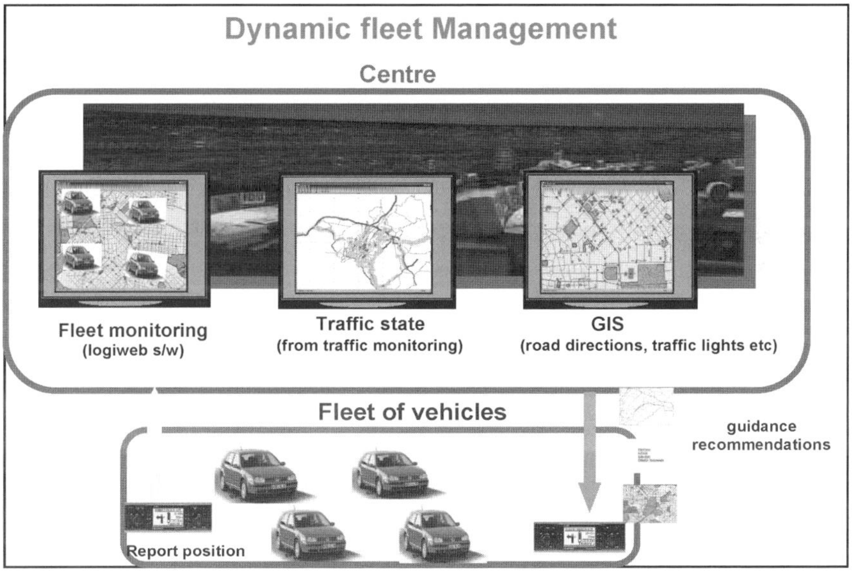

Figure 3. Dynamic fleet Management.

Customised Mobility Information

Based on the traffic information acquired by the synergy of airborne and terrestrial part customised mobility information is provided to the user which can be realised as a Traffic information service (Figure 4) or for the provision of Pre-trip internet information (Figure 4) or furthermore, for PDA applications.

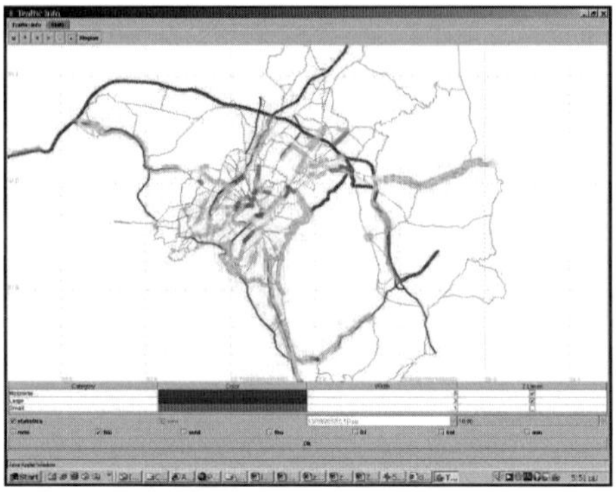

Figure 4. Sample of Pre-trip internet information.

SYSTEM'S COMPONENTS

The system can be broken down to a number of components. More specifically, the Airborne part consisted of: (a) the helicopter which provided a reliable and flexible platform for traffic measurements; (b) the camera on board the helicopter which provided digital imagery for traffic data acquisition; (c) the Video Transmission System which provided real time transmission of images from the helicopter to the Centre.

The Terrestrial part consists of: (a) the FCD (Floating Car Data) technology which enabled traffic diagnosis through a sample vehicle fleet moving throughout a road network; (b) the GIS technology which provided all the required spatial data for the operation of the applications; (c) the in-vehicle devices which sent position/speed information (for FCD vehicles) and could receive routing recommendations, and finally, (d) the GSM network which provided bi-directional transmission of information between the Centre and the vehicles.

Airborne Part

The technical characteristics of all the components of the Airborne Part can be found in the following tables (Tables 1 to 3). (See also Figures 5 to 8.)

Helicopter	
Type	Bell 206 L3
Technical characteristics	
Single engine	
Rotor diameter	11.27m
Length	10.13m
Height	3.6m
Weight	936.5kg (empty)
Fuel capacity	423.9L

Table 1. Helicopter Technical characteristics.

Figure 5. Helicopter view 1.

Figure 6. Helicopter view 2.

Video transmission System	
Airborne Antenna	
Polarization	circular
Azimuth Beamwidth	360°
Elevation Beamwidth	5°, 15°, 22°
Gain	14dBi, 9dBi, 6dBi
Ground Antenna	
GPS Tracking	
Polarization	circular
Azimuth Beamwidth	17°
Elevation Beamwidth	25°
Gain	14dBi
additional up-look antenna	elevation 0° - 90°

Table 2. Video Transmission System Technical characteristics.

Camera	
Black/white	10 bit/Pixel
Resolution	1980 x 1079 pixel
Pixel size	7.4 μm
Frame rate	6 images/sec max.
Weight	184g
Camera body size	56 x 56 x 56 mm3
Mounting with shock mounts	

Table 3. Camera Technical characteristics.

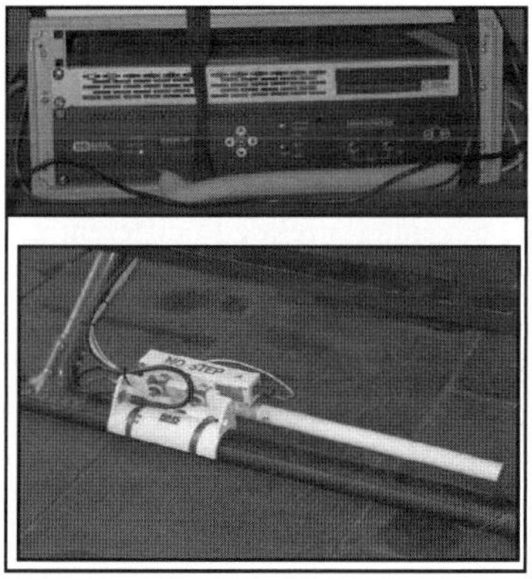

Figure 7. Video Transmission System.

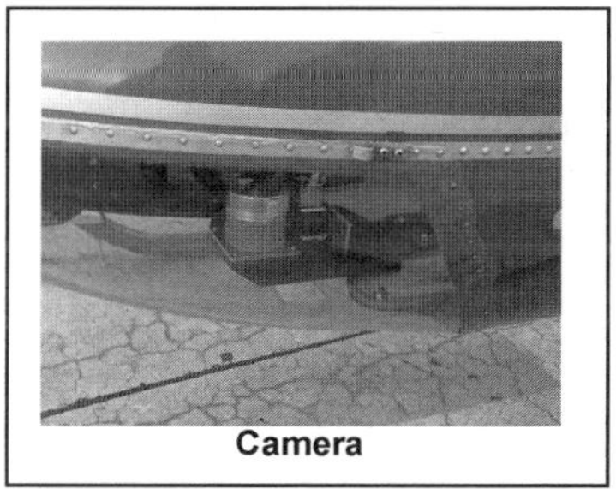

Figure 8. Camera.

Helicopter's equipment

The use of the camera on board the helicopter introduced several advantages. It provided real-time imagery which was fused with FCD data to improve traffic-data quality. It provided real-time imagery of areas not covered by the FCD fleet. It sent updated traffic status information from road segments that were last reported jammed by the Floating Cars and it provided additional visual view of the traffic situation to the Centre.

Terrestrial Part

FCD technology

The Terrestrial part was based on FCD methodology. Floating vehicles travelled throughout the road network and store the measured travel-time and speed profiles. The GIS database and especially the Traffic Information Network (TIN), provided all the required data for the road network. GPS was used to determine the position of the floating vehicles and GSM networks were used to transmit the information to the Centre where computer processing of the traffic data took place.
The FCD advantages relate to different aspects of the project. The Traffic data was generated using existing infrastructure (GPS/GSM).The FCD approach presented high area coverage as traffic statements were reported from the entire road network. The FCD algorithm was able to detect traffic jams and to provide a reliable travel time estimate. It provided swift and reliable traffic jam announcements and dissolution statements while the traffic congestion was detected according to scale and depends on road class.

Athens GIS

Another important component of the terrestrial part was the Athens GIS. The GIS provided all the required data for the operation of the applications as well as the necessary background for making all the services comprehensive. The area of implementation for the GIS was determined by several parameters and by taking into account all the characteristics of the applications and the provided services (Figure 9).

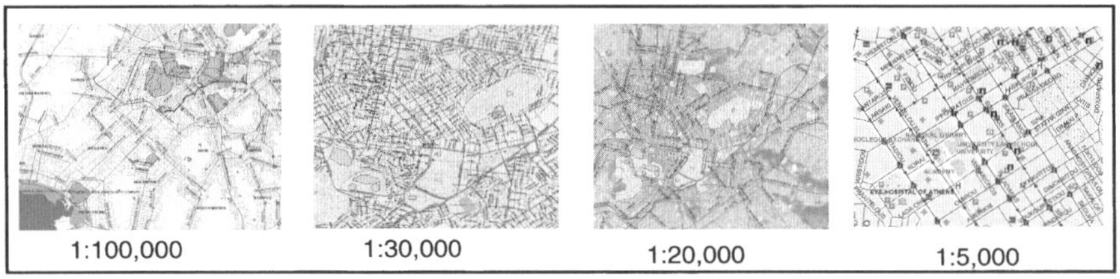

Figure 9. Samples of GIS maps in different scales.

The accuracy of the geometric information for the GIS database was 2m and the information consisted of several Layers which included city blocks, streets (names, numbering, zip codes, street type, number of lanes, directions, restrictions, etc.), Olympic venues, Olympic road

network, traffic lights, metro lines and stations, tram lines, Suburban railway, transportation gates and other Points Of Interest (POIs) (parks, hospitals, hotels, gas stations etc.)
The Athens GIS was used as a geo-reference for all relevant attributes of road elements required for FCD and multiple layers of relative information (road types, restrictions, topographic data and points of interest). The usage of the GIS was threefold: first it served as geo-reference for data-integration in the Centre, secondly it enabled location based services and thirdly it was required for generating the FCD-specific digital network of road-elements which was necessary for the operation of the in-vehicle-system.
Furthermore, the Athens GIS was used for visualization and data support for all the developed web services and the Dynamic Fleet Management. It was also used for data provision for all Location Based Services. It significantly contributed in the automatic traffic information extraction from images. It supported all Centre applications and in emergency support for Crisis Management

In vehicle devices

A number of devices were implemented in the FCD vehicles in order to realise the applications supported by the "Eye in the Sky" project.

EXPECTED ACHIEVEMENTS/IMPACT

As more than four-fifths of the European population dwell in urban areas, urban transport represents the most important aspect of mobility. The application of traffic monitoring and modelling techniques described in this project can create a positive impact on everyday urban transportation. Customised information can help individual drivers plan the fastest and safest route to their destination. The "Eye in the Sky" is fully aligned with EU policies regarding the framework of road transport and public passenger transport in Europe.
The implementation of the "Eye in the Sky" project is anticipated to have positive impact to economic and social aspects of the European community. Economic development is expected in the growth of companies acting as service providers using the proposed technology, as well as in the businesses that will benefit from these services. The proposed services can be provided by purely commercial entities or by private-public partnerships. In both cases growth of these services can generate new markets and employment opportunities. The successful implementation and growth of the "Eye in the Sky" services in Europe is expected to open market areas worldwide. In such case European innovation and expertise can place Europe in a competitive position in the global scheme.

Users

The project has the potential to facilitate a large number of different users and to be realised for many diverse applications. Firstly, the system is capable of providing traffic information to users at a fixed position, such as Public Authorities (Ministries, Public Transportation Organisation), Public Media (radio, TV, Internet portal) and the public in general (e.g. Citizens). (See also Figure 10.)

Secondly, it is capable of facilitating the provision of traffic information at mobile devices, for example drivers of private or public use vehicles (e.g. buses), closed user group, drivers of ambulances, police cars, etc. (See also Figure 11.) The same users with the in – vehicle devices can be recipients of fleet management services.

Figure 10. Example of providing traffic information to users at a fixed position

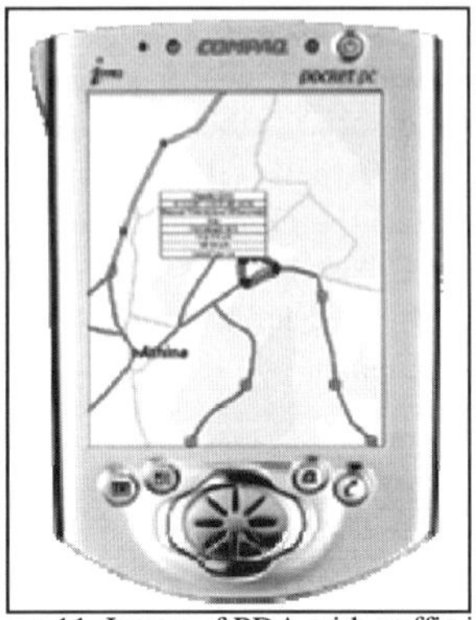

Figure 11. Image of PDA with traffic information. Provision of traffic information at mobile devices

LIST OF PARTICIPANTS

The consortium was comprised of organisations and companies of complementing technical profiles and expertise

GEOTOPOS S.A., Greece – Project coordinator
Technical University of Crete, Greece
Geosynthesis S.A. New Technologies Application , Greece
ND SatCom AG, Germany
JOINT RESEARCH CENTRE of the European Commission, EC
Deutsches Zentrum für Luft- und Raumfahrt e.V., Germany
gedas Deutschland GmbH, Germany
Fraunhofer-Gesellschaft zur Förderung der angewandten Forschung e.V., Germany
Robert Bosch S.A., Greece
Blaupunkt GmbH, Germany.

K.G. Goulias, editor

CHAPTER 9

ITS applications at Egnatia Odos Polimilos – Veria highway section

Konstantinos P. Koutsoukos, *Egnatia Odos AE, Greece*

Lefteris Koutras, *Thales Eng. & Consulting, Greece.*

Introduction

Egnatia Odos AE, the company responsible for the construction and operation of Egnatia Odos, a 680 Km highway, has recognized early the need for Intelligent Transportation Systems (ITS) applications for Egnatia. As a result, the company has conducted an ITS Architecture study for the whole axis, by examining both the user services requirements as well as the user needs, by taking into consideration the US and the European ITS architecture.

This paper describes briefly a) the results of the ITS Architecture study regarding the Traffic Management issues for the Egnatia highway, and
b) ITS applications and Traffic Management for a specific highway section (Polimilos - Veria section).

In addition, the paper highlights the difficulties, experiences and challenges involved in the ITS design and implementation process on the specific motorway section and also the solutions adopted to handle such difficulties, in order to render the design and implementation of the ITS systems feasible. Finally, the paper includes information about the implementation approach for ITS that was adopted by Egnatia Odos A.E. in the case Polimilos – Veria.

Description of Egnatia highway

Egnatia highway constitutes the most important modern infrastructure project for the development and transport of goods between North and Central European countries with South East Europe, the Balkans and the Middle East. The highway has a total length of 680 Km running from East to West in the Northern part of Greece. It is linked with 5 ports, 8 airports and 8 vertical axis that feed the transportation network of the Balkan countries.

Figure 1 shows Egnatia odos (in red) and how this network is placed in the North & Central European region.

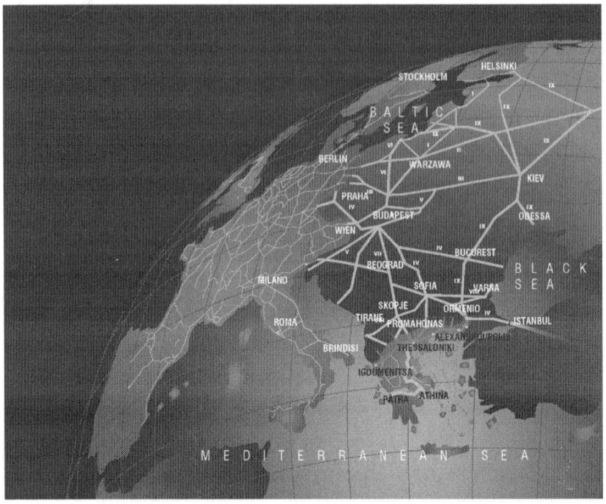

Figure 1. Egnatia Odos highway and European roadway network

The construction of this highway is by itself a very challenging project and its operation is anticipated to be equally difficult, if not more. The Greek Government has founded a private company named Egnatia Odos AE (EOAE), which is responsible not only for the management of the highway's design and construction but also for its operation and maintenance. Egnatia begins from the west (Figure 2 Hgoumenitsa) and covers many Kilometers with tunnels and bridges in order to cross the mountainous areas (Figure 3, 4). Central and Eastern sections of the highway have important structures as well. The highway is a dual carriageway with a central reserve , two traffic lanes per carriageway plus a hard shoulder.

Figure 2. Hgoumenitsa, western point of Egnatia Odos

Figure 3. Egnatia Odos bridges at western sections

Figure 4: Egnatia Odos bridges at western sections

Figure 5: Egnatia Odos tunnel at western sections

The highway network includes 80 Km of bridges and 90 Km of Tunnels (Figure 5). In addition Egnatia's network includes also 720 Km of service roads.
The cost analysis for the completion of Egnatia is 4.600 M€ (without VAT) and from that 7% is for the Project Management, Design 5%, Expropriations 8% and for the construction 80% as the **Figure 6** shows.

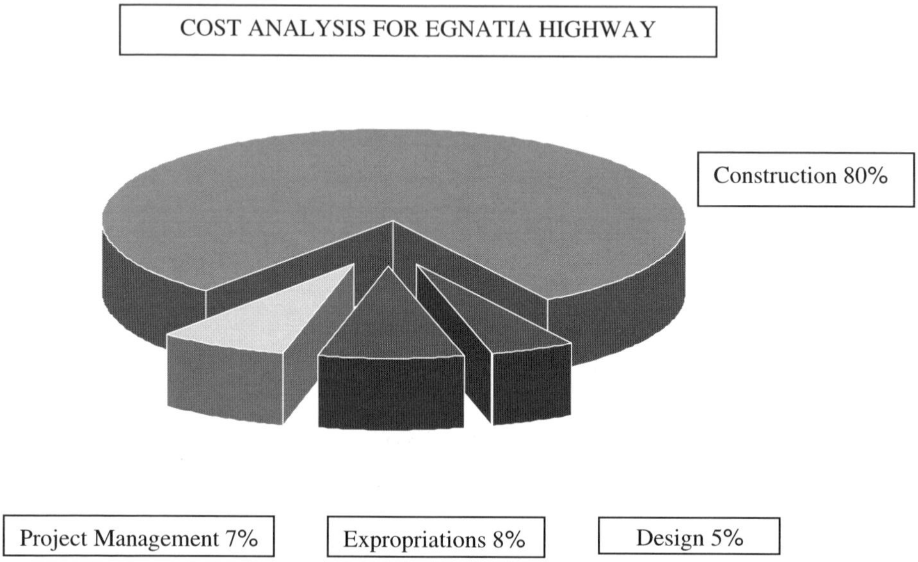

Figure 6: Cost Analysis for the project of Egnatia Odos

In order to support its planning decisions and the highway design including the telematics applications, Egnatia Odos AE has developed a traffic demand forecasting model. Based on that model, the traffic forecasts for the year 2010 are ranging from 10.000 veh. / day in western Egnatia sections trough over 40.000 veh / day in central sections. A map which depicts the travel forecast is shown in Figure 7.

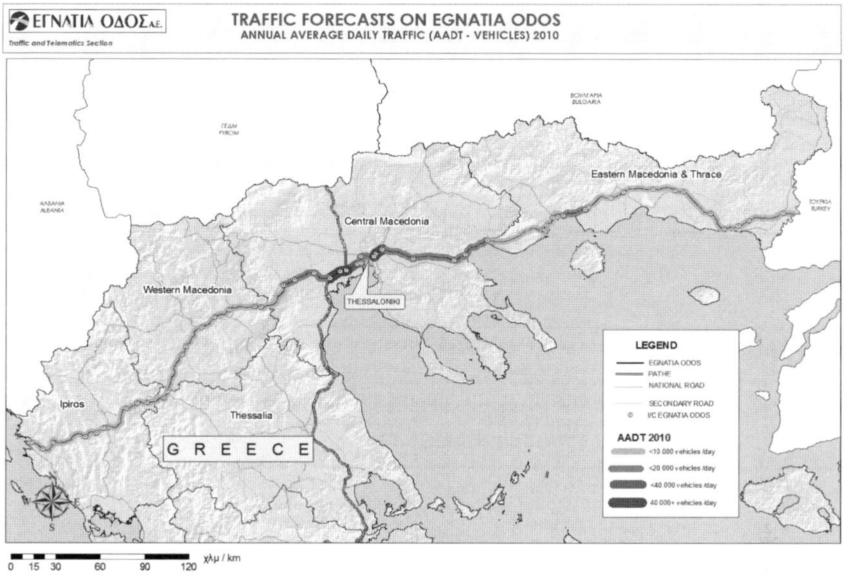

Figure 7: Traffic forecast for Egnatia Odos (Year 2010)

ITS Architecture Design for Egnatia Odos

EOAE, since the early stages of the project's design and construction, realized that the future operation of the Egnatia highway had to depend also on ITS applications. Thus, the design of the ITS Architecture for the whole Egnatia roadway network was assigned to a consultant (Delcan) in 1999.

The study provided a framework around the multiple design approaches that can be developed. The idea was that the systems goals should express the services that the transport stakeholders want to provide, in order to improve the movement of people and goods. Although this study had started for the whole Egnatia, the construction of the highway for the Polymylos – Veria was already underway since 1998. The ITS study for architecture study followed a number of phases in order to finally implement ITS applications. These phases are shown in Table 1.

Table 1. Phases of Egnatia's ITS Architecture study

Phases	Work / Steps
A. System Analysis	Targets of the Telematic system
	User Needs
B. System Design	User services
	System Operation – Logical Architecture
	Telematics Subsystems – Physical Architecture
. Implementation Strategy	Highway characteristics and segmentation
	Strategy of implementation – Cost model
	Procurement methods

Thus, the ITS architecture study started from phase A and examined the *needs assessment* in order to clearly identify the objectives and needs for telematics applications for the Egnatia Motorway. Another objective of this stage was to identify potential issues related to these needs, both in technical terms as well as in organisational terms, dealing with the co-operation between agencies and other stakeholders. The issues examined were a) Identification of stakeholders, b) Stakeholder interviews, c)Traffic needs and problems, d)Review existing operations, e) Review of agencies interface, f)Stakeholders Workshop, g)User objectives and needs.

Phase A and B concluded that 5 main ITS travel services should be offered to the users of Egnatia. These are :

a) Traffic collection data

b) Traffic Management – surveillance

c) Weather monitoring

d) Emergency management

e) Travel information

Following the phase for the implementation strategy (Phase C) the highway was categorized according to the area/terrain that was going through. Thus, three main categories were identified: 1) "open" highway through a flat terrain area, 2) urban highway 3) highway with tunnels and bridges through mountainous area. Table 2 shows the ITS services provided for the different roadway segments of Egnatia, while Figure 8 shows which sections of Egnatia belong to the above three different categories.

Table 2. ITS services by roadway segments at Egnatia Odos

ITS services at Egnatia	**Egnatia's Roadway segmentation**	
Traffic collection data Close circuit TV Traffic count stations Overweight vehicle detectors	Flat terrain and Urban highway	Tunnels and Bridges sections
Traffic Management – surveillance	Flat terrain and Urban highway	Tunnels and Bridges sections
Weather monitoring		Tunnels and Bridges sections
Emergency management Incident detection Lane control signs	Flat terrain and Urban highway	Tunnels and Bridges sections
Travel information Variable message Signs Blank out signs		

The ITS Architecture study, concluded that the traffic management for the highway should be handled by 5 Traffic Management Centers with the main TMC at the city of Thessaloniki. Figure 8 shows a map of Egnatia with the location of all the TMCs and Figure 9 shows an outline of all the elements of TMCs for Egnatia.

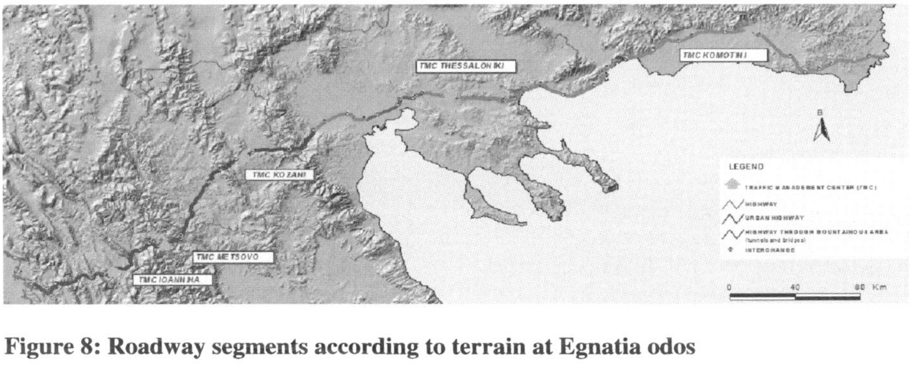

Figure 8: Roadway segments according to terrain at Egnatia odos

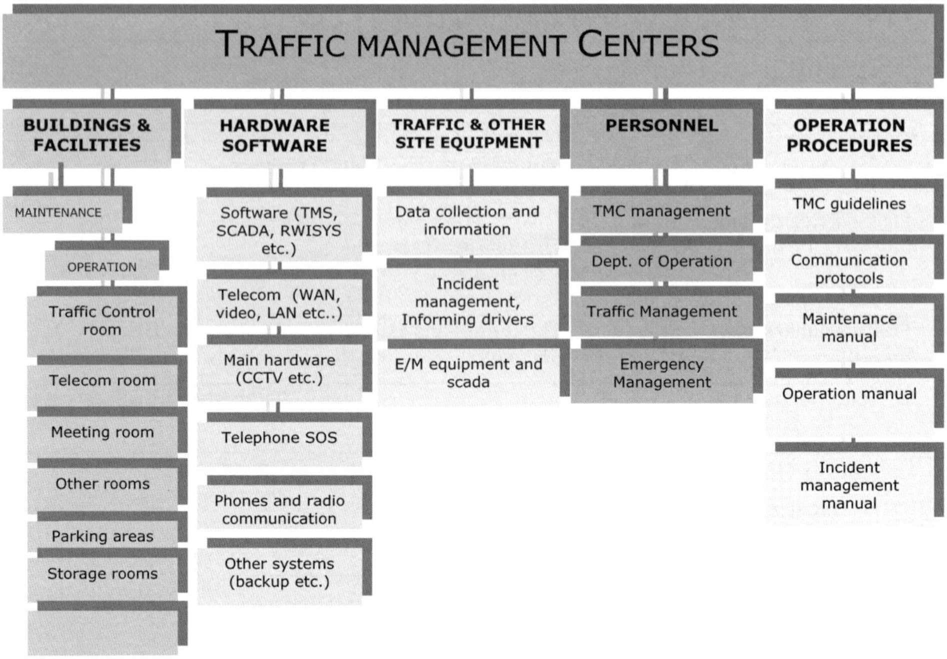

Figure 9: Elements of Traffic Management Centers at Egnatia

Description of the Egnatia highway section Polimilos-Veria

The highway section of Polimilos – Veria is one of the most challenging Egnatia's freeway section, in technical terms. It includes numerous tunnels and bridges that are successively constructed and closely spaced. For design purposes the section is divided in two subsections, Polimilos - Lefkopetra which is 12 km long and comprises 9.6 km of bored tunnels and 2.4 km of bridges and Lefkopetra - Veria which is 12 km long and the total length of bored tunnels and bridges is 4 km and 2.9 km, respectively. The highway is being constructed at a mountainous terrain with an elevation ranging from 200 m , at Veria I/C, to 800 m at Polymylos I/C. The region where the highway passes is characterized by frequent fog occurrences over extended road lengths. In addition, 17 Km out of the 25 Km of the section is constructed with split carriageways, a fact which hinders the safe operation of the road in case of emergency situations. Figure 10 depicts a ground plan of the 25 Km section with the positions of the Tunnels and Bridges in it.

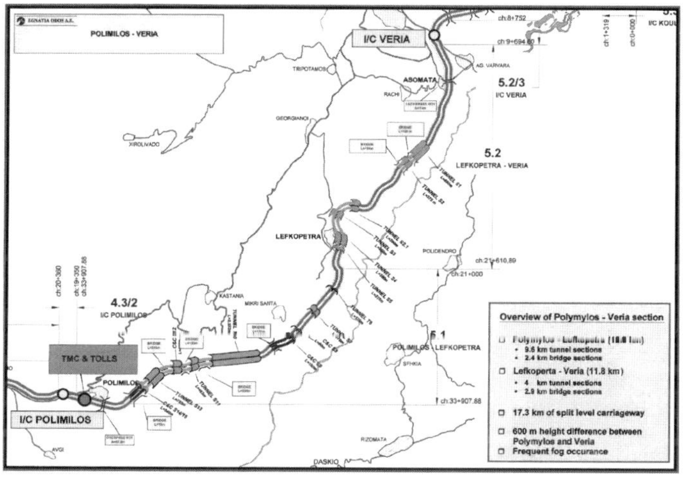

Figure 10: Polymylos – Veria highway section (Egnatia Odos)

Table 3 shows the major structures of Tunnels and Bridges of the section together with their associated length.

Table 3. Main structures of Polymylos section

Type of Structure	Length
Bridge 1	150m
Bridge 2	154 m
Bridge 3	130m
Bridge 9	170 m
Bridge 10	260 m
Bridge 11	300 m
Bridge 12	450 m
Tunnel 1	840 m
Tunnel 2	350 m
Tunnel 2.1	285 m
Tunnel 3	250 m
Tunnel 4	270 m
Tunnel 5	210 m
Tunnel 6	160 m
Tunnel 7	360 m
Cut & cover 8	112 m
Cut & cover 9	175 m
Tunnel 10	2240 m
Tunnel 11	440 m
Tunnel 12	470 m
Tunnel 13	770 m
Cut & cover 14	250 m

Based on the previous description it is clear that this highway section requires a special handling in terms of traffic control, surveillance and infrastructure issues as well as in terms of emergency management.

ITS applications and challenges at Polimilos-Veria section

Polymylos – Lefkopetra (12 Km) is one part of the Polymylos – Veria highway section; the other one is the Lefkopetra – Veria (12 Km). The longest Tunnel here has a length of 2.5 Km while the others range from 350m - 800m.

Construction for both sections of Polymylos – Veria have started in 1998, with a target date for completion in 2004. Construction contracts included the construction of Tunnels, Bridges, and the in between sections. In addition, the contracts included the electromechanical work, such as street lighting and ventilation as well as the installation of other equipment for Tunnels and the Tunnel service buildings.

One of the first challenges faced by the ITS study for the Polymylos project, was due to the fact that the implementation of traffic control was included in the general construction contracts without detailed traffic designs. In addition, the design for traffic related issues was not handled as a separate design subject, but rather was included in the other E/M designs for the highway. Thus, the whole design approach followed a segmented isolated approach that focuses on individual tunnels rather than on a macro traffic oriented approach for the whole

25 Km highway section. In order to overcome this difficult a separate ITS design study was performed by Thales Eng. & Consulting and Egnatia Odos SA, with subject the traffic management issues for Polymylos - Veria.

The differences between the segmented / isolated approach and the new ITS design approach can be seen if we recall the fundamental definition of ITS "that the system's goals should express the transport needs and the services of the transport stakeholders in order to improve the movement of people and goods". Designing ITS applications for an efficient traffic management from I/C of Polymylos to I/C of Veria, rather than simply installing equipment that cannot serve a complete traffic management system, served the previous definition.

Specifically the Polymylos – Veria ITS design included the following subjects:

1. Determination of Traffic Management requirements for the section
2. Definition of telematics systems and subsystems as well as interfaces between control equipment and procedures
3. Design of civil engineering infrastructure required to implement Telematics applications.
4. Selection of appropriate traffic technologies for the project specific needs in a cost effective manner. Applications proposed were for, traffic surveillance and recording, over height vehicle detection, meteorological stations, incident detection etc.
5. Drawings with the exact locations of Telematics systems
6. Telecommunications design, cabling and specifications

The ITS design for the Polymylos – Veria section concluded the following:

a) Different number of ITS traffic equipment for the highway,
b) A traffic oriented placement for the proposed equipment,
c) Updated traffic equipment specifications.

The total cost for traffic equipment for the Polymylos – Veria section was around 4 M € without including the cost for the telecommunications (ducts, fibre optic etc.)

The ITS study for Polymylos was designed and completed by taking into consideration the previously accepted ITS architecture study that was conducted by Egnatia Odos SA (1999). On the other hand, the ITS Polymylos study needed to adopt a design that could fit to an underway predefined construction project with a fix amount of money allocated for ITS equipment.

The highway section was already under construction since 1998 and the possible changes to construction schedule / budget were limited.
The ITS design for Polymylos concluded that the traffic control (Traffic Management center) should be handled centrally and not in segments from many tunnel service buildings located outside of every tunnel.

The design suggested that the main ITS services to be implemented were:

- Traffic management for the whole sections
- Incident detection for the whole section
- Trip Information to drivers

- Weather information to drivers
- Traffic data selection

In addition, for the E/M services and the tunnel service buildings for Polymylos – Veria the following were proposed :

- Control of E/M equipment (lights, ventilation etc.)
- Efficient maintenance for the E/M equipment

For the Traffic Management of the highway the ITS study proposed the following system architecture that Figure 11 shows. The figure shows the whole system Architecture that will operate at the Polymylos – Veria section. The system includes 7 subsystems, 6 of each are for traffic equipment and 1 for the scada system.

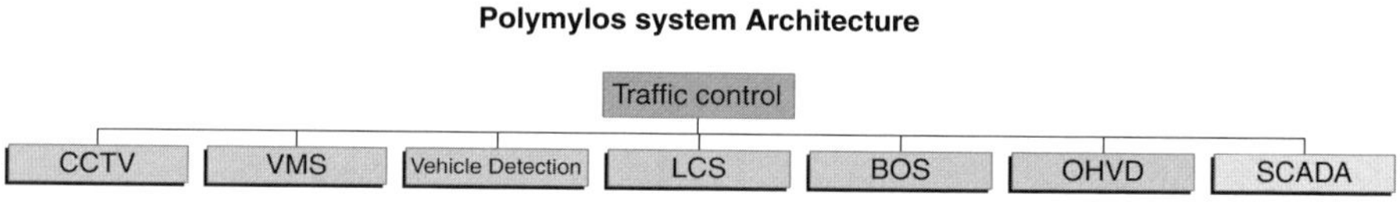

Figure 11: Polymylos – Veria Traffic control architecture

The ITS equipment were proposed in order to serve the previously mentioned services. Specifically the ITS equipment were:

1. **Lane Control Signs (LCS)**: The LCS equipment will play a key role for the traffic management of the highway section and will be installed inside tunnel areas as well as in other critical sections of the highway. The signs will be double-face in order to deviate traffic from one carriageway to another, in case of maintenance situations or incidents. The LCS is installed with 300 m spacing and have dimensions 600 X 600.
2. **Variable Message Signs (VMS)**: With the VMS, motorists will be informed in advance for any event that can cause non-recurring congestion for the highway. In that case, they either can reduce their speed or choose another path. In addition, the VMS can provide useful information for any programmed maintenance or abnormal meteorological conditions. For the Polymylos – Veria section, 4 VMS will be installed with each one having the capability of presenting messages (text) simultaneously in Greek and English languages in four lines, as well as a pictogram.
3. **Blank Out Signs (BOS)**: BOS are signs with smaller dimensions than the VMS. Each one can display up to three predefined text messages. They are used in pairs at locations of the highway where traffic can deviate to a different carriageway. BOS main purpose is to inform the drivers to reduce their vehicle speed for the upcoming traffic deviation point.
4. **Close Circuit TV (CCTV)**: There are two kinds of CCTV cameras that will be installed, one stable and one Pan/Tilt/Zoom. Their use is intended for traffic surveillance inside tunnel areas, as well as outside at major points.

5. **Inductive Loops**: They will be installed for traffic counting purposes as well as for incident management. They are installed in pairs in each traffic lane in order to identify the vehicle categories and speeds in addition to the number of vehicles. The spacing of the loops varies, every 500 m inside tunnels and every 900 m outside tunnels. The coverage is for the whole 25 Km of the highway in order to have traffic information at the TMC for the whole section.
6. **Over height Vehicle Detector (OHVD)**: OHVD detectors are placed at each interchange of Polymylos and Veria as well as in the main highway section. They can identify over height vehicles, which can be stop before entering the tunnel area with the use of VMS.
7. **Road Weather Information system (RWIS)** : The RWIS stations are placed in areas (close to bridges) where adverse weather conditions can create traffic problems and non-recurring congestion.

The total number of traffic equipment installed at Polymylos – Veria are shown in Table 4.

Table 4. Traffic equipment at Polymylos - Veria

Traffic Equipment	Number
LCS	82
VMS	4
BOS	20
CCTV	82
CCTV PTZ	16
Inductive Loops	92
OHVD	6
RWIS	4
Traffic Lights	8

The total number of the traffic equipment will be installed in two phases. First phase includes the installation of the basic equipment and second one the installation of additional equipment for a broader surveillance of the highway sections, which will result to a more efficient traffic management.

The traffic management for the Polymylos – Veria section will be handled by a Traffic Management Center located closer to the Polymylos I/C. For the first years of operation of the highway, the TMC will be collocated with the Tunnel Service Building of Tunnel 10 (S10). The TMC will control all the traffic equipment using the filed controllers installed at 11 Tunnel Service Buildings

Specifically for the case of Polymylos – Veria the TMC and the proposed traffic control room is shown in Figure 12. Two main areas for control are shown with the use of the appropriate software, one for the traffic management issues handled by a Traffic Management Software (TMS) and one for the electromechanical issues handled by SCADA software. The two systems operate on different PCs but exchange the data that are appropriate for the operation of the highway. TMS operators are located in front of a panel with dimensions 2,8 m x 6,05 m and includes a 2 m x 2.25 m Video wall, recorders, and 12 monitors of 28'' each.

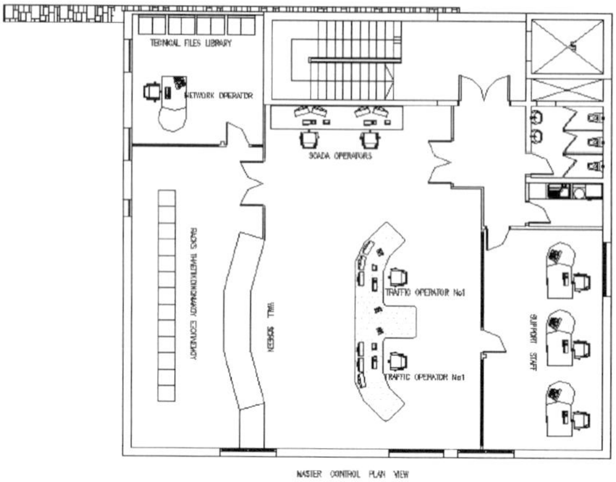

Figure 12 : Traffic Control room for Polymylos – Veria section

For the efficient traffic management of Polymylos – Veria highway, the ITS design study has taken into consideration operating instructions for incident management related issues provided by a French highway operator (Escota). According to the above instructions, the highway was divided into 3 sections for incident management as Figure 13 shows. These were, section A (9,4 Km), section B (10,7 Km) and section C (6,8 Km) while incidents were identified as major and minor, for areas inside and outside tunnels. Possible incidents are expected at areas of North or South carriageway. Incidents, include total physical blockage of the road following an incident (accident), passenger vehicle or HGV fire, incident involving dangerous goods vehicle, vehicle traveling in opposite direction to flow, smoke inhibiting visibility, serious weather (heavy snowfall, freezing rain) etc.

At the end, Egnatia Odos AE has developed an incident management plan for the Polymylos - Veria section, based on 800 traffic scenarios and incidents divided into four different categories according to the impact (severity) that they had on traffic flow.

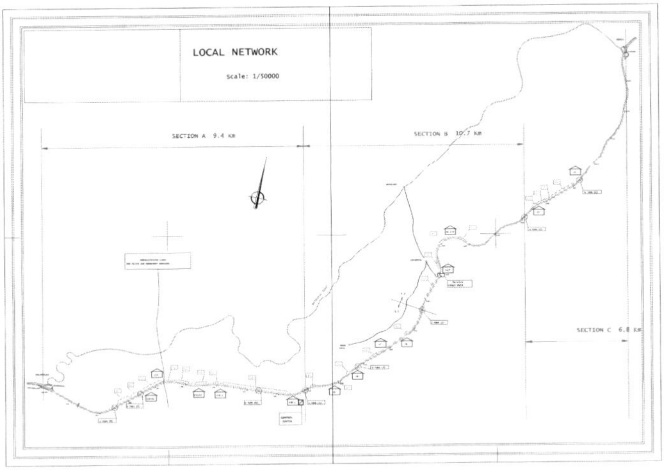

Figure 13 : Incident Management for Polymylos – Veria sections

ITS system integration for Polymylos – Veria

The implementation of an ITS system requires detailed specifications for the interface and logical connectivity of the various components, in addition to the normal functionality, performance and physical characteristics. Many times equipment suppliers are different form software suppliers, and the whole procurement needs to be integrated into one system.

There are three possible approaches in order to procure an ITS system: a) engineer/ contractor b) system management c) design – build approach. Egnatia Odos AE has identified the system management (SM) approach in order to implement ITS technologies on its highway. According to this, the SM is responsible for the system design and specifications, system integration, documentation, training, management of testing and system start-up. The system manager should be independent of manufacturer or supplier for any system components in order to avoid any conflict of interest. Contractors are responsible for the installation of traffic equipment while the system manager will provide the integration of all hardware and software. This approach allows the SM to act as an Engineer and system integrator on behalf of the owner, and the owner to be involved throughout the implementation phase. Another benefit is that the system is not linked to particular suppliers, which allows flexibility in the type of equipment to be procured. Last, a phased implementation of the project is possible with additions or changes introduced more easily than the other two approaches, especially in cases where detailed designs and specifications are not prepared in advance of the project's construction.

Acknowledgments

The authors would like to thank Mr. H. Kekis, Works Manager and the Engineers of the Works Management Dept. of West Macedonia – Egnatia Odos, for their cooperation and assistance for the implementation of ITS at the Polymylos – Veria highway section. Special recognition to Mr. Kyriakos Anagnostopoulos, Electrical Engineer of Egnatia Odos AE for his valuable assistance and contribution to the ITS Polymylos study.

K.G. Goulias, editor

CHAPTER 10

PROBLEMS OF ATTENTION DECREASES OF HUMAN SYSTEM OPERATORS

Mirko Novák, Technical University of Prague, Czech Republic

INTRODUCTION

The problems of non-satisfactory level of interaction between human subject and artificial system exist in almost all areas of human activity. Here we shall concentrate mainly on the case of the reliability of the interaction of the driver with car.

This is of course a very important area, while the volume and density of road transport rises every day and the number of road accidents reaches tremendous level. According the data from EU (presented e.g. on the conference of ERTICO, Prague, 2002), on European roads more than 42000 people per year are killed, which is estimated to losses of about 165 billions Euro. To this figure one has to add the price of non-mortal accidents, which are cheaper in average, of course, but much more frequent. Suppose that their total reaches about the same level.

[1)] Prof. Dr. Mirko Novák, Faculty of Transportation Sciences, Czech Technical University, Prague, 11000 Prague 1, Konviktská 20, Czech Republic, e-mail: mirko@fd.cvut.cz

This concerns the so-called primary losses. The secondary losses, involving the necessary medical care, social expenses, losses of work capacity etc. are hard to determine by statistics and the estimations differ. However as reasonable estimation the equality to primary losses can be taken. Very roughly speaking, one can therefore estimate the losses caused by accidents on EU roads and due the subsequent expenses to about 600 billions Euro per year. Without intensive and systematic preventive activity, this figure has the tendency to increase from year to year.

Similar situation is also as concerns other areas of transportation activities. The total figure of all these losses one can hardy estimate, but in any case this is extremely high.

Because of non complete statistics it is not easy to estimate, which part of this is caused by fatigue of the human subjects, as the methodology is not internationally standardized yet and differs significantly state to state. In literature, the values from 15 to 50% can be found. Nevertheless one can take the figure of 20% of the total volume of accidents being caused by the human subjects fatigue as very realistic. If one takes into account also the price, which we all have to pay for non-mortal accidents, we can speak of about 120 billion Euro per year lost due the decrease of attention of drivers below certain acceptable level.

The need to minimize these losses is the dominant motivation for activity in this area.

The progress in this respect could be reached by combination of the following 5 main approaches, which all need *very interdisciplinary* approach:

a) Improvement of the training the drivers with respect to their higher resistance to disturbing factors causing decrease of their attention,
b) Improvement of the interior of the car cockpit arrangement with respect to minimizing the influence of disturbing factors causing the decrease of drivers attention and enrichment of the set of installed car equipments by new active and passive tools allowing to improve the driving safety,
c) Development of the attention level and micro-sleep warning systems and their installation in car cockpit
d) Improvement of the traffic control systems with respect to wide scale detection of risky and aggressive driving and of its punishment,
e) Investigation of the influence of various drugs (including alcohol, nicotine etc) on human subject driving activity and development of new pharmatics improving the human attention

None of these 5 approaches is universal, but also no of them can be neglected.

As concerns the *drivers training*, much can be reached by the use of traditional methods, especially if they are completed by the systematic use of advanced driving simulators.

However, the progressive training methods based on the use of simulators equipped by *bio-feedback tools* (see Fig.1) seems to be very promissing, if the respective training is carried out in satisfactory number of repetitions and being controlled by skilled neurologist or psychologist. Such, considerably expensive training can lead to significantly improved resistance against both the fatigue and number of disturbing factors influencing the driver during his/her driving activity. Such enhanced state of the particular person resistance against fatigue can last considerably long, probably several months, may be sometime up to few years. In this period, the threat that his/her attention level falls down below acceptable level when driving is much less. Unfortunately, up to now, there is not enough knowledge about the percentage of population, which can be succesfully trained by this method and also about the possibility of succesfully repeated retrainings. Much more systematic measurements has to be done in this respect.

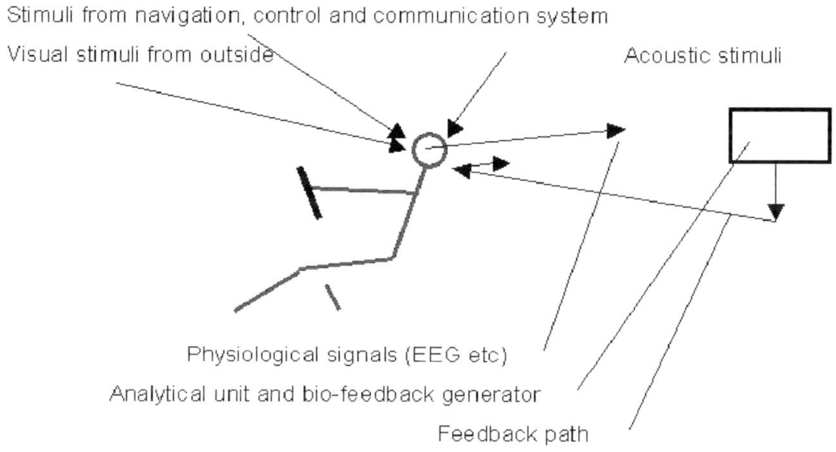

Figure 1 The basic principle of bio-feedback training

The education of new drivers (especially professionals) represents a very important part of transportation-oriented industry. To reach the main goal of its activity, i.e. the training a mass of people for to be able to operate as good drivers, with high efficiency and reliability is of course a very strong motivation, which projects also in significant economic gain.

As concerns the *car interior*, much can be done to optimize the shape, position and kind of use of the tools for driving control – i.e. the driving wheel, pedals, gear handle, instruments on the cockpit panel etc. This optimization has to be provided not only with respect to driver convenience and comfort, however before all to reliability and safety of his/her interaction with the system of car, especially with its driving control. One of most important aspect in this respect represents the optimization of the on-board mobile phone to minimize the negative influence of its use on driver attention. The development and design of so optimized cockpit stay in the focus of interest of various leading car manufacturers.

Much can be done also as concerns new kinds of electronic and information on-board tools, which will have positive influence on driving safety. Among them are e.g. the car radars, detecting not only the existence of nearest other car in the front and on the rear, but also controlling automatically the safe distance from it. Also the information systems, predicting on-board and in to driver acceptable form the most important weather parameters (temperature, wind, humidity, rain, fog, ice-on-road etc) along the expected car travel trajectory for prediction horizon of 1-2 hours can help very much.

In recent years there were realized several attempts how to design the on-board applicable system, which can automatically warn the driver against of his/her serious attention decrease and the advent of micro-sleep. Till now, however, no of them does reach the maturity for practical application. Those, which are based on the so called secondary markers (those, which are not directly derived from the tested person brain activity, like the electroocular signals, face grimaces, skin impedance etc) , face the problems with lower specificity and eventually long time delay between the real decrease of attention and significant change of respective parameter (in certain cases this delay can be several tens of seconds or few minutes). Those approaches, which deal with analysis of electromagnetic radiation from the driver brain (the so called primary markers) do not suffer from such problems, however they are not easy for practical application especially because of technical problems with measurement of very weak electromagnetic field on the driver head in moving car and of the very high individuality of each driver brain electromagnetic pattern. Nevertheless, there exist a very high motivation for further development of such warning tool.

All the above mentioned approaches how to diminish the losses in traffic accidents fail, if there is no good will of the driver to follow the respective recommendations of the on-board warning system. Because the drivers community consist unfortunately not only of the good willing people, but also of individuals of non-tolerant, careless, indolent, risky or aggressive nature, the *system of general supervision of drivers behavior* (and of the respective punishment of eventual aberrances from given standards) seems to be quite necessary (of course, the development and introduction into practical application of such system represents a very complicated problem, not only from the technical, but also from legal and juristic point of view).

The last mentioned approaches is based on recent developments, reached in the field of neuro-pharmacology as concerns the drugs, increasing the level of human subject attention, his/her speed of reaction and the probability of correct reactions, even in high physical and psychical load. Though much more research has to be done in this area, one can expect that in not to far future we shall know much more about possibilities how to use such medicaments, like methylphenidate (Ritalin) e.g. to prevent the possibility of fatal decrease of human subject attention and not causing a set of negative side-effects for particular person health. Also the possibilities of external attention stimulation (e.g. by suitable modulated magnetic or electric fields) represent a serious challenge, which needs intensive research.

THE STIMULI, INTERACTING WITH CAR-DRIVER

The main kinds of stimuli, which affect the driver behavior when driving car are sketched schematically in Fig. 2.

One can divide them into two main groups:

the external stimuli,
the internal stimuli.
Another division can recognize:
the natural stimuli,
the artificial stimuli.

Among the *external stimuli*, the visual onesare evidently of the main importance. Here one has to distinguish the visual stimuli, describing the position and move of the car on the road, the stimuli informing the driver about external situation, road signs and other traffic and the stimuli informing him/her about the car control, navigation and communication equipments. They differ nor only in shape, size, color and intensity, but also varies in time of appearance, length of existence and necessity of either periodic or permanent observation.

Beside the visual stimuli, also the acoustic signals play an interesting role. These can be of the warning kind (from outside traffic, or from the own car – skilled driver permanently listen the noise of his/her own car), of the disturbing nature (noise, communication with the car crew or listening radio – here much more research has to be done for to know, which kind of radio programs help and which destroy the driver attention).

Another important group of external stimuli is generated by the human subject mechanical sensors. The driver is exposed to the influence complicated mechanical forces, consisting of vibration components component of accelerations and decelerations and centrifugal components. This is of especial importance for skilled drivers, which very often analyze the driving situation out of their will just by these stimuli. On the other hand, the absence of such stimuli in simulator can lead to so called simulator sickness, which, especially for skilled divers can have the form of nausea. The furnishing of the simulator by a set for such stimuli simulation belongs to very expensive and laborious problems.

Almost all drivers operate the car having their hands on the driving wheel. The system driving wheel – drivers hands represent a very complicated and sensitive interface, where interact the mechanical stimuli coming from the move of the car on the road with the stimuli coming from drivers brain through his/her motoric system. The very careful analysis of hand reflections promises to be a good source of information about the level of the driver attention and his/her actual ability for safe driving.

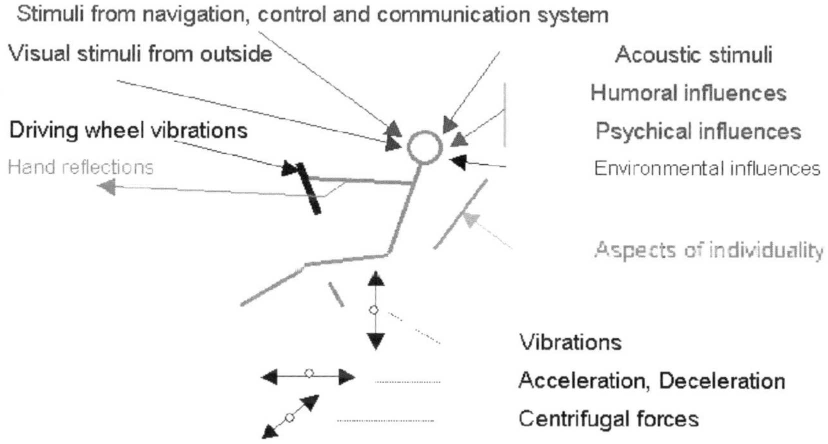

Figure 2: The main kinds of stimuli to which the car-driver has to face.

<u>*Internal stimuli*</u> come before all from the particular driver general physical and psychical conditions and of course also from the presence of eventual drug load of his/her organism. As concerns the drugs, the alcohol and nicotine load appears most frequently. While the negative influence of alcohol on particular driver ability for reliable and safe driving is widely known and in many countries respected by various legal limits for acceptable percent of alcohol in drivers blood, considerably few is known about influence of nicotine (and other drugs, including caffeine). This concerns especially their long lasting and combined exposition. Here also the factor of driver individuality must be taken into account.

All the above mentioned kinds of stimuli can be taken as <u>the natural</u>. However, besides these, the driver can be exposed also to <u>artificial stimuli</u>, like the external physical fields of drugs influence, which can have either the positive, but also negative influence on his/her level of attention and ability for safe control of the moving car. As it was already mentioned these need very intensive and systematic research, both from the preventive and also from the improving aspects.

All such investigation needs to be done on considerably large number of experimental persons (probands), especially because of very high level of human subject and namely his/her brain individuality. Here the development of the international data-base for neuroinformatics, organized in the range of the respective Global Science Forum OECD will be of very high significance.

RELIABILITY ASPECTS

As was already mentioned elsewhere (see Novak et al. 2003a, Noval et al 2003b), the ability for reliable and safe driving can be represented by some point in the multi-dimensional space {X} of the N parameters x_i representing the drivers attention level. In general various kinds of parameters x_i can be taken into account. However, because the determination of their values is very often loaded with considerably high level of fuzziness, the restriction of the number N to small values is recommendable. For practical investigations, one deals therefore before all with two main parameters, representing the level of attention, i.e. the driver reaction time RT and the probability P_{corr} of his/her correct or wrong response to certain external stimulus.

In the plane (RT, P_{corr}), the regions of acceptable attention are then restricted inside the gray shaded area, shown schematically in Fig. 3 (values of RF below 200 msec does not appear in practice, the RT above 1000 msec represent the fall into micro-sleep).

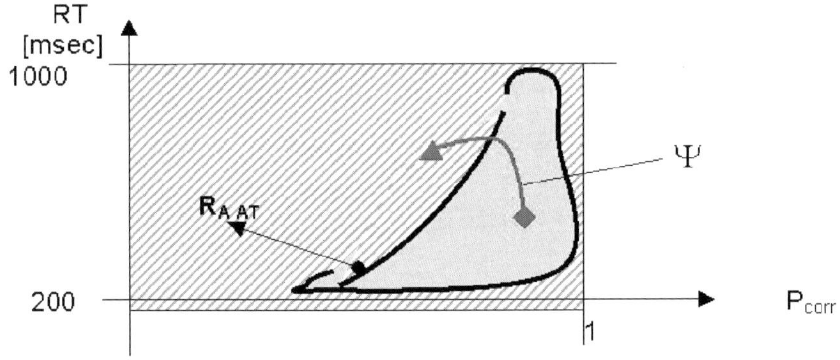

Fig. 3: The region of acceptable drivers level of attention and the respective life curve Ψ.

However, the investigation of the boundaries of $R_{A\ AT}$, even in such two-dimensional space represents a very laborious and complicated problem, especially because the various types of car, road, driving situation and especially also due the above mentioned drivers individuality has to be taken into account. Of course, often the boundaries of $R_{A\ AT}$ or some their parts have often more or less fuzzy character.

In the course of driving, the point X = {RT,P_{corr}}, representing the actual level of particular driver attention moves in the space {RT, P_{corr}}. It follows some curve, which in analogy to the technical system reliability theory can be called the "life curve" Ψ. This can be parametrized by the values of various independent variables, namely by time. If Ψ remains inside $R_{A\ AT}$, the driver is able to drive considerably reliable and safe. If it approaches the boundaries of $R_{A\ AT}$ or if it brakes it the situation becomes dangerous.

OPEN PROBLEMS

The above mentioned motivations stand before us some important actual problems, which also represent challenges for further research.

We shall try here to discuss some of them:

a) The creation of satisfactory large data-base of EEG data, measured on special selected set of human subject – probands, simulating the sample of drivers community. The respective laboratory measurements has to be made in a considerably dense grid of electrodes (at least 29) located according the wide accepted international standard (e.g. 10/20) during the appropriate long observation session (probably 30 to 45 minutes), during which the subject controls the car movement on simulator, observing some standard scene. In this artificial scene the rural and urban roads has to be simulated, both including the points, in which the probands reaction time and correctness of his/her reaction will be tested. While the filling out of such base – the Micro-Sleep Base MSB, which proposal was already made (see Novak et al. 2001a, Novak et al 2001b], is out of possibilities of one single laboratory, the necessity of coordinated international cooperation of several laboratories in different countries and different parts of world exist. These laboratories has to share a common methodology for to be able to produce compatible results, of course.

b) Mining of relevant hidden interrelations and knowledge from the sub a) created data-base.

c) The development of new, more selective and specific methods for analysis of quasi-periodic and quasi-stationary time series typical for EEG signals, which will be taken as part of the common recommended methodology (see Faber et al. 2002.).

d) The development of suitable electrodes for EEG recording applicable in moving car, not (or minimally) disturbing the driver. The application must be possible without any auxiliary help. Probably (as both recent experimental data and theoretical consideration show), only two pairs of electrodes located in the area of drivers head behind his/her ears will be satisfactory. The transmission of measured signals to the analytical equipment, installed in car panel must be wireless and satisfactory reliable.

e) The investigation of possibilities of contact-less measurements of electromagnetic radiation (either as electric potentials or eventually magnetic fields) emitted by human brain, which can be applied in moving car. In these investigations a special interest has to be given to the eventual possibility to use the specific parts of electromagnetic spectrum, for which the scalp and head are more transparent.

f) The investigation of the influences of special modulated and located either electric or magnetic fields on the drivers level of attention.

g) The investigation of influence of the set of external and internal disturbing factors, causing the diminishing of driver attention. Among such factors, the influence should be investigated both individual and in combination, such has to be included as:

- Temperature,
- Humidity,
- Air pressure,
- Illumination,
- Noise,

- Communication (including mobile phones),
- Alcohol,
- Drugs, including nicotine,
- Mental state diseases.

A special importance has to be given to the influence of mobile calls. Here also the analysis of the density and kind of use among drivers in different regions and time of year has to be done.

h) The investigation of possibilities to detect the driver fall into relaxant, somnolent or eventually micro-sleep stage by the use of suitable combination of some secondary factors, like the eye movement, face analysis etc., probably calibrated by the analysis of EEG signals (see Novak et al. 2003a, 2003b e.g.).
i) The development of special designed mobile set, which could be permanently inserted in car as a fixed part of its cockpit, designed to minimize the disturbing of drivers attention when driving.
j) The development of a set of recommendations for optimizing the car cockpit with respect to minimizing the subsequent degradation of driver attention. Here the special interest has to be given to the interaction of driver with navigation tools, radio (eventually TV) and communication systems (e-mail, Internet etc).
k) The development of auxiliary electronic and information tools, which can improve the car driving safety, like the car front and rear radars, on-board weather prediction systems etc.
l) The development of warning system, which can give to the driver satisfactorily in advance (at least few tens of seconds) the information, that his/her attention level is falling down near the boundaries of acceptability. The eventual warning has to be realized in the form, minimizing the possibility, that the somnolent driver either neglect it or on the other hand react panicky. Probably the artificial voice will be a good selection, combined with the set of other subsequently graduated warning signals, when the warned subject will not react adequate. As the last tool the automatic stopping of car movement has to be used.
m) The development of the satisfactorily reliable and safe classifiers and predictors of driver attention falling down, which will probably be highly individual for particular person. Investigation of the time (or other independent influences) for which they can be used (the time, or range of other independent influences, for which the image of the selected warning parameters of drivers attention – dominantly the EEG – is not significantly changed). Investigation of the possibilities to find among these individual classifiers and predictors some typical groups.
n) The investigation of the regions of minimal acceptable level of attention for different drivers, car and driving situation. The boundaries or their suitable approximations of these regions of acceptable attention has to be inserted in the driver attention analytic and warning system installed in the car cockpit together with the individual attention decrease predictors of the particular driver.
o) The development of the system, which allows to automatically investigate the actual drivers behavior on the selected dangerous parts of the road net, to detect the respective traffic situation, in which they eventually cross the limits of reasonable and safe driving and to start their necessary warning and subsequent punishment.
p) Development of the improved education and training system for drivers, which (e.g. on the base of advanced biofeedback) can enhance their resistance to fatigue and also diminish their eventual tendency to the risky and aggressive driving.
r) The investigation of drugs, supporting and improving the level of human being attention while driving a car.

The operation of driver in moving car is an example of very complicated interaction between several very heterogeneous systems. Some of them are artificial, i.e. the car, the road (tunnel, bridge), the traffic control system, some are of real nature (driver, passengers, surrounding community, the controllers of traffic control system, police, justice). All of them interact in very complicated manner, which we at present are not able to analyze with necessary accuracy and reliability. Even the relative simple interactions, like those between the driver and the moving car, sketched in Fig. 1 are not quite easy to understand.

Evidently, the solution of the above-mentioned challenges represents a very long research and development. Even if after much work some significant results will be reached, one can-not expect, that they will come very fast in practical use, even that there is evident strong need for it. This appears just because of natural conservatives of our human society.

Nevertheless, we can have the hope that subsequently it will be possible to reach also some success in this respect and that one can so contribute to minimization of those tremendous danger and losses, which are daily seen on our roads.

REFERENCES

Novák M., Faber J., Tichý T., Kolda T. (2001a): Project of Micro-Sleep Base Research Report No. LSS 112/01, CTU, Prague.

Novák M., Faber J., Votruba Z. (2001b) Project of International Cooperation in the field of Micro-Sleeps Research Report No. LSS 116/01, CTU, Prague.

Novák M., Faber J., Votruba Z (2003a) Theoretical and Practical Problems of EEG based Analysis of Human – System Interaction Proceedings of the International Conference on Mathematics and Engineering Techniques in Medicine and Biology Sciences, METMBS´03, Las Vegas, Nevada, June 23 - 26,2003, 247-255

Novák M., Votruba Z., Faber J. (2003b) Impacts of Driver Attention Failures on Transport Reliability and Safety and Possibilities of its Minimizing Lecture at conference SSGRR-2003, L´Aquila, Italy, July 27 – August 4, 2003.

Faber J., Novák M., Tichý T., Votruba Z.(2002) Problems of Quasi-stationary and Quasi-periodic Time-series Analysis in Human Operator Attention Diagnostics, Lecture at conference Diagnostika 2002, Brno, Czech Republic, October 1, 2002

Novák M. and Votruba Z.(2004) Challenge of Human Factor Influence for Car Safety Symposium of Santa Clara on Challenges in Internet and Interdisciplinary Research -SSCCII-2004, Santa Clara, Italy, January 29 – February 1, 2004

K.G. Goulias, editor

CHAPTER 11

CAN CREATIVITY BE RELIABLE?

Tomas Brandejsky, Faculty of Transportation Science, Czech Technical University in Prague, Konviktska 20, 110 00 Prague 1, Czech Republic, Brandejsky@fd.cvut.cz

WHY IS CREATIVITY SIGNIFICANT COMPONENT OF TRANSPORTATION PROCESSES ANALYSIS?

Creativity represents a significant face of human (and not only human) reasoning. The works in design theory field, namely of Gero (2002), underline this fact. But creativity is not limited only to artefact design. It affects the whole human reasoning including children games and a difficult problem solving (especially under uncertainty conditions). Of course, this case also includes such situations like route planning or solving of difficult situations like traffic jams. The role of creativity in routine control is less due to the need of large disposable reasoning (computational) capacity.

Because the creative part of drivers' reasoning is not studied, our predictions of traffic network varying impacts are ambiguous. It is difficult to say how drivers will solve contradictions in traffic signs or how they will orientate themselves in new roads area (in the first moments before the situation stabilisation).

We must also study creativity from a more practical viewpoint. It is the viewpoint of technical systems, control software and operation rules design. This fact brings the necessity to search an answer to a simple question: Can creativity be reliable? We also must search answers to two hidden question at least – under which conditions creativity is reliable and when creativity products are reliable. These two questions are more significant than the first question for practice!

CREATIVITY IN PROBLEM SOLVING – BOTH DESIGN AND COMMONSENSE PROBLEMS

We can presume that there exists the only creativity. Thus we need not to distinguish between transportation problem solving and, e.g., reliable system designing. It is possible to mention the paper of Harnad (unpublished manuscript) discussing particular creative and non-creative techniques used in creativity modelling. Creativity models differ from author to author and include a heterogeneous set of approaches from magic to genetic programming and other soft computing techniques. Harnad in his work recognises the following techniques as creative:

1. Analogies (analogical reasoning)

2. Anomalies (paradoxical reasoning)

3. Constraints (each useful reasoning must be limited by constraints, but creativity breaks some constraints!)

4. Heuristic strategies (and emergences concluding from parallel multiple heuristic use known in mobile robotics e.g.).

The role of analogies in creative reasoning is intensively studied by many authors from both theoretical and practical viewpoints. It is possible to mention herein the works of Bonnardel (2000), Ishikawa and Terano (1996), Pauen and Wilkening (1997) or Visser (1996). Paradoxical reasoning is studied sporadically and it is not studied from the creativity viewpoint usually. Constraint-based reasoning is known from deep history of Artificial Intelligence from the works on computer vision and qualitative simulation. Because analogical reasoning frequently uses abstraction and qualitative models present abstraction of differential equation models, we can find work joining these two distinguish disciplines (see Forbus, 2000).

Also different techniques are used within the area of conceptual design. These techniques are not creative according to Hanard and none of them can be used for human creativity modelling alone but they are in this moment well implemented and suitable for simple cases. These techniques include the following methods:

5. Production rules – expert systems and other systems based on logic cannot be creative due to the basic assumption of logic – closed world proposition. In such a system no unexpected discoveries can be done. Especially when the system works with a static set of rules (without learning).

6. Genetic programming algorithms – this modern technique is now frequently studied both by researchers in the field of soft computing (as Koza et al, 1999) and in the field of conceptual design (Gero et al., 1997).

Unfortunately, genetic programming is limited in large structure discovering. These limitations are made by the static character of the fitness function, inefficiency of long structures development and problems with using of design rules, meta-knowledge and standard tools.

The static character of the fitness function is problematic because in system design (design of artefact, device or procedural sequence) each new component included into the system brings additional constraints describing limitations and working conditions of this component. Non-genetic design techniques like rules, meta-knowledge and standard tools are frequently used by humans, but it is difficult to use them in combination with previously reviewed techniques in the computer.

APPLICATIONS OF CREATIVE REASONING

On the base of neurological research (e.g. Faber, 2003), and psychological studies, e.g. Pauen and Wilkening (1997), it is possible to recognise analogical reasoning as the most significant mechanism of human reasoning commonly and especially of creativity.

Many routine tasks neighbouring with the creative ones can be implemented as reactive (like conditional and unconditional reflexes known from living creatures). Reactivity implementation is studied in robotics (see e.g. Pfeifer and Scheier, 2000). Reactive reasoning is based on observation, satisfaction of proper reflexes conditions and their use. This type of reasoning is closed to analogical reasoning and this fact is useful for studying and implementing of them.

Creative reasoning is inherited in many tasks less or more related to robotics. From the viewpoint of penetration of mobile robots technology and transportation devices (and from the viewpoint of drivers reasoning, of course), the following fields are interested for us:

1. - learning

2. - reactive agents

3. - reasoning under uncertainty

Classic Artificial Intelligence studies successfully the problem of learning, but only from one side. It studies problem of memorizing prepared or observed data, recognising relationships in the data. But reasoning has also a second side. It is a problem of experiment preparation, problem of game construction. If the experiment or game might bring novel information, it must be able to answer new questions, it must be different from the previous ones. This is the sphere of creative reasoning.

Future intelligent agents will need to show more creativity in its behaviour than now if they could be applied in cars to help the driver or in the laboratory to model drivers.

CREATIVITY IN DIFFERENT KINDS OF TRANSPORTATION

It is interesting to study the relationship between the successive kinds of transportation and creativity. On the one hand we have railway transport with its strict rules and road transport with it freedom on the other hand. From this viewpoint the relation between reliability and creativity seems simple - creativity increases reliability and safety! But is such answer correct? Probably it is not because creativity enables us to design more reliable and safer systems than their predecessors. Does it mean that creativity is correct in device-design-time but erroneous in the time of use? Until we will not be able to study creative processes more precisely, such conclusion is correct.

REASONING BASED ON ANALOGIES AND METAPHORES

Analogical reasoning is now understood as fundamental mechanism of creative thinking not only in the above mentioned paper of Forbus (2000) but also e.g. in large book of Hofstadter (1995). This avouchment is supported by many interesting evidences from the field of psychology (Bonnardel, 2000; Pauen and Wilkening, 1997; or Visser, 1996). Within this context we cannot omit the presence of associative neurons in brain (Faber, 2003) physically forming "shortcuts" between distant brain centres.

We usually describe analogical reasoning by the following form (see Figure 1):

> **IF** object A has behaviours {Va} and A concludes C
> **THEN**
> **IF** B has a similar set of behaviours {Vb} **THEN** it is possible to expect that also B concludes C

Figure 1: Analogy example

Our model distinguishes between analogies and metaphors as two special kinds of analogical reasoning. The approach presented herein lets us to model the influence of real-time reasoning (and limited processing capacity) on complexity of the reasoning. This model is motivated by the simple fact that reasoning with metaphors is more computationally complex than the analogical one and thus it is used in different situations.

The model was presented in works (Brandejsky, 2003a, b). Model reasons analogies and metaphors as a special form of equivalencies. The difference is made by the presence or non-presence of this equivalencies on chains of analogies or metaphors respectively.

The presence of this equivalence is made by presence of behaviour sets equivalences (1):

$$\boldsymbol{D}_S(\mathrm{A},\mathrm{B}) \approx \boldsymbol{D}_S(\mathrm{B},C) \approx \boldsymbol{D}_S(\mathrm{A},C) \qquad (1)$$

If the condition (1) is satisfied, we speak about analogy, else about metaphor.

Scientific literature knows many methods how to work with analogies on the base of abstraction (Ishikawa and Terano, 1996), or abstraction reasoning models based on multi-agent approach like AMBR (Kokinov, 1994). Nevertheless, these models do not solve problems of work on the boundary of analogies and metaphors. Thus, the presented model was developed.

UNIFIED REPRESENTATION OF ANALOGIES AND METAPHORS MODEL

The key question from the viewpoint of metaphoric and analogical reasoning implementation is the representation of knowledge-base. The significance of this question increases at the moment when we speak about practical non-trivial application, when the size of the knowledge-base is gigantic.

Universal knowledge descriptions are usually based on ontologies and on frames, which are capable to let us produce metaphors dynamically. The frames have been introduced by M. Minsky (1975). They are flexible and enable describing wrongly structured knowledge. In the last decades the frames are also used in ontology description, like in the systems FrameLogic (Kifer at al, 1995) or Ontolingua (Gruber, 1993; Farquhar at al, 1997). Ontologies were developed within the frame of artificial intelligence research to enable sharing and reusing of knowledge. Ontologies are also capable to describe data meaning and knowledge representations (Fensel, 2001).

Implementation of analogies and metaphors look up unified algorithm

Reasoning objects (states, situations and related actions) as ontologies described in the form of frames we can connect their common behaviour and build a special case of metaphorical neural network. This network differs from the standard ANNs because it does not contain implicit mechanism of learning. To be able to work with contradictory input data (contradictory activation by objects with different magnitudes of the behaviour), it is necessary to insert special objects into network - behaviour arbiters - which will solve these conflicting situations. The architecture of the network is sketched in Figure 2:

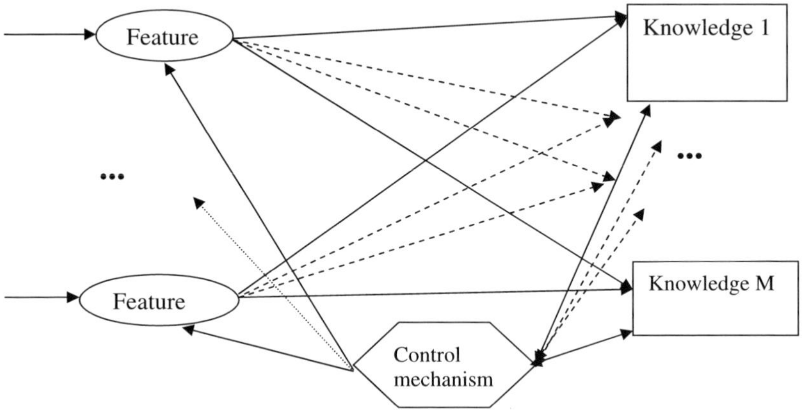

Figure 2: Communication structure of metaphorical and analogical network

The arbiters solve problems of multiple activation of a given behaviour and co-ordinate attributes of credibility to them. Knowledge objects get similarity measure of their inputs and incoming behaviours, from the weights of connections (expressing measure of equivalence of behaviour represented by related node and understanding of this behaviour in the knowledge object) and from evaluation of behaviour equivalence with a given pattern (e.g. equivalence with complicated structure describing device functions). The knowledge object then deduces new features and assigns them credibility calculated from the above mentioned similarities. Then the knowledge objects send information about their possible activations of behaviour nodes to the super-arbiter. The super-arbiter either allows or forbids these activations.

The super-arbiter plays a significant role in the system function. The super-arbiter reasons if to enable this propagation of deduced knowledge from the knowledge object to behaviour nodes. This architecture models periodical style of brain work. The super-arbiter also mines useful information from this communication (answers solved problem). The super-arbiter in addition stops this communication if the solution time is over - in robotics and transportation process modelling it is necessary to answer at the given time and then make reaction without looking if it is the best possible reaction or suboptimal one only. The super-arbiter is capable to restrict metaphorical reasoning to the benefit of analogical one by verifying newly initialised and activated nodes similarity with pattern.

The presented model is based on the idea of artificial (neural) network working with symbolic information. Now finishes the work on its implementation in order to verify correlation with experimentally measured data.

At the moment when the behaviour node is activated or re-activated by more than one knowledge object, the behaviour node must select one of them or record all accepted and to assert trust measure to each.

Control mechanism (super-arbiter) is capable to filter metaphors to the benefit of pure analogies and also to eliminate solutions that do not satisfy the given constraint (e.g. traffic rules). The arbiter can work stochastically or deterministically (in the case of strict application of rules). Stochastic filtering enables to producing a small number of metaphors and analogies not satisfying the given constraints (it gives a system chance „to have a brain-wave", solve a problem unsolvable within the frame of the constraint system and also produce erroneous solution. Such approach is closer to human thinking than a deterministic one which eliminates all mistaken solutions. Analogies and metaphors do not solve the problem of creative reasoning as the whole. As it is noted above, the reasoning based on analogies and metaphors solves only part of creative thinking (even if essential). None creative process is reasonable without processing a huge amount of routine operations. These operations can be modelled advantageously by various techniques (e.g. by expert systems working with mental models or by qualitative models).

THE USE OF THE MODEL IN RELIABILITY MODELLING

The presented model of analogical reasoning with control/discrimination unit is similar from many aspects to the brain structure regardless of its simplicity. It is closed to the formator-complex model of human brain explaining many neurological symptoms. This model is propagated by a neurologists Faber (2003) applying the Farley-Clark's model (Farley and Clark, 1954) on human brain. In the presented model of analogical and metaphorical reasoning we can see both basic components - formator (presented control unit) and complex (behaviour and knowledge nodes). We reason connections formed consciously; not Farley-Clark's random ones).

The problem of creativity safety is in above presented model the problem of relevant breaking mechanisms presence; the problem of existence and non-existence of relevant knowledge in knowledge nodes and discriminating mechanisms in the control unit.

The analogical reasoning is predicable and analysable more simply than metaphorical one due to the presence of equivalence between the initial and target states. Thus we must "only" find out if the subject (drivers) analogies do not contain an unsafe analogy link.

Metaphors are more creative, but there is no easy visible link between initial and final situations. Thus it is necessary to create powerful and easy applicable mechanisms (constraints, pre- and post-conditions) to eliminate "wrong" metaphors.

The presented model enables us to do more. The work (Halford and McCaredden, 1998) discusses a look at human thinking from the viewpoint of processing capacity. The processing capacity of the human brain from this viewpoint is not only limited by the capacity of short time memory, but also by ability to process grouping information into chunks. Halford and McCaredden recognize individual differences in experience, processing capacity and their interaction.

Applying this concept on above described drivers reasoning modelling we recognize that problems and collisions come when the number of solved tasks is greater than the processing capacity. Capacity decreases by tiredness, sleepiness, illness, drug influence and thus these factors increases the risk of collision and other events because they increase the risk of processing capacity excess.

Regardless of this influence, the risk of a wrong decision (and consequent collision) increases in situations with high number of stimuli and tasks or with complicated decision. Lack of time also increases requirements to processing capacity and this situation is known as stress.
In the presented model the processing capacity limitation is modelled by limitation of the number of metaphor/analogy steps - by limitation of discrimination unit capacity. The presented model enables us to form novel methods of operator's skills testing on the base of (non-)presence of relevant selective mechanisms of reasoning based on analogies and metaphors. The result is based on the ability of the selective mechanism to distinguish good and wrong analogies and metaphors (wrong from the defined viewpoint).

NOVEL VIEWPOINT TO MICROSCOPIC SIMULATION

The role of creativity can be studied within the frame of microscopic simulation tool. Microscopic simulation is ready to adopt more complex models of drivers' behaviour than nowadays ones. It means that study of creativity is strongly related to studies of drivers' mental models, drivers reasoning and decision processes.

The modelling of driver behaviour represents complicated problem especially due to the need to react on asynchronous events, whose arise causes the need to start additional processes, whose time of execution is sometimes longer than frequency of events arise. These processes start child ones, whose waits for raise of specified situation. E.g. they wait for change of signal on traffic lights. Whilst operator must solve unexpected events from programmer's viewpoint by event-driven method, expected situations are solved rather by goal-driven one. It is awkward because we must to join event-driven and classic approaches to programming. Out of this it is useful to solve questions of tasks priorities and problems of their terminating. On the opposite side goal-driven perceptive tasks enables to describe focusing of attention. This type of tasks is the sign of our expectations of some event or situation and enables to fix operator's attention in this direction and to decrease probability of his omit.

The operator must solve a lot of particular problems in sketched model. Usually we recognise tactical and strategic control. In the presented model the control system (operator model) musts solve on-line diagnostic of car (and to update its model in its mind), it musts to create model of environment for its purposes (to be able to provide strategic planning) and it musts also to form particular models of other drivers behaviour, to be able to react quickly on their future actions. In addition, the model with slower dynamics provides self-evaluation of its momentary health-state and its capabilities. Thus presented model differs from usual architectures used in robotics (Pfeifer and Scheier, 2000) by the use of sub-models describing particular aspects of more complex decision task.

The creativity is observed in complex problems solving content. Thus these problems are solved by strategic control unit. The tasks solved by the presented network are especially route planning, navigation in the city etc. Implementation of analogical-metaphorical network consists of connection of the network inputs (features) to car and driver behaviours and features. Knowledge then represents some actions, like turning to the left, deceleration or any else.

CONCLUSION

The presented model is applicable in road transport modelling, especially in microscopic simulation. The way of implementation of the model into microscopic simulator is sketched. The model also opens a new viewpoint of verification of the driver's education and testing.
The paper opens question if the road transport can be safe if cars are driven by human drivers or must be human drivers replaced by technical systems? Presented research opens novel ways to driver education and modelling of their reasoning. This research can also bring new viewpoints on human-car and human-out-car-traffic interactions.

ACKNOWLEDGEMENT

The work is supported by Grant Agency of Czech Academy of Science.

REFERENCES

Bonnardel N. (2000). Towards understanding and supporting creativity in design: analogies in a constrained cognitive environment. *Knowledge-based systems*, **13**, 505-513.

Brandejsky T. (2003a). Real-time analogical and associative reasoning machine. In: *ICCC 2003*, Vol. 1, pp. 763-766, TU Kosice, Slovakia,

Brandejsky T. (2003b). The Application of Analogical Reasoning in Conceptual Design System. In: *Artificial Intelligence and Applications*, Vol. 1, pp. 511-516, Acta Press, Anaheim.

Faber J. (2003). Isagogé to non-linear dynamics of formators and complexes in the CNS. The Karolinum Press, Prague.

Farley B. G. And Clark W. A. (1954). Simulation of self-organizing systems by digital computer. Trans. IRE 1954, **PGIT-4**, 76-84.

Farquhar A., Fikes R. and Rice J. (1997). The Ontolingua Server: A Tool for Collaborative Ontology Construction, International Journal of Human-Computer Studies, **46**,707-728.

Fensel D. (2001). Ontologies: A Silver Bullet for Knowledge Management and Electronic Commerce. Springer-Verlang, Berlin-Heidelberg-New York.

Forbus, K. (2000). Exploring analogy in the large. In: The Analogical Mind: Perspectives from Cognitive Science (Gentner, D., Holyoak, K. and Kokinov, B., eds), pp. 23-58, MIT Press, Cambridge, MA.

Gero J.S. (ed.) (2002). Artificial Intelligence in Design'02. Kluwer Academic Publishers, London.

Gero J.S., Kazakov V.A. and Schnier T. (1997). Genetic engineering and design problems, In: Evolutionary Algorithms in Engineering Applications (Dasgupta D. and Michalewicz Z. eds), , pp. 47-69, Springer-Verlang, Berlin-Heidelberg-New York.

Gruber T. R. (1993) A Translation Approach to Portable Ontology Specifications, Knowledge Acquisition, **5**,199-220.

Halford G.S. and McCredden J.E. (1998). Cognitive Science Questions for Cognitive development: the concepts of learning, analogy and capacity. Learning and Instructions, **8**, 289-308.

Harnad S. (unpublished manuscript): Creativity: Method or Magic? http://www.cogsci.soton.ac.uk/~harnad/Papers/Harnad/harnad.creativity.html

Hofstadter D. R. (1995). Fluid Concepts and Creative Analogies: Computer Models of The Fundamental Mechanisms of Thought. BasicBooks, New York, NY

Ishikawa T. and Terano T. (1996). Analogy by Abstraction: Case Retrieval and Adaptation for Inventive Design Expert Systems. Expert systems with applications, **10**, 351-356.

Kifer M., Lausen G. and Wu J. (1995). Logical Foundations of Object-Oriented and Frame-Based Languages, Journal of the ACM, **42**, pp. 741-843.

Kokinov B. (1994). A hybrid model of reasoning by analogy. In: Advances in connectionist and neural computation theory (Holyoak K. and Barnden J. eds.), Vol. 2, pp. 247-318, : Ablex, Norwood, NJ.

Koza J.R, Bennett F.H., Andre D. and Keane M.A. (1999), Genetic Programming III: Darwinian Invention and Problem Solving. Morgan Kaufmann Publishers. San Francisco, CA.

Minsky M. L. (1975). A Framework for Representing Knowledge, In: The Psychology of Computer Vision (P.H. Winston, ed.), pp. 211-277, McGraw-Hill, New York, NY.

Pauen S. and Wilkening F. (1997). Children's Analogical Reasoning about Natural Phenomena. Journal of experimental child psychology, **67**, 90-113

Pfeifer R. and Scheier Ch. (2000). Understanding Intelligence. MIT Press, Cambridge, MA.

Visser W. (1996). Two functions of analogical reasoning in design: a cognitive-psychology approach, Design Studies, **17**, 417-434.

K.G. Goulias, editor

CHAPTER 12

RELIABILITY OF INTERFACES IN COMPLEX SYSTEMS

Zdeněk Votruba, Mirko Novák, Jaroslav Veselý[1]

Abstract: **There is common, rather empirically supported knowledge within the body of the System Analysis that complex interfaces (for example "man – machine" interface within the hybrid system, or synapse in the human brain) susceptibly react both on the dimension of the task (i.e.: the number / type / domain of interface parameters / markers), and the level of uncertainty.**

In order to quantitatively evaluate this effect, the overview of the different concepts of interface is done first. Then the problem is analyzed on the background of geometrical considerations.

The results of the study indicate that even a low degree of uncertainty has significantly adverse effect on the interface regularity (consequently the reliability of systems processes, as well) if the dimension of the pertinent task is sufficiently high.

Practical implication of this result for system analytics is straightforward - keeping the dimension of the task as low as possible. The interface dimension higher than 5 is in the majority of tasks with moderate uncertainty considerably unfavorable. This result imposes serious constrain to the systems identification.

---

[1] Authors: Prof. Dr. Zdeněk Votruba (zdevo@ieee.org), Prof. Dr. Mirko Novák (mirko@fd.cvut.cz) and Dr. Jaroslav Veselý (vesely@fd.cvut.cz) are with the Czech Technical University in Prague, Faculty of Transportation Sciences, Konviktská 20, Praha (Prague) 1, CZ 110 00; Czech Republic;
Supported from grants: MSMT CR: MSM: 210000024 and AV CR :IAA 212 4301.

Key Words: **System interface, Hybrid system, Interface dimension, System identity Uncertainty, Regularity, Reliability, System Analysis,**

Complex system interfaces (IF), for example "man – machine" IF within the complex hybrid systems[2] or system alliances [3], are often recognized as the weakest points of the system from the reliability point of view. On the other hand, complex neuron synapse in the brain[3], seems to be quite reliable object. The aim of this contribution is to analyze the reliability of IF[4] within the framework of Systems Theory [2,1], taking into account the dimension of relevant IF and the degree of uncertainty. In order to provide this study, several concepts are to be detailed first.

1. Basic Concept of Interface (IF)

The concept of System Interface has been widely used for many years and in many areas of Science and Technology. Let us mention for example Systems Science, Computer Science, and Economic Management. The quite often used construction of this concept is based on specification of constrains on data structures conversion and compatibility [13].

1.1. The Simplest Model

Within the general systems the interface (IF) is most frequently introduced as a **fictitious cut** across the respective connection (relation) between the two parts (elements) of the pertinent system[5] (or between the system and the system neighborhood), described by two mutually corresponding pairs of sets:

- OUT and val OUT - at the output of the first part of a system under analysis, and
- IN and val IN - at the input of the second part of this system.

These pairs of sets consist of:

- sets of variables / parameters / parametric sentences[6]: OUT , IN respectively, and
- sets of intervals / domains of possible values of respective variables / parameters / parametric sentences: val OUT , val IN respectively.

[2] (respective interpreted system, for example of transportation or information nature)
[3] for many systems analytics the prototype of complex interface
[4] I.e. the probability that analyzed IF does not change the "run" of the chosen process against reference. The process (in the system) is further defined as a sequence of events. The event is either transition of (any) system element, or the change of the system structure (within synchronous systems the event could also be the elementary step of time – in fact the transition of the system specific element – clock)
[5] see Fig.1, (parts 1 and 2)
[6] The terms Variables and Intervals, respectively are usually used within the real or interpreted systems, mostly of the continuous nature in the systems base, while terms Parameters, Parametric Sentences and Domains, respectively, are mostly utilized in systems of a higher degree of abstraction, usually of discontinuous nature in systems base.

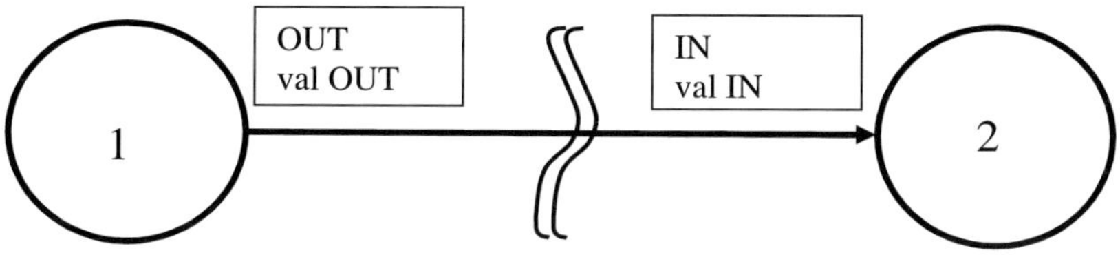

Fig.1. Schematic sketch to the basic definition of interface

The IF is regular if and only if both OUT = IN and val OUT =val IN

One can hardly over-estimate how clever this basic concept of IF is. The power of the basic concept results from both the universality and the fictitious nature of interface. Such an IF posses no "real world" qualities, as for example time – delays or resources consumptions are. It is pure quasi–entity that could be measured by cardinalities of respective sets and by the binary – valued parameter of regularity.

Unfortunately, the basic concept of IF is not well suited for specific deeper studies of the reliability of systems owing a substantial degree of uncertainty and complex interfaces, as well as for both analyses and control of the processes of complex interfaces regularization.

1.2. Interface as a fictitious system element

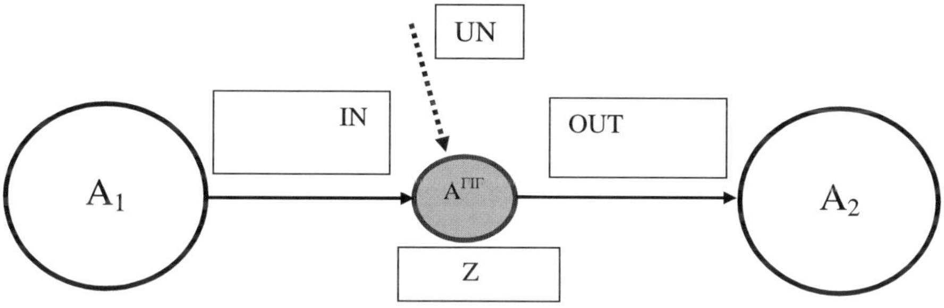

Regular interface: A^{FIF}: $Z_0 = \{1\}$; α: $Z_0 \rightarrow Z$; β: (IN x Z) = OUT;dim(IN) = dim(Z) =dim(OUT)

Fig.2. Schematic sketch of the regular interface identified as a fictitious system element.

For a complex IF seems to be advantageous the introducing the IF as a <u>fictitious system element</u> (**A^{FIF}**). The advantage of this approach is anchored in the richness of the concept of the system element, which is generally defined as an <u>automaton</u>. For the sake of simplicity the finite deterministic automaton (FDA) is usually chosen. FDA can be described by a <u>triple of sets</u> **IN, Z, OUT** – inputs, internal states and outputs, respectively (within the set of internal states, **Z** is further defined a specific subset – initial internal state **Z_0**.), and a <u>double of</u> (mapping) <u>functions</u>: **α, β**. Function α transforms the Cartesian product (IN x Z) into the set of internal states Z. Function β transforms Cartesian product (IN x Z) into output set OUT.

$$\textbf{FDA:=(IN,Z, } \mathbf{Z_0}\textbf{,OUT, } \boldsymbol{\alpha}, \boldsymbol{\beta}\textbf{)}$$

$$\boldsymbol{\alpha}: (\mathbf{IN} \times \mathbf{Z})^{7} \rightarrow \mathbf{Z}, \quad \boldsymbol{\beta}: (\mathbf{IN} \times \mathbf{Z}) \rightarrow \mathbf{OUT})$$

The fictitiousness of the IF element reflects its important features:

- No demands for systems resources, and
- No transform of base variables or parameters, i.e. no consumption of time to carry out the functions, as well.

It is worth mentioning that these features are strictly valid for regular IF, while any disturbances of regularity can harm these features, as well.

1.2.1. Probably, the simplest introduction of a **regular** IF as a fictitious systems element[8] $\mathbf{A}^{\mathbf{FIF}}$ could be:

Z is an empty set,

$\boldsymbol{\alpha}$ is any arbitrary function without demands for system resources (in fact α is meaningless),

$\boldsymbol{\beta}$ transforms IN into OUT, $\boldsymbol{\beta}$**: IN→OUT**, the transform being an equivalence for all the parameters (components) of the sets (vectors) IN and OUT respectively :

OUT = IN.

In detail:

Let OUT = $\{a_i^{OUT}\}$ be a set of parameters a_i^{OUT} , Let IN = $\{a_j^{IN}\}$ be a set of parameters a_j^{IN}, (i, j, being natural numbers), then interface IF is regular if and only if :

$$(i = j) \text{ AND } (\wedge_i(\mathbf{a_i^{OUT}} = \mathbf{a_i^{IN}})) = \mathbf{1} \text{ (true)}$$

The regular function of $\mathbf{A}^{\mathbf{FIF}}$ therefore means plain instantaneous transition of IN into OUT.

1.2.2. To describe the impact of irregularities and uncertainties, a slightly modified model of the IF is more suitable:

Let OUT = $\{a_i^{OUT}\}$ be a set of parameters a_i^{OUT} ; Let IN = $\{a_j^{IN}\}$ be a set of parameters a_j^{IN};

Let Z = $\{a_k^{Z}\}$ be a set of parameters a_k^{Z} ;

$Z_0 = \{z_{0k}\}$; (i, j, k, natural numbers),

$\boldsymbol{\alpha}: \mathbf{Z} := \mathbf{Z_0}$

$\boldsymbol{\beta}: \mathbf{OUT} := (\mathbf{IN} \times \mathbf{Z})$

For regular IF the respective $\mathbf{A}^{\mathbf{FIF}}$ has, of course, the following features:

$i = j = k$; $Z_0 = \{z_{0k}\} = \{1\}$;(i.e.: $z_{0k} = 1$ for $\forall$ k);

In Chapter 3 we discuss how to express irregularities and uncertainties within this model.

1.3. Interface identified as a conversion element (CA)

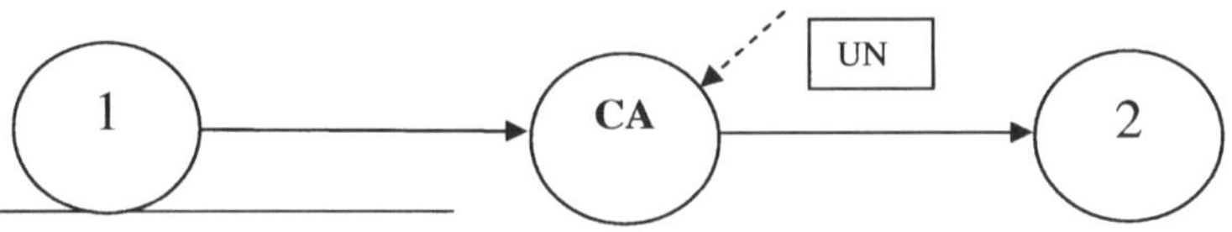

[7] (IN × Z), etc. means Cartesian product of respective sets.

[8] (i.e. fictitious finite deterministic automaton)

Fig.3. Schematic sketch of the interface of the system elements 1 and 2, being regularized via the conversion element CA.

In specific cases, especially within the interpreted systems (scarcely in the abstract ones) it should be meaningful to identify the IF as a (full-valued) real or virtual system element[9]. This approach is within the System Analysis quite familiar, the respective task being known as the "Construction of the Conversion IF Element (CA)" . A significant advantage of this approach is that the "well constructed" CA is able to regularize the respective IF dynamically[10], or to optimize the IF holding certain (pre-defined, goal seeking, respectively) criterion. On the other hand, this approach has a serious disadvantage, namely being often very difficult[11].

An introduction of the system element in the role of IF implies full identification of all the element[12] definition components[13]. [12]

1.4. Language description of IF

The well known equivalence of FDA and the syntax of certain finite language [11] opens the possibility to study this type of IF utilizing artificial language methodologies. This is probably a very promising approach, unfortunately the irregular IF (which is in the focus of interest) is not an equivalent of deeply studied and quite well understood Chomski languages, but it is rather an equivalent of a till not fully understood class of incomplete or pragmatic languages [3,4].

1.5. System Alliance Interfaces

The concept of the Systems Alliance has been recently introduced [3] to cope with the emergence of synergic effects even for the groups of systems that do not share Common System Identity[14].

The principle of the Systems Alliance[15] (for which the role of the IF seems to be crucial) has been explained utilizing the concepts of Information Power (IP) and multilingual translation efficiency, respectively [4].

A simplified illustration of the basic phenomena[16] resulting in the emergence of Alliance could be based on the concepts of Interface Sharing, and Irregularities

---

[9] For the alternative of virtual element, the twins of functions α,β are not associated with the strictly specified FDA, there are "pools" of element sets as well as element functions, and the association is processed dynamically.

[10] (during the "run" of processes)

[11] It is a frequently very difficult synthetic task. The solution often assumes suboptimal intelligent searches, the "hard" algorithmic attempts are scarcely of any practical value due to their trans-computability (non-polynomial algorithms). That is why the prevailing solutions of this task still utilize heuristic or soft methodologies.

[12] (i.e. FDA)

[13] (i.e. **IN, OUT, Z, Z_0, α, β**)

[14] (as is true for the category of Hybrid Systems)

[15] and the emergence of synergic effect, as well

[16] (within the Alliance and especially in Alliance IF)

Conjugation, as well. The case could be illustrated on a simple constructive example from digital electronics environment[17].

1.5.1. Construction of the Example

Let A and B be two synchronous binary digital systems. Both A and B consist of three elements (see Fig. 4.)

The elements with subscript 1 perform either a logic function OR, or a logic function AND of max. 4 inputs. The choice of a respective logic function is controlled by the (binary) parameter R[18].

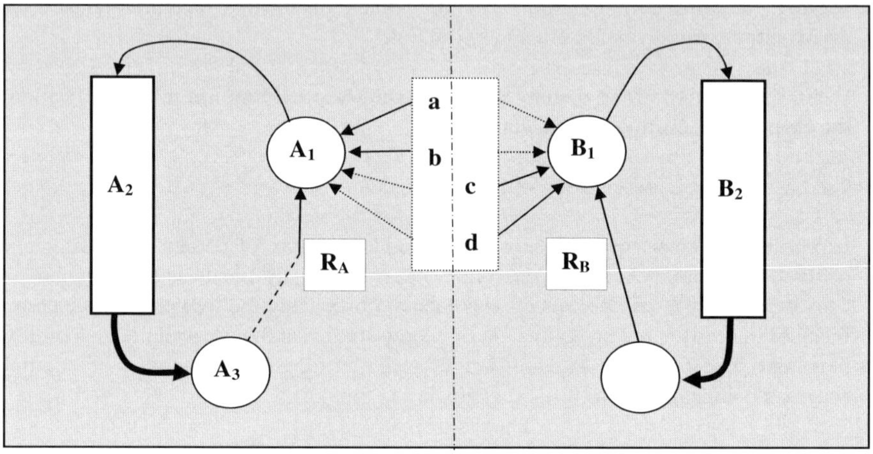

Fig.4. An example of IF sharing and irregularities conjugation

The elements with subscript 2 are shift registers of the pre-defined length.

The elements with subscript 3 identify the total number of zeros in respective shift-registers, and eventually generate control parameters R.

- The goal of system A is to fill dynamically the shift-register $\mathbf{A_2}$ with ones only.
- The goal of system B is to fill dynamically the shift-register $\mathbf{B_2}$, 1:1 with ones and zeros.

Goal (seeking) processes are evidently different for A and B. As a result, the system Identities are different, as well. Consequently, any composition of these (sub)systems A, B cannot be a Hybrid System.

1.5.2. Regular IF:

(1) At the beginning of our consideration, let $\mathbf{A_1}$ utilizes inputs **a** and **b**, while $\mathbf{B_1}$ utilizes inputs **c** and **d**[19]. There is no a priori information about the state of any input.

[17] Such a presentation is also fruitful and important from the epistemological point of view. It is always productive within the frame of Systems Science, to demonstrate that some complex phenomenon (which can be usually identified in social or biological systems) emerges also in systems recognized on hard and in principle not too complex physical objects. In fact such a demonstration is also verification of the Systems nature ("Systemhood") of this phenomenon.

[18] (i.e. R_A, R_B, respectively; R = 0 means function OR, et vice versa)

[19] IF is **not** shared.

It is possible to prove that the optimal choice of the logic function of element $\mathbf{A_1}$ has to be permanent **OR**, and therefore no control $\mathbf{R_1}$ has to be generated. That is not the case of **B**, where $\mathbf{R_2}$ must be zero for more than 50% of ones inside $\mathbf{B_2}$, et vice versa. Both systems A and B can dynamically seek their goals, but generally neither is able to reach the goals permanently.

(2) The situation quantitatively changes if both systems A and B start to share all the inputs a-d[20]. In such a case the frequency of reaching both goals dynamically probably arises[21].

1.5.3. Irregular IF:

More significant changes occur if some (of many possible) irregularities of respective IF are taken into account.

(1) Consider again the original case 1.5.2 (1), but now let a = (permanent) 0 and let c = (permanent) 1. System A is in this case dynamically far from reaching the goal, because the long term probability of the content of $\mathbf{A_2}$ is the same as the probability of "1" at **b**[22]. A slightly better result is obtained for system **B,** but also in this case the appearance of IF irregularity (c = 1) worsens the result in comparison with the original situation 3.3.(1).

(2) Significant improvement of the goal seeking ability of both systems occurs if the (irregular) IF is shared between systems A and B.

Then system A reaches the goal (by chance) absolutely[23] and the goal – seeking ability of system B improves[24], as well. It is the result of (partial) IF irregularities conjugation.

1.5.4. Discussion

The implication of this simple example is straightforward: The presented constructive example shows us that there is a nonzero chance to find the doubles of systems[25] for which the sharing of IF improves the efficiency of the goal seeking processes[26]. Such a double of systems can constitute System Alliance, if either self-ordering or controlled ordering[27] processes occur within the respective systems or systems environment.

An analogical result could be found when taking into account another important component of the system identity - strong processes, as well[28]. Therefore, it is reasonable

[20] IF is shared.

[21] There is a higher probability that for mutually independent and a priori unknown inputs **a, b, c, d** the function **OR (a, b, c, d) = 1**, in comparison with the function **OR (a, b) = 1** (and similarly the same is valid for: **AND (a,b,c,d) = 0**) .

[22] Taking into consideration the Laplacean principle of insufficient evidence, one can expect probability ½.

[23] OR (0,b,1,d) = 1

[24] OR (0,b,1,d) = 1; AND (0,b,1,d) = 0; ⇒ $R_B = 0$ for more than 50% of ones inside B_2 and $R_B = 1$ for less then 50% of ones inside B_2; the dynamic error of the goal-seeking process is of the (binary) order of L^{-1}, where L is the length of respective shift-register B_2.

[25] (even with different identities)

[26] There is of course also a nonzero chance to find the doubles of systems for which the sharing of IF worsens the efficiency of goal seeking processes. This fact is not any real objection against our result, as the process of IF sharing (the emergence of the conjugate irregularities) originates only if it actually has the positive global (for the Systems Alliance as a whole) effect.

[27] (diagnostics and repair subsystems, e.g.)

[28] This case is significantly more complicated. To construct the model rigorously means for example also to identify the double of systems with mutually different goal-seeking asa well as strong processes. An example from the transportation environment may be the double of microcontrollers within the same interlocking system. These microcontrollers are of mutually different both HW and SW (due to the strict demands on security and reliability), but they participate on the common task: control of railway traffic. They also share common IF. A detailed analysis is beyond the scope of this study.

to study the shared IF (with conjugate irregularities) in Systems Alliances as a distinguished case, owing to new both quantitative and qualitative aspects.

More general elaboration and deeper understanding of these phenomena could bring just planned study of IF interaction.

1.6. Specific IF models

Models of neuron synapse could be occasionally used for studying some aspects of IF. From the methodological point of view these models are analogical to cases described in Chapter 1.2. or 1.3. While the models of "electrical synapse" are quite simple ones, the models of "chemical synapse" are generally too complex.

Chaotic and fractal models of IF will probably be in foreseeable future utilized for to tackling with certain emerging effects in very complex and / or unstable systems [4,5,14].

2. System Uncertainty

An uncertainty (in complex systems substantial and almost omnipresent) [1,2,10,12] has many "resources" and aspects. Any effective analysis of complex systems reliability can hardly be made without any careful evaluation of the impact of uncertainty to the system[29]. Thus, there is no question that uncertainty is in the focus of contemporaneous system science.

An the beginning of the further study methodological problems arise:

How to "incorporate" uncertainty into the system[30]?

The significant majority of authors localize uncertainty into:

- Systems (or system elements) **functions / processes** or
- Systems **structure**, eventually
- Systems **neighborhood**.

The localization of uncertainty into the IF is not frequent[31]. Nevertheless, the authors believe that it is just this approach that could help us to illustrate some nontrivial aspect of the task. To model IF we have primarily chosen an attempt described in Chapter 1.2.2. (Interface as a fictitious system element) in which uncertainty "enters" the initial state Z_0.

3. Specification of the Task

The aim of this study is to analyze the combined effect of the dimension and uncertainty of chosen IF within the system with respect to the reliability of defined[32] processes.

The task is structured to the following main steps:

---

[29] The problem has some analogies with the process of system identification [2].

[30] (i.e. specific model of object)

[31] One of the reasons of this situation is probably certain semantic proximity of the concepts of interface irregularity and interface uncertainty, and consequently the possibility of misinterpretations. The most frequent concept of interface as some fictitious entity and the concept of uncertainty as a shortage of information or knowledge (and also the reciprocal relation: information ⇔ removed uncertainty) [10] makes it difficult to imagine how uncertainty (i.e. quasi-entity) enters the interface (i.e. system quasi-object).

[32] (usually strong or goal – seeking)

A. Reliability of a single (non-interacting) IF
B. Reliability of interacting interfaces.

3.1. Reliability of the non-interacting IF

is directly connected with the regularity of this interface. The respective relation is as follows:

Reliability of regular IF is equal to 1.

Rel (Reg (IF)) = 1

To specify the impact of irregularity we have to turn back to the chosen model of interface (see Chapter 1.2.2.). Assume further the same dimension of sets IN, OUT and Z,(i = j = k).

To simplify the following discussion let us suppose that Z_0 is a vector the components of that can be either 1 or 0, as well. For a regular IF the vector $\mathbf{Z_0 := \{1,1,1,....1\}}$. The impact of uncertainty (resulting in possible IF irregularity) could then be expressed in the simplest possible way as the existence of zero components in Z_0.

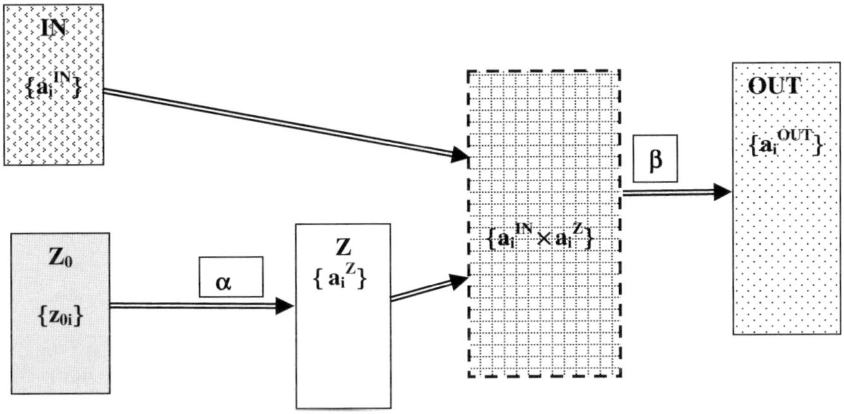

Fig. 5. Model of IF function (IF represented by fictitious finite deterministic automaton)

α: $\{a_i^Z\} := \{z_{0i}\}$ **i.e.** $\mathbf{Z := Z_0}$

β: $\mathbf{a_i^{OUT} := u_i\, a_i^{IN};\ u_i = 1}$ **for** $\forall i$, for that $\mathbf{a_i^Z = 1}$, else $u_i = \aleph$, where $\aleph$ is the number of undefined real value[33] in interval $0 \le \aleph \le 1$.

<u>Verbally</u>: All the components of input vector IN for which the corresponding components of the initial internal state Z_0: $z_{0i} = 1$ are directly mapped into the respective components of output vector OUT: $\mathbf{a_i^{IN}|_{(z_{0i} = 1)} \rightarrow a_i^{OUT}}$, while these components of input vector IN for which the corresponding components of Z_0 : $z_{0j} \neq 1$ are mapped into the component $\mathbf{a_j^{OUT}}$ which has value $\aleph$, uncertain without any a priori knowledge within the interval $\langle 0,1 \rangle$.

---

[33] $\aleph$ is a real number with undefined value in the strong sense, i.e. neither the probability density function nor the membership function within the interval <0,1> are known.

For example: For IN:= {a_1, a_2, a_3... a_n}, and Z_0:= {0,1,0....1} ⇒ OUT={ℵa_1, a_2, ℵa_3.... a_n}.

3.1.2. Assuming that the length of vectors IN, OUT, Z, Z_0 is **n** and that there is **m**[34] components of Z not equal to 1, **m** could naturally be an absolute measure of IF irregularity, while the relative measure of IF irregularity can then be introduced as **rir := m/n.**

Reliability of irregular IF could be expected to be a monotonous non-increasing function of the rir. As the reliability from its definition is probability, it must be defined within the interval ⟨0,1⟩.

Rel (Irreg (IF) = Rel(rir)

3.1.3. This consideration does not take into account (for pragmatic reasons quite important) concept of "**acceptable degradation of IF**" which is quite often used within the Systems Analysis. This concept reflects the experience of analytics that the minor irregularities of IF could often have (in real or interpreted systems) no measurable effect on the reliability of respective processes. The nature of this phenomenon can be linked with the redundancy of input parameters/variables (IN), and consequent possibility to reconstruct the correct values of the disturbed vector components in OUT. To introduce this aspect of the task into the model the threshold parameter ξ[35] can be defined and the impact of uncertainty is then quantitatively expressed assuming $\mathbf{z_{0i}}$ ⟨0,1⟩[36] . The function α in the model is also modified:

α: **If (z_{0i} +ξ)≥1 then a_i^Z:=1, else a_i^Z:= (z_{0i} +ξ),** while β remains unchanged:

β: **a_i^{OUT} : = $u_i a_i^{IN}$; u_i =1 for** ∀i, for which **a_i^Z=1,** else u_i = ℵ, where ℵ is an undefined real number[37] in interval 0≤ℵ≤1.

The further results of the previous chapter (3.1.2.) remain unchanged.

3.2. Generalization of the model for interacting interfaces

This model of IF could be generalized assuming that interfaces within the respective system interact. The interaction means for our purposes that the measure of irregularity **rir** of the IF under study can be modified by irregularities of other[38] system interfaces[39].

The generalization is based on the idea that instead of taking into consideration only the initial state vector $\mathbf{Z_0}$ of the IF under study (as is done in the Chapter 3.1.) the analogical vectors of the neighboring IF in the same system are to be considered as well. In this case the function α is of a significantly more complex nature, mapping Cartesian product of initial internal states vectors of all the interacting IF into the internal state Z of IF under study.

---

[34] (m≤n; m is a natural number or zero, n is a natural number)

[35] **$\xi \in$(0,1).** An obvious generalization could be done assuming the vector character of ξ (ξ_1, ξ_2 ,.. ξ_i.. ξ_n), but this generalization is not reasonable for our purposes.

[36] (not only the binary values 0/1)

[37] ℵ is a real number with undefined value in the strong sense, i.e. neither the probability density function nor any membership function within the interval <0,1> are known.

[38] (neighboring)

[39] An example can be seen in Chapter 1.5.

Let the index of IF under study be **p** $\in\langle 1,q\rangle$ and e =1,2....q

Then:

α_p: $\Pi(Z_0)_e \rightarrow Z_p$; where Π_q means a Cartesian product of q sets and the arrow "→" means certain mapping rule.

i.e.: $(z_{01} \times z_{02} \times \ldots \times z_{0q})' \rightarrow z_1$, etc.

A simplified version of this case should illustrate the complex nature of the generalized task:

Assume two interacting interfaces, IF_1 and IF_2. The first one let be under study.

Z_{01}:= (1,0); Z_{02}:= (1,0,1)[40]; The relations are two-valued ones.

Respective (degraded) Cartesian product $(Z_{01} \times Z_{02})$ = ((1-1), (0-1), (1-0), (0-0), (1-1), (0-1));

Let us further define α_1: {(1-1):= 1, (0-0):= 0, (0-1):= 0, (1-0):= 0}, then $(Z_{01} \times Z_{02})'$:= (1,0,0,0,1,0), and $(Z_{01} \times Z_{02})''$:= **max|** $_{\dim Z1}$ **(comp(($Z_{01} \times Z_{02}$)´)**[41] = (1,1).

Then Z_1= (1,1) and therefore this IF is regularized.

For slightly different definition of α_1: Z1:= Z_{01}AND(max (comp($Z_{01} \times Z_{02}$)´) the IF remains irregular one.

This generalization makes it possible to utilize the proposed model of IF for both interacting and externally controlled interfaces. This feature is important especially in complex hybrid systems and system alliances.

4. Problem of IF identification

An intuitive approach how to increase the reliability of any IF[42] is:

To identify and tightly control all the interface variables / parameters.

This, at first sight quite natural and smart approach turns to be quite counterproductive if the significant **uncertainty** enters the playground[43]. It could be exemplified in simple geometrical re-interpretation of the model presented.

5. Geometric re-interpretation of the model

- Let the analyzed IF consist of **n** mutually independent variables / markers. Then its dimension is **n.**
- Let all the variables of IF be renormalized[44]. Then, in the geometrical view, IF could be supposed to form **n** dimensional compact body[45].

[40] (Therefore the dimension of IF_1 is 2 and the dimension of IF_2 is 3.)

[41] vector of the length Z_1 the components of which are the maximum ones of the vector $(Z_{01} \times Z_{02})'$

[42] i.e. how to tune regularity of the respective interface and subsequently to increase reliability of the system as a whole

[43] A similar problem has recently been both theoretically and experimentally (computer experiments) discussed within the framework of fuzzy sets [9]. The results are exciting, but all the procedure presented is a bit complicated. The complications result from the very nature of fuzzy sets, as different membership functions are to be properly analyzed and proceeded. On the other hand, we are convinced that origin of the problem is not in the shape of the membership function, but it has more general, simple geometrical roots. Nevertheless, the results of the computer experiments presented in this article cast light into the important aspects of the problem. From our point of view so called "curses of dimensionality", the outcomes of the increase of both uncertainty and dimensionality of the task seem to be of high importance.
Analogical results, based on the concept of shadowed sets and geometric consideration backgrounds, are presented in the course of analysis of application potential of predictive diagnostics [5].

[44] (to the interval <0,1>, occasionally taking into account the weights of the respective variables)

[45] cube or sphere (or, taking into account the weights of variables, an n – dimensional cuboid or ellipsoid)

- Let <u>uncertainty</u> enter solely the studied IF, not the system as a whole. It modifies **m** variables/parameters of Z_0.

These assumptions should be expressed geometrically as a reduction of the n - volume of the IF, extracting from the core[46] the outer shell in which the OUT is totally uncertain[47].

For the sake of simplicity, the same coefficient **rir := m/n** of the reduction for any dimension of the IF n – volume is then utilized.

This re-interpretation builds a bridge to the consequent utilization of the results of [5][48] in Chapter 6:

6. MODEL ANALYSIS

Our problem is now reduced to a purely geometric task:

- Let the "n-volume" of the **IF** be $\mathbf{V_{IF}}$
- Let the "n-volume" of the core be $\mathbf{V_{CORE}}$

Then the constructed variable $\mathbf{v = V_{CORE} / V_{IF}}$, being a function of **n**, $(\mathbf{v = v(n)})$ is an effective measure of the "weight" of the (regular) core for given **n**.

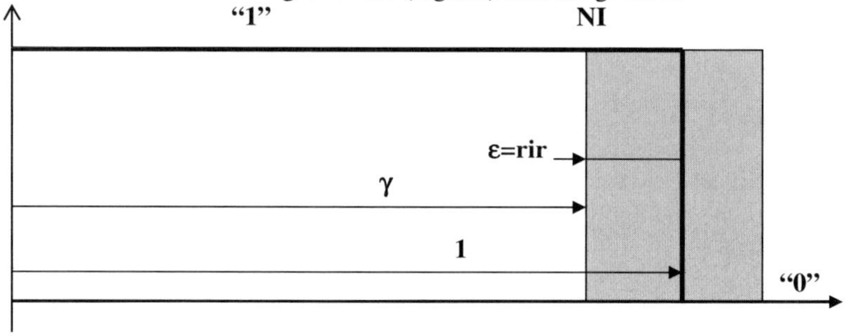

Fig. 6. Central cut through renormalized IF. "1" denotes full regularity, "0" denotes full irregularity of IF and NI denotes non-identified (i.e. totally uncertain) area of IF. ε=rir ; γ≤1; ε + γ= 1

The expressions for **v(n)**, for Euclidean space, typical shapes of IF and fixed **γ, ε,** respectively are as follows:

<u>Sphere</u> (with the radius r):

$$v(n) = \frac{\alpha_n (r-\varepsilon)^n}{\alpha_n r^n} = \left(1 - \frac{\varepsilon}{r}\right)^n \approx 1 - n\frac{\varepsilon}{r} + O\left(\varepsilon^2\right),$$

α_n is a constant for given n

<u>Cube</u> (with the edge a):

$$v(n) = \frac{(a-2\varepsilon)^n}{a^n} = \left(1 - \frac{2\varepsilon}{a}\right)^n \approx 1 - n\frac{2\varepsilon}{a} + O\left(\varepsilon^2\right)$$

Remark: The expressions for the cube and sphere are the same for $a = 2r$.

---

46 (Inner, fully certain and reliable part)

47 (factor ℵ, no membership functions are introduced, i.e. uncertainty means total lack of knowledge)

48 These results were obtained for substantially different, but to some extent analogical task from the area of predictive diagnostics.

<u>Cylinder</u> (bases are (n –1) dimensional spheres with radius r and with the height a):

$$v(n) = \frac{\alpha_{n-1}(r-\varepsilon)^{n-1}(a-2\varepsilon)}{\alpha_{n-1}r^{n-1}a} = \left(1-\frac{\varepsilon}{r}\right)^{n-1}\left(1-\frac{2\varepsilon}{a}\right) \approx 1-\frac{2\varepsilon}{a}-(n-1)\frac{\varepsilon}{r}+O\left(\varepsilon^2\right),$$

<u>Cuboid</u> (with the edges $2a_1, 2a_2, \ldots 2a_n \ldots$):

$$v(n) = \frac{(2a_1-2\varepsilon)(2a_2-2\varepsilon)\ldots(2a_n-2\varepsilon)}{2a_1 \cdot 2a_2 \ldots 2a_n} = \left(1-\frac{\varepsilon}{a_1}\right)\left(1-\frac{\varepsilon}{a_2}\right)\ldots\left(1-\frac{\varepsilon}{a_n}\right) \approx 1-\varepsilon\sum_{i=1}^{n}\frac{1}{a_i}+O\left(\varepsilon^2\right)$$

<u>n - Ellipsoid</u> (with axes $a_1, a_2, \ldots a_n$):

<u>Simplification</u>: Region RA is an n-ellipsoid with axes $a_1, a_2, \ldots a_n$, the core being an ellipsoid with axes: $a_1-\varepsilon$, $a_2-\varepsilon, \ldots a_n-\varepsilon$. (The core is not bounded with an equidistant surface in the distance ε, as was previously supposed by the assumption: the volume of an uncertain shell

RA = (surface of RA x ε. *Instead of an equidistant surface a minor ellipsoid was chosen. This simplification results in inaccuracy $O(\varepsilon)$).*

$$v(n) = \frac{\alpha_n(a_1-\varepsilon)(a_2-\varepsilon)\ldots(a_n-\varepsilon)}{\alpha_n a_1 a_2 \ldots a_n} = \left(1-\frac{\varepsilon}{a_1}\right)\left(1-\frac{\varepsilon}{a_2}\right)\ldots\left(1-\frac{\varepsilon}{a_n}\right) \approx 1-\varepsilon\sum_{i=1}^{n}\frac{1}{a_i}+O\left(\varepsilon^2\right)$$

(Where $O(\varepsilon)$ is an inaccuracy resulting from the simplifications made).
<u>Remark</u>: The expressions for cuboid and ellipsoid are de facto the same.

<u>n - Sphere</u> of radius 1.

$$v(n) = \gamma^n$$

Let us show some values for $\gamma = 0.9$ (quite moderate value):

n	0	1	2	3	4	5	10	20	30	50
v(n)	1	0,9	0,81	0,729	0,656	0,590	0,349	0,121	0,042	0,005

Generalization should be done:

The n-volume of the core of IF for non-zero uncertainty is a significantly decreasing function of the IF dimension n.

7. Discussion

From this model it should be concluded:

a. In presence of uncertainty the increase of the dimension of the IF significantly reduces the relative weight of its regular **core**. This effect impairs the conditions for the reliable function of IF. For n > 10 the IF (for a quite moderate level of uncertainty $\gamma = 0.9$) has a relatively too high proportion of potential irregularity for practical purposes, e.g. That is probably the reason why system analytics on the experience and intuitive background try to keep the number of markers (and therefore the dimension of the IF) on a as low as possible value[49,50].

---

[49]They can do it (to some extend), utilizing some known simplification methods [2].

b. Another way how to improve the conditions for achieving the regularity of IF is to (re)construct it as robust as possible. It means that the acceptable degradation of the regularity of respective IF, (expressed by coefficient ξ) is to be sufficiently high. Within the artificial part of system it should be done quite easily. The redundancy in codes or artificial system elements should be utilized, as well as time redundancy or sophistical means of predictive diagnostics. But this way is of controversial value for "man – machine" IF, as it often assumes in fact the reconstruction[51] of both interfacing parts of the respective system.

c. A specific discussion is needed for the class of interacting IF. In this case the effect of IF conjugation could emerge. This effect could either worsen or improve the regularity of respective IF.

d. Another possible approach is the construction of combined IF variables which may help to reduce consequently the dimensionality of respective IF. This approach is promising if the variables of IF are mutually dependent. A problem then arises with **geometric interpretation** that is important for the presented approach to the IF regularity task. The authors decided to evaluate the potential of **fractal geometry** for this purpose, but till now no satisfactory results has been obtained.

8. References

1. Vlček J.: *System Engineering*, ed. CTU in Prague 1999. (in Czech)
2. Klir G. J.: *Facets of System Sciences*, 2nd edition, Kluver Academic / Plenum Publ., New York, 2001
3. Vlček J. et at.: Reliability of Hybrid System, *Research report No.: K 620 163/02 (in Czech), CTU, Faculty of Transportation Sciences*
4. Votruba Z., Novák M.: An Approach to the Analysis and Prediction of the Complex Heterogeneous Systems Evolution, *Proc. conf.: Technical University Košice, Slovak Republic, October 2000*
5. Votruba Z, Novák M, Voráčová Š.: Problem of dimensionality in predictive diagnostics, *Neural Network World*, **4/03**, 2003
6. Novák M.: *Theory of system tolerances*, (in Czech), Academia, Prague, 1987
7. Novák M., Šebesta V., Votruba Z.: *Safety and reliability* (in Czech) ed CTU in Prague, 2003
8. Novák M.: Theory of Reliable Systems Based on Tolerance Prediction, Dexa'93, Prague, Czech Republic, September 6-8, 1993
9. Mitaim S., Kosko B.:The Shape of Fuzzy Sets in Adaptive Function Approximation *IEEE Trans. on Fuzzy Systems*, Vol. 9, No. 4, Aug. 2001
10. Klir G. J.: Uncertainty in System Science, (invited lecture) CTU in Prague, Faculty of Transportation Sciences, Nov. 11, 03.
11. Hopcroft, J., Ullman, J.: *Formal Languages and their Relations to Automata*, Addison-Wesley, Reading, Mass., U.S.A., 1969.

[50] This approach has also some remarkable links with epistemology, for example with the old, famous and in science very useful principle of "Occam's razor": "Frustra fit per plura, quod potest fieri per paociora", or later his successors: "Entia non sunt multiplicanda praeter necessitatem" William Occam (Ockham) (1285(?)– 1349)

[51] The "reconstruction "of human component of man – machine IF implies for example demanding specific training of the operator, or multiplication of the number of human operators. Such measures could be only rarely met.

12. Votruba, Z. et al.: *System analysis* (in Czech - Systémová analýza) CTU 2004
13. Veselý, J: *System Interface* (in Czech) Prague, Academia 1983
14. Veselý, J: Chaos theory and Synergetics in Transport Engineering Informatics (Research Report CTU, Faculty of Transportation Sciences, May 2003, in Czech)

Transport Science and Technology
K.G. Goulias, editor

CHAPTER 13

OBSERVED AND MODELLED BEHAVIOURAL CHANGES CAUSED BY THE COPENHAGEN METRO

Goran Vuk and Tine Lund Jensen

INTRODUCTION

The Copenhagen metro is the largest urban public transport infrastructure improvement in the last decade in Denmark. The chief reasons for introducing a new public transport system in the capital were to increase market share for public transport, and accordingly reduce car traffic and environmental impact as well as enhance urban development, especially on the island of Amager, which was previously only served by bus. The establishment of the metro is linked to the development of a new urban area south of Copenhagen, where land prices are expected to rise due to the high-quality infrastructure improvements and thus provide financing for the metro.

Over the past years, the Metro has generated a lot of interest among city traffic planners resulting in the initiation of two projects, among other things. The first was the building of an operational traffic model for Copenhagen and the second the metro impact study. The aim of the article is to compare changes in travel behaviour caused by the metro as observed in the available data and as predicted by the traffic model. Traffic growth, induced traffic and modal split are the specific targets of the analysis.

The traffic effects of the metro were measured in the harbour corridor, a narrow strait between the islands of Zealand and Amager. The corridor can be crossed only on the Langebro and Knippelsbro bridges in the city area, which makes it an obvious candidate for the analysis.

METRO IN COPENHAGEN

The Copenhagen Metro

As early as 1961, plans existed to build an S-train connection between the islands of Zealand and Amager, on which Copenhagen is located (Steen Eiler Rasmussen, 2001). The alignment was a straight line between west Amager and the city. At that time west Amager was unpopulated and the plans proposed building some 40,000 new dwellings to ease the pressure on the fast-growing city centre.

In 1992, the Danish Parliament passed the Ørestad Act permitting the construction of a new railway infrastructure in Copenhagen. The government thus endorsed the regeneration of public transport and the reduction of road congestion in the Danish capital. Of the three suggested public transport modes – metro, light rail and tram – the first option was chosen. Metro construction started in 1994 and the first phase was opened in October 2002. The construction costs of phase 1 of the metro amounted to DKK 6.7 billion (approximately EUR 900 million). Figure 1 shows the metro's alignment. The 11-km route consists of two metro lines connecting the island of Amager in the south with the city terminus at Nørreport on the island of Zealand. Metro line 1 (M1) runs to a new town Ørestad in west Amager while metro line 2 (M2) runs to Lergravsparken in east Amager. The two lines meet on the Amager side of the harbour corridor, just before Knippelsbro bridge where the line goes underground.

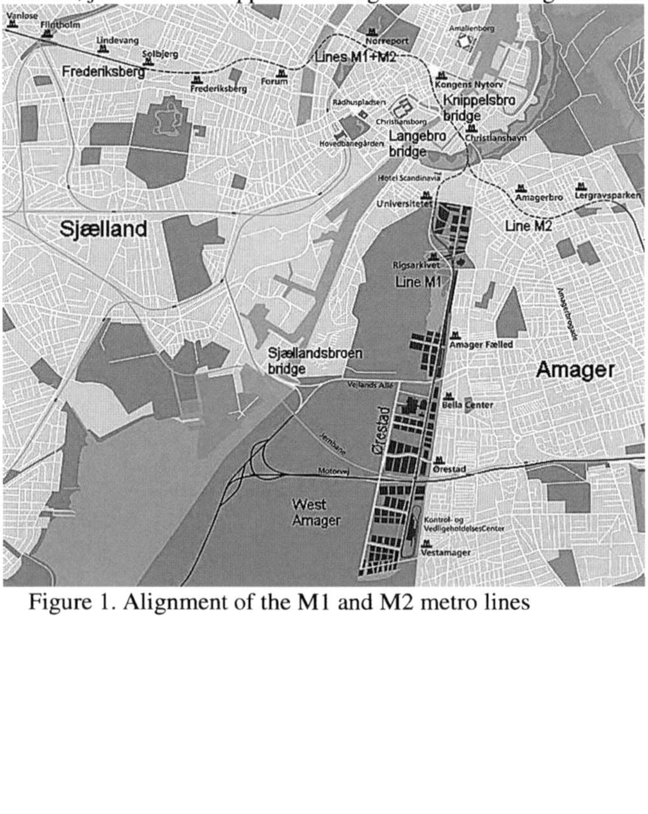

Figure 1. Alignment of the M1 and M2 metro lines

The alignment of the M1 metro line in west Amager largely corresponds to the 1961 plan. The new town of Ørestad will expand over the next 20 years to an area of 310 hectare, providing 60,000 jobs, 20,000 education places and 20,000 dwellings. Some major companies have built new offices in Ørestad: Telia, Copenhagen Energy, Keops and Ferring. The Danish Broadcasting Corporation is due to transfer all its activities to a new television centre in Ørestad in 2006. The University of Copenhagen has been enlarged in the area where construction of the IT University has been finalised.

Phase 2 of the metro opened in October 2003, linking another part of the city, Frederiksberg, with the city centre. Both metro lines continued their alignment in this phase from Nørreport to Vanløse. With phase 2 completed, the metro system has a total length of 16 km and consists of 17 stations, of which eight are underground.

Phase 3, the last projected phase of the Copenhagen metro, will continue line M2 from Lergravsparken station to Copenhagen's international airport at Kastrup. This phase opens for operation in 2007. The full metro system will be a 22-km network, of which 11 km will be underground.

The next potential phase of the metro, the so-called Metro City Ring, with construction costs amounting to DKK 12 to 15 billion (approximately EUR 1.6 to 2.0 billion), is at the planning stage, and a final political decision is yet to be made on whether to go ahead with the project. Connecting areas close to the city centre, the Metro City Ring will provide good opportunities for interchanging at major metro and S-train stations across the existing lines of public transportation. Various alignments are being proposed at this preliminary stage. The Metro City Ring will not be built until 2015 at the earliest.

The Copenhagen metro is fully automated and operated from a computer centre in Ørestad. The M1 and M2 lines operate with a four-minute headway between trains during peak periods, which gives a two-minute headway in the city centre. The operational system allows a minimum headway of only 85 seconds between trains. The metro operates in a self-contained network.

As the metro network has been expanded, changes have been made to the bus service in the capital, including the introduction of a so-called A-line bus network. A-line buses are high-frequency city buses that operate without timetables. They cross the metro network frequently and therefore serve as an access/egress mode to/from the metro.

DATA COLLECTION

Data from two sources are analysed in the article. These are traffic counts for private cars and public transport modes, and panel data.

Traffic counts

Road counts on the Knippelsbro and Langebro bridges were available from the Copenhagen metro impact study for the years 2000, 2001, 2002 and 2003. The counts were done manually once a year on an average workday (typically Tuesday or Thursday) in March in the time frame 6.00 a.m. to 10.00 p.m.

Road counts were also performed at checking posts outside the research area in order to measure the general fluctuations in traffic over time. The Copenhagen Municipality and the Danish Road Directorate collected the data for the period 1998-2003.

Bus passenger counts in the corridor were available from the city bus company for the period 1996 to 2003. Finally, metro counts were available from the Copenhagen Metro Company for the period December 2002 to April 2003.

Panel data

Panel surveys for the metro impact study were conducted the first time six months before the opening of the metro's phase 1 in autumn 2002 and the second time six months after the opening. The 'before' sample consisted of 1,111 respondents while the 'after' sample consisted of 1,056 respondents. The survey included 862 persons on both occasions. Respondents were recruited via postcard surveys conducted on the two bridges prior to the panel surveys. Interviews were conducted by telephone, and respondents reported all trips made the day before they were interviewed. Each trip was described by departure and arrival times, travel purpose and travel mode.

TRAFFIC MODEL FOR COPENHAGEN

Passenger demand model

The Copenhagen traffic model covers the Greater Copenhagen area with 1.8 million inhabitants. The analysis area was split into 601 zones and the surrounding area into 17 port zones. The model predicts traffic for an average working day, i.e. weekend traffic is omitted from the model.

The passenger demand model is a state-of-practice model built in a nested logit structure in which the generation, distribution and mode choice models are connected via logsums. The model includes four segments: business, commuter, education and leisure. The business segment is trip-based while the other three are tour-based. In the model, a tour is defined as a sequence of a simple trip from home to destination and a return simple trip from destination to home.

Inputs to the demand model are zone data and files of level of service (LOS). The zonal data describe the distribution of the population, workplaces, education places, shopping areas, car ownership and parking costs in the 601 zones. Zone-to-zone travel times as well as travel costs for each available mode are represented in the LOS files. These files are produced in the car and transit assignment models, and hence the levels of service variables are exogenous to the demand model. The distribution model is conditional on the generation model, and the mode choice model is conditional on the distribution model. The models are connected in the opposite direction via the measure of accessibility, i.e. logsums. Demand for the metro, a new mode of transport, is modelled in the mode choice model by the application of Stated Preference (SP) data.

The model applies base 1992 observed matrices in a pivot point procedure consequently throughout the demand model as a last step of both the generation, distribution and the mode choice models. Day matrices for four journey purposes, i.e. business, commuter, education and leisure, and four modes, i.e. car, transit, bicycle and walk, are produced after the execution of the mode choice model. They are then split into three time-of-day matrices, namely matrices for morning peak (7a.m.-9a.m.), afternoon peak (3p.m.-5p.m.) and out-of-peak (the rest of the day) based on the observed time split existing in the base matrices. Simultaneously, the home-work, home-education and leisure tour matrices are transformed into trip matrices. A detailed description of the Copenhagen passenger demand model is given in Jovicic and Hansen (2003).

Assignment models

The Copenhagen traffic model includes assignment models for all four modes, where walk and bicycle trips are assigned by the all-or-nothing procedure based on travel times. A probit-based stochastic user equilibrium model is used in the car assignment, applying the principles developed by Danganzo & Sheffi (1977), and Sheffi & Powell (1982).

In the public transport assignment, trips are split between the bus, train and metro modes, and routes. It is a simple assignment procedure based on a frequency-aggregated network. The model parameters are estimated by the SP data, which do not differentiate between travel purposes.

OBSERVED VS MODELLED BEHAVIOURAL CHANGES

Traffic growth and induced traffic

Traffic counts

Based on the counted data there was an increase of 7,300 person day trips over the harbour screen line in 2003 relative to 2002, corresponding to a general increase in traffic of 4.2%. The Copenhagen traffic model predicts that 1,000 trips occur as a result of socio-demographic and zonal changes. This results in 6,300 trips in the corridor related to positive-induced traffic, an increase of 3.6%.

Public transport traffic in the corridor increased by 10,300 trips from 2002 to 2003, a general traffic increase of 14.9%. A minimum of 3,000 and a maximum of 5,000 of those trips shifted from car traffic (see 5.2.2 for more details). Furthermore, the Copenhagen traffic model predicts that without the changes in public transport infrastructure, public transport traffic would increase by 700 day trips from 2002 and 2003. This means that the increase in public transport traffic in the corridor that is related to the metro infrastructure (i.e. public transport induced traffic) is in the range of between 4,600 and 6,600 day trips. This corresponds to an increase in public transport induced traffic in the corridor of 6.7% to 9.6% in 2003 relative to 2002. Also, assuming that 4,600 to 6,600 new public transport trips across the corridor all relate to the metro, positive-induced traffic accounts for 12.6% to 18.1% of metro traffic in the corridor. Table 1 summarises the conclusions based on traffic counts.

Table 1. Traffic growth and induced traffic across the Knippelsbro and Langebro bridges from 2002 to 2003

	Trips	Percentage values
General traffic growth	7,300	+4.2%
Public transport traffic growth	10,300	+14.9%
General induced traffic	6,300	+3.6%
Public transport induced traffic	4,600 to 6,600	+6.7% to +9.6%
Metro induced traffic	4,600 to 6,600	+12.6% to +18.1%

Source: The Copenhagen metro impact study and the Copenhagen traffic model

Panel data

Of the before sample, 88.9% had conducted at least one trip on the day before the interview. In the after sample, 87.8% had travelled on the day before. For these respondents, the average trip rate was calculated to be 3.53 trips per person per day in both surveys. Thus the average trip rate for the samples did not change from 2002 to 2003. Note that the trips reported include both trips across the corridor and trips that took place exclusively on either Zealand or Amager.

In the data, the main travel mode for each trip conducted was defined on the basis of travel length. This means that if a respondent travelled between origin and destination by bus and then by train, the main travel mode for the trip was defined as the mode with the longer travel length. In Table 2 we show trip rates (number of trips per person per day) for the two samples split by travel modes. To sum up the average day trip rate, we also refer in the table to trips conducted by slow modes (i.e. walking and cycling).

Table 2. Trip rates for different travel modes in the panel survey

	Before panel	After panel
Modes	Trip rate	Trip rate
Slow modes	1.68	1.45
Car	1.35	1.38
Bus	0.29	0.24
Train	0.21	0.34
Metro	-	0.12
Total	3.53	3.53

Source: The Copenhagen metro impact study

The car trip rate was almost the same for the two panels, i.e. 1.35 trips per day in the before panel and 1.38 in the after panel. However, the trip rate for public transport modes increased from 0.50 trips per day in the before panel to 0.70 in the after panel. This happened at the expense of the slow modes, whose trip rate decreased by 0.23 trips per day from 2002 to 2003.

Within public transport modes, the bus trip rate decreased slightly between the two surveys, possibly due to the shift to the metro. Furthermore, the metro trip rate was lower than either the bus or the train trip rate. A possible explanation for this is the current rather limited extent of the metro network compared with the train and, in particular, the bus network. As a main travel mode, the metro offers a relatively modest number of travel destinations compared with bus and train. Accordingly, the metro appears to serve as an access/egress mode for many long trips where train is the main travel mode, and this would explain why the train trip rate increased significantly in the after panel relative to the before panel.

Model results

For the 2003 socio-economic data (population, employment and car ownership statistics), the Copenhagen model predicts 6.66 million day trips in the greater Copenhagen area in 2003. That is 60,000 trips more, or 0.9%, than in 2002. The model predicts a person trip rate of 3.55 in both 2002 and 2003.

Modelled total traffic across the harbour corridor in 2002 is 192,600 day person trips. In 2003, modelled traffic equals 206,680 day person trips. In conclusion, the Copenhagen traffic model predicts an increase of 14,080 trips from 2002 to 2003, which gives a traffic growth of 7.3%. Two thousand trips resulted from zonal changes from 2002 to 2003. Therefore, the modelled induced traffic across the two bridges is 12,080 trips or 6.3%.

Modelled public transport traffic across the harbour corridor in 2002 is 73,320 day person trips. In 2003, modelled public transport traffic equals 83,900 day person trips. In conclusion, the Copenhagen traffic model predicts an increase of 10,580 public transport trips from 2002 to 2003, which gives a traffic growth of 14.4%. One thousand five hundred trips resulted from zonal changes from 2002 to 2003. Therefore, modelled public transport induced traffic across the two bridges is 9,080 trips or 12.4%.

Discussion

Traffic counts from the Copenhagen metro impact study show a general traffic growth across the harbour corridor in the period March 2002 to March 2003 of 7,300 day person trips or 4.2%. In 2003, 79,300 public transport day trips were counted on the screen line, an increase of 10,300 trips relative to 2002, or 14.9%. The Copenhagen traffic model overestimated the total traffic growth in the corridor (7.3%), while public transport traffic growth is the same (14.4%). Table 3 shows the changes in observed and modelled traffic growth and induced traffic across the Knippelsbro and Langebro bridges from 2002 to 2003.

Table 3. Observed and modelled traffic growth and induced traffic across the Knippelsbro and Langebro bridges from 2002 to 2003, in %

	Observed	Forecasted
Total growth	+4.2	+7.3
Total induced traffic	+3.6	+6.3
Public transport traffic growth	+14.9	+14.4
Public transport induced traffic	+6.7 to +9.6	+12.4

Source: The Copenhagen metro impact study and the Copenhagen traffic model

The 14.9% public transport growth in the harbour corridor is lower than that observed in other European cases. Knowles (1996) reported that Manchester's light rail system (Metrolink) generated about 20% of new traffic, while Monzon (2000) found that 25% of the trips in the new subway system in Madrid were newly generated trips. It is, however, rather difficult to compare the impact of metro systems across countries, since the effect of the metro on transport behaviour in a given city is influenced by a number of parameters such as the extent and geographical location of the metro lines, socio-economic variables and level of congestion.

Six thousand three hundred observed trips (3.6%) in the harbour corridor can be directly attributed to positive-induced traffic caused by the metro from 2002 to 2003. A minimum of 4,600 trips and a maximum of 6,600 trips are attributed to positive-induced public transport traffic, depending on how car traffic development in the corridor without the metro is perceived. This equals a percentage increase of at least 6.7% and at most 9.6%. The traffic model over-predicted both total induced traffic and public transport induced traffic in the corridor. In conclusion, the period of adaptation to the new transport mode was longer than predicted by the model, possibly due to the long period of unreliable metro operation immediately after the opening of phase 1.

The existing literature shows that one year after the opening of the railway fixed link across the Great Belt in Denmark, public transport induced traffic was 10% (Danish Transport Council, 1997, and Danish Transport Council, 1998). Public transport induced traffic was 12% after introducing the Supertram system in Sheffield (WS Atkins, 2000), and 13% one year after a reorganisation of the bus system in Jönköping (Johansson & Svensson, 1999; and Holmberg, Johansson & Svensson, 1999).

The increase in traffic in the harbour corridor, both general growth and induced traffic, is not related to the increase in trip rate. According to both the panel analysis and the model forecasts, the trip rate stayed constant between the two years, i.e. the observed trip rate was 3.53 trips per person per day while the modelled trip rate was 3.55 trips per person per day. Additionally, the car trip rate from the panel data decreased slightly from 2002 to 2003 while the public transport trip rate increased in the same period.

Choice of transport mode

Traffic counts

According to the statistics from the Danish Road Directorate, car traffic in the peripheral districts of the Copenhagen and Frederiksberg municipalities (checking posts into the city) has increased since 1998.

Table 4 shows the development of car traffic across the Langebro and Knippelsbro bridges on an average workday in the period 2000-2003 in absolute figures as collected in the metro impact study. Car traffic increased by 1.8% in 2001 relative to 2000, and then again by 1.9% in 2002 relative to 2000. In 2003, after the opening of the metro, car traffic decreased almost to the level of 2000. From 2002 to 2003, car traffic in the corridor decreased by 2,200 car trips on a daily basis, a decrease of 3%.

Table 4. Observed car traffic across Langebro and Knippelsbro bridges on an average workday in the period 2000-2003

Year	Observed	Change relative to 2000

2000	72,400	0%
2001	73,700	+1.8%
2002	75,100	+3.7%
2003	72,900	+0.7%

Source: The Copenhagen metro impact study

Until the metro opened, the only form of public transport across the Langebro and Knippelsbro bridges was the city bus service. Table 5 shows the counted bus traffic in the corridor since 2000 when the bus market share declined according to the figures up to 2003. In 2003, 26,200 fewer bus trips occurred in the corridor relative to 2002, a decrease of 38%.

Table 5. Observed bus passenger traffic over the Knippelsbro and Langebro bridges for an average workday in the period 1996-2003

Year	Observed	Change relative to 2000
2000	71,600	0%
2001	71,600	0%
2002	69,000	-3.6%
2003	42,800	-40.2%

Source: The Copenhagen bus operating company

Since the opening of phase 1 of the metro, screen line counts have been conducted once a week, rotating between weekdays. In most cases, counts were located on the stretch between Nørreport and Kongens Nytorv stations (sample counts), and estimates of metro passengers computed from these sample counts. Total counts (all metro stations) of boarding passengers have only been conducted a couple of times to date owing to the high costs associated with conducting this type of count. The observed metro traffic in the harbour corridor in 2003 was 36,500 passenger trips.

Discussion

Table 6 shows changes in car person trips and public transport traffic from 2002 to 2003 on an average working day in both counted statistics and model forecasts. Counted car person traffic is obtained from Table 4, and the car occupancy rate of 1.4 persons per vehicle in 2003 (aggregated car occupancy for all travel purposes) as applied in the Copenhagen traffic model.

Table 6. Observed and modelled car and public transport traffic across the Knippelsbro and Langebro bridges in 2002 and 2003; person trips on an average workday and shares

Modes	Counted 2002	Counted 2003	Forecasted 2002	Forecasted 2003
Car traffic	105,100 (60%)	102,100 (56%)	62%	59%
Bus traffic	69,000 (40%)	42,800 (24%)	38%	20%
Metro traffic	-	36,500 (20%)	-	21%

Total	174,100 (100%)	181,400 (100%)	100%	100%

Source: The Copenhagen metro impact study and the Copenhagen traffic model

According to the figure counts in the table, 3,000 car person trips have shifted to the metro, a decrease of 2.9% in car traffic relative to 2002. Accordingly, the car modal share in the corridor dropped by 4% from 2002 to 2003. The Copenhagen model predicted a decrease of 3%.

The 2.9% modal shift is based on the assumption that car traffic in 2003 would have been of the same magnitude as in 2002. If we assume that the observed increasing trend in car traffic in recent years also continued throughout 2003, we would expect around 76,500 day car trips across the two bridges in 2003, or 107,100 day car person trips. In that case, the metro infrastructure produces a drop in 2003 equal to 5,000 day car person trips (or 3,600 day car trips). In conclusion, between 3,000 and 5,000 car trips in the corridor shifted to the metro from 2002 to 2003, a decrease of between 2.9% and 4.7%.

In the after-metro situation, 79,300 public transport day trips were counted in the corridor, giving an increase of 10,300 trips relative to 2002. The share of public transport trips across the corridor rose by 4% from 2002 to 2003. Bus traffic in the corridor decreased by 26,200 passenger trips from 2002 to 2003. Assuming that bus traffic would have remained constant in 2003 relative to 2002 if the metro had not been not built, the share of bus trips that shifted to the metro would be 38.0%. However, by extrapolating the bus traffic trend shown in Table 5, we calculate that without the metro, bus traffic in the corridor in 2003 would be 68,200 trips. In that case, 25,400 bus trips shifted to the metro, a decrease of 36.8%. In conclusion, between 25,400 and 26,200 bus trips in the corridor shifted to the metro from 2002 to 2003, a decrease of between 36.8% and 38.0%. According to Table 6, the model forecasted a bigger share of metro traffic than bus traffic in the corridor, while the counted data show the opposite. Between 8% and 14% of metro passengers in the corridor are former car users while between 70% and 72% are former bus users.

The modal shift from other public transport modes due to the metro in Copenhagen approximates that of similar European projects while the modal shift from car is somewhat lower. In the Madrid subway project, the observed modal shift was 50% from bus and 26% from private cars. The new metro in Athens has attracted 56% of former public transport users (53% bus passengers and 3% train passengers) and 16% of former car users (Golias, 2002). The Croydon Tramlink impact study shows that the majority of the Tramlink passengers are former bus users (69%) while 19% of passengers formerly undertook the trip by car (Copley at al., 2002). Finally, one half of the Metrolink's passengers (Manchester) are former bus and train users, while about 27% used to travel by car.

LESSONS LEARNT

The generation model accurately predicts the person trip rate compared to the observed trip rate from the panel. However, it over-predicts total traffic growth in the corridor from 2002 to 2003 (7.3%) compared to the observed growth (4.2%). There is also a difference between observed total induced traffic (3.6%) and modelled total induced traffic (6.3%). The differences may be due to certain model characteristics, but the metro additionally experienced serious problems in the initial implementation phase, which caused delays for many passengers. This may indicate that we have not yet seen the full impact of the metro on traffic growth.

The aggregate application of the generation model can sometimes result in an 'explosion' of some zones in the forecasts. In particular, if a zone includes only a few persons of different employment profiles in the base matrix, but many in later years, the zone may generate far too many trips in the forecasts. One way of dealing with this problem would be first to determine the 'wrong zones' and then reduce their forecasting power by setting a correction factor. A similar result would possibly be obtained by not pivoting for those zones, but this has not been verified. Finally, another suggestion would be to include both disaggregate estimations and disaggregate forecasts in the generation model.

It is even more important to recognise that the metro's impact on a person's travel behaviour may differ if the person's combination of daily activities changes between the before-metro and after-metro situations. A model for daily activity pattern is therefore needed to highlight this problem.

The modal split forecasts accurately reflect the observed changes from 2002 to 2003, i.e. the forecasted car traffic decrease was 3% while we observed a 4% car share decrease, and the forecasted public transport increase was 4% while we observed a 3% transit share decrease. The majority of metro passengers were former bus travellers (70-72%), while a smaller share were former car users (8-14%). This is similar to the findings of surveys conducted in other European cities such as Athens, Madrid and Manchester.

REFERENCES

Copley, G., Thomas, M. and Georgeson. N. (2002). Croydon Tramlink impact study. European Transport Research Conference. September 9-11. Cambridge, UK.

Danganzo, C.F. and Sheffi, Y. (1977). On Stochastic Models of Traffic Assignment. Transportation Science 11 (3). Pages 253-274.

Danish Statistical Bureau (2004). The databank available at the Statistical Bureau's website.

Danish Transport Council (1997). Øst-Vest Trafikken – vurdering af persontrafikken efter åbning af Storebæltforfbindelsen. Report nr. 97-10.

Danish Transport Council (1998). Storebælt i en overgangsperiode – interviewanalyse blandt togpassagerer. Report nr. 98-01.

Golias, J.C. (2002). Analysis of traffic corridor impacts from the introduction of the new Athens Metro system. Journal of Transport Geography. Volume 10, Issue 2, June 2002. Pages 91-97.

Holmberg, B., Johansson, S. and Svensson, H. (1999). Evaluation of the Reorganization of Public Transport in Jönköping. Institutionen för trafikteknik, Tekniska Högskolan i Lund, Lund University, Sweden.

Johansson, S. and Svensson, H. (1999). Har kollektivtrafikomläggningen påverkat resvanorna i Jönköping. Institutionen för trafikteknik, Tekniska Högskolan i Lund, Lund University, Sweden.

Johansson, S. and Svensson, H. (1999). Vad tycker resenärerna i Jönköping om trafikomläggningen?. Institutionen för trafikteknik, Tekniska Högskolan i Lund, Lund University, Sweden.

Jovicic, G. and Hansen, C.O. (2003). A Passenger Travel Demand Model for Copenhagen. Transportation Research, Part A. Volume 37. Pages 333-349.

Knowles, R. (1996). Transport impacts of Greater Manchester's Metrolink light rail system. Journal of Transport Geography. Volume 4, Issue 1, March 1996. Pages 1-14.

Monzon, A. (2000). Travel demand impacts of a new privately operated suburban rail in the Madrid N-III corridor. European Transport Research Conference. September 7-11. Cambridge, UK.

Rasmussen, S. E. (2001). Steen Eiler Rasmussens København: Et bysamfunds særpræg og udvikling gennem tiderne. Gads Forlag. Copenhagen, Denmark.

Sheffi, Y. and Powell, W.B (1982). An algorithm for the Equilibrium Assignment Problem with Random Link Times. Networks vol. 122. Pages 191-207.

WS Atkins (2000). Supertram (Sheffield) monitoring study. Final report.

Transport Science and Technology
K.G. Goulias, editor

CHAPTER 14

QUALITATIVE TECHNIQUES FOR URBAN TRANSPORTATION

Pat Burnett
Department of Economics
Massachusetts Institute of Technology
Cambridge, MA 02142
USA.

There has recently been interest in qualitative methods in the travel behavior literature. Goulias (2001) and Weston (2004) are among those to state the case for qualitative research. Examples are seen from the August 2003 10th International Conference on Travel Behavior Research in papers on focus groups by Mokhtarian et al. and Garling et al., and in work by van der Waerden et al. using a descriptive analysis. However, the emphasis of research work in the travel behavior area is still on the formulation and calibration of quantitative models.

This paper reviews the arguments for further developing qualitative methods. First, the definition and purposes of qualitative methods are briefly reviewed: it is shown they may have an affinity for current travel behavior work. Secondly, the paradigm or "scientific worldview" of mathematical modeling is examined and some reasons advanced from philosophy of science as to why it is challenged. Then some other paradigms related more to qualitative methods are advanced. Finally, a selection of qualitative methods will be referenced, and the paper will conclude with a beyond the "quantitative and qualitative divide" section.

DEFINITIONS OF QUALITATIVE METHODS

Standard reference works on qualitative methods emphasize that the social world is complex and that qualitative methods were developed in part to deal with this complexity (Strauss, 1998). And again, "by the term 'qualitative research,' we mean any type of research that produces findings not arrived at by statistical procedures or other means of quantification. It can refer to research about persons' lives, lived experiences, behaviors, emotions and feelings as well as about organizational functioning, social movements, cultural phenomena and interactions between nations" (Strauss, 1998, 10-11). With this breadth of application and application to complexity, qualitative procedures seem well suited to, say, current complex activity-travel questions.

Some change the definition of qualitative to include work which contains some low level statistical summaries of results (Golledge and Stimson, 1997). Be that as it may, Weston cites one specific well-developed research problem in intra-urban travel for qualitative analysis:

"To elicit and reveal subjectively experienced time-space constraints in everyday urban travel, to delineate those experiences and to learn more about how people construct and live their travel routines in different cultural settings" (Weston, 2004, 2).

This does show how in-depth interviewing, focus groups, ethnographies and other methods (qualitative procedures) may suit travel behavior research: and not just in the form of clarification, before, during or after a quantitative analysis.

PARADIGMS AND PROCEDURES

The worldview of quantitative travel behavior analysis is logical positivism. This originated as a philosophical-scientific schema with the Vienna Circle in the 1920s. Many philosophical problems were found with it to WWII. "Post World War verdicts on logical positivism have been numerous: some have been sympathetic but others have been rather less so.... Like many modern commentators Ian Hacking finds 'the success of the verification principle amazing...for no one has succeeded in stating it.' John Earman is less charitable, stating baldly that the extreme forms of positivism...are not a suitable basis on which to found an adequate epistemology (way of knowing)" (Ray, 2000, 250).

And again: for some fields in the social sciences, for example, Geography, "...logical positivism had by the turn of the century ceased to be a significant philosophical force. In many contexts, logical positivist now functions chiefly as a term of abuse, while post-positivist has become a widely used term of approbation" (Salmon, 2000, 233-34).

So what are the chief characteristics of logical positivism that are still in use in the field of urban travel behavior? What are the challenges to them leading to other possibilities as forms of analysis?

Goulias (2001) and Lincoln and Guba (2003) are among those who describe in depth the customary features of the paradigm in use in the urban transportation domain. In summary, the ontology is naïve realism: there is a real apprehendable reality. The epistemology is objectivist and findings are believed to be true. The methodology is experimental/manipulative with hypothesis-verification; quantitative methods are in principal use. The nature of knowledge is verified hypotheses established as facts or laws. There is knowledge accumulation through "building blocks," generalizations, and cause-effect linkages. Conventional benchmarks of rigor are used (reliability and internal and external validity, containing premises from which the conclusion may logically be derived). Ethics are extrinsic. The inquiry aim is explanation, prediction and control. Values are excluded and their influence in all forms denied. Finally, propositional knowing about the world is an end in itself, is intrinsically valuable. (Lincoln and Guba, 2003, 256-263).

Post-positivism challenges these beliefs. It does not mark a transition to new probabilistic forms of theory and modeling in the travel behavior field in the sixties and seventies as Goulias (2001) suggests. Rather it marks a rupture with logical positivism altogether on the grounds summarized by Zammito and listed by Hooker (2004, 14).

1. Theories cannot be reduced to observations;
2. Scientific method is not merely logical entailment;
3. Observation is not theory-neutral;
4. Theories do not cumulate historically;
5. Facts are theory-laden;
6. Science is not isolated from human individuals;
7. Science is not isolated from society;
8. Method is not timelessly universal;
9. Logic should not be privileged;
10. There is no gulf between fact and value.

The rupture with logical positivism on these grounds paves the way for new paradigms and new methods. There is insufficient space to trace the origins of all of Hooker's objections, but let us take one to illustrate: 5 facts are theory-laden. The succinct argument is that in the choice of facts to test a theory, we can't but help use theory in selection. Therefore observations and theories are not independent, as assumed, in the verification process. This is one source of presumed difficulty with objectivity for the logical positivist paradigm.

Two other paradigms which became widespread at the end of the last century as a result of some or all of these objections and helped promulgate qualitative procedures, were social constructivism and critical theory. These will be presented as alternatives, briefly, for some future endeavors in urban travel behavior.

Social constructivism has an ontological belief not in a "real" reality "out there," but in relativism, in socially constructed individual and contextual realities. The epistemology or ways to knowledge are transactional (interacting with subjects)/subjectivist; they are created findings. The methodology is hermeneutic (interpretive, explanatory); dialectic (balancing of contradictions in data, theory). The nature of knowledge is reconstructing individual realities, coalescing around consensus. Knowledge accumulation occurs through more informed and sophisticated reconstructions and vicarious experience. Goodness criteria are trustworthiness and authenticity. The inquirer posture is one of a passionate participant. Propositional, transactional knowing is instrumentally valuable as a means to social emancipation, which as an end in itself is intrinsically valuable; inquiry is often incomplete without action on the part of participants. Criteria of rigor have been developed. Reflexivity (the projection of the author's values and social positions into the study) and some textual representation practices may be problematic—however, methods of overcoming these are available. (Lincoln and Guba, 2003).

Critical theorists hold that social sciences have normative aims and this makes them different from the natural sciences. A social science searches for emancipatory legibility, not laws or generalizations, or even rules. Its methods are criticism of ideologies or belief systems masquerading as fixed truths, which are really social constructs. Since theory can influence behavior, the social scientist has the responsibility of framing and disseminating theory in a way that will emancipate people from social institutions, and emancipate their communities. (Rosenberg, 2000, 457). This is a long way from planning, prediction and control! Qualitative instead of quantitative procedures are used in the social theoretic project, for example, the exploration of beliefs, or for empowerment. (Dialogic/Dialectic methods, Lincoln and Guba, 2000).

Clifton and Handy (2003) review many of the studies using qualitative methods in travel behavior (2003). These complement other work or stand alone within the positivist paradigm. It would be timely to look beyond this paradigm to others for alternative sources of constructs and methods for the study of travel behaviour. Some of these methods come up in the review which follows.

METHODS

Qualitative Interviewing

This section will concentrate on features of qualitative interviewing that aren't generally known. This form of interviewing is suggested to replace the current form of data collection in travel behaviour surveys. Current surveys not only frame the questions, they also frame the responses (Clifton and Handy, 2003). In-depth interviewing enables better access to a person's (or organization's) beliefs, desires, emotions, mental processes or contexts, or 'workings.' The following is based on Rubin and Rubin, Qualitative Interviewing: The Art Of Hearing Data (1995). However, Strauss and Corbin, Basics of Qualitative Research, Techniques and Procedures for Developing Grounded Theory, (1998) and Berg, Qualitative Research Methods for the Social Sciences, (Chapters 4 and 11, developing Content Analysis, 2004) show the variety in the field.

To illustrate the conduct and richness of data from a qualitative interview, first we note some main questions, follow-ups and probes may be designed for it: or the latter may "flow" in the course of the interview itself. For example, a form of qualitative interview is the verbal protocol method which has been used in travel behavior. Here "think of the alternative malls from which you choose and select the one you prefer" was the main question. The follow-ups and probes were "tell me what is going on in your mind" and "is that all?" This elicited the following affective response from one participant: "Oh God I hate malls. When I step inside it I want to run out within 5 minutes. I feel like my soul is getting sucked right out of me. If I had to go to a mall (run in and out in 10 minutes to pick up something I need) I'd go to X because it's close to school and my house." In this study, the prevalence of affective responses could lead to new conceptualizations of travel (Burnett, 2004). The richness and nuances of these responses could not be picked up with traditional measurement procedures, leading to their inclusion as an extra variable in behavioral models.

The analysis of qualitative data is often regarded as inductive, though Strauss and Corbin (1998) consider it a deductive process yielding "grounded theory." There is no standard way to analyze the data from interviews. For example, Rubin and Rubin (1995) use, first, the discovery of themes" "offering descriptions of how people do or should behave" in the data. (234). They proffer up four pages as to how to identify these themes throughout the interviews. Then the themes are coded, "The process of grouping interviewees' responses into categories that bring together...the themes you have discovered." (238). This generates theory. After the interviews have been marked up with coding categories, the researcher puts all material with the same code together. This used to be done by hand but now there are variety of computer programs that will handle coding and data analysis (Weston, 2004).

The final step in the Rubin and Rubin (1995) data analysis is "to compare material within the coding categories (placed together) (and) compare material across categories." (251). This should permit a "clear explanation of a topic." As noted earlier, low-level statistics and frequency counts are often used to clarify. We illustrate with Table 1 from Burnett (2004) showing the number of phrases by decision making category code in a protocol: comparisons between codes and within codes can be made to show differences in travel rules in importance, and differences in travel rules between contexts. The Table as a whole suggests an explication of travel as a decision making process.

We are going into some depth here on data analysis and results since it is here that difficulty is experienced over threatened subjectivity. However, the preceding review shows that the logical positivist approach (nomological-deductive) is not free of contamination, especially of its "objectivity" and that travel response data is highly constrained. The grounds for accepting qualitative methods are authenticity and reliability. The conscientious presentation of data analysis methods step-by-step here demonstrates this rigor. Considerable time is spent by authors on presentation so the procedures can be replicated for interview analysis. However, enough has been said to show that a qualitative approach is a different way of thinking about collection and analysis of data through interviews.

Urban Ethnography

Urban ethnography also offers the prospect of revealing much that is hidden or not well understood in current research. It is little used in current travel behavior work.. (This contrasts with some current interest in focus groups—see a review in Clifton and Handy, 2003; and more recent work in Mohktarian et al., 2003 and Garling et al. 2003. For this reason focus groups will not be further discussed here).

Berg (2004) describes the various meanings of ethnography, borrowed from anthropology and sociology. However, he concludes "that the various ways researchers speak about ethnography may amount to little more than terminological preferences:" (148). Van Maanen (1982, 103) gives the most relevant definition of the most recent variant for our purposes: "(it) involves extensive fieldwork (in urban settings) of various types including participant observation, formal and informal interviewing, document collecting, filming, recording and so on." The concentration on individuals and small purposely sampled groups which the study of urban travel behavior would seem to entail places it in the field of micro-ethnography. For example, travel behavior questions could be asked by

I. visiting selected homes and observing and participating in how travel schedules are formed;
II. Visiting selected institutions (workplaces, colleges) and observing how selected travel 'cultures' are created and function;
III. Participating in the travel decisions and behaviors of members of oppressed groups: the elderly, the working class, those on social welfare, women.

Topics two and three also involve a shift in paradigm. Two suggests individual realities instead of a 'true' one and thus is social constructivist; three belongs in the realm of critical social theory. Ethnography can come from many traditions.

Unless this seems too far-fetched, Clifton and Handy (2003) review a study which delivered unusual findings using the ethnographic method. Neimeir (2001) studied the travel of welfare mothers through surveys at job fairs and traveling with several throughout the day. The finding was that survival depended on schedule flexibility, for example, one woman changed her schedule four times in forty minutes.

Table 1

SUMMARY: # PHRASES BY DECISION MAKING CATEGORY AND CONTEXT

	Counting	Lexicographic	Conjunctive	EBA	Add. Diff	Compromise Effect	Affective	Take the Best	Compensatory	Other
TOUR	18	28	0	0	0	0	35	0	1	7
GROCERIES SHOPPING	32	12	4	2	0	0	9	3	28	5
CLOTHING SHOPPING	30	13	6	8	0	1	9	1	15	5
REGIONAL SHOP. CTRS	22	8	1	2	1	3	16	1	29	4
LOCAL SHOP CTRS	21	13	0	0	0	1	12	2	18	9

subjects (r) = 55

Source: Burnett (2004, Table 8, 27)

Other Qualitative Procedures

It is not the purpose of this paper to cover the battery of qualitative procedures that are in place. That has been done elsewhere (Goulias, 2001; Clifton and Handy, 2003; and an extensive literature on qualitative methods). The focus of this paper has been on further developing the arguments for qualitative research in the travel behavior field. Therefore we only note here some remaining methodologies from Berg (2004), before turning to unobtrusive measures and archival research. The remaining methodologies are action research, historiography and oral traditions, and case studies. Unobtrusive research is chosen here because it also breaks with the travel behavior field's paradigm and research habits—yet it also could yield some insights into some new questions.

Unobtrusive Measures in Research

These are an interesting and innovative way for gathering and examining data. One thing they are meant to do is access aspects of social settings and their inhabitants" that are simply unreachable by any other means" (Berg, 2004, 209). Travel behavior researchers rarely ask in-depth questions about the social worlds which are the realities of the human beings so often treated as the objects of their enquiry. These are subsumed under socio-economic indicators. Yet it is now accepted that travel may have a complex genesis. Isn't it time to use an appropriate method to explore these worlds?

One strategy of exploration involves the use of public archival records (running accounts). These can cover, for example, not only libraries but also graveyard tombstones, hospital admittance records, computer-accessed bulletin boards, and credit companies' billing records. There are three categories, accounts in the commercial media (including that in photographed or video or audio forms), actuarial records (e.g., demographic or residential types of records such as application forms held by credit companies), and official documentary records (such as the files of schools, social agencies, retail establishments and the like).

Another strategy of exploration involves the use of private archival records—memoirs, diaries and letters, home movies and videos, and artistic and creative artifacts. In both the first and the second strategy the practice of video ethnography can be used, the collection and analysis of some of the data in video or still photo form.

Data triangulation is recommended for use with unobtrusive measures. That is, more than one type of qualitative procedure is used simultaneously to check results agree. For example, this copes with missing data.

Unobtrusive measures can take us into an area of descriptive research questions about inhabitants and their worlds which are far from the current predict/control orientation of the transportation profession. Work is only constrained by the imagination: for example, such as that of the well-known researcher who found the popularity of radio stations in Chicago by "having automobile mechanics record the position of the radio dial in all the cars they serviced" (Webb et al., 1981).

BEYOND THE QUALITATIVE AND QUANTITATIVE DIVIDE

Previously, the suggestion has been that either qualitative or quantitative methods be followed: that these are mutually exclusive either/or entities. Recent work however transcends the divide, of which we will mention three streams. The first is the research shift to a pragmatic paradigm in which the philosophical background details are less important than choosing an appropriate set of tools (quantitative and/or qualitative) for the research question (e.g. Corbin and Strauss, 1998; Snape and Spencer, 2003). This move should be appealing to those in the transportation field. It has been formalized into the mixed method approaches which are overviewed in Tashakkori and Teddlie (2003). Triangulated designs are recommended: that is, the quantitative and qualitative sections should check each other. There is no license to mix methods any way the investigator wants, in this formalization. Such mixed method studies are computer assisted (Bazeley, 2003).

The second stream to transcend the divide occurs in Geography, with a history of using mathematics and logical positivism (now in disfavor), and qualitative methods in other paradigms. Kwan (2004) advocates a hybrid geography as a way forward. This should prove ameliorative in the unlikely event that conventional ways of undertaking travel behavior research prove resistant to new suggestions.

The third stream of research takes us back out of the qualitative altogether. Sheppard (2001) shows that there is no necessary connection between mathematics and logical positivism. He outlines mathematical (and GIS) techniques which belong in a post-positivist world. These include chaos theory, disequilibrium theory, fuzzy sets, fractals, and the new Bayesian probability theory. There is plenty to work with and to develop if you aren't inclined to take up qualitative research.

CONCLUSION

This paper has had the aim of elaborating on recent work in the travel behavior literature which points to an interest in qualitative methods. The nature of qualitative work is first defined as dealing with the complexity of social acts and social life which now seems the nub of travel behavior research. Then it is shown that qualitative methods are used with 'scientific worldview' paradigms other than the logical positivist one which currently informs transportation modeling. Thus the use of qualitative methods could broaden the views of the world, as well as the procedures, found in the literature.

Next, qualitative interviewing, urban ethnography and unobtrusive measures are discussed in depth. Particular care is used to give points of analysis to show how rigor and reliability are obtained. It should be clear however that the use of qualitative procedures requires a shift in thinking about data—its collection, handling and analysis. Hopefully from the portrayal, the benefits of additional insights not available from other methods will be obtained.

The paper concludes with something like a compromise. Mixed method and hybrid studies proffer a way ahead for an area that has long used quantitative methods. For those who eschew qualitative procedures, there are alternative mathematics. Altogether, hopefully, the work here will help open up new prospects in worldview and method in the travel behavior field.

REFERENCES

Bazeley, P. (2003). Computerized data analysis for mixed methods research. In: *Handbook of Mixed Methods in Social and Behavioral Research* (A. Tashakkori, and C. Teddlie, eds.), pp. 385-422. Sage, Thousand Oaks.

Berg, B.L. (2004). *Qualitative Research Methods for the Social Sciences*. Pearson, Boston.

Burnett, K.P. (2004). *Variable Decision Strategies, Rational Choice and Travel Demand.* Mimeo, Dept. of Economics, MIT, Cambridge, MA, USA.

Clifton, K. and S. Handy (2003). Qualitative methods in travel behavior research. In: *Transport Survey Quality and Innovation* (D. Jones, and P. Stopher, eds.). Elsevier, New York.

Golledge, R.G. and R.J. Stimson (1997). *Spatial Behavior: A Geographic Perspective.* Guilford, New York.

Goulias, K.G. (2001). On the role of qualitative methods in travel surveys. Workshop on Qualitative Methods, International Conference on Transport Survey Quality and Innovation. Kruger National Park, South Africa.

Handy, S.L., L.M. Weston and P.L. Mokhtarian (2003). Driving by choice or necessity? Paper presented at the 10th International Conference on Travel Behaviour Research. Lucerne, Switzerland.

Kwan, M.-P. (2004). Beyond difference: from canonical geography to hybrid geography. 100th Annual Meeting of the Association of American Geographers. Philadelphia, PA.

Lincoln, Y.S. and E.G. Guba, (2003). Paradigmic controversies, contradictions and emerging confluences. In: *The Landscape of Qualitative Research: Theories and Issues* (N.K. Denzin and Y.S. Lincoln, eds.), pp. 253-291. Sage, Thousand Oaks.

Lofland, J. (1996). Analytic ethnography: features, failings and futures. *J. Contempo. Ethno.*, **24**, 30-67.

Loukopoulos, P., C. Jakobsson, T. Garling, C.M. Schneider and S. Fujii (2003). Car-user responses to travel demand management measures: goal intentions and choice of adaptation alternatives. Paper presented at the 10th International Conference on Travel Behaviour Research. Lucerne, Switzerland.

Ray, C. (2000). Logical positivism. In: *A Companion to the Philosophy of Science* (W.H. Newton-Smith, ed.), pp. 243-251. Blackwell, Oxford.

Rubin, H.J. and J.S. Rubin (1995). *Qualitative Interviewing: The Art of Hearing Data.* Sage, Thousand Oaks.

Salmon, W.C. (2000). Logical empiricism. In: *A Companion to the Philosophy of Science* (W.H. Newton-Smith, ed.), pp. 233-242. Blackwell, Oxford.

Sheppard, E. (2001). Quantitative geography: representations, practices and possibilities. *Envt and Plan. D: Society and Space*, **19**, 535-554.

Snape, D. and L. Spencer (2003). The foundations of qualitative research. In: *Qualitative Research Practice* (J. Ritchie and J. Lewis, eds.), pp. 1-23. Sage, Thousand Oaks.

Strauss, A. and J.Corbin, J. (1998). *Basics of Qualitative Research: Techniques and Procedures for Developing Grounded Theory*. Sage, London.
Tashakkori, A. and C. Teddlie, eds. (2003). *Handbook of Mixed Methods in Social and Behavioral Research*. Sage, Thousand Oaks.
van der Waerden, P., H. Timmermans, and A. Borgers (2003). The influence of key events and critical incidents on transport mode switching behavior: a descriptive analysis. Paper Presented at the 10th International Conference on Travel Behaviour Research. Lucerne, Switzerland.
Van Maanen, J. (1988). *Tales of the Field: On Writing Ethnography*. University of Chicago Press, Chicago.
Weston, L.M. (2004). The case for qualitative methods in transportation research. Paper Presented at the 83rd TRB Annual Meeting, Washington, DC, January.
Zammito, J.H. (2004). *A Nice Derangement of Epistemes: Post Positivism in the Study of Science from Quine to Latour*. University of Chicago Press, Chicago.

Transport Science and Technology
K.G. Goulias, editor

CHAPTER 15

Toll Modelling in Cube Voyager

Tor Vorraa
Regional Director, Citilabs Ltd.,
London, United Kingdom

Abstract

Toll systems are installed on highways and around cities with an increasingly pace. To help planning the toll system, the capacities they need, find the right locations and get a reasonable estimate of revenues, advanced toll modelling must be implemented in the travel demand model. Cube Voyager is a member of the Cube transport planning system and offers tools for handling any type of toll modelling that is and will be required.

Overview of Road Toll Systems

Road charging is becoming increasingly 'popular' throughout the world to support transport infrastructure developments. In many places road toll collection is also used in policy measures such as congestion charging schemes.

Toll collection is organised in a variety of ways including:

- Single toll collection links
- Multiple toll collection links
- 'Network' of toll collection links
- Toll cordons around towns

The toll collection is in many places still based on manual payment systems but automatic collection with various electronic systems is becoming standard. The toll cost can be fixed, vary by time of day and may also depend on entry-exit points, time between entries etc.

The planner is faced with many challenges when dealing with toll systems. The designing and planning of the toll collection itself needs information about expected flows through the system for finding the best possible locations for toll collection points and for calculating expected revenues.

This paper describes how toll systems can be handled in travel demand forecasting tools.

Modelling Tolls and Road Charging

Generalised Costs

The travel demand forecasting methodologies uses various forms of costs calculations along routes through the transportation network to determine which routes are chosen between origins and destinations. The general form of these cost calculation formulas is:

GenCost = Time * TimeFac + Distance * DistFac + OtherCost * OtherFac

The time and distance factors may be set to monetary units and in this way the general costs equation can accommodate toll costs.

The primary purpose of this cost calculation is to find realistic routes through the network for different types of vehicles, for people with different travel purposes and for different time of day. This means that the generalised cost might be different for the different categories of travel. The more detailed travel demand model, the more detailed these cost calculations should be.

How to include toll cost in models

The toll cost is applied to the link (s) with toll stations. It is added to the time and distance based costs in the 'OtherCosts' element of the formula above. Obviously, the toll cost varies with type of vehicle and therefore the formula must reflect that.

The toll **price elasticity** will also vary by vehicle type and most probably also by travel purpose and even time of day. This elasticity represents the tendency to choose the toll link or not. In a simple 'all-or-nothing' path building exercise this will represent the 'OtherFac' element in the equation above, but might be included in logit choice formulations in more advanced toll modelling calculations.

It is not given that the toll link is used, even for the most 'obvious' route through the toll corridor. The driver will make a choice whether to use the toll way or not. **Driver's choices** will vary between vehicle groups, travel purposes and time of day for origin-destination trips where alternative routes can be found.

In recent toll studies logit choice models have been used to split the trips between toll and none toll routes. The first step is to identify all 'reasonable' routes that a driver might choose between an origin-destination pair. Based on the price elasticity for the trip category, the driver will choose to go through the toll station(s) or not.

The choice function might take the form:

$$P_{toll} = \frac{\exp(-\lambda C_{toll})}{\exp(-\lambda C_{toll}) + \exp(-\lambda C_{no-toll})}$$

The choice between using the toll road or not depends on the cost difference between the alternative routes and how quickly people respond to these differences. This cost sensitivity or elasticity is illustrated in the figure below:

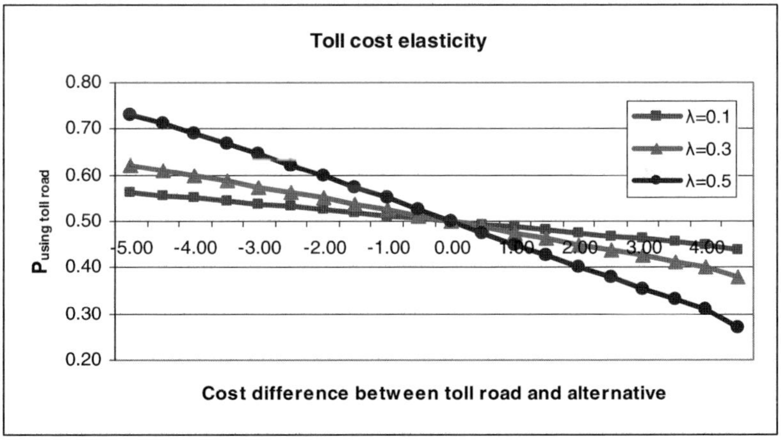

Figure 1: Example of logit choice function for determining the split between toll road and avoiding the toll road. The graph shows logit model sensitivity which may represent price elasticity (example only and not necessarily showing realistic values)

How to represent different toll systems

The typical path building algorithms in travel demand forecasting models will in most cases be able to accommodate the toll costs and any weights associated with these. However, toll costs will often and increasingly so, depend on the actual travelled distance on the toll way or simply based on the entry and exit points. The challenge for the modellers is to be able to add the correct toll cost to any possible route between origin-destination pairs and to find the best ones(s) that people will choose for their journey.

So, the requirements for a travel demand model for handling road tolls, are:

1) Allow for toll costs in generalised costs calculations for path building
2) Add exact toll costs to the journey for alternative routes through the network
3) Be able to handle different user classes, e.g. vehicle types , with different toll costs
4) Include price elasticity for calculating the split between the toll road/system and avoiding the toll road for any origin-destination pair

The more complex these calculations are, the longer it takes for the model to run. Strategic models are usually running until a convergence between assignment iterations have been found and in congested urban areas, there is a need for balancing out the travel demand to the overall network capacity that is offered. The time it takes to do this modelling is crucial in many projects and handling advanced toll calculations increases the run times.

Cube Voyager is used as an example of how advanced toll modelling can be handled properly for finding the correct paths through the network and at the same time allow for correct calculations of toll revenues.

Toll Modelling in Cube

Overview of the Cube Planning System

Cube is a transportation planning software system designed for forecasting of passengers and freight movements. Cube offers advanced and flexible tools for the generation, distribution, mode split and assignment of personal and freight transport as well as detailed analysis of environmental issues. Cube Voyager is a new generation travel demand forecasting software which is based on legacy products TRIPS and TRANPLAN. The modelling is supported by efficient and user friendly data and scenario interfaces allowing for seamless integration with GIS.

Cube Voyager

Cube Voyager combines the latest technologies for the forecasting of personal travel. Cube Voyager uses a modular and script-based structure allowing the incorporation of any model methodology ranging from standard four-step models, to discrete choice to activity-based approaches. Advanced methodologies provide junction-based capacity restraint for highway analysis and discrete choice multi-path pubic transport path-building and assignment. Cube Voyager includes highly flexible network and matrix calculators for the calculation of travel demand and for the detailed comparison of scenarios.

Cube Voyager was designed to provide an open and user-friendly framework for modelling a wide variety of planning policies and improvements at the urban, regional and long-distance level. Cube Voyager brings together these criteria with a comprehensive library of planning functions applied under the general Cube framework.

Through its flexible scripting system any type of cost function can be used and for any number of user classes. This allows for very advanced modelling including practically any type of implementation of toll models. In the next few chapters, some examples are covered.

Toll Modelling in Cube Voyager

The Toll Modelling Process

Cube Voyager uses a flow charting system for specifying models. Functions can be put together in any meaningful way and the user has full control over this process. An example model flow is presented in figure 2 below.

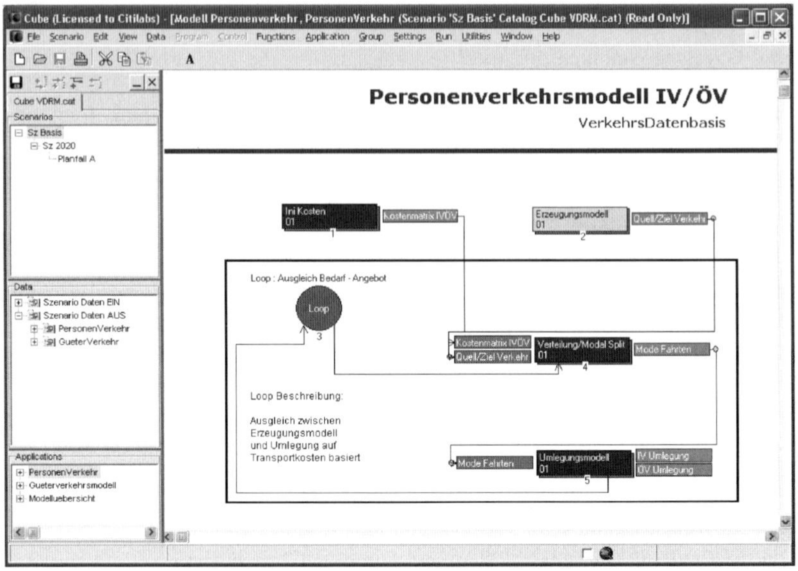

Figure 2: Interactive flow chart showing model structure of a typical 4-stage model

The example shows a four stage modelling process with the distribution, modal split and assignment stages in a loop to achieve a balance between demand and supply. In this way, the real costs in a congested situation will be used along with any toll costs in path calculations. The toll modelling itself is done within the assignment stage and is controlled with a script as shown in the example in figure 3.

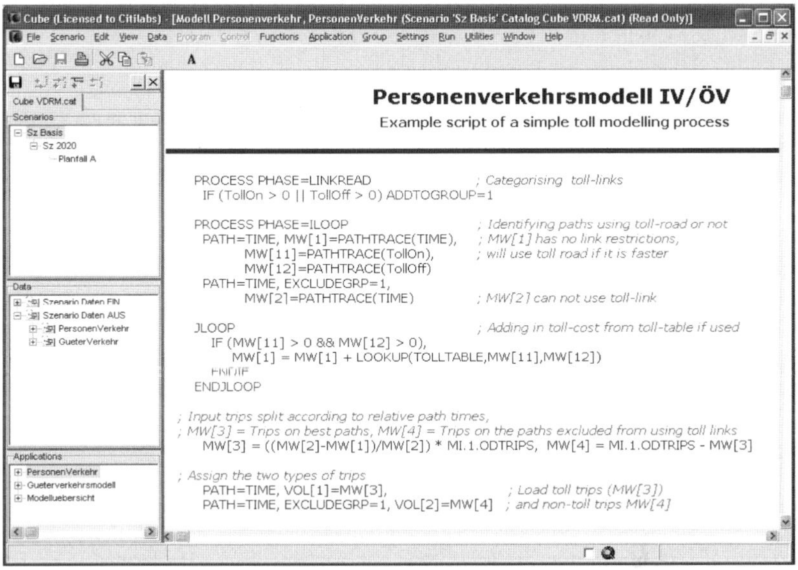

Figure 3: Cube Voyager script for implementing the demand model (simplified)

The script consists of a series of modelling processes:

Process 'linkread':
This process is identifying links with toll costs; these will be the toll stations where vehicles either pay directly or are registered when entering or exiting the toll system. Toll links are added to a special group of links that will later be used when finding paths that avoid the toll systems.

Process 'iloop':
This process finds paths through the network for any origin-destination pair. Two sets of paths are identified, the first with no restrictions and where the paths will go through the toll system when it's faster/less costly without adding the tolls. The other path excludes the toll links and will represent the paths that avoid the toll system.

Process 'jloop' (process inside the 'iloop' process):
Here toll costs are added to the travel cost for any origin-destination pair. Costs are in the example held in a lookup table that represents the cost for any entry-exit combination through the toll system.

Process trips (process inside the 'iloop' process)
Establishing which trips (o-d relations) use the best path according to total cost including toll cost and the ones that do not use the toll links even if this would be best if no toll cost was added. The example above represents a mathematical diversion process but this can alternatively be substituted with a logit choice function with the alternative routes, toll or no toll, in the choice-list.

Exploring and Evaluating Results

When applying the toll model, the user defines a scenario which includes the toll system and the costs associated with it, runs the model to forecast how many and which trips will use the toll system for the scenario and analyses the results. These tasks are supported by a flexible and intuitive, map based user interface.

The figure below shows the results of a select link analysis for the link without and with the toll station. The trips passing are shown as bandwidth diagrams along the roads and also as desire line plots representing the trip distribution for these particular trips. One can study the impact of the toll station in terms of reduced traffic on the toll link.

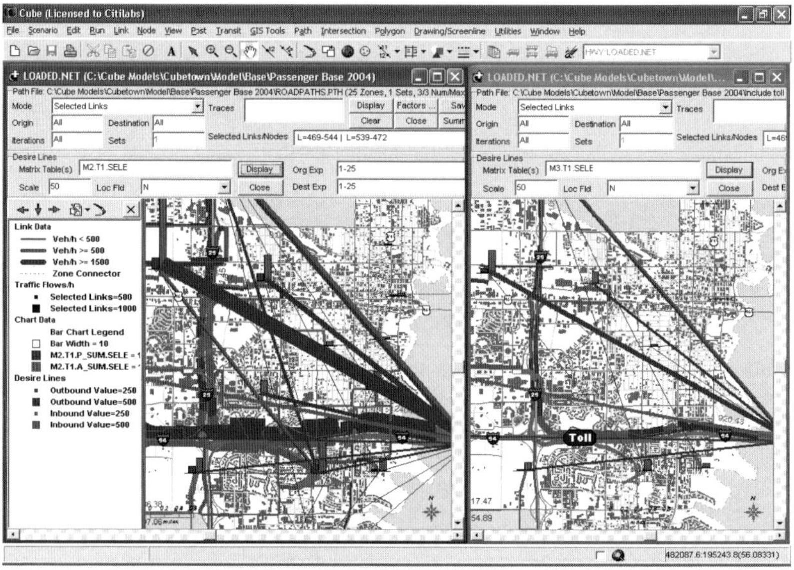

Figure 4: Map based bandwidths and desire lines showing traffic flows and trip distribution for trips using toll system

The next step in the analysis of the impact of the toll station is to look at what alternative routes the rejected traffic is likely to take. The figure below shows these trips which chose one major alternative route because of the toll cost.

As one can imagine, such analyses are vital for being able to predict and to avoid unwanted effects of the toll system.

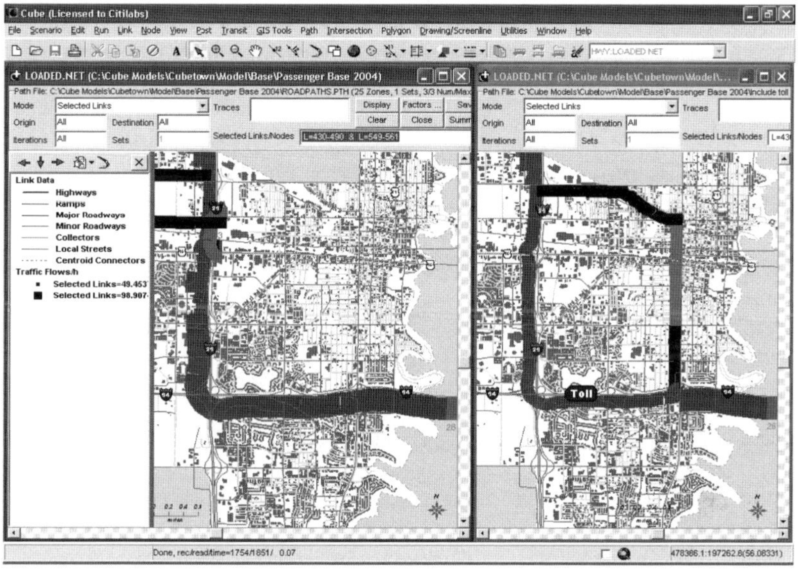

Figure 5: Traffic flow bandwidths showing traffic 'leaking' from the toll system

Testing toll scenarios

As indicated earlier, modelling a toll system will help with the planning of the system itself, to find the optimal locations for toll collection points, to avoid unwanted effect from the toll system and more. A road charging scheme might be put into action to reduce congestion in peak periods, to help with modal shifts from car to public transport or simply to be able to fund the road infrastructure.

It is clear that there might be a lot of combinations possible and the demand on the modelling is high. Getting the model right and calibrated in order to get sensible and reliable results is vital and the need to test multiple scenarios and scenario combinations will make it essential to be able to generate and run scenarios easily. The Cube system has been designed with this in mind and lets the model developer customise a scenario run menu easily. Results from the model are structured customised to the user's needs and customised reports including scenario comparisons are 'falling' out. An example of a customised run menu for a hierarchical scenario structure is shown in the figure below along with an example of a user specified report of expected toll revenues.

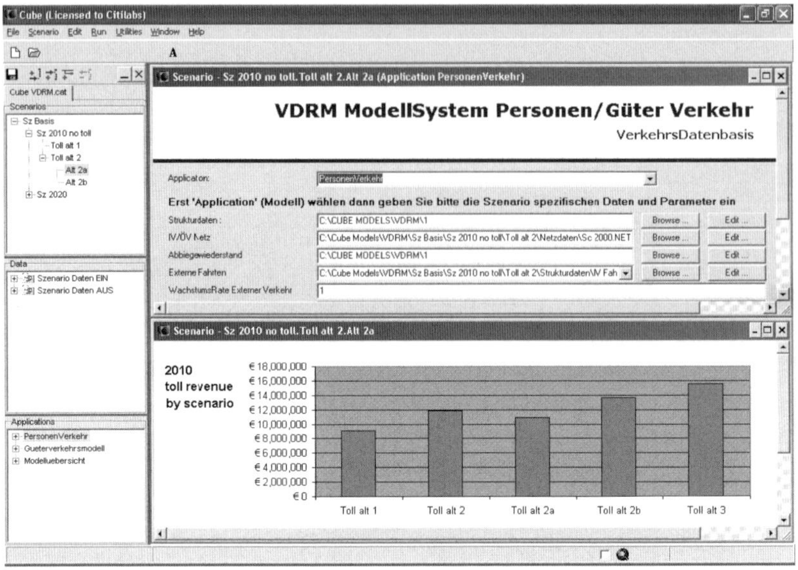

Figure 6: Customised scenario run menu and modelling reports in Cube

References

Citilabs 2001-2004: Cube Voyager help system

Transport Science and Technology
K.G. Goulias, editor

CHAPTER 16

SIMULATION MODELLING IN THE FUNCTION OF INTERMODAL TRANSPORT PLANNING

Ass. Prof. Natalija Jolić, D. Sc., Head of International Cooperation Board

Ass. Prof. Zvonko Kavran, D. Sc, Vice Dean

Faculty of Transport and Traffic Engineering, Vukeliceva 4, University of Zagreb, Croatia

ABSTRACT

This paper presents simulation modelling in the function of the intermodal transport system planning especially intermodal terminals. The simulation modelling based on the functional model of intermodal environment defines a model (among several possible meta-models) of the information flow in the integrated environment. The model of intermodal transport system can be represented by a matrix, map and a table of connectivity and the initial conditions about the maximum and minimum number of channels are selected according to the number of connections of the interest groups of intermodal transport system.

INTRODUCTION

Transportation, land use and development decisions are best addressed in a comprehensive planning process. Transportation planners need an integrated understanding of future transport scenarios to make informed policy decisions and debate the investment trade-offs on mobility, productivity and environmental consequences. Transportation planning research initiatives help decision makers identify solutions to address economic, social, environmental, land use and technical developments.[1]

---

[1] U. S. Department of Transportation, web site http://www.fhwa.dot.gov/rnt4u/init_planning.htm

Intermodal environment has developed from the need to improve the organisation and co-ordination of the transport subjects and to accelerate the traffic flows. The simulation modelling of the intermodal environment of the transport system has enabled a detailed study of the influence of single system entities on the functioning of the continuous intermodal chain. The intermodal characteristic of the transport system – intermodal terminal is complex and the simulation research represents an important tool in planning and developing the system design. In accordance with the Wymore theory, characteristics of the intermodal system – intermodal terminal (IS-IT) can be defined through the implementability cotyledon (cross-section of the functionability cotyledon and the buildability cotyledon) and represents the first step of the research activities of this work.
Mathematical defining of the IS -IT is shown as:

KTLI (IS -IT) = KTLF (IS -IT) & KTLB (IS -IT) = KTLF, B (IS -IT) =
{ID(IS -IT) : ID(IS -IT) = (Z, DSZ, TSZ, PSZ, SCR) (1)

Where:
KTLI (IS -IT) – implementability cotyledon of IS -IT
KTLF (IS -IT) – functionability cotyledon of IS -IT
KTLB (IS -IT) – buildability cotyledon of IS -IT
ID (IS -IT) – imlementability design of IS -IT
Z – example system Z
DSZ – design system Z
TSZ – time system Z and PSZ – place system Z
SCR – system coupling recipe.

MODELLING THE INTERMODAL TERMINAL REQIREMENTS

Model of the modern IS-IT is strongly influenced with the many developments in the specific area of the maritime logistics, supply chain and information flows tool. Therefore the functional areas are to be defined. Functional areas are defined in such a way that they completely (or almost completely) satisfy the needs - requirements of the users. Therefore, the first step in defining functional areas of the intermodal transport system – intermodal terminal is to define the requirements of the port system users. Functional areas include functions, flow of data and databases. The definition of a functional area consists of a number, name and description and it is best illustrated by a table. Functionality of each area is divided into functions.

There are two types of functions:
1. High-level functions – very complex functions. For the purpose of understanding and description they have to be divided into low-level functions, and thus divided some may again feature as high-level functions. The description of high-level function consists of an overview and list of component function

2. Low-level functions – simple functions which can be described without being divided into smaller parts and represent the lowest level of functionality of each area. The description of low-level functions consists of an overview, description of the flow of input and output data and detailed functional requirements.

In determining of the intermodal terminal users' requirements it is necessary to predict the possible future development and requirements that do not exist today, but could appear with the development and advancement of technology. Therefore, an increase of cargo in terminals, greater number of users and an increased number of the means introduced by the users into the intermodal system may well be expected. The means introduced by the users into the system are considered from the organisational, and not from the ownership point of view, resulting in the existence of groups of means and elimination of the participants belonging to the same horizontal level. The engagement of a single group depends on the functioning of the rest of the sub-system and restrictions in space and time that depend on the continuity of flow.

The identification of the users' requirements depends on the answers to the following questions:
1. Who are the interested subjects and what are the existing *inputs* (users' requirements) and to which extent do they influence the *outputs* (offered services)?
2. Are there any interaction relationships between single subjects and single inputs they initiate?
3. What are the requirements of *output* (effectiveness, traveling time, delays, energy consumption, accidents, environmental protection, etc)?
4. What are the expected potential inputs (users' requirements for which it may be assumed that they will occur in the future), how much they differ from the existing inputs and whether they are qualitatively and quantitatively matched by the outputs defined in the previous step?

Requirements of the intermodal terminal are defined as follows:

1. CARGO HANDLING
1.1. The system shall provide cargo operations for all types and sizes of operators.
1.2. The system shall be able to provide information about cargo during traveling and during stay in the port, obligatory loading status, contents, and delay.
1.3. The system shall provide the shipper with information about: destination, cargo properties data.
1.4. The system shall be able to transfer any important – necessary information about the cargo, e.g. in case of dangerous cargo, relevant to the government.
1.5. The system shall be able to control physical and administrative status of cargo during transport and in the intermodal terminal.
1.6. The system shall be able to reconstruct the traveling route and predict the time of cargo passing through the intermodal terminal gate.
1.7. The system shall be able to locate, identify, and control cargo at any moment in the terminal.

2. CONTROL OF THE INTERMODAL TERMINAL GATE
2.1. The system shall be able to record data on vehicles, traveled route, equipment, and sensors on the cargo unit, for later analysis.
2.2. The system shall provide control of transport units and cargo units in the terminal.
2.3. The system shall be able to control the vehicle and cargo unit by detecting faulty operation (e.g. open door, etc.) and alert the center about the identified irregularity.
2.4. The system shall provide planning of the terminal capacity according to the traffic volume.
2.5. The system shall be able to send information to the shipper, cargo owner, forwarder, and road and railway operators about their respective cargo status.
2.6. The system shall be able to provide information about future cargo at terminal to all the interested parties.
2.7. The system shall be able to store the gathered and processed data in the database which can be partially accessed, depending on authorization, by all the interested parties.

3. HANDLING OF TRANSPORT EQIPMENT
3.1. The system shall provide operations with fleet for all types and sizes of operators.
3.2. The system shall provide operations of loading, contents, delay, delivery status to the fleet management centre in real-time.
3.3. The system shall be able to control the physical status of traffic means in real-time conditions.
3.4. The system shall be able to locate, identify, and control the status of vehicles, cargo or equipment at any moment.
3.5. The system shall be able to plan, monitor, control and evaluate the fleet operation.
3.6. The system shall be able to control vehicles, and identify faulty operation (e.g. open doors, etc.) and alert the centre about the identified irregularity.
3.7. The system shall be able to provide management of the combined transport interface.

Functional areas of intermodal terminal derived from the defined users' requirements, are presented as follows. The name of the area has been formed so as to illustrate the functions unified by the respective functional area. Functional architecture has been divided into five functional areas that contain functions (that have not been divided into high-level and low-level functions - if they exist) data flow and databases.
The description of each functional area is represented by means of **shall provide** thus facilitating the checking whether this area provides functionality expected in the intermodal transport system modeling.

1. System of port cargo handling mechanization - this area shall provide functionality which allows performing of functions of cargo handling mechanization. Functions of this area will be connected to the functional area of the system of automatic identification and the functional area of fleet monitoring and management.

2. System of automatic identification - this area shall provide functionality which allows automatic identification of equipment, cargo, vehicle and driver, i.e. management and control of the maritime and land port system gates. This area will have defined interfaces with the functional area of the port cargo handling mechanisation system, vessels system and fleet monitoring and management system.

3. System of vessels - this area shall provide functionality which allows integration with port system subsystems so as to connect this area with the port cargo handling mechanization system, automatic identification system, and fleet monitoring and management system. In this area it will be possible to use the advantages of ITS applications on the vessels in order to improve the port system functioning.

4. System of navigation management - this area shall provide functionality which allows traffic management in front of the port, and will be connected with the area of automatic identification system, vessel system, fleet monitoring and management system.

5. System of fleet monitoring and management – this area shall provide functionality which allows management of cargo and fleet management. Functionality will include cargo, i.e. vessel detection, control and monitoring, creating and maintenance of a database providing the possibility of port traffic planning. The functionality will be realized through connectivity with the system of vessels, system of navigation management and system of automatic identification that is terminators (cargo ownership and shipper) in order to provide accurate and timely information about the cargo and ship.

INTERMODAL TERMINAL MODELLING DESIGN

The model of intermodal terminal consider the functional characteristics – functional requirements and information flow defined by them. Intermodal terminal model that will be used in simulation experiments with the conditions of priority arrivals, and the results of the classical intermodal system – intermodal terminal (C IS-IT) and the parallel intermodal system – intermodal terminal (P IS-IT) will be quantified. The information and statistical concept of research means the choice of simulation variables by studying the ITS models with FIFO service (Z = FIFO), and in performing simulation experiments, the models with non-prompt (relative) PRI service discipline with two or three priority classes (Z = relative PRI) are researched.

The research will analyse the multi-channel Markov systems which include c ($c > 1$) identical service places with the same μ and the information arrivals are exponentially distributed with mean time interval $1/\lambda$. The transport system model of type *M/M/s* ($s = c$) is the simplest type of the multi-channel Markov system without capacity limitation ($Y = \infty$) and with FIFO service, as well as with the condition of normal system functioning such that the average information arrival rate is always less than the maximal possible service rate of the whole system - relation $((\lambda / c\mu) < 1)$ represents the same condition expressed in the form of non-equality. The simulation tools MathProg were selected for the simulation, and the results of the simulation modelling. After having completed the research of enhancing the service system by simulation modelling and experimental simulation, the problem of intermodal transport system capacity dimensioning can be regulated. The research results show the advantages of implementing parallel transport system and give guidelines from the scientific point of view, for the future system architecture with the aim of establishing a continuous, efficient and cost-efficient intermodal transport system.

PRE-RESEARCH RESULTS

In this part of research, motivated by the objective of improving **(IT-IS functions of meta-model) the I system**, the simulation modelling was used, and through various simulation modelling of diverse multi-channel models of type ***M/M/s***, the problem of rational dimensioning and organisation of the IS-IT capacity was studied, with the aim of avoiding the possible standstills in the IS-IT model. The behaviour of IS-IT model ***M/M/s*** was analysed with regard to various traffic intensities (Table 1) of the considered system (i.e. changes of the factors $(\rho_s = \lambda / s\,\mu)$ of system utilisation), and according to all combinations for the values: $\lambda = \{2, 3, 4\}$, $\mu = \{4, 5, 6\}$, and for the models: M/M/2, M/M/3, M/M/4, M/M/5 and M/M/6. Such new approach to the study of advanced ITS *M/M/s* models is in accordance with the condition of parallel processing of the advanced digital information of the IS-IT entities, or with the P IS-IT property: $s_{CITS} = n$, $\Rightarrow$ $s_{PITS} = n + r$, where $r \geq 2$ $(s_{PITS\ min} = n + 2)$ and it is possible to see the calculation of the considered models as well as the differences between their behaviours (their static values) when the difference in the number of servers is at least two, i.e. for the minimal value r = 2 (= Δs). The cases when s = {3, 5} are interesting, but since this concerns a research of (complex) IS-IT issues, then multi-channel systems C IS-IT and P IS-IT are considered.
Therefore, the research here focused on the behaviour of different IS-IT M/M/s models with the minimal difference in the number of servers in the amount of two servers, for s = {3, 5}.

$\rho_s = \lambda / (s \cdot \mu)$	μ=4	μ=5	μ=6
λ = 3 and s = 2	*0.375*	*0.300*	*0.250*
λ = 4 and s = 3	*0.333*	*0.267*	*0.222*
λ = 2 and s = 2	*0.250*	*0.200*	*0.167*
λ = 3 and s = 3	*0.250*	*0.200*	*0.167*
λ = 4 and s = 4	*0.250*	*0.200*	*0.167*
λ = 4 and s = 5	*0.200*	*0.160*	*0.133*
λ = 3 and s = 4	*0.188*	*0.150*	*0.125*
λ = 2 and s = 3	*0.167*	*0.133*	*0.111*
λ = 4 and s = 6	*0.167*	*0.133*	*0.111*

$\rho_s = \lambda / (s \cdot \mu)$	μ=4	μ=5	μ=6
λ = 3 and s = 5	*0.150*	*0.120*	*0.100*
λ = 2 and s = 4	*0.125*	*0.100*	*0.083*
λ = 3 and s = 6	*0.125*	*0.100*	*0.083*
λ = 2 and s = 5	*0.100*	*0.080*	*0.067*
λ = 2 and s = 6	*0.083*	*0.067*	*0.056*

Table 1. Different usage of ITS system (ρ_s) of different simulation models of type ***M/M/s***

SIMULATION OF IS-IT MODEL OF TYPE M/M/S WITH RELATIVE SERVICE PRIORITIES

Based on the obtained measures of the IS-IT success (C IS-IT and P IS-IT), and for further research of the relative priority service, the following proposition (in the form of a premise, i.e. hypothesis) had to be made:

"In P IS-IT always at least one more relative document service priority can be used than in the case of CITS".

Further research is oriented to simulation experimenting of the elementary models ***M/M/s*** for P IS-IT and C IS-IT (***M/M/5*** for P IS-IT and ***M/M/3*** for C IS-IT) with individual service time exponentially distributed from the average $1/\mu$ and according to priorities, limitless capacity of receiving information (documents) and serving with exponential distribution of arrival times. Only those cases are considered where there is no congestion of the multi-channel model of the IS-IT of relative priority service and where $\rho_s < 1 \Leftrightarrow \lambda < s\mu$. If we consider initially (for the sake of presentation simplicity) the elementary model of the M/M/1 system with two relative priority service classes, then the information with higher priority of relative serving arrive exponentially distributed at the average speed of λ_1, and the information with lower relative service priority arrive at the speed of λ_2. The parameter of common input exponential distribution λ represents their sum, i.e. $\lambda = \lambda_1 + \lambda_2$. The assumed service is the same for both types of information (classical and digital), according to speed μ.

The solution of the IS-IT model of type M/M/3 in C IS-IT (N = 2) with the same mean information arrival time according to relative priority classes, means that, with μ = 5 and λ_{kCITS} = 4, the value for λ_1 = 2, and for λ_2 = 2. The solution of the model of type M/M/5 in P IS-IT (N = 3) with the same (uniform) mean information arrival time according to relative priority classes, means that, with μ = 5 and λ_{kPITS} = 4, the value for λ_1 = (1,4), for λ_2 = (1,3), and for λ_3 = (1,3). The parameters of monitoring success in solving the systems P IS-IT and C IS-IT are presented in Table 2.

	k=1	k=1	k=2	k=2	k=3	k=3
M/M/s	P IS-IT	C IS-IT	P IS-IT	C IS-IT	P IS-IT	C IS-IT
W_k	0.200925	0.199602	0.201044	0.199716	0.199835	0
L_k	0.144688	0.099476	0.154078	0.099606	0.154086	0
$W_q(k)$	0.000006	0.000069	0.000008	0.000098	0.000005	0
$L_q(k)$	0.000005	0.000034	0.000006	0.000049	0.000004	0

Table 2. Comparison presentation of parameters CITS and PITS

According to the parameters in Table 2 the expected value of the time the information of the first class of relative priority service (including the service time) spends in the whole system is less in the P IS-IT (W_k = 0.200925) than in C IS-IT (W_k = 0.199602), and this means that the first class information stay shorter in the whole system in P IS-IT than in C IS-IT.
The expected value of time that the information of the first relative class of priority serving spends in the queuing system ("queue"), excluding the service time, is less in the P IS-IT ($W_q(1)$ = 0.000006) than in C IS-IT ($W_q(1)$ = 0.000069) by 91,3 percent, which means that the first class information spend in total less time queuing in P IS-IT than in C IS-IT. The expected number of information for the class of the first order of relative priority service in the queuing system, excluding those that are being served, is smaller in P IS-IT ($L_q(1)$ = 0.000005) than in C IS-IT ($L_q(1)$ = 0.000034) by about 85,29 percent. The throughput capacity of the information of the first class of relative service with priorities in P IS-IT is far greater than in C IS-IT, with the same mean information arrival rate according to relative priority classes, and under the condition of equal (0.2) mean exponential service time distribution.
The first experimental simulation in a series of 100,000 information arrivals was used to study the behaviour of the described IS-IT models when the mean information arrival time is the same according to relative priority classes, which means that for P IS-IT $\lambda_{kPITS} = \lambda_1 + \lambda_2 + \lambda_3 = 1.4 + 1.3 + 1.3$ (since N = 3 and $\lambda_{kPITS} = 4$) and that for C IS-IT $\lambda_{kCITS} = \lambda_1 + \lambda_2 = 2 + 2$ (since N = 2 and $\lambda_{kCITS} = 4$), and with the same (0.2) mean exponential distribution of service time. The results of simulations for C IS-IT and P IS-IT are presented in Table 3.

	k=1	k=1	k=2	k=2	k=3	k=3
M/M/s	PITS	CITS	PITS	CITS	PITS	CITS
W_k	0.201627	0.000197	0.200712	0.200442	0.199949	0
L_k	0.100388	0.066372	0.200882	0.199228	0.201244	0
$W_q(k)$	0.000009	0.000184	0.000008	0.000187	0.000008	0
$L_q(k)$	0.000004	0.000066	0.000008	0.000188	0.000008	0

Table 3. Comparison presentation of parameters CITS and PITS

Based on the obtained simulation results with equal information arrival for relative priority service classes one can notice the difference between the behaviours of the P IS-IT and C IS-IT models. According to Table 3, the total expected value of the waiting time of the first priority service class information (including the service time) in the system is greater in the P IS-IT system (W_k = 0.201627) than in C IS-IT (W_k = 0.000197), which means that in the P IS-IT system the total average time needed for information service is longer. Also, there is a greater number of information which stay in the queuing system and in the whole service system in P IS-IT (L_k = 0.100388) than in C IS-IT (L_k = 0.066372). The total expected value of the waiting time of the first priority service class information (excluding the service time) in the queuing system ("queue") is less in P IS-IT ($W_q(1)$ = 0.000009) than in C IS-IT ($W_q(1)$ = 0.000184) by 95,1 percent, and this means that in total the information are delayed less in queues in P IS-IT. There is a smaller number of information which stay in the queuing system (without the whole service system) in P IS-IT ($L_q(1)$ = 0.000004) than in C IS-IT ($L_q(1)$ = 0.000031) by 93,93 percent.
The solution of the M/M/3 type model in C IS-IT (N = 2) with different (non-uniform) mean information arrival times according to relative priority classes means that with $\mu = 5$ and $\lambda_{kCITS} = 4$, the value for $\lambda_1 = 3$, and for $\lambda_2 = 1$. The solution of the M/M/5 type model in P IS-IT (N = 3) with different (non-uniform) mean information arrival times according to relative priority classes, means that, with $\mu = 5$ and $\lambda_{kPITS} = 4$, the value for $\lambda_1 = 3$, for $\lambda_2 = 1$, and for $\lambda_3 = 1$. The success parameters in solving the systems C IS-IT and P IS-IT are presented in Table 4.

	k=1	k=1	k=2	k=2	k=3	k=3
M/M/s	P IS-IT	C IS-IT	P IS-IT	C IS-IT	P IS-IT	C IS-IT
L_k	0.400127	0.613008	0.200072	0.205913	0.200079	0
W_k	0.200064	0.204336	0.200072	0.205913	0.200079	0
$L_q(k)$	0.000127	0.013008	0.000072	0.005913	0.000079	0
$W_q(k)$	0.000064	0.004336	0.000072	0.005913	0.000079	0

Table 4. Success parameters of PITS and CITS systems

Based on the obtained results of the simulation modelling of PITS and CITS with different (non-uniform) information arrivals for relative priority service classes one can notice the difference between the behaviours of the PITS and CITS models. According to Table 4 the total expected waiting time value of the information of the first priority service class (including the service time) in the system is less in PITS (W_k = 0.200064) than in CITS (W_k = 0.204336), which means that in total the information are served within a shorter average time period in the PITS system. Also, there is a smaller number of information which stay in the queuing system and in the whole service system in PITS (L_k = 0.400127) than in CITS (L_k = 0.613008). The total expected waiting time value of the first priority service class information (excluding the service time) in the queuing system ("queue") is less in PITS ($W_q(1)$ = 0.000064) than in CITS ($W_q(1)$ = 0.004336) by 98.52 percent, which means that in total the information stay shorter queuing in PITS. The number of information that are delayed in the queuing system (without the service system) is smaller by 99.02 percent in the considered system PITS ($L_q(1) = 0.000127$) than in CITS ($L_q(1) = 0.013008$).

CONCLUSION

Simulation modelling in the function of the intermodal transport system planning as presented in this paper is based on the functional model of intermodal environment defines a model (among several possible meta-models) of the information flow in the integrated environment. The intermodal characteristic of the transport system – intermodal terminal is complex and the simulation research represents an important tool in planning and developing the system design. The model of intermodal terminal consider the functional characteristics – functional requirements and information flow defined by them. Intermodal terminal model that was used in simulation experiments with the conditions of priority arrivals, and the results of the classical intermodal system – intermodal terminal (C IS-IT) and the parallel intermodal system – intermodal terminal (P IS-IT) was quantified. Intermodal transport planning and simulation modelling continue to be important issues in the 21st century. Intermodal transport modelling requires mathematical techniques in order to make predictions, which can then be utilised in planning and design. This is the basis for improved decision-making and planning in the transport area.

LITERATURE

Maletić, N. (2000). Development of the Intelligent Transport Systems in Water Transport, Master thesis, Faculty of Transport and Traffic Engineering, Zagreb, Croatia.

Ortuzar, J., Willumsen L. G. (2001). Modelling Transport, J. Wiley,

O'Flaherty, C. A. (1996). Transport Planning and Traffic Engineering, Butterworth-Heinemann.

U. S. Department of Transportation, http://www.fhwa.dot.gov/rnt4u/init_planning.htm

Klir, G. J. (1972). Trends in General Systems Theory, Wiley Interscience, New York.

Giannopoulos, G. A. (2000). Innovative Approaches and Telematics for Ports and Transport Chain Management, 7th World Congress on Intelligent Transport System, Power Point Presentation, Turin, Italy.

Brandt, S. (1999). Data Analysis: statistical and computational methods for scientists and engineers - 3rd ed., Springer-Verlag New York Inc., New York.

Hiller, F. S. , Lieberman, G. J(1995). Introduction to Operations Research - 6th ed., McGraw Hill Book Co. - Singapore, Singapore.

Klir, G. J. (1972) Trends in General Systems Theory, Wiley Interscience, New York.

Ottjes, J. A., Veeke, H. P. M. , Duinkerken, M. B. (2002) Simulation Studies of Robotised Multi Terminal Systems, Faculty of Marine Technology, Delft.

Review of Maritime Transport 2001, (2001). Chapter 1, Development of international trade, United Nations Conference on Trade and Development, Geneve.

Wymore, W. (1993) Model –based Systems Engineering, CRC Press, London. Brunne, D. et al.(1998). Toward Increased Use of Simulation in Transportation, Proceedings of the 1998 Winter Simulation Conference, Washington.

K.G. Goulias, editor

CHAPTER 17

USE OF MOBILE COMMUNICATIONS TOOLS AND ITS RELATIONS WITH ACTIVITIES

Kuniaki Sasaki, University of Yamanashi, Kofu, Japan

Kazuo Nishii, University of Yamanashi, Kofu, Japan

Ryuichi Kitamura, Kyoto University, Kyoto, Japan

Katsunao Kondo, University of Marketing and Distribution Sciences, Kobe, Japan

INTRODUCTION

Recent developments and innovations in information technology have most certainly brought changes in our daily life (e.g. Graham and Marvin, 1996, Zimmerman et al.; 2001). It is easily seen that many kinds of innovative communication tools have effects on the individual's communications and consequently on his daily life, because communication is one of the fundamental activities in our life. Shopping online and communication through the Internet have become familiar acts to ordinary people. In addition, in the transportation field, it has been said that telecommunications has three effects on travel behaviour: substitution, supplement, and induction of travel. Especially, substitution effects have been researched from the early 1990s in the transportation field, because that effects are expected to reduce transportation problems such as traffic congestion (Koppelman et al., 1991, Koenig et al., 1996). However, the innovation in the information and telecommunications technology (ICT) and its use are so rapid that it is difficult to capture and to forecast the effects correctly. In this study, we focus on the effect of mobile telecommunications devices such as cellular phones.

Mobile telecommunications tools have become prevalent in contemporary daily life. The total number of cellular phone contracts in Japan is over 93 million the spring of 2005, while its population is 127 million. Most of mobile phone users in Japan use this tool not only for verbal communication but also for character-based telecommunications and collecting information through the Internet. The share of 3G cellular phones, which enable high-speed connection to the Internet, is 40% and is increasing rapidly. In Japan, SMS (short message services) have almost been displaced by the e-mail system.

We focused on the effects of mobile telecommunications with the anticipation that its use affects joint activity engagement by households. Generally, engagement in many types of activities is based on communication and interaction with other individuals. For example, the availability of a family car may cause schedule adjustments by household members if the number of license holders who want to drive is larger than the number of cars available to the household. In this case, the family car must be allocated based on the communication among household members. Picking someone up is a typical household joint activity that requires communication.

Mobile telecommunications facilitates flexible schedule adjustments. The rapid penetration of mobile telecommunications devices may have produced essential changes in the relationship between communication and behaviour, because mobile telecommunications diminishes time and space constraints on communication (Mohkatarian and Salomon, 2002). Before the prevalence of mobile phones, the conventional telephone was a representative communication tool for a household which was assigned to the household collectively, not to each individual household member. Cell phones have released people not only from the location of the stationary telephone, but also from the household by providing direct person to person connetion. Socially, communication has become more personal with mobile telecommunications devices. From this perspective, contacts among family members may be one of the activities affected by mobile telecommunications. It is possible that both communication and activities among household members are changing with mobile devices (Zumkeller, 2000).

Many studies have been accumulated on household decision making as group decision making (e.g. Davis, 1975; Golob and McNally, 1997; Zhang et al., 2003). In some studies, telecommunications is indicated as an important factor. However, incidents of communication via telecommunications are not treated explicitly in most studies because of difficulties in measurement involved.

This study is based on a survey of telecommunications and activity-travel behaviours of about 150 families. We asked every member of the households to participate in the survey to capture telecommunications incidents as episodes of interactions among household members. The survey is designed to capture detailed information on the relations between telecommunications and joint activity engagement by respective household members of each household; hence it is hosehold-based rather than individual-based.

Trendsetters on mobile phones that have became prevalent in the last 10 years or so are the younger generations and business people. It is often reported that there exist generation divides in the use of mobile telecommunications tools. To capture such phenomena, households are grouped by life cycle stage in the study.

The Framework of the Research

The conceptual framework on mobile telecommunications and joint activity engagement is described in Figure 1. The figure, which is a version of the time-space path diagram in activity analysis, shows two joint activities. One is a business meeting involving husband and business partner, and the other is the picking up of husband by wife. Suppose telecommunications is needed to coordinate these activities. The figure shows a conventional telecommunications (TC) incident and a mobile telecommunications (MC) incident. The purpose of this research is to gain empirical evidence of the link between mobile telecommunications and joint activity engagement by household members as illustrated in the figure.

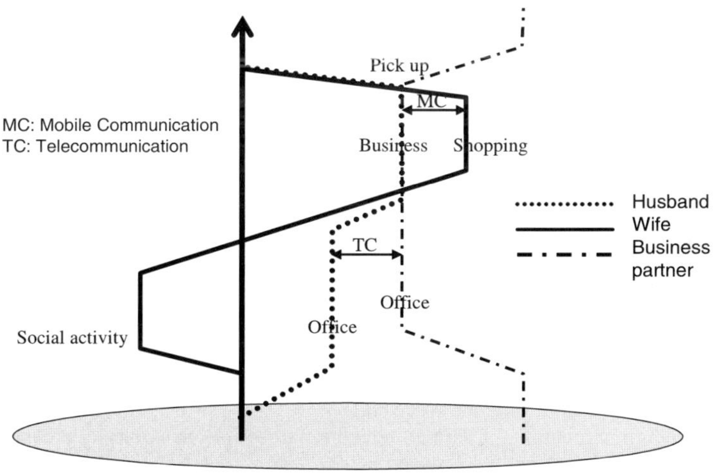

Figure 1. The framework of this research

OUTLINE OF THE SURVEY

The survey, named SCAT2 (Survey on Communication and Activity-Travel 2), involved a self-administrated questionnaire and was conducted in Kofu City, Japan, in November 2003 (Sasaki et al., 2005), on two successive days of which on was a week day and the other was a weekend day (i.e., Friday and Saturday, or Sunday and Monday). The sample households included 158 households, of which 70 were Kofu University students' households and 88 were government employees' in the city. The sample is not based on random sampling from a population, and the results of this study may not be indicative of trends in the population. Nonetheless it is hoped that the survey results will reveal prominent relationships between mobile telecommunications and household behaviour.

The contents of this survey consist of mainly two parts as below.

1. ***Activity diary***: All in-home and out-of-home activities and trips were recorded in bar-chart type format. The time and place of joint activities by household members were recorded in the record of each participating member, with a flag attached on each joint activity. In-home activities were categorized into five types, and out-of-home activities into seven types in the questionnaire. The trip is classified as a category of out-of-home activity.

2. ***Telecommunications diary***: Respondents were asked to record all telecommunications incidents with their times, purposes, partners, media and activities involved.

Individual and household attributes were also collected as part of the survey.

Prior to this survey, a preliminary survey (SCAT1) was conducted in 2002 with 150 university students from five universities in Japan (Nishii et al., 2005). The purpose of the preliminary survey was to test the efficiency and effectiveness of the survey design, especially the bar-chart format of the activity diary. The results showed the survey well captured activities and telecommunications incidents. The information items in the survey are summarized in Table 1.

Table 1. Questioned Items and Contents of SCAT2

Questioned Items	Contents
Activity and travel patterns in two consecutive days (Activity Diary Survey)	The entire records of in-home activities, out-of-home activities, and trips in two days. A joint activity was so marked in the bar-chart.
Mobile telecommunications on the survey day	The records of all telecommunications incidents on either weekday or weekend: Purposes, time, partners, activities involved, and media (receive or send, e-mail or voice call)
Mobile telecommunications in relation to activity plans of the surveyed day.	Role of telecommunications and activity plans of the surveyed day
Individual attributes and household attributes	Individual attributes such as sex, age, employment status and driving license holding. Attributes of households such as car ownership, the number of household members and residential type.

OUTLINE OF INDIVIDUAL CHARACTERISTICS AND TELECOMMUNICATIONS INCIDENTS

The Basic Characteristics of the Sample

In this section, we show the sample attributes in detail. Note that mobile telecommunications behavior is influenced by sex and age (Sasaki, 2000). The total number of respondents is 322 from 148 households. Ten households were excluded because of defective answers. There are about the same numbers of men and women in the data. Figure 2 shows the distribution of age in the data. Since about half of the respondent households are from university students' households, 20~29 and 40~49 age cohorts are over-represented in this sample. The composition of employment status is shown in Figure 3. Since the half of selected households were from workers in the Kofu City government, full time employees have a larger share in the data. The composition of household members is one of the important factors in this analysis, because we are to categorize the household into life cycle stage. Figure 4 shows the number of members in a household. The average is 3.8 persons per household. The university students' households have more members than other households since those households include at least one student. Household car ownership is shown in Figure 5. There is no sample household without a car available. The average of number of cars available is 2.6, which compares with the average household size of 3.8. Kofu City is located 100 km west of Tokyo and is surrounded by mountains. The population of the city is about 200,000. The use of public transport in this city is decreasing and the share of car trips is about 80%. This is typical of smaller metropolitan areas of Japan. Figure 6 shows the cell phone ownership of each sample individual. The penetration rate of cellular phones in the sample is almost 90%. Most of non-owners are over 60 years of age. The cell phone ownership in the sample is higher than the average in Japan.

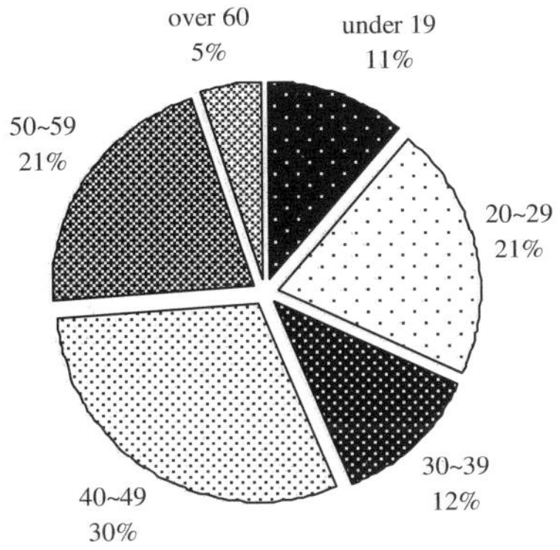

Figure 2. Distribution of Age

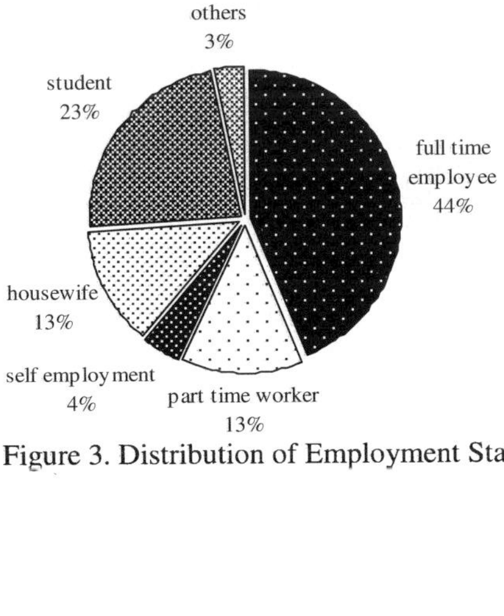

Figure 3. Distribution of Employment Status

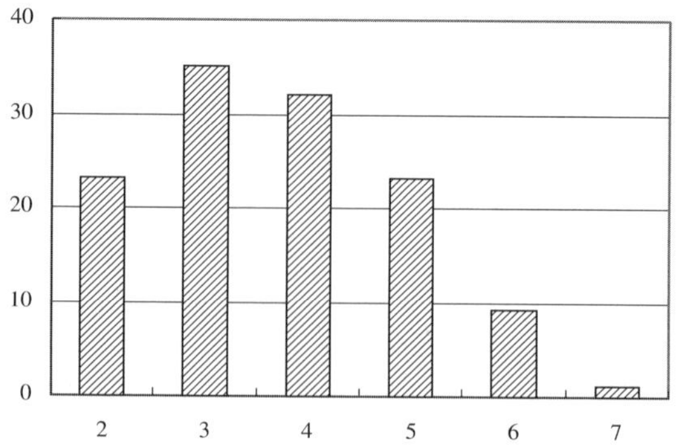

Figure 4. Distribution of the Number of Household Members

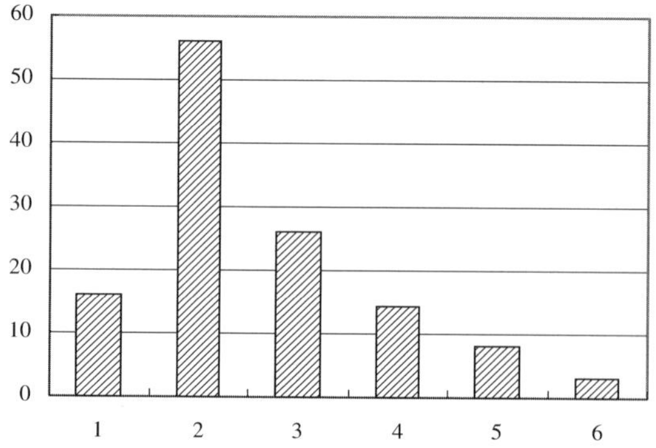

Figure 5. Distribution of the Number of Cars Available to Household

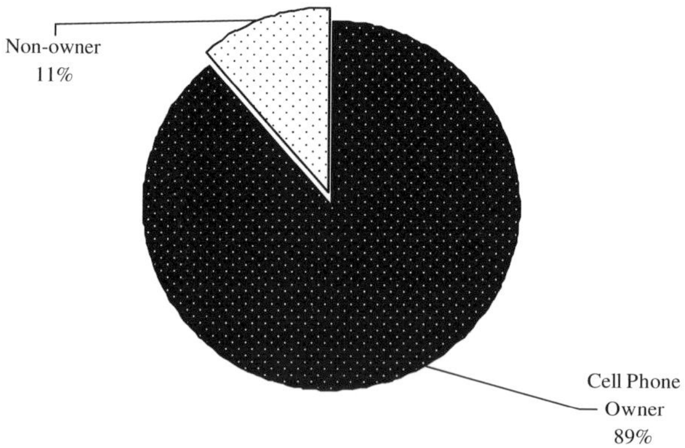

Figure 6. Cellular Phone Ownership of Sample Individuals

Descriptive Statistics: Telecommunications Diary

The distribution of the total frequency of telecommunications incidents by all media is shown in Figure 7, on weekday and weekend respectively. The averages frequency is 3.0 on a weekday and 3.2 on a weekend day. The distribution is similar in shape between weekdays and weekend days. The averages for university students from our previous study are 5.1 on a weekday and 5.5 on a weekend day (Nishi et al., 2005). The distribution curves are different between university students and this sample and indicate that the telecommunications frequency, in other words, the demand for communication, is different between generations.

Figures 8 and 9 show the distribution of telecommunications incidents by media, by direction (send and receive), by type (voice call and e-mail, including SMS). There are some differences in the choice of telecommunications media between weekends and weekdays. The frequency of "phone calls" on weekdays is smaller than on weekend days, while the use of e-mail is almost the same. Since most of the sample individuals worked or studied on weekdays, time available for voice telecommunication may be limited on weekdays. The telecommunications diary includes the purpose of communication categorized into six types:

1. chat (no relations or consequences to activity engagement)
2. to confirm an appointment
3. to make an appointment for the day
4. to confirm an appointment for a later day
5. to make an appointment for a later day
6. urgent contact

Figure 10 shows the distribution of purposes by media on weekdays. Chatting is most frequent on both voice calls and e-mail. This is especially on the case with younger generations. This is a piece of evidence that cellular phones has penetrated as an everyday communication tool. This is followed by confirming or making appointments of the day or a later day. We investigated the detail of these two purposes. An appointment on the day tends to be made by a voice call, while that for a later day tends to be processed by e-mail. This tendency can be seen also on weekends (Figure 11). The results indicate that the choice of telecommunications media is affected by the characteristics of communication contents, especially their immediacy. E-mail is less immediate than a phone call, though e-mail on mobile phone is convenient because it can be sent and read at convenient times. Consequently e-mail is not used in urgent contacts as the figure indicates.

Figure 12 shows the distribution of types of activities for which appointments were made by mobile phones. The distribution is different between weekdays and weekends. For example, going back home has the largest share on weekdays, while personal shopping is most frequent on weekend. Details of the relation between activity and telecommunication are shown in the next section.

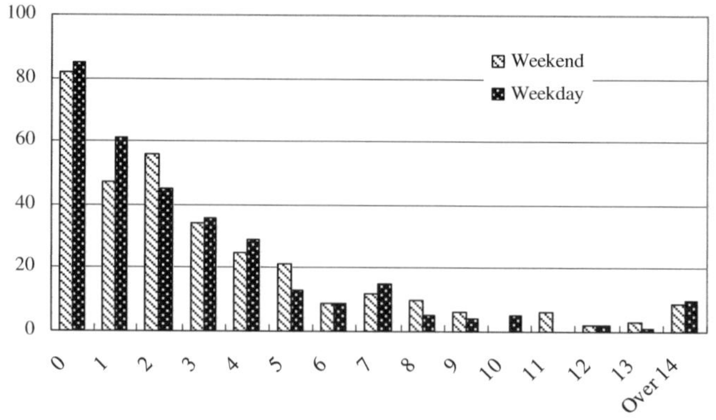

Figure 7. Frequency Distribution of Telecommunications Incidents

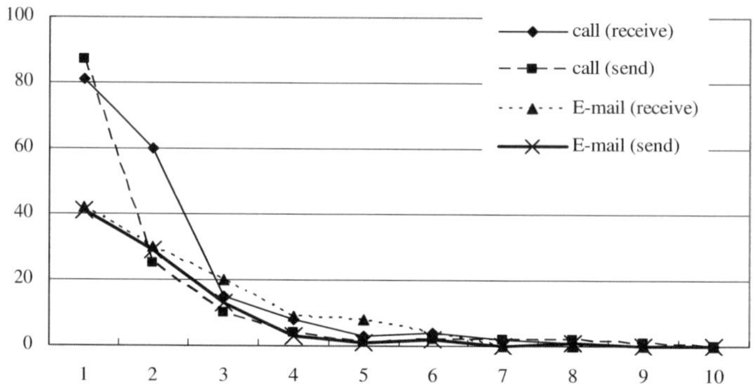

Figure 8. Frequency of Telecommunications Incidents by Media: Weekend

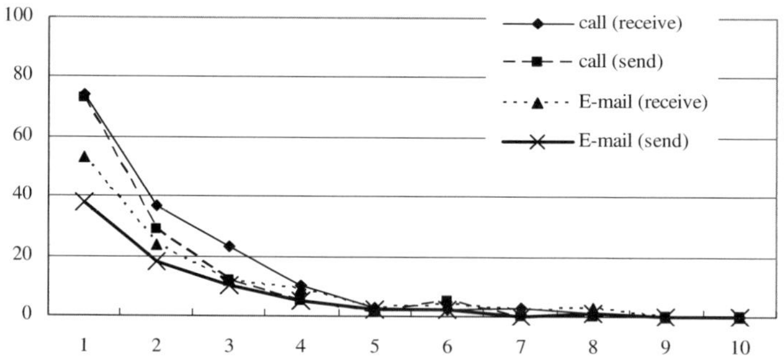

Figure 9. Frequency of Telecommunications Incidents by Media: Weekday

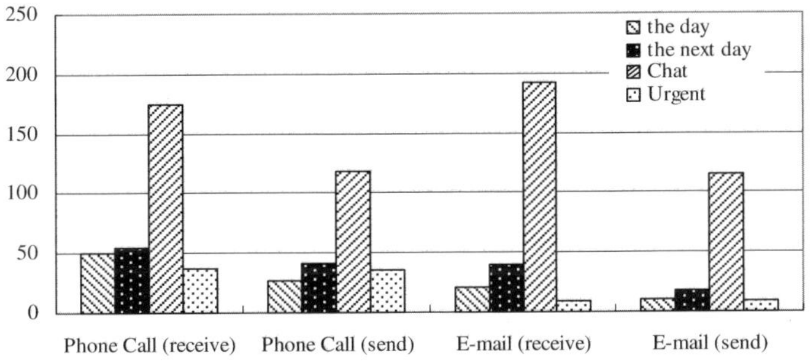

Figure 10. Purpose of telecommunication by Media: Weekday

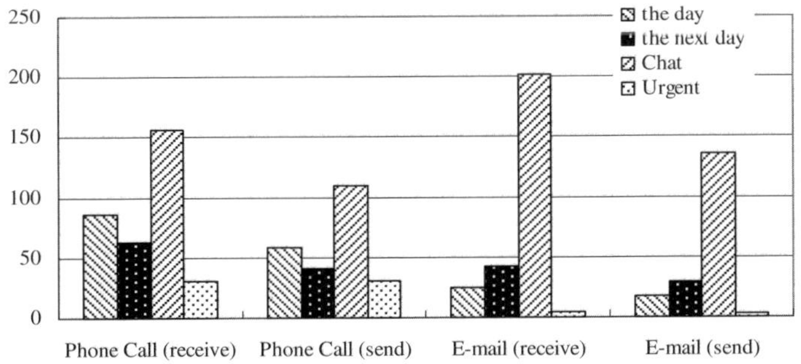

Figure 11. Purpose of telecommunication by Media: Weekend

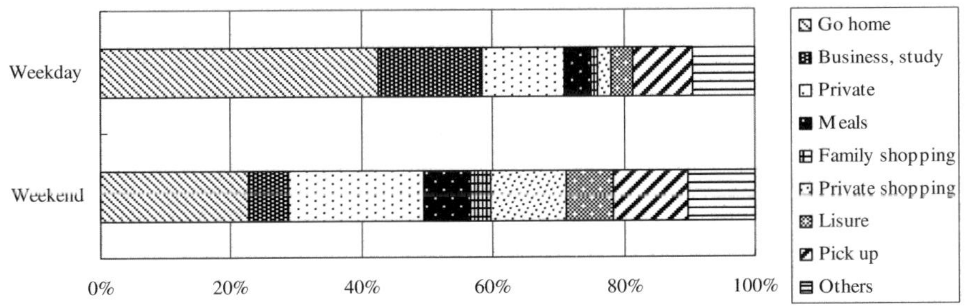

Figure 12. Distribution of Activity Types: Activities with Appointment (?)

THE RELATION BETWEEN TELECOMMUNICATION AND TRAVEL-ACTIVITY

A hypothesis is formed based on the framework of Figure 1 that people can spend time more efficiently by using telecommunication, because they can arrange for and accomplish joint activities in any situation. However, this hypothesis is not concerned with how extra time made available may be used. Therefore, trips, activities and trip chains are statistically investigated in this section.

The average number of recorded activities per day is 6.36, while the average number of trips is 2.64, and average telecommunications incidents is 3.10. The simple correlation coefficients between telecommunications and activity-travel indexes are summarized in Table 2. The result shows no obvious relationship. However, Figure 13 describes the relation between the frequency of telecommunications incidents and the number of trips on a weekend day, after grouping sample individuals by the number of telecommunications incidents. Though the direct simple correlation between telecommunications incidents and trips is 0.089, the figure shows that there is positive correlation between telecommunications frequency and the number of trips. The reason of the small simple correlation is that there is non-direct proportion in intermediate telecommunications category. However, the difference in trip frequency between the group with no telecommunications incident and the group with over 10"incidents is statistically significant. Additionally, the bar charts in Figure 13 shows similar relations between average time per trip and total telecommunication.

The figure also shows the relation between total travel time and telecommunications incidents. The trip time and total travel time of the group without telecommunications incident are significantly larger than those of the group with over 10 telecommunications incidents. These results prompt a conjecture that a long-trip maker does not have very much telecommunications incidents on weekend.

We now focus on activities and investigate their features on weekend. Figure 14 shows the fraction of sample individuals engaging in each type of out-of-home activity tabulated with 10-minute intervals for those without telecommunications incident. Figure 15 shows the same tabulation with those over 10 telecommunications incidents. By comparing the two figures, the samples in the no-telecommunication group more often engaged in obligatory activities such as work/study or family shopping and came back home around noon and after 5:00 P.M. The rate of engaging in discretionary activities such as meal or private activity out-of-home is low in this group. The frequency of telecommunications incidents becomes high when obligatory activities are not on the weekend day.

These results suggest another conjecture that telecommunication is related to discretionary activities. Generally, obligatory activities are pre-determined and a routine. The results in the previous section showed that telecommunication is mainly used to confirm or to make appointments just before activities. Additionally, considering the results that frequency of telecommunication is related to travel characteristics, telecommunication is possibly used for out-of-home discretionary activities with other people, i.e., joint activities.

To verify this conjecture, the timing of telecommunication for the activity of the day and the timing of the out-of-home discretionary activiti are shown together in Figure 16. The figure shows the aggregate number of telecommunications incidents concerning activity engagement for each hour, and the graph of out-of-home discretionary activities is the rate of activity engagement in each 10 minutes. This graph clearly shows the difference in the location of the peaks between the two. The peak of telecommunications incidents is at 11 A.M., while that of discretionary activities is at around 3 P.M. The difference is about 4 hours. Telecommunication is increasing rapidly at 10 A.M. and decreasing gradually in the afternoon. On the other hand, the ratio of out-of-home discretionary activities starts to decrease in the evening but remains high until late evening. This represents situational evidence that telecommunication is used for making and confirming appointments for joint activities in the afternoon. Obviously there is a complementary relationship between telecommunication and joint activity engagement.

Table 2. The Correlation Coefficients between Trip, Telecommunication, and Activity Indexes

Weekend		Weekday	
Total telecomm-trip	0.089	Total telecomm-trip	0.094
Total telecomm-average trip time	0.218	Total telecomm-average trip time	-0.026
Total telecomm-activities	-0.103	Total telecomm activities	0.068
Total telecomm-out-of-home acts.	0.104	Total telecomm-out-of-home acts.	0.111

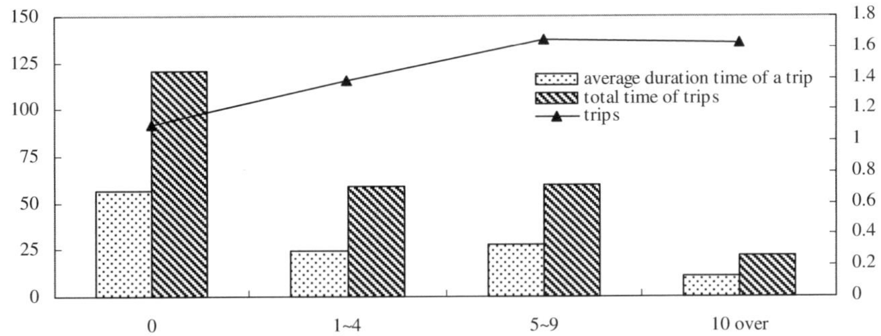

Figure 13. Average Number of Trips, Trip Duration, and Total Travel Time by the Number of Telecommunications Incidents

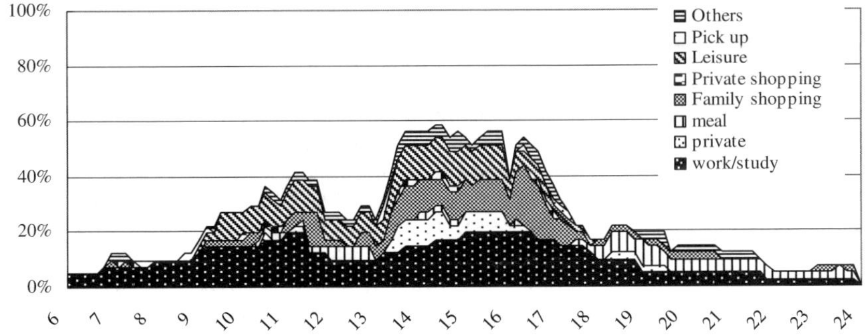

Figure 14. Fraction of Sample Individuals Engaged in Activity by Type: No Telecommunications Incident

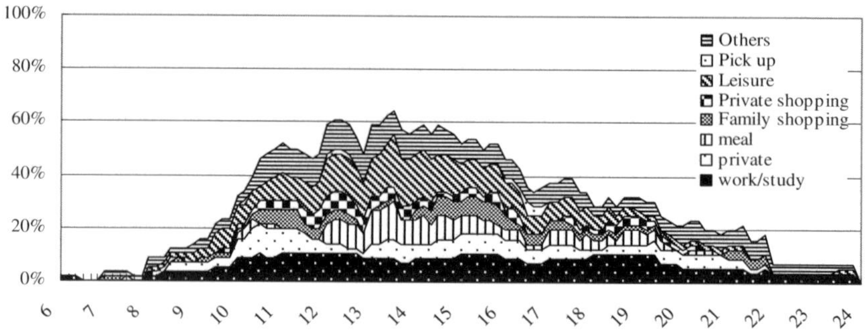

Figure 15. Fraction of Sample Individuals Engaged in Activity by Type: Over 10 Telecommunications Incident

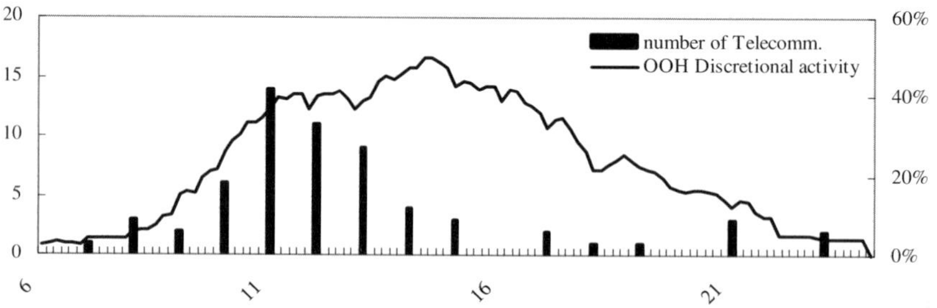

Figure 16. Timing of Telecommunication and Out-of-home Discretionary Activity

HOUSEHOLD CHARACTERISTICS AND JOINT ACTIVITY

In this section, we focus on the household activities to capture the interaction of individuals in the household. The analysis takes into consideration household characteristics because the use of ICT is significantly affected by life cycle stage (Bhat et al., 2003). Household activity is usually decided by group decision making by members. This has been analyzed by some researches (e.g. Recker, 1995). The interaction of household members is important part in this kind of decision. Consequently, joint activity engagement is an area where the influence of mobile communication can be manifested. By investigating this type of activity, we expect the role of communication in decision making will become evident. Since the feature of household activity is usually affected by the level of independence of individuals in the household, we first categorize the samples by life cycle stages (LCS), and then we analyze the aggregate and average characteristics of household activities. Life cycle stage is defined as follows taking, among others, mobility restrictions of household members into consideration.

I. Householder head is less than 40 years old, and is living with children under 18 years old.

II. Head is more than 40 years old, and is living with children under 18 years old

III. Hean is more than 40 years old, and is living with children over 18 years old

IV. Head is less than 50 years old, no child in the household

V. Head is more than 50 years old, no child in the household

The frequency of joint activities and time expenditure for them, both on weekend and on weekday, are summarized in Table 3. Both total time expenditure and the number of joint activities are large for LCS-I, LCS-IV and LCS-V, especially on weekend. However the duration per joint activity is between 78 and 100 minutes for every LCS on weekend. The difference among LCS becomes small for average duration on weekend. This means that the number of joint activities is more for LCS-I and LCS-IV, and LCS-V on weekend. On weekday, LCS-I has a long total time for joint activity and other LCSs have a short total time on weekday.

Joint activities on a weekday are affected by the time and space constraints of the householder, because the sample households in this study had at least one full time worker. We are to investigate particularly weekend joint activities because this class of activities are frequently carried out on weekend. The distribution of joint activity durations is shown by activity type on weekend in Figure 17. This figure shows that almost 80% of joint activity time is spent for family gathering (in-home). LCS-III and LCS-V have relatively high shares of sightseeing or leisure, but differences are not statistically significant. These LCSs are without children or with children over 18 years old. Consequently, it is probable that joint activities involve only the parents. Time for trip is extremely small in LCS-V as compared with other LCS. The reasons are that this LCS represents the most mature stage and is less inclined toward out-of-home activity engagement.

To see the effect of telecommunications on joint activity engagement, the average frequency of telecommunications incidents among household members (line graph) and the average total time (bar chart) by LCS in Figure 18. From this figure, we can see that, the more progressed LCS becomes, the smaller the average frequency of telecommunications incidents among the household members becomes. However, the number of joint activities does not have the same trend. Both the number of telecommunications incidents and the number of joint activities are large in LCS-I, which is characterized by the presence of young children whose mobility is low, and a younger head. We can derive a conjecture that low mobility is one of causes of frequent telecommunications. However, LCS-II has low mobility persons in it, too. Younger householder and low mobility might have synergetic effects on the use of mobile telecommunication. There is correlation between total joint activity duration and mobile telecommunications incidents among household members. This can be another support for the conjecture that mobile communication is used as a supplemental tool for joint activity.

Table 3. Average Frequency and Total Duration of Joint Activity per Household

LCS	Weekend	Weekday	Weekend	Weekday
	The average number of joint activities	The average number of joint activities	The average total time of joint activities (per joint activity)	The average total time of joint activities (per joint activity)
I	7.5	2.8	589 (78.5)	308 (110.0)
II	3.9	2.4	325 (83.3)	153 (63.57)
III	3.1	1.5	317 (102.3)	143 (95.3)
IV	4.5	1.2	471 (104.6)	135 (112.5)
V	4.5	1.3	476 (105.7)	171 (131.5)

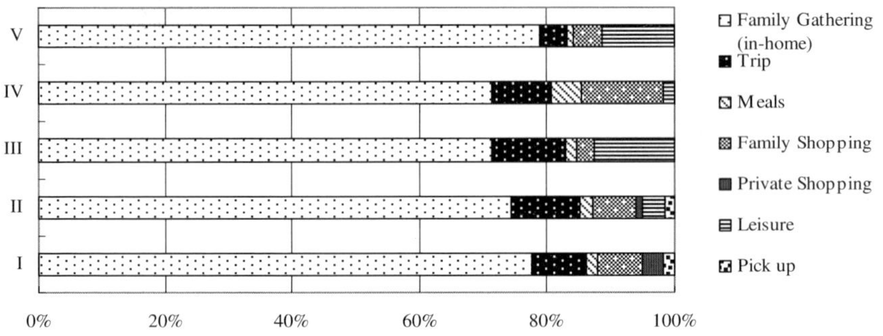

Figure 17. Distribution of Joint Activity Durations by Type by LCS

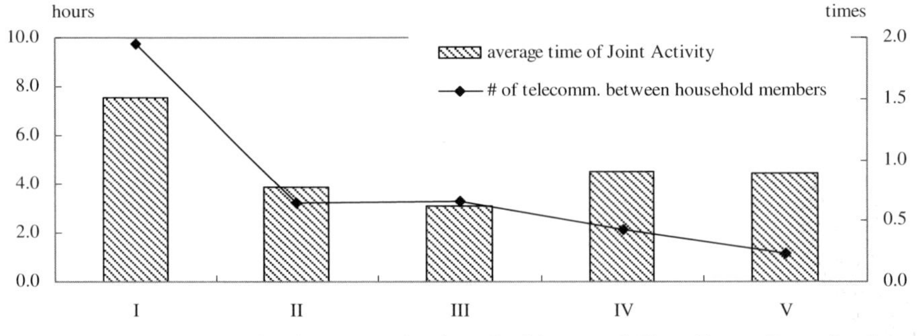

Figure 18. Frequency of Telecommunications Incidents and Time Expenditure for Joint Activity by LCS

DISCUSSION AND CONCLUSION

This research is based on a survey on communication, activity and travel with the purpose of clarifying the relation between joint activity engagement and telecommunication among household members. The survey made it possible to determine the role of mobile telecommunication in household. Using the data, we investigated the features of telecommunication, activity and travel. Some pieces of evidence have been obtained to show that telecommunication is used to supplement travel and out-of-home activity engagement as:

1. The frequency of telecommunication is statistically significantly associated with trip characteristics
2. Trip characteristics are related with out-of-home discretionary activity engagement on the day
3. The peak of mobile communication incidents concerning activity engagement is located about four hours before the peak of out-of-home discretionary activities
4. Households with a young householder and low mobility children spend more time on joint activities and use mobile telecommunication more.

These pieces of evidence imply that the joint activity may be activated by mobile telecommunication as a supplementary tool. That is, the effect of mobile telecommunication on activity-travel is not only substitution effect when focusing on the household. In the sociology field, it is said that a household member has a role in daily life. From this point of view, we can conclude that the penetration of mobile telecommunications devices has opened a new way for flexible forms of household activities, coordinating activities between family members. However, in this study, the interdependence of activities among household members was not examined. This is our next important research theme.

Acknowledgment

This research was supported by a Grant-in-Aid for Scientific Research (KAKENHI) of the Ministry of Education, Culture, Sports, Science and Technology, Japan. The members of the Mobile Market Study Group also gave us a number of suggestions for this research.

References

Anderson, B. (2004): Everyday Time Use in Britain in 1999 -Implications for telecommunication strategy-, Working Paper, University of Essex.

Bhat, C. R., A. Sivakumar and K Axhausen (2003): An analysis of the impact of information and communication technologies on non-maintenance shopping activities, Transportation Research Part B, 37, pp. 857–881.

Koppelman, F., I. Salomon and K. Proussaloglou (1991): Tele-shopping or store shopping? A choice model for forecasting the use of new telecommunications-based services, Environment and Planning B, 18, pp.473–489.

Davis, H.L. (1976): Decision Making within the Household, Journal of Consumer Research, vol. 2, pp.241-260.

Golob, T.F. and M.G. McNally (1997): A model of household interactions in activity participation and the derived demand for travel. Transportation Research, 31B, pp. 177-194.
Graham, S., Marvin, S., 1996. Telecommunications and the City: Electronic Spaces, Urban Places. Routledge, London.

Koenig, B.E., Henderson, D.K., Mokhtarian, P.L. (1996): The travel and emissions impacts of telecommuting for the State of California Telecommuting Pilot Project. Transportation Research C, pp.13–32.

Mokhtarian, P.L., Salomon, I. (2002): Emerging travel patterns: do telecommunications make a difference. In: Mahmassani, H.S. (Ed.), In Perpetual Motion: Travel Behavior Research Opportunities and Application Challenged. Pergamon, Elsevier Science, Oxford, UK, pp.143–182.

Nishii, K., K. Sasaki, R. Kitamura and K. Kondo (2005): Recent Developments in Activity Diary-Based Surveys and Analysis: Some Japanese Case Studies, In: Timmermans, H. (ed.), Progress in Activity-Based Analysis, Elsevier, Oxford, 2005, pp.335-354.

Recker, W.W. (1995): The Household Activity Pattern: General Formulation and Solution, Transportation Research, 29B, pp. 61-77.

Sasaki, K. (2000): The relationship between ICT use and behaviour in a train station, Presented paper in JEKI Competition Conference, Tokyo, Japan (in Japanese).

Sasaki, K., K. Nishii, R. Kitamura and K. Kondo (2005): An Analysis of Urban Activities with Focus on the Relationship between ICT and Activity-Travel, presented paper in ERSA2005, Amsterdam, CD-ROM.

Zhang, J., Timmermans, H.J.P., and Borgers, A.W.J. (2002): Utility-Maximizing Model of Household Time Use for Independent, Shared, and Allocated Activities Incorporating Group Decision Mechanisms, Transportation Research Record, 1807. pp.1-8.

Zumkeller, D. (2000): The impact of telecommunication and transport of spatial behavior, Periodica Polytechnica Series, Transportation Engineering, Vol., 28, pp.23-38.

Zimmerman, C.A., Campbell, J.L., Cluett, C. (2001): Impact of new information and communication technologies on transportation agencies: a synthesis of highway practice. NCHRP Synthesis 296, National Cooperative Highway Research Program, Transportation Research Board.

Transport Science and Technology
K.G. Goulias, editor

CHAPTER 18

A MULTIVARIATE MULTILEVEL ANALYSIS OF INFORMATION TECHNOLOGY CHOICE

Tae-Gyu Kim, North Carolina Department of Transportation, Raleigh, North Carolina, USA

Konstadinos G. Goulias, University of California, Santa Barbara, California, USA

INTRODUCTION

During the last decade, the rapid advance and explosive growth of information and communications technology (ICT) has given rise to new paradigms in the manner in which people conduct their everyday affairs and also the way in which businesses are conducted. Under these circumstances, our traditional concept of accessibility may no longer be valid (Golob, 2001; Golob and Regan, 2001). In fact, through ICT people can get virtual accessibility to a rapidly growing range of activities without the more traditional spatial and temporal limitations and constraints. Consequently, people have more flexibility to arrange their schedules, and eventually change their activity and travel patterns. These substantial impacts of ICT motivate the need for research on the present and future impacts of telecommunication on activity and travel behavior.

The first step to forecast impacts of ICT on activity and travel behavior is to develop a technology choice model that identifies user groups for each ICT device and estimate the number of its users. To accomplish this task in this paper, trends of people's ICT adoption during the period of 1997-2000 are first examined using Puget Sound Transportation Panel (PSTP) data collected in 1997 and 2000. In addition, a multivariate multilevel technology choice model is developed to identify user groups for different ICT devices while accounting for a strong correlation within household members as well as a high degree of person heterogeneity in technology adoption. In this model, the ownership of a variety of ICT devices in the year 2000 such as desktop computers, Internet, cellular phones, pagers, and laptops, is modeled as a function of cross-sectional and longitudinal information.

The rest of this paper is organized as follows. First, a brief description of Puget Sound Transportation Panel data and sample characteristics used for this paper are provided. This analysis is followed by multivariate multilevel model formulations, which identify ICT user groups. In addition, the result of the model estimation is discussed. Lastly the paper ends with a summary and conclusions.

DATA DESCRIPTION

The database used in this study is from Puget Sound Transportation Panel (PSTP). It is the first general purpose, urban household panel survey designed for transportation analysis in the United States (Murakami and Watterson, 1990; Goulias, Kilgren, and Kim, 2003).

The PSTP was initiated in the four counties (King, Kitsap, Pierce, and Snohomish) of the Puget Sound region including Seattle in the fall of 1989 by the Puget Sound Regional Council (PSRC, then the Puget Sound Council of Governments) in partnership with transit agencies in the region, and continues until today. Unlike traditional cross-sectional surveys, the PSTP is a panel or longitudinal survey in which similar measurements are made repeatedly on the same households and their members over time. A survey conducted at each point in time is called a wave. Each wave of the PSTP collects three groups of data that are household demographics, person's social economic information, and reported travel behavior. Trip information is collected using a travel diary.

Since 1997 (Wave 7), the survey participants above 15 years of age were asked some additional questions about their personal use and attitudes towards existing and potential new information sources. In addition, respondents were asked about their familiarity with, and use of, electronic equipment and information services. For example, respondents provided information regarding their use of a desktop computer at home or work with access to the Internet at least once a week on average. Other questions asked if the respondents carried a personal cellular phone, pager, laptop computer (with modem) and a personal digital assistant (PDA) at least ten times a month. The survey also asked if respondents are aware of and use a variety of traffic and transportation information over the radio, television, telephone or the World Wide Web (WWW). Therefore, the PSTP dataset offers a unique opportunity to study people's dynamics in ICT adoption.

For the analysis in this study, we used the data from 1480 persons in 866 households who provided detailed information in both waves 7 (1997) and 9 (2000) for all the variables used in this analysis. Table 1 shows the number of persons and households and a few social and demographic characteristics of the sample. It also contains information about technology ownership and use with focus on the more contemporary technologies. Computer and Internet at home are the two fastest growing technologies. Computers at work appear to be stabilizing at 50 percent of the sample, and although Internet use at work is increasing, it did not reach the level of market penetration of computers at work. Cellular phone ownership is increasing even faster than computers and the Internet, causing a potential negative impact on pagers. Laptop computers and personal digital assistants (PDA) are used by a small sample segment as reported.

Table 1 Sample Characteristics

Characteristics			Wave 7 (1997)	Wave 9 (2000)
Number of persons in the sample			1480	1480
Number of households in the sample			866	866
Person & Household	Percent of males		46.9	46.9
	Number of employed persons		859	881
	Number of persons in household		2.5	2.5
	Number of cars per household		2.2	2.2
Information & Communication Technology	Number of persons who use computer regularly	At home	751 (50.7 %*)	982 (66.4%*)
		At work	700 (47.3 %*)	717 (48.4 %*)
	Number of persons who use the Internet regularly	At home	458 (30.9 %*)	896 (60.5 %*)
		At work	438 (29.6 %*)	544 (36.8 %*)
	Number of persons having cellular phones		419 (28.3 %*)	685 (46.3 %*)
	Number of persons having pagers		156 (10.5 %*)	129 (8.7 %*)
	Number of persons having laptops		71 (4.8 %*)	79 (5.3 %*)
	Number of persons having PDAs		6 (0.4 %*)	38 (2.6 %*)

** % over total number of persons (1480) in the sample*

MODEL FORMULATION

Variables Used

As shown in Table 2, two groups of ICT variables are used as the dependent variables in the model. The first group of the dependent variables is whether or not a person uses computer or the Internet at work/school or home in the year 2000. The second group is whether or not a person carries personal mobile technologies in the year 2000, such as cellular phones, pagers, and laptops. PDAs are excluded in the model due to its usage by a small sample segment. All the dependent variables are binary: if persons use a technology, the dependent variable is coded as 1; otherwise, it is coded as 0.

Table 2 List of Dependent Variables

ICT Groups	Dependent Variables	Descriptions
Computer and Internet	CW	Computer use at work/school
	NW	Internet use at work/school
	CH	Computer use at home
	NH	Internet use at home
Personal Mobile Technology	CL	Cellular phone use
	PG	Pager use
	LP	Laptop use

For independent variables, two groups of variables are used: household-level variables and person-level variables. We used cross-sectional information in the year 2000 as well as longitudinal information between the years 1997 and 2000 about household-level and person-level social, economic, and demographic characteristics.

To account for task allocation and roles within the household, number of adults, number of children by age group as well as vehicles owned and income representing resource availability are included as independent variables. In addition, in order to account for the sampling stratification of the panel participants, the county of residence and sample indicators (TRANSIT, CARPOOL) are also included as independent variables. Appendix contains a comprehensive description of the independent variables used here.

Using information between 1997 and 2000 a large set of variables is defined to capture the changes in social and economic circumstances experienced by each respondent and include them as explanatory variables of ICT adoption in the year 2000. In other words, indicators for changes in household and person characteristics, such as increase or decrease in car-ownership, changes in household composition, and changes in employment, were used to examine the effect of the changes on technology choice and use.

Multivariate Multilevel Technology Choice Model

Since in PSTP information on technology choice is collected from a few household members within a household, it may be viewed as data with a nested hierarchical structure and gives an opportunity to explore household interactions in technology adoption.

The use of multivariate multilevel modeling techniques is employed in this study, due to their capability to explicitly formulate the hierarchical data structure in the model, consider observed and unobserved household interactions in technology use, relax the usual regression assumption of independent observations, and account for the correlations among the observations in the model formulation.

The multivariate multilevel technology choice models are defined for seven ICT dependent variables with the following specifications:

$$\begin{aligned}
\text{logit}(\pi_{ij}^{CW}) &= \beta_{ij}^{CW} + \gamma_1^{CW} x_{1ij}^{CW} + \ldots + \gamma_{k_{CW}}^{CW} x_{k_{CW}ij}^{CW} \\
\text{logit}(\pi_{ij}^{NW}) &= \beta_{ij}^{NW} + \gamma_1^{NW} x_{1ij}^{NW} + \ldots + \gamma_{k_{NW}}^{NW} x_{k_{NW}ij}^{NW} \\
\text{logit}(\pi_{ij}^{CH}) &= \beta_{ij}^{CH} + \gamma_1^{CH} x_{1ij}^{CH} + \ldots + \gamma_{k_{CH}}^{CH} x_{k_{CH}ij}^{CH} \\
\text{logit}(\pi_{ij}^{NH}) &= \beta_{ij}^{NH} + \gamma_1^{NH} x_{1ij}^{NH} + \ldots + \gamma_{k_{NH}}^{NH} x_{k_{NH}ij}^{NH} \qquad (1) \\
\text{logit}(\pi_{ij}^{CL}) &= \beta_{ij}^{CL} + \gamma_1^{CL} x_{1ij}^{CL} + \ldots + \gamma_{k_{CL}}^{CL} x_{k_{CL}ij}^{CL} \\
\text{logit}(\pi_{ij}^{PG}) &= \beta_{ij}^{PG} + \gamma_1^{PG} x_{1ij}^{PG} + \ldots + \gamma_{k_{PG}}^{PG} x_{k_{PG}ij}^{PG} \\
\text{logit}(\pi_{ij}^{LT}) &= \beta_{ij}^{LT} + \gamma_1^{LT} x_{1ij}^{LT} + \ldots + \gamma_{k_{LT}}^{LT} x_{k_{LT}ij}^{LT} \\
\beta_{ij}^{m} &= \gamma_0^{m} + v_j^{m} + u_{ij}^{m}, \text{ where m= CW, NW, CH, NH, CL, PG, LT}
\end{aligned}$$

π_{ij}^{CW} is the probability that person i in household j uses a computer at work/school (with i =1,2, ... number of people in household j, j = 1,2, ... number of households in the sample). Similarly, we define the other six dependent variables (π_{ij}^{NW}, π_{ij}^{CH}, π_{ij}^{NH}, π_{ij}^{CL}, π_{ij}^{PG}, and π_{ij}^{LT}) in Equation (1).

The first term (β_{ij}^{m}) in the right hand side of each equation in the multilevel model is a random intercept. The term u_{ij}^{m} is a random person–to–person variation (also called within household variation) and is a deviation of travel expenditure around $\gamma_0^m + v_j^m$. The term v_j^m is a random household-to-household variation and it is a deviation around γ_0^m. These are also called random error components, and are assumed normally distributed with E(u) = E(v) = 0, and $Var(u)$ = σ_u^2 and $Var(v)$= σ_v^2. The random components (u and v) and their variances represent unobserved heterogeneity at the person and household levels, respectively. They can be interpreted in the usual way as random error terms in linear and non-linear regression. In Equation (1) observed heterogeneity at different levels is included in the model when level-specific variables are included in the matrix **X**. Their effect on the probability of choosing a given technology is estimated by estimating the vector γ. In equations all the γ coefficients are defined in a manner similar to a typical regression model. Although all the coefficients of explanatory variables are defined as fixed in the model specification above, the coefficients can be defined as random with a mean and a variation around their means γs. In this way we can define a more general model at each of these levels to represent heterogeneous technology choice either due to personal or household variation.

In multilevel models, all the fixed and random parameters can be estimated by iterative generalized least squares (IGLS). This approach separates estimation of the fixed parameters from the random parameters at different steps in sequence repeatedly as follows until subsequent estimates of the parameters changes are very small. First, an estimate of the fixed (non-random) coefficients can be obtained using generalized least squares as: $\hat{\beta} = \left(X^T V^{-1} X\right)^{-1} X^T V^{-1} Y$ with a covariance matrix $\left(X^T V^{-1} X\right)^{-1}$ in which V is a function of the random parts at the two levels in the model. Then, the estimates of random parameters (θ) can be calculated by using a generalized least squares as follows: $\hat{\theta} = \left(Z^{*T} V^{*-1} Z^*\right)^{-1} Z^{*T} V^{*-1} Y^{**}$, $V^* = V \otimes V$, where $\otimes$ is the Kronecker product. The covariance matrix of $\hat{\theta}$ is given by $\left(Z^{*T} V^{*-1} Z^*\right)^{-1} Z^{*T} V^{*-1} \operatorname{cov}\left(Y^{**}\right) V^{*-1} Z^* \left(Z^{*T} V^{*-1} Z^*\right)^{-1}$. Z^* is the design matrix of the random part in the model and Y^{**} is a vector stacking of the residuals. These procedures iterate using the current estimates of fixed and random parameters until a very small change is observed in the estimates in subsequent steps. However, IGLS procedures produce biased estimates in general, especially in the small sample. Goldstein (1995) has improved the IGLS algorithm by taking account of the sampling variation of the $\hat{\beta}$ leading to restricted iterative generalized least squares (RIGLS).

The models here are estimated using a first order marginal quasi-likelihood (MQL) model with extra binomial distributional assumption. Goldstein (1995) provides a discussion of some additional detail of the log-odds model in Equation (1) and a variety of estimation methods that have been implemented for the model using a linearization technique.

MODEL RESULT

The first model presented here contains no explanatory variables (Table 3) which is in essence an error components model. It is called the null or fully unconditional model and it is used as a benchmark to assess other model specifications that include explanatory variables and regression coefficients (fixed and/or random) at each level. The estimation method used is Goldstein's RIGLS. Table 3 shows the proportion of variance across persons within a household (person level) and across households (household level) in terms of the probability of ICT usage. There is a greater variance in ICT usage across households (50.1% to 80.0%) than between persons within a household (20.0% to 49.9%), especially in the home-based technologies and cellular phone usage. It is because decisions of ICT ownership and usage are most likely to be joint decisions among members of the same household. Key finding here indicates that the household level variance in ICT usage is too large to be neglected and should be accounted for in a model. The multilevel specification is therefore justified. It also indicates that it is necessary to specify models using explanatory variables capturing and/or depicting factors affecting a person's and household's technology choice.

The bottom half of Table 3 contains the estimated covariances (below the diagonal) and the estimated correlation coefficients (above the diagonal) within each of the levels for the combination of the seven dependent variables. In terms of correlations among ICT usage, there is a strong, positive correlation between computer usage and Internet usage at both the household and person levels, regardless of the location of the technologies. In general, the correlations among ICT usage are higher across households than between persons within a household, except for laptop.

Table 3 Multivariate Multilevel Technology Choice Error Components Models

Model Component	CW	NW	CH	NH	CL	PG	LT
Fixed Effect	Coef. [S.E.]	Coef. [S.E.]	Coef. [S.E.]	Coef. [S.E.]	Coef. [S.E.]	Coef. [S.E.]	Coef. [S.E.]
Intercept	-0.079 [0.056]	-0.557 [0.060]	0.633 [0.063]	0.383 [0.062]	-0.162 [0.060]	-2.363 [0.096]	-2.881 [0.119]
Random Effects	σ^2 [%]	σ^2 [%]	σ^2 [%]	σ^2 [%]	σ^2 [%]	σ^2 [%]	σ^2 [%]
Person variation within households (u_{ij})	0.794 [48.8]	0.750 [40.5]	0.616 [26.9]	0.509 [20.0]	0.626 [29.2]	0.938 [49.9]	0.958 [49.8]
Between households variation (v_j)	0.833 [51.2]	1.101 [59.5]	1.675 [73.1]	2.038 [80.0]	1.517 [70.8]	0.943 [50.1]	0.965 [50.2]
Total	1.627 [100]	1.851 [100]	2.291 [100]	2.547 [100]	2.143 [100]	1.881 [100]	1.923 [100]
-2LogL				7236.76			

Variance / Covariance Matrices (upper triangle correlations)

		CW	NW	CH	NH	CL	PG	LT
Between Persons	CW	0.794	0.709	0.200	0.123	0.122	0.176	0.246
	NW	0.547	0.750	0.213	0.153	0.137	0.218	0.307
	CH	0.140	0.145	0.616	0.656	0.156	0.079	0.127
	NH	0.078	0.095	0.367	0.509	0.120	0.001	0.108
	CL	0.086	0.094	0.097	0.068	0.626	0.107	0.155
	PG	0.152	0.183	0.060	0.001	0.082	0.9381	0.138
	LT	0.214	0.260	0.098	0.075	0.120	0.131	0.958
Between Households	CW	0.833	0.931	0.570	0.633	0.201	0.506	0.020
	NW	0.892	1.101	0.474	0.495	0.176	0.259	0.029
	CH	0.673	0.644	1.675	0.924	0.338	0.233	0.344
	NH	0.825	0.742	1.708	2.038	0.363	0.251	0.288
	CL	0.226	0.227	0.539	0.638	1.517	0.208	0.168
	PG	0.449	0.264	0.293	0.349	0.249	0.943	-0.373
	LT	0.018	0.030	0.438	0.404	0.203	-0.356	0.965

Tables 4 and 5 give the estimates along with t-statistics for the simultaneous equation model of all seven dependent variables as a function of household-level and person-level cross-sectional and longitudinal variables

Cross-sectional Effects

As expected, working and attending a school are major factors to use computers and the Internet at work/school. Persons employed as managers have a higher propensity to use computers and the Internet at work/school than their counterparts employed in other occupations. The higher household income is accompanied by the higher probability of using both technologies at work and at home. As expected, old people are less likely to use computer and the Internet at both places. A person with a driver's license is more likely to use most of ICT, except for pager and laptop. Respondents with more children tend to make more use of computers and the Internet at home. People living in Pierce county are less likely to use computers and the Internet at home than those living in the other counties

In terms of personal mobile technology usage, cellular phones seem to be more popular among people with household income higher than $75,000 and multi-cars in the household. People in secretarial occupations and males are less likely to use cellular phones. Managers are more likely to use pagers while professionals are less likely to use them. Males and students also tend to make greater use of pagers. Persons with household income more than $75,000, men, students, or managers are more likely to use laptop computers.

A likelihood ratio comparison between the error components model and the final model yields a difference in likelihood of 5448.49 with 98 degrees of freedom, which indicates that the explanatory variables give a statistically significant improvement over the error components model.

Table 4 Multivariate Multilevel Technology Choice Models (Fixed Cross-sectional Effects)

	CW	NW	CH	NH	CL	PG	LT
Variables	Coef. [t-stat]	Coef. [t-stat]	Coef. [t-stat]	Coef. [t-stat]	Coef. [t-stat]	Coef. [t-stat]	Coef. [t-stat]
constant	-3.646 [-5.956]	-5.608 [-8.034]	-2.060 [-4.852]	-0.662 [-1.932]	-1.900 [-4.989]	-5.682 [-8.537]	-5.884 [-12.980]
transit	0.589 [3.281]				-0.526 [-3.145]		
Pierce			-0.346 [-2.140]	-0.349 [-2.156]			
totadult			0.203 [2.415]				
tot6-17			0.389 [3.495]	0.229 [2.237]			
tot1-5			0.604 [2.856]	0.940 [4.294]			
midinc	0.369 [2.062]	0.459 [2.341]				0.510 [1.687]	
highinc	0.877 [4.203]	1.144 [5.358]	0.456 [2.592]	0.574 [3.375]	1.033 [6.613]	0.532 [1.695]	0.818 [3.128]
mhinc					0.395 [1.772]		
dkinc			-0.551 [-2.177]	-0.593 [-2.272]			
car1				-0.435 [-2.700]			
car2			0.574 [3.388]		0.503 [2.891]		
car3+_			0.492 [2.593]		0.809 [4.373]		
male		0.352 [2.971]	0.063 [0.762]		-0.434 [-4.381]	0.613 [2.806]	0.921 [3.535]
midage	-0.641 [-2.558]					0.693 [1.490]	
old	-1.367 [-4.180]	-1.090 [-3.632]	-0.791 [-4.817]	-0.850 [-5.730]		0.975 [1.580]	
wk5	0.742 [3.952]	0.465 [2.686]				0.526 [1.802]	
prof	1.225 [6.417]	1.343 [7.858]				-0.501 [-3.404]	
manag	1.610 [4.971]	1.707 [6.785]				0.757 [2.840]	0.997 [3.337]
secre	1.237 [4.228]	0.859 [3.540]			-0.738 [-3.249]	-0.763 [-1.633]	
dlicen	1.820 [3.462]	1.811 [2.868]	1.719 [4.678]	1.151 [3.506]	0.964 [2.609]		
dbpass	0.815 [3.232]	1.035 [4.894]	0.283 [2.360]			-0.455 [-1.462]	
dpupil	1.108 [2.338]	1.941 [4.440]				2.221 [3.423]	1.664 [2.316]
-2LogL				1788.27			
Deviance from Error Component Model (Table 3) = 5448.49 with d.f.=98							

Longitudinal Effects

Table 5 shows the effect of social and demographic changes at the person and household levels on the probability of each ICT use. An increase in the number of children (6-17 years old) shows a lower likelihood of computer and Internet use at home, while a decrease in the number of children (6-17 years old) shows a higher likelihood of Internet use at home. One explanation for this could be the roles younger individuals play in the household as a computer and Internet expert. A decrease in the number of employed persons in the household is also more likely to increase the probability of computer and Internet use at home. Interestingly, a decrease in the number of driver's license holders is accompanied by a lower likelihood of pager usage. A decrease in the number of vehicles and an increase in number of employed persons in the household are more likely to increase the probability of laptop usage. In general, the employed people are more likely to use most of ICT with a higher propensity for those who are employed in both years 1997 and 2000, than for those who are employed in 2000 only.

Table 5 Multivariate Multilevel Technology Choice Models (Fixed Longitudinal Effects)

Variables	CW Coef. [t-stat]	NW Coef. [t-stat]	CH Coef. [t-stat]	NH Coef. [t-stat]	CL Coef. [t-stat]	PG Coef. [t-stat]	LT Coef. [t-stat]
inkid			-1.434 [-2.852]	-1.210 [-2.428]	0.417 [1.753]		
dnkid				0.360 [2.186]		0.361 [1.306]	
dnbaby			1.111 [2.123]	1.068 [2.092]			
dnveh	-0.292 [-1.925]			-0.184 [-1.621]	0.307 [1.953]		0.796 [2.815]
inemp							0.622 [2.169]
dnemp			0.541 [2.679]	0.524 [2.671]			0.664 [1.866]
dnlicen						-0.948 [-2.483]	
inbpass	-0.455 [-2.040]						0.455 [1.245]
dnbpass							-2.413 [-2.153]
expemp	1.955 [7.858]	2.144 [6.878]	0.245 [2.142]		0.628 [4.354]	1.969 [3.964]	1.989 [4.875]
novemp	0.751 [2.612]	1.560 [4.551]	0.462 [1.854]	0.618 [2.744]	0.656 [2.902]	1.768 [3.235]	
quitemp		0.712 [1.941]					

Table 6 shows the variances for the multivariate multilevel model for ICT use and ownership when explanatory variables are included in the model. The table also shows the covariances (below the diagonal) and correlations (above the diagonal). From Table 6, it can be seen that there is again a greater variance in ICT usage across households as compared to between persons within household.

Table 6 Multivariate Multilevel Technology Choice Models (Random Effects)

Variance / Covariance Matrices (upper triangle correlations) [Std. Error]								
		CW	NW	CH	NH	CL	PG	LT
Between Persons	CW	0.905 [0.043]	0.484	0.147	0.087	0.119	0.111	0.071
	NW	0.387 [0.030]	0.705 [0.034]	0.180	0.158	0.150	0.018	0.100
	CH	0.120 [0.029]	0.129 [0.263]	0.728 [0.039]	0.644	0.128	0.068	0.126
	NH	0.065 [0.028]	0.103 [0.024]	0.427 [0.030]	0.605 [0.033]	0.100	-0.004	0.079
	CL	0.093 [0.029]	0.104 [0.025]	0.089 [0.027]	0.064 [0.025]	0.671 [0.037]	0.098	0.097
	PG	0.109 [0.032]	0.016 [0.029]	0.060 [0.031]	-0.003 [0.029]	0.082 [0.030]	1.068 [0.482]	0.031
	LT	0.063 [0.029]	0.079 [0.026]	0.102 [0.028]	0.058 [0.027]	0.075 [0.027]	0.030 [0.031]	0.896 [0.039]
Between Households	CW	0.351 [0.248]	1.180	0.371	0.624	-0.118	-0.208	0.322
	NW	0.803 [0.199]	1.315 [0.237]	0.113	0.112	-0.071	0.947	0.297
	CH	0.271 [0.178]	0.160 [0.176]	1.526 [0.217]	0.902	0.227	-0.279	0.027
	NH	0.531 [0.173]	0.184 [0.171]	1.599 [0.188]	2.059 [0.210]	0.278	-0.112	0.019
	CL	-0.087 [0.160]	-0.101 [0.157]	0.348 [0.146]	0.495 [0.143]	1.539 [0.187]	0.046	0.067
	PG	-0.040 [0.239]	0.355 [0.227]	-0.113 [0.241]	-0.052 [0.232]	0.019 [0.214]	0.107 [0.401]	-0.061
	LT	0.202 [0.295]	0.360 [0.281]	0.035 [0.298]	0.029 [0.285]	0.088 [0.261]	-0.021 [0.357]	1.120 [0.547]

SUMMARY AND CONCLUSIONS

In this paper, data from two waves of the Puget Sound Transportation Panel (PSTP) are analyzed to study trends in ICT adoption by persons and their households during the periods of 1997-2000. The fastest growing ICT are the home-based that have a practical ceiling equal to the number of computer users in the household. Computers at work appear to be stabilizing at 50 percent of the sample. Cellular telephony use is also on the rise in this sample during the period 1997 to 2000, causing a potential negative impact on pagers. However, laptop computers and personal digital assistants (PDA) are still used by a small sample segment. In addition, using a multivariate multilevel model specification, technology choice models are developed. When compared to more traditional regression models, the multivariate multilevel model provides additional insights on components of variance and heterogeneous behavior of persons and their households.

The results of the analysis show that there is more variability between households when compared to the variability contributed by persons within a household, which implies that decisions of ICT ownership and usage are most likely to be joint decisions among members of the same household. It is found that income level, employment, and age are important factors in ICT choice and usage. For example, the employed and persons with household income more than $75,000 are more likely to use most of ICT types. Older persons tend to use computers and internet less than their younger counterparts. In addition, the length of employment seems to have positive effects on ICT usage.

REFERENCE

Goldstein, H. (1995). *Multilevel Statistical Models*, Edward Arnold, New York.

Golob, T.F. (2001). Travelbehaviour.com: Activity Approaches to Modeling the Effects of Information Technology on Personal Travel Behaviour. In: *Travel Behaviour Research, The Leading Edge* (D.A. Hensher, ed), pp. 145-184. Pergamon, Kidlington.

Golob, T.F. and A.M. Regan (2001). Impacts of Information Technology on Personal Travel and Commercial Vehicle Operations: Research Challenges and Opportunities. *Transportation Research Part C,* **10**, 87-121.

Goulias K.G., N. Kilgren, and T. Kim (2003). A decade of longitudinal travel behavior observation in the Puget Sound region: sample composition, summary statistics, and a selection of first order findings. Presented at the 10th International Conference on Travel Behaviour Research, Moving through nets: The physical and social dimensions of travel, Lucerne, 10-14 August 2003.

Murakami, E. and W.T. Watterson (1990) Developing a household travel panel survey for the Puget Sound Region. *Transportation Research Record*, **1285**, 45-46.

APPENDIX: INDEPENDENT VARIABLES USED IN MULTIVARIATE MULTILEVEL TECHNOLOGY CHOICE MODELS

Household level	SOV	Indicator, 1= household is sampled from SOV class; 0=otherwise
	CARPOOL	Indicator, 1= household is sampled from carpool class; 0=otherwise
	TRANSIT	Indicator, 1= household is sampled from public transit class; 0=otherwise
	KING	Indicator, 1= living in King County; 0=otherwise
	KITSAP	Indicator, 1= living in Kitsap County; 0=otherwise
	PIERCE	Indicator, 1= living in Pierce County; 0=otherwise
	SNOHO	Indicator, 1= living in Snohomish County; 0=otherwise
	TOT1-5	Number of children in the household who are less than 6 years old
	TOT6-17	Number of children in the household who are between 6 and 17 years old
	TOTADULT	Number of adults in the household who are 18 years old or older
	LOWINC	Indicator, 1= household income < $35,000; 0=otherwise
	MIDINC	Indicator, 1= $35,000 ≤ household income < $75,000; 0=otherwise
	HIGHINC	Indicator, 1= $75,000 ≤ household income; 0=otherwise
	MHINC	Indicator, 1= $35,000 ≤ household income; 0=otherwise
	DKINC	Indicator, 1=household income is unknown; 0=otherwise
	CAR0	Indicator, 1= no car household; 0=otherwise
	CAR1	Indicator, 1= one car household; 0=otherwise
	CAR2	Indicator, 1= two car household; 0=otherwise
	CAR3+	Indicator, 1= three or more car household; 0=otherwise
	INBABY	Indicator, 1=an increase in the number of children < 6 years in the household between waves; 0=otherwise
	DNBABY	Indicator, 1= a decrease in the number of children < 6 years in the household between waves; 0=otherwise
	ENBABY	Indicator, 1= no change in the number of children < 6 years in the household between waves; 0=otherwise
	INKID	Indicator, 1= an increase in the number of kids whose age is 6-17 in the household between waves; 0=otherwise
	DNKID	Indicator, 1= a decrease in the number of kids whose age is 6-17 in the household between waves; 0=otherwise
	ENKID	Indicator, 1= no change in the number of kids whose age is 6-17 in the household between waves; 0=otherwise
	INADULT	Indicator, 1=an increase in the number of adults in the household between waves; 0=otherwise
	DNADULT	Indicator, 1=a decrease in the number of adults in the household between waves; 0=otherwise
	ENADULT	Indicator, 1= no change in the number of adults in the household between waves; 0=otherwise
	INLICEN	Indicator, 1=an increase in the number of drivers license holders between waves; 0=otherwise
	DNLICEN	Indicator, 1=a decrease in the number of drivers license holders between waves; 0=otherwise
	ENLICEN	Indicator, 1= no change in the number of drivers license holders between waves; 0=otherwise
	INBPASS	Indicator, 1=an increase in bus pass holders in the household between waves; 0=otherwise
	DNBPASS	Indicator, 1=a decrease in bus pass holders in the household between waves; 0=otherwise
	ENBPASS	Indicator, 1= no change in bus pass holders in the household between waves; 0=otherwise
	INVEH	Indicator, 1=an increase in the number of cars in the household between waves; 0=otherwise
	DNVEH	Indicator, 1=a decrease in the number of cars in the household between waves; 0=otherwise
	ENVEH	Indicator, 1= no change in the number of cars in the household between waves; 0=otherwise
	INEMP	Indicator, 1=an increase in the number of employed persons in household between waves; 0=otherwise
	DNEMP	Indicator, 1=a decrease in the number of employed persons in household between waves; 0=otherwise
	ENEMP	Indicator, 1=no change in the number of employed persons in household between waves; 0=otherwise
Person level	MALE	Indicator, 1=male; 0=female
	YOUNG	Indicator, 1=18≤ age ≤ 34; 0=otherwise
	MIDAGE	Indicator, 1=35≤ age ≤ 64; 0=otherwise
	OLD	Indicator, 1=65≤ age; 0=otherwise
	PROF	Indicator, 1=having professional occupation; 0=otherwise
	MANAG	Indicator, 1=having managerial occupation; 0=otherwise
	SECRE	Indicator, 1=having secretarial occupation; 0=otherwise
	SALES	Indicator, 1=having sales occupation; 0=otherwise
	UNEMP	Indicator, 1=unemployed; 0=otherwise
	WK5	Indicator, 1=working outside of home for 5+ times a week; 0=otherwise
	DPUPIL	Indicator, 1= student; 0=otherwise
	DLICEN	Indicator, 1= having driver's license; 0=otherwise
	DBPASS	Indicator, 1= having bus pass; 0=otherwise
	EXPEMP	Indicator, 1= employed outside home in both waves; 0=otherwise
	NOVEMP	Indicator, 1= started getting employed outside home after wave 7; 0=otherwise
	QUITEMP	Indicator, 1= employed outside home in wave 7 but not in wave 9; 0=otherwise
	NOTEMP	Indicator, 1= unemployed in both waves; 0=otherwise

Transport Science and Technology
K.G. Goulias, editor

CHAPTER 19

A DYNAMIC ANALYSIS OF DAILY TIME USES, MODE CHOICE, AND INFORMATION AND COMMUNICATION TECHNOLOGY

Tae-Gyu Kim, North Carolina Department of Transportation, Raleigh, North Carolina, USA

Konstadinos G. Goulias, University of California, Santa Barbara, California, USA

INTRODUCTION

As activity-based approaches to travel demand modeling are recently emerging as a viable alternative to the more traditional four-step travel model, a need arises to examine the patterns used by people in allocating time to activities and travel. In addition, repeated observations of the same people over time are also emerging as powerful tools to study changes in activity and travel behavior. The advent of increased use of Information and Communication Technology (ICT) and its relationship with time allocation and travel is one phenomenon that can be understood better by adopting an activity based framework and studying changes over time. To accomplish this in this paper, we developed a structural equation model implementing a broader perspective that includes time use indicators and modal split indicators in a system of equations viewed as dependent variables jointly. The complex relationships among social and demographic change, ICT ownership change, daily time allocation, and mode choice are then studied using traditional hypotheses testing of regression coefficients and an examination of the magnitude of effects an indicator has on another indicator considering both the direct and indirect effects of one factor on another. The study here creates a system of simultaneous equations of activity and travel demand and builds on more recent structural equation modeling (SEM) applications in travel behavior that are used as guidance (Golob, 1998; Chung and Ahn, 2002).

Figure 1 shows the conceptual framework of the model. Using the Puget Sound Transportation Panel (PSTP) data, especially the data collected in Wave 7 (1997) and Wave 9

(2000), this study attempts to examine a variety of relationships between ICT and activity and travel behavior within comprehensive conceptual model systems that examine correlation patterns from both cross-sectional and longitudinal viewpoints. More specifically, the correlation among the amount of time allocated to subsistence, maintenance, leisure, and travel, as well as the total number of trips by the most important modes (drive alone, car sharing, public transportation, walking, biking, and all other modes used) can be estimated. The model system is designed to extend and parallel other past studies, and for this reason, variables at the household and person level are used. In addition, some specific variables to the database in PSTP are also employed to account for other factors such as stratification sampling and potential self-selection bias.

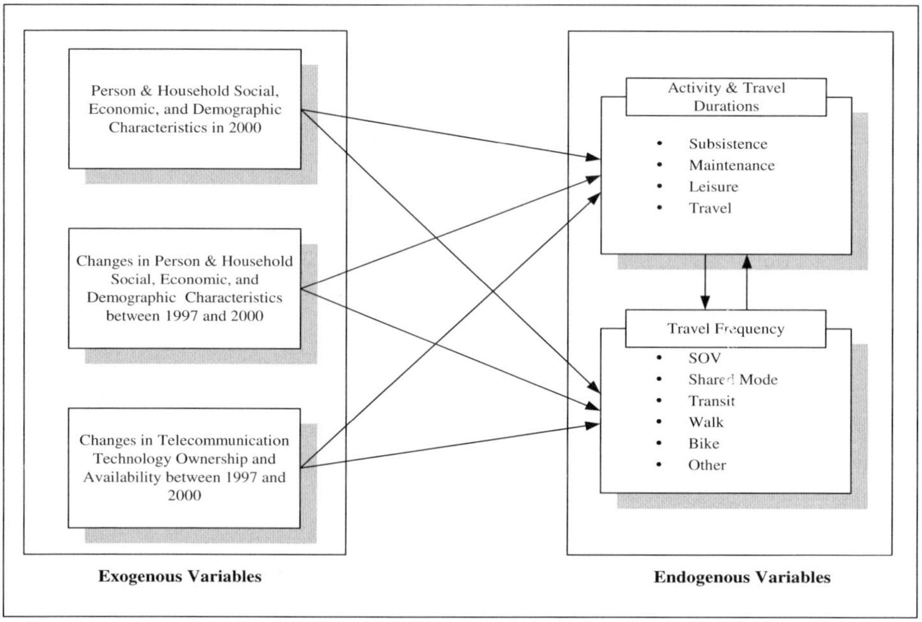

Figure 1 Model System Overview

A key aspect of this model system that sets it aside from other studies is the inclusion of "experience" variables. Using information between 1997 and 2000, a large set of variables is defined to capture the changes in social and economic circumstances experienced by each respondent and include them as exogenous variables of the behavior in the year 2000. The second element that characterizes the research work here is the study of ICT on activity and travel behavior considered jointly with all the other determinants of activity and travel behavior. Within this framework, we are addressing questions such as: What are some of the factors that influence activity and travel behavior? What is the relationship among different activities and travel in terms of time use and mode choice when we control for many other exogenous factors? What is the role played by ICTs and what changes do we observe?

The remainder of this paper is organized as follows. In the next section, a brief description of the data used is provided, followed by model formulation and estimation results. The paper ends with summary and conclusions.

DATA DESCRIPTION

The database used in this study is from the Puget Sound Transportation Panel (PSTP), which is the first general-purpose urban household panel survey for travel analysis in the four-county region including Seattle in the United States. In PSTP a household questionnaire and a two-day travel diary are administered repeatedly on the same households and their members (15 and older) over time. Each survey occasion is called a wave. PSTP started in the fall of 1989, and it continues until today with sample size of approximately 1700 households per wave. To date, there have been ten waves in PSTP. Each survey includes three groups of data that are household demographics, persons' social and economic information, and reported travel behavior. Trip information was collected using a travel diary. In the travel diary each driving age person reports every trip made during two consecutive weekdays, which remained approximately the same throughout the panel years. Each trip contains information about trip purpose, type, mode, starting and ending time, origin and destination, and distance. Based on the reported trips, activity engagement information is derived from the trip purposes. The durations of activities are computed by the difference between the start time of the next trip and the end time of the current trip giving the sojourn time at an activity location. Total activity duration includes the amount of time each person spends in activities, including in–home activities (between the first and last out–of–home activity), during a day but neither before the first trip in the morning nor after the final return to home in the evening. Total travel time is the total amount of time spent by a person traveling during the day.

Since 1997 (wave 7), PSTP has included additional questions on respondents' traveler information system use and computer and telecommunications ownership. This is to gain some insight on how people use traveler information to make their transportation decisions. Additional details about the panel can be found in Murakami and Ulberg, 1997 and Goulias, Kilgren, and Kim, 2003.

In the original trip data of PSTP, the mode chosen for each trip has been classified into 17 different types: car, carpool, vanpool, bus, para-transit, taxi, walking, bicycle, motorcycle, school bus, drive-on ferry, walk-on ferry, monorail, boat, train, airplane, and other. In this case, car indicates a single occupant vehicle trip mode by car/truck/sport utility vehicle (SUV), while carpool and vanpool implies an official carpool or vanpool as well as an informally shared trip mode by car/truck/suv or van. The public transportation trips are mostly by bus and taxi. To make this analysis tractable, the modes have been grouped into: single occupant vehicle (car), shared ride (carpool and vanpool), transit (bus, para-transit, and taxi), walk, bike, and others (all other categories). In addition, the trip purpose for each trip has been classified into 9 different types at Wave 7 (work, school, college, shopping, personal business, appointment, visiting, free-time, and home) and 21 different types at Wave 9. The use of these different schemes for trip purposes between the waves makes it difficult to compare the two waves. To make the data for the two waves comparable and this analysis tractable, activity types have been grouped into: Subsistence (work, school, and college), maintenance (shopping, personal business, appointment, and errand/picking-up/dropping-off), and leisure (free time, recreation/exercise, visiting, and home staying).

Table 1 Sample Characteristics

Characteristics			Wave 7 (1997)	Wave 9 (2000)
Number of persons in the sample			1480	1480
Number of households in the sample			866	866
Person & Household	Percent of males in the sample		46.9	46.9
	Number of employed persons in the sample		859	881
	Number of persons in household		2.5	2.5
	Number of cars per household		2.2	2.2
Activity & Travel	Total amount of time in subsistence activities (min.)	Day 1	275.0	235.3
		Day 2	265.2	224.9
	Total amount of time in maintenance activities (min.)	Day 1	43.3	46.4
		Day 2	39.4	46.9
	Total amount of time in leisure activities (min.)	Day 1	109.8	93.5
		Day 2	106.9	93.0
	Total amount of time traveling (min.)	Day 1	84.7	75.0
		Day 2	79.0	73.8
Travel Mode*	Total Number of trips per person	Day 1	4.57	4.08
		Day 2	4.30	3.98
	Number of trips driving alone per person (%*)	Day 1	2.52 (55.1)	2.31 (56.6)
		Day 2	2.37 (55.1)	2.25 (56.5)
	Number of trips by shared modes per person (%*)	Day 1	1.57 (34.4)	1.34 (32.8)
		Day 2	1.52 (35.3)	1.34 (33.7)
	Number of trips by transit per person (%*)	Day 1	0.19 (4.2)	0.17 (4.2)
		Day 2	0.15 (3.5)	0.15 (3.8)
	Number of trips by walking per person (%*)	Day 1	0.20 (4.4)	0.20 (4.9)
		Day 2	0.16 (3.7)	0.18 (4.5)
	Number of trips by biking per person (%*)	Day 1	0.02 (0.4)	0.02 (0.5)
		Day 2	0.02 (0.5)	0.02 (0.5)
	Number of trips by other modes per person (%*)	Day 1	0.06 (1.3)	0.05 (1.2)
		Day 2	0.07 (1.6)	0.05 (1.3)
Information & Communication Technology**	Number of persons who use computer regularly (%**)	At home	751 (50.7)	982 (66.4)
		At work	700 (47.3)	717 (48.4)
	Number of persons who use the Internet regularly (%**)	At home	458 (30.9)	896 (60.5)
		At work	438 (29.6)	544 (36.8)

	Number of persons having cellular phone (%**)	419 (28.3)	685 (46.3)
	Number of persons having pager (%**)	156 (10.5)	129 (8.7)
	Number of persons having laptop (%**)	71 (4.8)	79 (5.3)
	Number of persons having PDA (%**)	6 (0.4)	38 (2.6)

** % over total number of trips per person*

*** % over total number of person in the sample*

For the analysis in this study, we used the data from 1480 persons in 688 households who provided detailed information in both waves 7 (1997) and 9 (2000) for all the variables used in this analysis. Table 1 provides the number of persons and households as well as a few social and demographic characteristics of the sample. It also shows a summary for some key variables used in the analysis; the amount of time dedicated to various activities and traveling as well as travel frequency of each mode during the two interview days in wave 7 (1997) and wave 9 (2000). As expected, people show a similar pattern of activity and travel behavior between the two days and between the two waves, except for activity and travel durations between the waves. The rather large discrepancy in activity and travel durations is most likely a result of genuine change but also the use of different schemes for trip purposes in the travel survey between the waves. In the year 2000, the respondents spent an average of 230.1 minutes in subsistence activity, 46.7 minutes in maintenance activity, 93.3 minutes in leisure activity, and 74.4 minutes in travel per day. The respondents are heavily dependent on cars/truck/SUV for their travel mode. Traveling alone with car/truck/SUV is the most popular mode accounting for 56.6 percent of total trips, while bike is the least used mode accounting for only 0.5 percent. The bottom of Table 1 also contains information about technology ownership and use with focus on the more modern technologies.

MODEL FORMULATION

Variables Used

As shown in Table 2, two groups of activity-travel variables were used as the endogenous variables in the model. The first group of the endogenous variables is total amount of time dedicated to a specific activity (subsistence, maintenance, and leisure) and traveling in a day. All four are measured in minutes per day. The second group is the frequency of trips by a specific mode (SOV, shared mode, transit, walk, bike, other modes) a person makes in a day, and it is measured in number of trips per day.

Table 2 List of Endogenous Variables

Endogenous Variables		Descriptions
Activity and Travel Durations	Sdur	Total subsistence activity duration per day (min)
	Mdur	Total maintenance activity duration per day (min)
	Ldur	Total leisure activity duration per day (min)
	Ttime	Total travel time per day (min)
Travel	Ssov	Number of trips per day by driving alone

Frequency by Mode	Shared	Number of trips per day by shared ride
	Transt	Number of trips per day by transit
	Walk	Number of trips per day by walking
	Bike	Number of trips per day by bike
	Others	Number of trips per day by other modes

For exogenous variables, four groups of variables were used: household-level variables, person-level variables, time-related variables, and ICT variables. Appendix provides an inventory of all the exogenous variables used for this study. We used cross-sectional information in the year 2000 as well as longitudinal information between the years 1997 and 2000 about household-level and person-level social, economic, demographic characteristics and ICT ownership and availability.

To account for task allocation and roles within the household, number of adults, number of children by age group as well as vehicles owned and income representing resource availability were included as exogenous variables. In addition, in order to account for the sampling stratification of the panel participants, the county of residence and sample indicators (TRANSIT, CARPOOL) were also included as exogenous variables.

Indicators for changes in household and person characteristics, such as increase or decrease in car-ownership, changes in household composition, and changes in employment, were used to examine the effect of the changes on time use and mode choice.

Two types of time variables were also included. The first type is person-level time elapsed in panel since the first time participating and it can capture two effects: a) a genuine change in activity and travel behavior by a person during the time of her/his panel participation; and b) possible travel diary and/or panel fatigue in reporting trips. The second type is the set of day of week indicators to account for different activity and travel behaviors among weekdays (weekends are not targeted in PSTP) and a correlation between the first and second diary day.

To examine the effect of changes in information and telecommunication technology between the years 1997 and 2000, we defined indicator variables for four groups of persons for each ICT:

- Persons that started using these technologies some time after 1997 and they are using them in 2000 (*new users*);
- Persons that stopped using these technologies since 1997 (*past users*);
- Persons that never used them (*non users*); and
- Persons that started some time before 1997 and never stopped (*experienced users*).

Structural Equation Model (SEM)

The use of structural equation modeling techniques is employed in this study, due to their capability to estimate a set of simultaneous equations capturing the interrelationship among a large number of endogenous and exogenous variables. The general SEM with latent variables consists of two parts: 1) measurement model and 2) structural model. The measurement model specifies how latent variables are indicated by the observed variables, while the structural model specifies the causal relationships among the latent variables and describes the causal effects of the exogenous variables on the endogenous variables. However, since no latent variables are involved in the SEM for this study, the SEM with observed variables takes the following form:

$$y = \mathrm{B}y + \Gamma x + \zeta \tag{1}$$

where $y = p \times 1$ vector of observed endogenous variables.
$x = q \times 1$ vector of observed exogenous variables.
$\mathrm{B} = p \times p$ matrix of coefficients of the y-variables.
$\Gamma = p \times q$ matrix of coefficients of the x-variables.
$\zeta = p \times 1$ vector of equation errors.

SEM is a covariance–based model, because structural equations systems are estimated by covariance analysis. In the procedure, the difference between the sample covariances and the covariances predicted by the model is minimized, instead of minimizing the difference between observed and predicted individual values. The underlying theory of this estimation procedure is that the population covariance matrix of the observed variables (Σ) is a function of a set of parameters:

$$\Sigma = \Sigma(\theta) = \begin{bmatrix} \text{covariance matrix of y} & \text{covariance matrix of y and x} \\ \text{covariance matrix of x and y} & \text{covariance matrix of x} \end{bmatrix}$$

$$= \begin{bmatrix} (I-\mathrm{B})^{-1}(\Gamma\Phi\Gamma' + \Psi)[(I-\mathrm{B})^{-1}]' & (I-\mathrm{B})^{-1}\Gamma\Phi \\ \Phi\Gamma'[(I-\mathrm{B})^{-1}]' & \Phi \end{bmatrix} \tag{2}$$

where Φ = covariance matrix of x.
Ψ = covariance matrix of ζ.

The matrix $\Sigma(\theta)$ basically consists of three covariance matrices. The unknown parameters $\mathrm{B}, \Gamma, \Phi, \text{and } \Psi$ are simultaneously estimated by finding the parameters such that the covariance matrix ($\hat{\Sigma}$) implied by the model is as close as possible to the sample covariance matrix (S). To know when the estimates are as close as possible, a fitting function that is to be minimized is defined.

There are several estimation methods for SEM The choice of the estimation method depends mainly on the assumption of the probability distribution, the type of variables, and sample size. ML estimation method assuming a multivariate normal distribution was employed for this study, becasue it is kmown that ML estimation is fairly robust to deviation of multivariate normality and sample size commonly used in transportation research (Golob, 2003). The ML fitting function that is minimized is:

$$F_{ML} = \log|\Sigma(\theta)| + tr(S\Sigma^{-1}(\theta)) - \log|S| - (p+q) \tag{3}$$

In SEM, there are three types of effects of one variable on another: direct, indirect, and total effects. The direct effects, which are estimated as B and Γ, are the influence of one variable on another that are not mediated by any other variable, while the indirect effects are ones mediated by at least one intervening variable. The total effects are the sum of the direct and indirect effects. It should be noted that interpreting a model with the direct effects only provides misleading conclusions when the direct and the total effects are very different. It is the total effects that should be used in interpretation. As implied in equation (2), the decomposition of effects for SEM with obseved variables is reported in Table 3.

Table 3 Direct, Indirect, and Total Effects

Decomposition of Effects		Effects on Y
	Total Effect	$(I-\mathrm{B})^{-1}\Gamma$
Effect of x	Direct Effect	Γ
	Indirect Effect	$(I-\mathrm{B})^{-1}\Gamma-\Gamma$
	Total Effect	$(I-\mathrm{B})^{-1}-I$
Effect of y	Direct Effect	B
	Indirect Effect	$(I-\mathrm{B})^{-1}-I-\mathrm{B}$

MODEL RESULTS

In this section, the discussion of model results focuses on the complex relationships among the activity and travel indicators and the total effects of the ICT variables on the activity and travel indicators. A complete inventory of all the parameter estimates for this model is available in Kim and Goulias, 2004.

Relationships among Endogenous Variables

Table 4 provides an overview of the complex relationships found among the endogenous variables: the amount of time allocated to various activities in each day by each person and the number of trips made by each mode. The sum of the number of trips by mode is also the frequency of activity episodes that involve the change of an activity location. In the section with label "goodness-of-fit indices" we see that every indicator shows a model with excellent fit to the data.

Travel time does not influence any other variable and therefore implicitly treated as the outcome of all the other indicators. One way of looking at the relationships among daily subsistence duration (sdur), daily maintenance duration (mdur), daily leisure duration (ldur), and daily travel time (ttime) is to consider the possible effect of a minute more in one type of activity on another. Subsistence duration's effect is very clear: persons that work longer are more likely to spent less time in leisure (-0.129) and maintenance (-0.062). Golob (1998) claims this as evidence of time budgeting by individuals and their households. In another analysis, however, Goulias, Kilgren, and Kim (2003) find extreme variation in "budgeting" and the existence of multiple groups with very different budgets. Note also the circular effect through the indirect effects. The direct effect of the subsistence on leisure, however, is the highest. As expected, persons that spend more time working/studying are also more likely to travel longer (0.032). Time expenditure on maintenance (e.g., shopping) has also a negative total effect on leisure and subsistence, and it is considerably larger than the two effects of subsistence before. The total effect of leisure on maintenance is positive and small, while the total effect of maintenance on leisure is negative (inhibiting) and large. None of the effects is larger than one (no substitution one for one minute) and the largest appears to be the "trade-off" between maintenance and leisure (1 minute of maintenance leads to a third less leisure).

Moving to the effects of activity daily durations on the number of trips by mode a striking finding is the overall lack of influence time allocation has on the modal frequencies. As expected it is more likely that the number of episodes determine activity duration and not the other way around. In addition, persons with longer durations are more likely to make more trips driving alone and with others in the same private car than to take the bus. Leisure duration appears to have a positive effect on bicycle trips but very small, while the use of most "others" modes are inhibited by longer durations of any activity type.

When we consider the number of trips effect on time allocation we see a few systematic relationships. The frequency of traveling by transit and other modes are accompanied by a positive and fairly large effect on the amount of time allocated to activities on maintenance (34.7 minutes per day for transit) but also leisure (18.8 minutes) and as expected travel time (17.9 minutes). Walking has a similar effect on maintenance and exactly the opposite on leisure duration. Using a car either alone or with others is accompanied by longer daily allocations to maintenance, leisure, and travel but not subsistence. The difference between direct and total effect in car sharing lead us to believe that using a simultaneous equation technique is a worthwhile exercise leading to better understanding of the effect one variable has on another. Overall, the effect of mode frequencies on duration of activities appears to be much stronger than the effect of activity duration on the trips made.

Effects among the use of travel modes generally seem to be quite different in magnitude as well as in signs by direction. These different effects by direction may be due to the fact that each mode has unique characteristics, thus each pair of modes cannot be a complete alternative to each other. For example, SOV has positive and small total effects (0.121) on transit, while transit on SOV has negative and large effects (-0.691). So the net effects of transit on SOV (-0.570) indicate that transit could be a good substitute to SOV, but not true vice versa. In the case between shared modes and transit, shared modes has small, negative total effects (-0.039) on transit, while transit has relatively large, positive effects (0.381) on shared modes, resulting in the net effect of (0.342). It indicates that transit has enhancement effects on share mode. However, it should be noted that caution is required in interpretation, because trip frequency by each mode in this study was treated as continuous variables rather than counting variables, which may influence the magnitude of the coefficient.

Table 4 Total, Direct, and Indirect Effects among Endogenous Variables

Causal Endogenous Variables		Resulting Endogenous Variables									
		sdur	mdur	ldur	ttime	ssov	shared	transt	walk	bike	Others
	total	-0.019	-0.062	-0.129	0.032	0.002	0.000	-0.000	0.000	0.000	-0.000
sdur	direct	0.000	-0.094	-0.225	0.020	0.002	0.000	-0.000	0.000	0.000	-0.000
	indirect	-0.019	0.033	0.096	0.012	0.000	0.000	0.000	0.000	0.000	0.000
	total	-0.041	-0.181	-0.311	0.024	0.000	0.004	-0.002	-0.004	0.000	-0.000
mdur	direct	0.000	0.000	-0.383	0.058	0.000	0.006	-0.003	-0.003	0.000	-0.000
	indirect	-0.041	-0.181	0.072	-0.034	0.000	-0.002	0.001	-0.001	0.000	0.000
	total	-0.003	0.003	-0.120	-0.012	0.000	0.001	-0.001	0.000	+0.000	-0.001
ldur	direct	0.000	0.000	0.000	0.015	0.000	0.001*	-0.001	0.001	+0.000	-0.001
	indirect	-0.003	0.003	-0.120	-0.027	0.000	0.000	0.000	-0.001	0.000	0.000
	total	0.000	0.000	0.000	0.000	0.000	0.000	0.000	0.000	0.000	0.000
ttime	direct	0.000	0.000	0.000	0.000	0.000	0.000	0.000	0.000	0.000	0.000
	indirect	0.000	0.000	0.000	0.000	0.000	0.000	0.000	0.000	0.000	0.000
	total	-10.491	13.452	30.448	11.654	-0.069	0.121	0.026	-0.002	-0.007	-0.002
ssov	direct	-10.074	11.813	30.497	9.456	0.000	0.000	0.094	0.000	-0.019	0.021
	indirect	-0.417	1.639	-0.049	2.198	-0.069	0.121	-0.068	-0.002	0.013	-0.023
	total	-5.644	3.934	17.626	9.538	-0.206	0.037	-0.039	-0.023	-0.004	0.003
shared	direct	-7.330	7.917	24.923	11.543	-0.218	0.000	0.000	0.000	-0.017	0.020
	indirect	1.686	-3.984	-7.298	-2.005	0.012	0.037	-0.039	-0.023	0.013	-0.017
	total	4.059	34.685	18.765	17.877	-0.691	0.381	-0.243	0.365	-0.003	-0.012
transt	direct	0.000	31.913	60.697	21.346	-0.815	0.216	0.000	0.596	-0.021	0.000
	indirect	4.059	2.772	-41.932	-3.469	0.123	0.165	-0.243	-0.231	0.018	-0.012
	total	-1.918	31.807	-10.625	6.346	0.124	0.121	-0.189	-0.210	0.006	0.048
walk	direct	0.000	44.333	0.000	5.453	0.000	0.000	-0.153	0.000	0.000	0.067
	indirect	-1.918	-12.525	-10.625	0.893	0.124	0.121	-0.036	-0.210	0.006	-0.020
	total	29.075	22.214	-121.959	-7.157	0.072	-0.041	-0.005	-0.149	-0.132	-0.293
bike	direct	33.985	35.926	-79.274	5.494**	0.000	0.000	0.000	0.000	0.000	-0.407
	indirect	-4.910	-13.711	-42.685	-12.651	0.072	-0.041	-0.005	-0.149	-0.132	0.115
	total	7.519	12.001	89.963	38.195	-0.051	0.193	0.029	0.044	0.248	-0.167
others	direct	0.000	0.000	133.056	38.460	0.000	0.000	0.176	0.000	0.249	0.000
	indirect	7.519	12.001	-43.094	-0.265	-0.051	0.193	-0.147	0.044	-0.002	-0.167
Goodness-of-fit Indices		Chi-Square =305.287, d.f.=471, P-value=1.000 Chi-Square/d.f =0.650 NFI=0.998 TLI =1.007 CFI =1.000 RMSEA =0.000, 90 Percent C.I. of RMSEA=(0.000 0.000), Probability(RMSEA <=0.05) =1.000 CN=5049									

Note: A direct effect value of 0.000 for a variable indicates that the variable was constrained to 0 in the model, because of its insignificance at 90% level. However, the values of –0.000 or +0.000 indicates that the effects are less than 0.0005 in magnitude, but significant at 95 % level.

** Significant at 90% level; all others are significant at 95% level, except for***

*** insignificant at 90 % level*

ICT Effects on Time Use and Mode Choice

One of the advantages in using panel data is our ability to measure change in a variable and concomitant effect on another. From a travel behavior viewpoint it is also interesting and useful to know if there is behavioral symmetry when these changes happen. For example, is the difference in activity participation the same when a person looses access to a computer at work with when she/he gains one? Or what is the effect of becoming a mobile technology user?

Table 5 reports the total effects of ICT availability on time allocation to activities and travel giving evidence of lack of symmetry and linearity. The ICT effect, however, is not always asymmetrical. For example, acquiring access to a computer at work is accompanied by 34 more minutes of work. In contrast, loosing access of computer at work is accompanied by 41 less minutes working. The net difference is a small number of about 7 minutes per day. When we consider the two ICT variables and maintenance duration, gaining access of computers at work shows a negative 6.368 and loosing access a positive 6.398 demonstrating an almost perfectly symmetric effect. Then, turning to leisure duration we see that gaining access of computers at work has an effect that is more than three times (-14.063) that of loosing access (4.135).

Table 5 Total Effects of Changes in ICT on Time Allocation to Activities and Travel

ICT Groups			Total Effects on Endogenous Variables			
			Sdur	Mdur	Ldur	Ttime
Computer	Work/ School	Experienced user (39.9%)	80.446	-10.356	-20.034	-1.284
		New user (8.6%)	34.121	-6.368	-14.063	-2.848
		Past user (7.4%)	-41.395	6.398	4.135	-0.575
	Home	Experienced user (45.5%)	-2.831	6.622	22.699	10.719
		New user (20.8%)	1.916	6.191	5.470	11.996
		Past user (5.2%)	-0.555	-4.741	-2.565	-2.444
Internet	Work/ School	Experienced user (23.7%)	5.690	-4.318	-8.281	5.581
		Past user (6.0%)	0.324	-5.371	1.794	-1.072
	Home	Experienced user (28.6%)	-16.479	1.039	2.164	-0.534
		New user (32.0%)	-0.845	-2.784	1.123	-7.318
		Past user (2.4%)	5.452	4.165	-22.869	10.004
Cellular Phone	Experienced user (23.8%)		-1.499	-0.024	-4.238	7.743
	Past user (4.5%)		4.452	-5.708	-12.919	-4.945
Pager	Experienced user (4.7%)		-30.531	1.925	4.009	-0.989
	New user (4.0%)		-0.511	6.506	30.782	0.821
Laptop	Experienced user (2.0%)		-91.257	5.754	11.983	-2.956
	New user (3.2%)		-1.515	-7.502	-12.755	2.877
PDA	New user (2.6%)		42.700	-2.693	-5.607	1.383

The ICT effect on activity and travel behavior seems to depend on the location and type of technology. Comparing the experienced information and telecommunication users to the new users and to the non-users, we find the overall effect of computers and Internet at work to be an increase in subsistence participation and a decrease in maintenance and leisure participation. However, computers and Internet at home have the opposite effect on the three activities. The more specific example is the amount of time allocated to leisure by the persons that are experienced users of computers at work and at home. The experienced users of computers at work spend 20 minutes less in leisure per day, while the experienced users of computers at home spend 22.7 more. As the average values of the ICT user indicators show we have an increase in technology users over time at home but not a substantial increase at work/school. The new computer users at home are 20.8% and new Internet users at home are 32%. The first group spends more time in all activities and travels more and the second group spends less time in subsistence and maintenance activities and travels less. Laptop users increase rapidly if one considers that experienced laptop users are 2% of the sample and new users are 3.2%. Experienced mobile technology users are also diverse but the experienced mobile technology users spend less time in subsistence and traveling, except that experienced cellular phone users travel more.

Table 6 reports the effects of ICT on number of trips by mode. New users of computers at home use public transportation and bike more often, but exactly the opposite happens to the new users of the Internet at home but at lower levels because of less trip making in general.

Table 6 Total Effects of Changes in ICT on Mode Frequencies

ICT Groups			Total Effects on Endogenous Variables					
			ssov	shared	transt	walk	bike	others
Computer	Work /School	Experienced user (39.9%)	-0.274	-0.083	0.004	0.018	0.006	0.036
		New user (8.6%)	-0.246	-0.055	-0.003	0.007	-0.001	0.001
		Past user (7.4%)	-0.067	0.034	-0.030	0.086	0.005	0.005
	Home site	Experienced user (45.5%)	0.228	0.079	0.054	0.029	0.001	-0.011
		New user (20.8%)	-0.095	0.064	0.104	0.048	0.042	0.049
		Past user (5.2%)	0.095	-0.052	-0.103	-0.050	0.000	0.002
Internet	Work /School	Experienced user (23.7%)	-0.181	-0.423	0.010	0.013	0.023	0.064
		Past user (6.0%)	-0.021	-0.020	0.032	-0.133	-0.001	-0.008
	Home	Experienced user (28.6%)	-0.033	0.008	-0.003	-0.003	0.002	0.000
		New user (32.0%)	0.045	-0.025	-0.051	-0.022	-0.017	0.007
		Past user (2.4%)	0.014	-0.008	-0.001	-0.028	0.163	-0.055
Cellular Phone	Experienced user (23.8%)		-0.036	0.138	0.003	-0.001	-0.025	0.015
	Past user (4.5%)		-0.395	-0.051	-0.011	0.001	0.003	0.001
Pager	Experienced user (4.7%)		-0.060	0.015	-0.006	-0.006	0.003	0.000
	New user (4.0%)		0.040	0.063	-0.067	0.164	0.010	-0.016
Laptop	Experienced user (2.0%)		-0.180	0.044	-0.017	-0.019	0.009	-0.001
	New user (3.2%)		0.130	-0.089	-0.139	-0.070	-0.025	-0.085
PDA	New user (2.6%)		0.084	-0.021	0.008	0.009	-0.004	0.001

Table 6 also allows us to perform some additional calculations. Regression coefficients defined for a group of indicators are relative to the excluded group (implicitly assumed to have a zero coefficient) and in this case is the non-user group. If a person is a heavy ICT user that has always (in 1997 and 2000 and we call them experienced users) made use of all the technologies available we can compute the number of driving alone trips s(he) makes in day as a difference from the trips a non-user would make and they are: (-0.274+0.228-0.181-0.033-0.036-0.060-0.180) = -0.536, which means on average half a drive alone trip less than non-users. Similar calculations lead to a –0.222 less sharing a car trips than non-users. Given the market penetration of ICT is also interesting to see what happens when we perform the same calculations for new users in all ICT categories leading to –0.042 for driving alone and–0.063 for sharing a car with others. Examining travel time, however, we see that fewer trips do not necessarily mean less traveling. For example, these same users that make less frequent trips may also travel long distances or in congested conditions for longer times. Additional analysis is required to also examine the interaction in the use of these technologies.

SUMMARY AND CONCLUSIONS

In this paper, a plethora of relationships among the most popular activity and travel behavior indicators are studied. A system of equations is first defined that includes as endogenous variables the amount of time allocated to subsistence, maintenance, leisure, and travel, as well as the total number of trips by the most important modes (drive alone, car sharing, public transportation, walking, biking, and all other modes used). The model system is designed to parallel other past studies and for this reason variables at the household and person levels are used. In addition, some specific variables to the database in PSTP are also employed to account for stratification and potential participation fatigue.

A key variant of the model system that sets it aside from other studies is the inclusion of "experience" variables. Using information between 1997 and 2000 a set of different level of ICT users is defined to capture the changes in information and telecommunication technology experienced by each respondent and included them as exogenous variables of the behavior in 2000. In this way we can detect if the opposite events in ICT ownership and use have the same but of opposite sign effect on behavior. Most ICT changes have asymmetric effects.

The second element that characterizes the research work here is the study of ICT on activity and travel behavior considered jointly with all the other determinants of travel behavior. The ICT effect on activity and travel behavior seems to depend on the location and type of technology. Comparing experienced information and telecommunication users to the new users and to the non-users, the overall effect of computers and Internet at work shows an increase in subsistence participation and a decrease in maintenance and leisure participation. However, computers and Internet at home have the opposite effect on the three activities. As the average values of ICT variables show, there is an increase in technology users over time at home but not a substantial increase at work/school. The new computer users are 20.8% at home and new Internet users are 32% at home. The first group spends more time in all activities and travels more and the second group spends less time in subsistence and maintenance activities and travels less. New users of computers at home use public transportation and bike more often but exactly the opposite happens to the new users of the internet at home but at lower levels because of less trip making in general. Heavier experienced users of technologies make fewer trips but do not spend less time traveling. Similar findings are reported for the new heavy ICT users but at a lower level.

The model system presented in this paper proves to be a very powerful tool in understanding activity and travel behavior in a social, economic, and technological context and allows one to examine behavioral aspects in unprecedented detail for hypotheses testing. This model system, however, has also some limitations due to some fundamental assumptions. All the dependent variables were assumed to be multivariate normally distributed and continuous. This may influence the values of the effects and their significance (although testing and experimentation with single equation models lead to similar conclusions). For this reason one potential expansion of the work here is to use a limited dependent variable formulation for the time allocation indicators (to account for the large concentration of persons at zero minutes per day) and to use a count data regression formulation for the frequencies by mode.

REFERENCE

Chung, J. and Y. Ahn (2002). Structural Equation Models of Day-to-Day Activity Participation and Travel Behavior in a developing Country. *Transportation Research Record*, **1807**, 109-118.

Golob, T.F. (1998). A model of household choice of activity participation and mobility. In *Theoretical Foundations of Travel Choice Modeling* (T. Gärling, T. Laitila, and K. Westin, eds.), pp. 365-398. Pergamon, Amsterdam.

Golob, T.F. (2003). Structural Equation modeling. In *Transportation Systems Planning: Methods and Application* (K.G. Goulias, ed.), pp 11.1-11.23. CRC Press, Boca Raton, FL,

Goulias, K.G., N. Kilgren, and T. Kim (2003). A decade of longitudinal travel behavior observation in the Puget Sound region: sample composition, summary statistics, and a selection of first order findings. Presented at the 10th International Conference on Travel Behaviour Research, Moving through nets: The physical and social dimensions of travel, Lucerne, 10-14 August 2003.

Kim, T. and K.G. Goulias (2004). Cross-sectional and Longitudinal Relationships among Information and Telecommunication Technologies, Daily Time Allocation to Activity and Travel, and Modal Split using Structural Equation Modeling. Presented at the 83rd Annual Meeting of the Transportation Research Board, Washington D.C., January 2004.

Murakami, E. and C. Ulberg (1997). The Puget Sound transportation panel. In *Panels for transportation Planning methods and Applications*, (T.F. Golob, R. Kitamura, and L. Long, eds.), pp.159-192. Kluwer, Boston.

APPENDIX: EXOGENOUS VARIABLES USED IN STRUCTURAL EQUATION MODELS

Household level	CARPOOL	Indicator, 1= household is sampled from carpool class; 0=otherwise
	TRANSIT	Indicator, 1= household is sampled from public transit class; 0=otherwise
	KITSAP	Indicator, 1= living in Kitsap County; 0=otherwise
	PIERCE	Indicator, 1= living in Pierce County; 0=otherwise
	SNOHO	Indicator, 1= living in Snohomish County; 0=otherwise
	TOT1_5	Number of children in the household who are less than 6 years old
	TOT6_17	Number of children in the household who are between 6 and 17 years old
	TOTADULT	Number of adults in the household who are 18 years old or older
	MIDINC	Indicator, 1= $35,000 ≤ household income < $75,000; 0=otherwise
	HIGHINC	Indicator, 1= $75,000 ≤ household income; 0=otherwise
	MHINC	Indicator, 1= $35,000 ≤ household income; 0=otherwise
	DKINC	Indicator, 1=household income is unknown; 0=otherwise
	CAR1	Indicator, 1= one car household; 0=otherwise
	CAR2	Indicator, 1= two car household; 0=otherwise
	CAR3_	Indicator, 1= three or more car household; 0=otherwise
	INBABY	Indicator, 1=an increase in the number of children < 6 years in the household between waves; 0=otherwise
	DNBABY	Indicator, 1= a decrease in the number of children < 6 years in the household between waves; 0=otherwise
	INKID	Indicator, 1= an increase in the number of kids whose age is 6-17 in the household between waves; 0=otherwise
	DNKID	Indicator, 1= a decrease in the number of kids whose age is 6-17 in the household between waves; 0=otherwise
	INADULT	Indicator, 1=an increase in the number of adults in the household between waves; 0=otherwise
	DNADULT	Indicator, 1=a decrease in the number of adults in the household between waves; 0=otherwise
	INLICEN	Indicator, 1=an increase in the number of drivers license holders between waves; 0=otherwise
	DNLICEN	Indicator, 1=a decrease in the number of drivers license holders between waves; 0=otherwise
	INBPASS	Indicator, 1=an increase in bus pass holders in the household between waves; 0=otherwise
	DNBPASS	Indicator, 1=a decrease in bus pass holders in the household between waves; 0=otherwise
	INVEH	Indicator, 1=an increase in the number of cars in the household between waves; 0=otherwise
	DNVEH	Indicator, 1=a decrease in the number of cars in the household between waves; 0=otherwise
	DNEMP	Indicator, 1=an increase in the number of employed persons in household between waves; 0=otherwise
	DNEMP	Indicator, 1=a decrease in the number of employed persons in household between waves; 0=otherwise
Person level	MALE	Indicator, 1=male; 0=female
	YOUNG	Indicator, 1=18≤ age ≤ 34; 0=otherwise
	MIDAGE	Indicator, 1=35≤ age ≤ 64; 0=otherwise
	PROF	Indicator, 1=having professional occupation; 0=otherwise
	MANAG	Indicator, 1=having managerial occupation; 0=otherwise
	SECRE	Indicator, 1=having secretarial occupation; 0=otherwise
	SALES	Indicator, 1=having sales occupation; 0=otherwise
	WK5	Indicator, 1=working outside of home for 5+ times a week; 0=otherwise
	DPUPIL	Indicator, 1= student; 0=otherwise
	DLICEN	Indicator, 1= having driver's license; 0=otherwise
	DBPASS	Indicator, 1= having a bus pass; 0=otherwise
	EXPEMP	Indicator, 1= started getting employed outside home in both waves; 0=otherwise
	NOVEMP	Indicator, 1= employed outside home in wave 9 only; 0=otherwise
	QUITEMP	Indicator, 1= employed outside home in wave 7 but not in wave 9;0=otherwise
Time related	PELAP	Time duration (No. of year) for person in panel
	PELAP2	Square of time duration for person in panel
	TUE	Indicator, 1=diary on Tuesday; 0=otherwise
	WED	Indicator, 1=diary on Wednesday; 0=otherwise
	THU	Indicator, 1=diary on Thursday; 0=otherwise
	FRI	Indicator, 1=diary on Friday; 0=otherwise
Information and Communication Technology	EXPCW	Indicator, 1= using computers at work/school in both waves; 0=otherwise
	NOVCW	Indicator, 1= started using computers at work/school after wave 7; 0=otherwise
	QUITCW	Indicator, 1= stopped using computers at work/school after wave 7; 0=otherwise
	EXPCH	Indicator, 1= using computers at home in both waves; 0=otherwise
	NOVCH	Indicator, 1= started using computers at home after wave 7; 0=otherwise
	QUITCH	Indicator, 1= stopped using computers at home after wave 7; 0=otherwise
	EXPNW	Indicator, 1= using the Internet at work/school in both waves; 0=otherwise
	QUITNW	Indicator, 1= stopped the Internet at work/school after wave 7; 0=otherwise
	EXPNH	Indicator, 1= using the Internet at home in both waves; 0=otherwise
	NOVNH	Indicator, 1= started using the Internet at home after wave 7; 0=otherwise
	QUITNH	Indicator, 1= stopped using the Internet at home after wave 7; 0=otherwise
	EXPCEL	Indicator, 1= using cell phones in both waves; 0=otherwise
	QUITCEL	Indicator, 1= stopped using cell phones after wave 7; 0=otherwise
	EXPPAG	Indicator, 1= using pagers in both waves; 0=otherwise
	NOVPAG	Indicator, 1= started using pagers after wave 7; 0=otherwise
	EXPLAP	Indicator, 1= using laptop computers in both waves; 0=otherwise
	NOVLAP	Indicator, 1= started using laptop computers after wave 7; 0=otherwise
	NOVPDA	Indicator, 1= started using PDAs after wave 7; 0=otherwise

Transport Science and Technology
K.G. Goulias, editor

CHAPTER 20

TRANSPORT COMPANY INFORMATION SYSTEM: A TOOL FOR ENERGY EFFICIENCY ENHANCEMENT

Vladimir Momčilović, Vladimir Papić, Olivera Medar and Aleksandar Manojlović
Faculty of Transport and Traffic Engineering, University of Belgrade

INTRODUCTION

Since the goal of managing the transport company is the highest possible profit, one direction is in the business broadening and engaging more transport resources, while another is in better usage of resources and by this, the transport costs decrease. The energy efficiency has proved itself as an important issue in decreasing transport related costs, while its role in decreasing the impact on the environment is also important. This is especially the case with large transport companies whose vehicle fleets comprise large number of vehicles with a different structure, technology and age. Transport company management responsibility is to recognize important influences under its control and to define the ways of action. Therefore, the energy efficiency increase should play an important role within their business strategy.

According to the economic evaluation of emissions reductions in the transport sector of the EU (*Bates et al., 2000*), within the operational measures with the purpose of increasing the energy efficiency – *good operating and fuel management practices* comprise: (a) effective monitoring of fuel use, (b) driver awareness, training and incentive schemes, and (c) preventive maintenance. As a result of our own researches and others' experiences, the problem to which is paid attention in this paper, has been defined as follows: how to enable transport company management to make and carry out high-quality decisions with respect to the company operation, which will equally lead to energy efficiency increase in further sustainable development. Long-term cooperation with transport companies has pointed out that managers are often overloaded by information not giving sufficient and high-quality elements for good decision making process.

During the problem solving process, the authors have been devoted to separate important influences on the energy efficiency being under company management's control. In that manner, all indispensable information that make possible "good" decision making have been defined. A special emphasis has been given to the solution of the correlation between energy efficiency and vehicle condition. Implementation of a method for managing vehicle fleet energy efficiency and a software package have as a consequence improved energy efficiency and reduction of its impact on the environment.

In this paper have been presented some of the results of the survey on a transport company vehicle fleet performance that has been completed within the project "Increasing the technical condition level of the transport company vehicle fleet with the purpose of raising its energy efficiency" financed by the Ministry of science and environmental protection of the Republic of Serbia, where the developed method has been applied.

ENERGY EFFICIENCY WITHIN TRANSPORT COMPANIES

The actions with the goal of energy efficiency increase on the level of a transport company could be educational, controlling or investment measures. Since driving skills and habits influence importantly fuel and lubricant consumption, therefore purpose of educational measures is to diminish their influence to a reasonable extent (*DfT Research Database, 2002*). Furthermore, controlling measures comprise two sorts of activities: consumption monitoring and vehicle condition monitoring. Good management requires a modern information system with high-quality input data processed into updated information for the decision makers, which must clearly indicate the values of indicators that surpass defined boundaries, i.e. alarm deficiencies and initiate certain activities. The last but not least are investment measures concerning the latest technology vehicles and equipment, which will by themselves facilitate the energy efficiency increase and in the long term are the most efficient and profitable. Although they might look as the simplest solution in developed countries, this unfortunately is not the case in developing countries because of required investment.

The energy efficiency issue is considered from the aspect of management measures dealing with vehicle fleet operation and maintenance. Besides technical measures (applied by vehicle manufacturer) there are so-called operational measures for energy efficiency increase, out of which on the transport company level are: fuel and lubricant consumption monitoring, preventive maintenance, logistic activities and many others.

Transport companies record large quantity of data, among which fuel and lubricant consumption related data, nevertheless the manner of their presentation does not allow high-quality efficiency analysis. Managers have an insight only to global indicators: consumed fuel and lubricants in the defined period and total costs per vehicle. The problem lies in data preparation to allow managers to use them timely and appropriately in the decision making process as well as monitoring effects of those implemented measures.

Energy efficiency indicators

Until recently (*McKinnon, 1999*), efforts to improve fuel efficiency in road freight operations have focused on engine performance and vehicle design. These reduce the ratio of fuel consumed to distance traveled, which is only one of three ratios which influence total fuel

consumption in this sector. Account must also be taken of vehicle utilization (i.e. the ratio of vehicle-km to ton-km) and transport intensity (measured by ratio of ton-km to sales). Fuel savings accruing from improved vehicle technology can be offset by declining load factors or increases in the average distance each unit of freight is transported.
According to definition of ODYSSEE indicators (*Enerdata s.a., 2002*) the most important macro indicator is energy intensity – a ratio between energy consumption measured in energy units (toe, Joule, etc) and an indicator of activity measured in monetary units (GDP, value added, etc.). Based on the analysis of energy intensity, it can be noticed that in developing countries this indicator is importantly unfavorable in comparison to developed countries, which means that there are significant possibilities for consumption improvements. Meanwhile, indicators related to fuel efficiency of road vehicles are: unit vehicle consumption (ratio of total consumption and realized annual mileage), test specific consumption of new vehicles (technical efficiency of new vehicles derived from fuel consumption test), and specific vehicle consumption (ratio of total consumption and effected passenger-km, ton-km etc.).
In this paper, by constant monitoring of fuel and motor oil consumption indicators per vehicle, the following indicators have been appointed as relevant: (a) unit consumption (fuel [L/100 km], lubricant [L/1000 km]); (b) specific efficiency (usage efficiency in kilometers traveled per 1L of fuel or lubricant [km/L]).

Vehicle Fleet Management needs and requirements

The management needs updated, clear and single-meaning data, i.e. an actual state that represents a base for the decision making, which means that management must have all indispensable data on time in the adequate form. Therefore it is not sufficient just to provide the data, but also that those clearly stress possible causes of energy "inefficiency" and ways to improve it.
Since managers, with different education, experience, capacities and interests, are often obliged to make ad-hoc decisions about issues of different nature, besides the defined goal (energy efficiency increase) they also need a high-quality Decision Support System (DSS). DSS operation is facilitated by implemented information system which should provide and process information, distribute decisions, monitor effects of the decisions, etc.
The manager must have overview of all critical vehicles, at least to the level of those that request interventions with several most important activities, define decisions in order to increase energy efficiency, forward decisions to responsible persons, obtain information about the reception and beginning of decision implementation, and obtain the results of its implementation. Accordingly, manager must posses a developed DSS, which defines what to observe, what to measure, how and when to do it, as well as to provide the link between the measured parameters, i.e. the implemented activities and their effects. Therefore the role of the information system can be presented by three main activities: (a) providing to DSS the data upon defined protocol and corrects the criteria upon methodology, (b) distributing decisions between management and all other participants within the business information system, and (c) making the links between vehicle fleet energy efficiency indicators – criteria – maintenance interventions.

Current conditions in developing countries

Transport company vehicle fleets in developing countries are technologically outdated. The average age of vehicle fleets in the Republic of Serbia is around 13 years. Within transport companies, vehicle fleets are heterogeneous regarding types as well as regarding their age. Therefore, to efficiently monitor energy efficiency indicators, vehicle fleets have to be segregated in the first place to homogeneous groups – so-called Technical Construction Operation (TCO) groups – vehicle sets with same technical features operating in similar conditions.

In most transport companies, fuel consumption norms by TCO groups are implemented, including variations depending on climatic, road and traffic conditions, but as for the motor oil consumption, the norms are only sporadically implemented.

On the other hand, drivers in transport companies are on different levels of training and education. Their training, as well as obligatory knowledge testing most frequently consists of vehicle driving skills from the aspect of traffic safety, but rarely from the aspect of economical ways of driving.

METHODOLOGY FOR INCREASING ENERGY EFFICIENCY

Among the most cost efficient measures for increasing vehicle fleet energy efficiency are operational measures that comprise fuel and motor oil consumption monitoring, transport routes optimization, logistic methods implementation, etc. The first intervention (*Coyle, 2000*) that any transport operator must take is to obtain accurate fuel and motor oil consumption data. Whilst this might seem like common sense, examination of the fuel records supplied by transport operators in the research project (*Coyle, Murray, and Whiteing, 1998*) showed that up to 20% of the records had a field containing erroneous data. Until the data is of reasonable quality it is not practicable to introduce most interventions (because neither the effects will be reliable).

Secondly (*Kaes, 2000*), the driver's influence on the energy efficiency and on decreasing the greenhouse gas emissions shouldn't be neglected. His influence is perceived through driving behavior (low revs at maximum torque, anticipation, no unnecessary idling, selection of optimal speed, etc.), observation of unusual vehicle behavior (excessive fuel and motor oil consumption, noise, black smoke from the exhaust, etc.), and regular and systematical application of servicing and maintenance activities (even with driver carrying out small technical jobs himself).

Therefore, the authors have started from the hypothesis that constant monitoring of vehicle fleet energy efficiency and activities on vehicle condition are the most efficient among cost efficient measures. The idea is to timely determine energy "inefficiency" by continuous monitoring, then to determine its cause, inspect the vehicle condition to eliminate causes and follow up monitoring to quantify the maintenance effectiveness. This method is based on the algorithm shown on the figure 1.

The method comprises two parallel processes: selected energy efficiency indicators monitoring, and vehicle maintenance. As a first monitoring measure, everyday follow-up of vehicles' and drivers' energy efficiency is introduced. To reach higher efficiency and decrease consumption deviations, maximum two drivers are in charge of each vehicle, which leads to higher liability for each vehicle, in terms of energy efficiency, as well as of vehicle care and maintenance. Vehicle fleet operation manager defines preliminary allowed deviations – "critical" intervals for unit consumptions. At the very beginning of method implementation, defined norms for TCO groups or entire vehicle fleet can be adopted as a starting point. In terms of "critical" intervals, following impacts must be taken into account: climatic irregularities, passenger flows peak loads, and human factor.

Figure 1 - Algorithm of methodology implementation for measuring maintenance interventions' influence to vehicle fleet energy efficiency

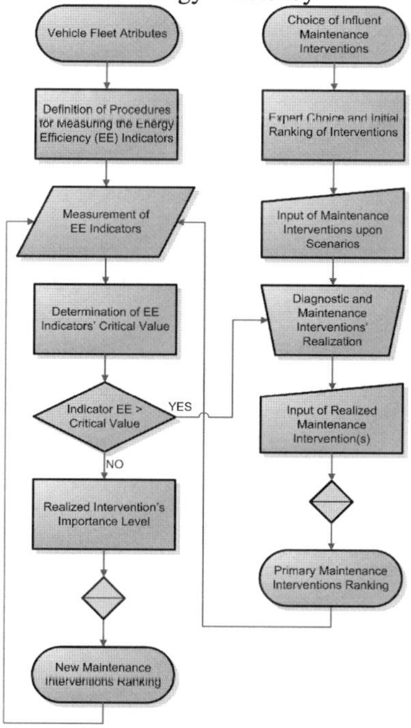

Regarding increase of fuel and lubricants consumption, three scenarios have been created:

(a) First scenario – increased fuel consumption,

(b) Second scenario – increased motor oil consumption, and

(c) Third scenario – a combination of two previous scenarios, i.e. increased consumptions of both fuel and lubricants related to TCO group "critical" intervals.

Recorded energy efficiency data are observed periodically, on monthly basis, and compared to TCO group average, while data for the current period are monitored on daily basis and compared to preliminary "critical" intervals for each scenario. This interval may be additionally enlarged or reduced for predefined vehicle percentage to be included in the "allowed" zone. For all vehicles definitely in the "unallowed" zone, the manager must decide what actions to take to increase their efficiency. The manager has following three options: do nothing (eventually driver warning), consumption test, or maintenance intervention.

If manager chooses "do nothing" strategy, in the following period vehicle is treated as if nothing had happened. If manager decides to undertake the consumption test, it is realized by a test driver and consequently compared to the results of regular vehicle drivers. If test consumption is equal or worse than the starting point, a maintenance intervention can be initiated, while if it is better, the drivers are penalized. As a special monitoring measure, drivers can be observed individually by full refueling on drivers' shifting. This measure could be difficult if drivers' shifting takes place on "field" instead in the garage.

A complete set of maintenance interventions influencing energy efficiency is then determined. Those interventions are ranked by experts assigning them corresponding "importance indexes". The experts that participate in interventions' ranking must be from the practice – maintenance field (from the particular transport company and/or other companies from the region) and from scientific and research institutions dealing with energy efficiency and vehicle maintenance. The experts are interviewed for their evaluation, individually quantifying interventions' influence to each scenario by respective weight factor called "importance index" (in the span from 1 to 10) assigning the highest index (10) to the intervention or defect influencing the most analyzed consumption scenario taking into account its maximum negative impact and so on. Maintenance interventions may be preventive or corrective, although focus must remain on diagnostic measures. After collecting expert "opinions", interventions are ranked upon overall "importance index". Results of this expert evaluation are entered to the database to obtain the initial ranking list of interventions. Such initial ranking lists of interventions differ by scenarios. The experts also indicate what, upon their opinion, represents a small, a medium and a large increase in unit consumption in percentage [%], as well as in absolute values [L], so that small increase could be tolerated, medium requires at least a consumption test, while large requires maintenance intervention.

After the alarm on surpassed "critical" value of one or more indicators, initially "recommended" maintenance intervention is checked. While performing the interventions, it is crucial to check them upon exact order from the ranking list of "recommended" interventions. This practically means that maintenance worker must check two things: that suggested intervention has been checked, and that the intervention has been completed. This procedure is introduced to facilitate monitoring and measurement of each intervention importance. Upon completed intervention, its influence to respective consumption is observed in following time period. New consumption is registered and compared to the previous as well as to the average for the current period. This result is memorized as this intervention attribute in the given case, which will enable current ranking list of the most "attractive" maintenance interventions leading to significant consumption decrease. If its effects on consumption indicators in the following period are shown as especially good and intervention's frequency grows, its "importance index" rises towards the top of the current ranking list of interventions. Influent interventions take their respective place in time, but some interventions fall off the list (insignificant appearance frequency and/or neglected impact on indicators). This process practically represents "learning" and method improvement process.

In short, featured methodology comprises the following steps:

(a) Vehicle sample selection and division into TCO groups,

(b) Initial criteria definition by vehicle fleet managers for the DSS ("critical" deviations of energy efficiency indicators, monitoring period, etc.),
(c) Consumption averages determination,
(d) Disclosure of "problematic" vehicles (comparing consumptions to TCO group average),
(e) Maintenance interventions determination and implementation (according to energy "inefficiency" scenario),
(f) Repeated fuel and lubricant consumption daily monitoring,
(g) Maintenance intervention's "importance index" correction,
(h) Regular correction of "critical" interval criteria (with gradual decrease of deviations),
(i) Influent interventions ranking with purpose of shortening the ranking list of interventions.

SOFTWARE FEATURES

For the implementation of DSS, special software was designed and developed based on the algorithm shown on the figure 1. The software synthesizes large quantity of data into high-quality and "synoptic" reports for managers. Based on those reports the most energy inefficient vehicles and drivers in the period are indicated and sent to maintenance and/or testing.
The software is composed of the following sections: (a) vehicle and operation data processing, (b) consumption indicators monitoring and alarming, and (c) maintenance interventions ranking.

Vehicle and operation data processing section

In this segment input of all essential elements for vehicle registration (vehicle age monitoring), program of preventive maintenance, operation program etc. is performed. While opening "vehicle file" all important data for undisturbed vehicle operation are entered. The vehicle is afterwards assigned to drivers and classified into TCO group by which it will "inherit" its respective program of preventive maintenance. During vehicle operation, from trip logs relevant data are inputted to the software. The software provides data memorizing, analysis and processing with purpose of generating required reports. All data are linked to corresponding mileage.

Consumption indicators monitoring and alarming section

Based on the input of each vehicle refueling and oil addition, average values of fuel and lubricant consumption are calculated. When considered period average value for TCO group is determined, all periodical values are compared to it. Manager at the beginning must define the initial "critical" interval for the first monitoring period as allowed deviation or as percentage from TCO group average. This alarm or "critical" interval is adopted as a preliminary interval for the forthcoming period, but with possibility to be modified upon analysis of results from the period. After the insight to the periodical report, he defines the definitive "critical" interval by efficiency indicators, which should be preferably lower or at least equal to the previous, although there is a possibility for increase – especially in case of seasonal (climatic) changes – with the warning that the value is higher than previous (manager can add a comment which will help him in future analysis to clarify reasons for such "inefficient" tendency).

Maintenance interventions ranking section

The starting point is initial intervention ranking list from the database, resulting from expert evaluation. For each appearance of a particular intervention, in case of energy "inefficiency", its importance grows in view of appearance frequency. In case of determining that energy efficiency indicator has been enhanced in comparison to the "critical" value, after intervention, it obtains extra "points" for efficiency. Whether its monthly average in the forthcoming month interval has importantly decreased compared to TCO group "critical" value then intervention obtains further extra "points" for efficiency. Maximum number of points that an intervention could reach is set, so when any surpasses previously mentioned limit, the number of points exceeding the limit will be subtracted from all the interventions on the list (so that none of the intervention's index could be negative). All the interventions with index 0 will be erased from the list.

CASE STUDY

The research had been realized within a transport company, where as a sample has been chosen a homogeneous group of vehicles. This group is made of 17 articulated buses of the model IK 202 (IKARBUS, engine RÁBA - EURO 1, automatic transmission), with average from 5 to 9 years of age. Those vehicles have the annual average mileage of around 90 000 km, so that total mileage of all vehicles up to date is lower than 750 000 km. One part of these vehicles (around 25%) has already completed the engine overhaul in recent years.
A special type of intensive monitoring of the chosen vehicles' energy efficiency has been applied on monthly basis. The illustrative graph resulting from fuel efficiency monitoring from 2002 to middle of 2004 is shown on the figure 2. The authors have opted for an initial "critical" interval deviation on entire monitoring period, which is e.g. for the fuel consumption +15% of the average, as also illustrated on the figure 2. Regarding the mentioned interval, vehicles that exceed it are determined and a special attention is dedicated to them.
The experimental research has been performed on the most energy "inefficient" vehicles in the given period. A "quick" consumption test has been realized on vehicles in usual operation conditions but with different drivers. On vehicles that exceeded defined "critical" interval upon any scenario on the consumption tests, the need for interventions is checked (by order from the ranking lists upon scenarios). From eight vehicles participating in the "quick" tests, three "problematic" vehicles were taken into further consideration (5145, 5146 and 5161) and all of them were "inefficient" by fuel, as well as by lubricant consumption – Scenario 3. On two of them (5146 and 5161) have been selected and implemented interventions followed by

a new consumption monitoring (until the next tanking/refueling), which gave positive effects regarding registered values. Both vehicles have been "treated" by first intervention from the ranking list of interventions, which led to increased energy efficiency in the forthcoming period. However, for vehicle 5145 the cause of increased consumption was not defect but driving manner, which has been established by another consumption test. The new testing was manager's decision shown as efficient because its consumption in following period has normalized and fitted into allowed "critical" deviation interval.

Figure 2 – Monthly unit fuel consumption in [L/100 km] with the average for the entire TCO group beginning from 2002

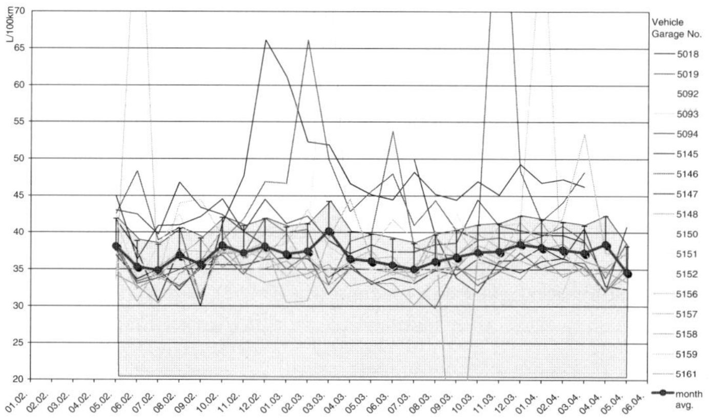

The values of operation and energy efficiency indicators for the entire vehicle sample are shown in the Table 1, on the chosen time period (monthly values) during the first year of monitoring.

Table 1. Overall vehicle sample operation indicators during 2002

Month	Total motor oil (incl. oil change) in Liters	Motor oil consumption in Liters	Unit motor oil consumption in L/1000 km	Mileage in km	Total Fuel in Liters	Unit fuel consumption in L/100 km
01	588.00	346.00	2.475	139,823	N/A	N/A
02	538.50	433.50	3.090	140,306	N/A	N/A
03	606.00	396.00	2.667	148,506	N/A	N/A
04	648.00	504.00	3.679	137,003	N/A	N/A
05	775.50	578.50	4.777	121,091	46120	38.09
06	626.50	508.50	3.494	145,532	51400	35.32
07	536.00	391.00	2.736	142,914	49813	34.86
08	561.50	393.50	2.746	143,305	52821	36.86
09	610.50	396.50	2.764	143,426	51100	35.63
10	589.00	397.00	2.781	142,752	54581	38.23
11	656.50	461.50	3.317	139,136	51761	37.20
12	566.00	520.00	3.884	133,883	51023	38.11
Σ	**7302.00**	**5326.00**	**3.175**	**1,677,677**	**408620**	**36.75**

RESULTS

The research have pointed out that the implementation of monitoring procedure with processed and adequately presented results (fluctuating "critical" values) that have been used in the managing process, has an outstanding positive effect – stabilizing fuel and lubricant consumption on a lower level. The DSS implementation demonstrated its importance combined with monitoring in lowering energy consumption level as well as in decreasing deviations from TCO group average value and the following results were obtained:

(a) the deviations of fuel efficiency indicators have been lowered from approximately 20% to 10% relative to the monthly average (figure 2),
(b) reduced fuel consumption for a total of 8.83%, from 38.04 L/100km (6.18 mpg) to 36.12 L/100km (6.51 mpg) in the first year, and then finally to 34.68 L/100km (6.78 mpg) in the second year (figure 2),
(c) energy savings of 2,342,615.66 MJ during 2002 and 2003,
(d) minor financial expenditures, e.g. with the total kilometers driven from about 3,205,000 km the savings are approximately 56,500 US$, and
(e) reduced emission of greenhouse gases for approximately 5% in the first year of monitoring and 4% in the second year, especially important for the CO_2 with the total savings of 565.0 t.

It is important to emphasize that all the suggested measures are followed by relatively small total investments in respect to the accomplished savings in the considered time period.

With the software assistance by which interventions effects and frequency have been analyzed, the ranking list of interventions for each scenario of energy "inefficiency" has been shortened. Also, by timely and adequately selected maintenance intervention extra savings in fuel and lubricant consumption have been reached (up to 2%), especially regarding motor oil consumption and decreasing corrective maintenance working time (time savings).

CONCLUSIONS

The developed method and the designed and implemented information system, based on resulting requirements, represent the outcome of the long theoretical and empirical researches. This paper presents a part of the research performed until the middle of 2004. The implementation in the Transport Company "ASP Lasta Beograd" has shown positive results regarding the energy efficiency (around 8%) and greenhouse gases emission (around 5%), which is a confirmation that on described manner, by implementing the method supported by featured information system, especially in developing countries, with minor investments, might reach prominent positive effects.

Here, as presented, the leading role had the developed software as part of the IS which is completed upon defined requirements and highly integrated into the BIS of the company where the research has been performed. The intention of the research project, whose part has been presented in this paper, is to supervise all possibilities and effects of the preventive interventions implementation. By this, the circle would be closed: monitoring – actions – effects – actions – preventive interventions – effects – Program of Preventive Maintenance correction, and all this with the purpose of increasing energy efficiency. Considering the first presented results, a further increase in the energy efficiency is expected. The link between fuel and lubricant consumption monitoring and appropriate timely interventions in case of major deviations is emphasized: drivers "justifications" for deviations will be slowly eliminated. Complementary driver training with the purpose of better energy usage and much efficient and economical behavior is the only activity left. Finally, it is important to notice the secondary effect: everyday vehicle (and drivers) operation monitoring system influenced drivers' behavior, therefore slight fuel consumption decrease (2-3%) has been observed as a consequence of the monitoring itself.

REFERENCES

Bates J., Brand C., Davison P., Hill N., (2000) Economic Evaluation of Emissions Reductions in the Transport Sector of the EU, Final Report, AEA Technology Environment, Abingdon, UK

Coyle M. (2000), Fuel saving interventions: do they really work?, Hong Kong Institute of Engineers/Society of Automotive Engineers (HKIE/SAE), Transport and Logistics Research Unit (TLRU), University of Huddersfeld, UK

Coyle M., Murray W., Whiteing A.E. (1998), Optimising fuel efficiency in transport fleets, Logistics Research Network Annual Conference, Cranfield University, UK

DfT Research Database (2002), Project: Fuel Saving Tips Guide, Reference: RHMF 001, AEA Technology Environment, ETSU Harwell, Didcot, Oxfordshire, UK

Enerdata s.a. (2002), Definition of ODYSSEE indicators, Monitoring Tools For Energy Efficiency In Europe, http://www.odyssee-indicators.org/Publication/PDF/def-ind.pdf

Kaes P. (2000), The importance of Maintenance and ensuring it gets done, Smart CO2 Reductions, Non-product Measures for Reducing Emissions from Vehicles, Joint Conference ACEA - ECMT – OICA in Turin, Italy

McKinnon A. (1999), *A Logistical Perspective on the Fuel Efficiency of Road Freight Transport*, Improving Fuel Efficiency in Road Freight: The Role of Information Technologies, Joint OECD/ECMT/IEA Workshop, Heriot Watt University School of Management, Edinburgh, UK

Transport Science and Technology
K.G. Goulias, editor

CHAPTER 21

AN IMAGE PROCESSING BASED TRAFFIC ESTIMATION SYSTEM

Hartwig Hetzheim and Wolfram Tuchscheerer, German Aerospace Centre, Berlin, Germany

INTRODUCTION

Road planning and traffic-management and control need traffic information like flow rates, share of cars and lorries, velocity etc. Digital video image based traffic data collection becomes more and more popular. Cameras images comprise a bigger observation area than other detectors do. Area related traffic characteristics like lane changing frequency, traffic density, length of traffic jam and vehicle behaviour can be obtained. Drawbacks are occlusion, influence of weather conditions and sudden changes of lighting. Moving non-traffic objects - rotating panels, trees and windows - can lead to ghost-vehicle detection. Solving these problems, traffic analysis by image processing over a larger road area can give valuable results for control and understanding of traffic. Named problems can be reduced by exploring the texture, which is nearly independent on brightness. Vehicle's texture differs from road, bushes, buildings or trees. Texture is described here by a system of coupled non-linear stochastic differential equations. Nonlinearities and the number of equations are estimated by an adaptive process. Other properties to distinguish vehicles from other objects are generated by logical and algebraic combinations of grey values. The generated properties are combined to obtain better adapted properties to distinguish the vehicles from the road. A binary representation of shape and size of vehicles is the basis to characterise their type, velocity and direction. Different kinds of properties are represented by fuzzy measures and functions. Different properties are fused by fuzzy integral to generate better adapted properties for decision making.

THE MATHEMATICAL BASIS FOR TRAFFIC ESTIMATION

Mathematical basis is the detection of different kinds of properties, their fusion and the generation of new kind of properties for decision making.

Determination of traffic active areas

To detect road users in real time, the searching area has to be reduced. This can be achieved by summation of differences of subsequent images taken at the time t_k and t_{k+1} . These difference images $b(t_{k+1})$-$b(t_k)$ are summed over a longer time. The algorithm for estimation of the active area S_{act} is described by

$$S_{act}(t_{k+1}) = \langle (s_{act}(t_{k+1}) + S_{act}(t_k))/2 \rangle_m \quad \text{where} \quad s_{act}(t_k) = f(b(t_k) > s_1 - b(t_{k+1})) > s_1)$$

The mean values are written by $\langle\ \rangle_m$ and realised by smooth or rank operations used over the area m. The extension by mean values fills the gaps created if vehicles are fast and very small, e.g. cyclists, and help to mask a larger area for the occlusion by bigger vehicles, such as buses or lorries. The gaps are closed also by application of an erode function. The threshold s_1 is determined by the mean grey value of the image. The function f is e. g. a quadratic function. By use of different thresholds on S_{act}, different widths of traffic flow areas are obtained. If the threshold is near the maximal grey value, the traffic lanes can be distinguished. The areas, where the road is occluded by masts or advertising panels, are excluded from the active area. Using a threshold near the maximal value of S_{act} and using the operation thinning, the change of direction of the vehicles can be measured and given as a mean value in a time interval. This is important for understanding the reaction of drivers in selected traffic situations, such as lane changing and overtaking. From this the angles of the vehicles on intersections are obtained as well.

Background estimation for enhancement of vehicles in the image

Background estimation is important to detect vehicles stopped at an intersection. The vehicles with a grey and texture value similar to the road are enhanced by background subtraction from the image. The background is successively estimated in a pre-processing. Usually, the road is not free of traffic and changes, so that all kinds of mean values give a bad result for the background. For the estimation of the background two basic properties are used - the difference in the texture between vehicles and the road, and the sensitivity for motions in difference images. The algorithm begins with the detection and marking of image parts, where texture is different from the road texture, i.e. the homogeneity is more violated. The difference images $b(t_k)$-$b(t_{k+1})$ are mapped in the so marked image parts. If this difference is greater than a given threshold (about maximum of grey value/3), the vehicle is moved and the end of the moved part (s_2-s_3) is replaced by the road as background. The obtained algorithm is

$$s_{mask} = (f(b(t_k)) - f_{smooth}(b(t_k))) > const_1 < (const_1 + 1) \tag{1}$$

$$s_2 = [(f(b(t_{k+1})) - f(b(t_k))) > const_2 < (const_2 + 1)] - const_2 \tag{2}$$

$$s_3 = ((f(b(t_{k+1}))/const_2) > 0 < 1) \ AND \ ((f(b(t_k))/const_2) > 0 < 1) \tag{3}$$

$$s_4 = \{ [s_2 - s_3] \ AND \ (s_{mask} - \min(s_{mask})) \} * b(t_{k+1}) \tag{4}$$

The property s_4 is replaced in the masked area. After some time, in all areas with moving vehicles, this replacement is done by s_4. Areas with lawn, bushes or trees are more inhomogeneous. For these areas abs(s_2-s_3)*s_{mask} is added for each image. If this sum after a longer observation exceeds a given threshold, this area is replaced by the grey values of the last image in this area, because there are no vehicles

Masking of inactive areas of traffic surrounding the road

For masking the road surrounding, 3 properties are combined for better decision. The first property s_6 within an image b is described by their stochastic and geometric structures coupled with the brightness. Rank or smooth operations, given by $\langle\ \rangle$, are used over different areas m_2 or m_3 to characterise the structure s_6.

$$s_6 = f\left(\langle b\rangle_{m_2} - \langle b\rangle_{m_3}\right) \tag{5}$$

The non-traffic objects, such as buildings, trees and road surfaces, have another structural uniformity and other textural structures than vehicles. The geometrical structures of buildings are simpler and can characterised by edge enhancement s_7

$$s_7 = f\left(\, dilate(b) - erode(b)\,\right) > const. \tag{6}$$

Another property is generated from car flow during a time interval, generated by the difference of subsequent images. By this property, the road is marked by observation of moving vehicles. By the third property, masts are isolated by morphological edge detection and following comparing with a line character, represented by equal gradients at different points:

$$s_8 = \left(s_7(x_1, y_1) - s_7(x_2, y_2)\right) / \left(s_7(x_3, y_3) - s_7(x_4, y_4)\right) \tag{7}$$

where s_8 is nearly equal 1 for masts. The masking reduces the area where vehicles are searched. Non-parametric methods are applied if the parametric information about the surroundings or vehicles are not or incomplete known. They are described by rank with different thresholds. For grey values $x_{i,j}$ at the pixel points (i, j) the rank $R_{i,j}$ is given by

$$R_{i,j}(M,N) = \sum_{l=-M}^{M} \sum_{k=-N}^{N} u\,(x_{i,j} - x_{l,k}) \quad \text{with the step function} \quad u(z) = \begin{cases} 0 & \text{for } z<0 \\ 1 & \text{for } z \geq 0 \end{cases} \tag{8}$$

To reduce the search areas, the area of expected road users is isolated by masking of the areas, where traffic is not possible such as houses, lawn areas, bushes or trees. Moved trees are detected by comparing differences of smoothed and original image over a longer time. The areas, where the trees are strongly moved or occlude the road, are masked for a better decision.

Mathematical representation of texture properties

The texture is nearly independent of sudden change in lighting and the basis for detection of surrounding objects. The mathematics for this representation is given in a reduced form here. Texture is described by a stochastic sampled at discrete points. If parametric information about the stochastic in the image does not exist a priori, non-parametric methods, see Bickel et al. (1993), Hetzheim, Dooley (1995), are applied. With a non-parametric algorithm, different areas with hidden properties are isolated. The surrounding of a pixel-point x_{ij} are compared by logical or arithmetical operation like signum-relationship $s_{i,j}$ or difference-relationship $d_{i,j}$ given by

$$s_{i,j} = \sum_{k=-m}^{m} \sum_{l=-n}^{n} \operatorname{sgn}(x_{i,j} - x_{i+k,j+l}) \quad \text{or} \quad d_{i,j} = \sum_{k=-m}^{m} \sum_{l=-n}^{n} abs(x_{i,j} - x_{i+k,j+l}) \cdot \tag{9}$$

Where point $x_{i,j}$ is related to each point in this area. Point $x_{i,j}$ is moved over the entire image. In this case a new image with generalised properties is generated. Another efficient method to describe the stochastic within an image is the rank description of eq. (8) with the surrounding rectangle with k={i-M,...,i+M}and l={j-N, ,...,j+N}. The rank shows how many of the pixels in the area 2N*2M have a grey value less than the selected value $x_{i,j}$. A generalised image is generated by shifting $x_{i,j}$ over the entire image. A stochastic component $S_{i,j}(g)$ with threshold g is obtained if rank values with $x_{i,j} > g$ are selected. Different textures are characterized by changing the threshold g :

$$S_{i,j}(g, M, N) = \sum_{l=-M}^{M} \sum_{k=-N}^{N} u\left(\left|x_{i,j} - g\right| - \left|x_{l,k} - g\right|\right) u(x_{i,j} - x_{k,l}) \tag{10}$$

As $u(x)=\frac{1}{2}sgn(x)+\frac{1}{2}$ and $u\left(x_{ij} + x_{kl} - 2g\right)=1$, this equation is fulfilled iff $\left|x_{ij} - g\right| \geq \left|x_{kl} - g\right|$ and $x_{ij} > x_{k,l}$ or $\left|x_{kl} - g\right| \geq \left|x_{ij} - g\right|$ and $x_{kl} > x_{i,j}$ and with a threshold g can be written:

$$S_{i,j}(g, M, N) = \sum_{k=-N}^{N} \sum_{l=-M}^{M} u\left(x_{ij} + x_{k,l} - 2g\right) \tag{11}$$

$S_{i,j}(g)$ is a new value for the pixel point (i,j) and creates an image with reduced fluctuations of the textural component. Based on the rank 3 types of stochastic can be separated for (i,j):

- $S_{i,j}(g\ )$ is calculated for different values of g
- $S_{i,j}(g_m)$–$S_{i,j}(g_n)$ is a calculated difference for different levels
- $S^{(N)}{}_{i,j}(g, M, N)$ is calculated over the direct neighbourhood values

If a parametric representation is possible, then different superposed kinds of elementary textures can be represented by a system of stochastic differential equations. The number of equations, the non-linearity and the gain factors of the noise $n(x)$ characterise the different superposed structures of the textures. The first s equations describe the non-stochastic properties, whereas the $m-s$ equations with the gain factors $G_{s+1}, \cdots, G_m$ of the noise describe the stochastic effects. For the coupled system of equations can be written

$$
\begin{aligned}
y_1(x_{l+1}) - y_1(x_l) &= \mathbf{F}_1(\mathbf{y}(x_l))(x_{l+1} - x_l) \quad where \quad \mathbf{y}(x_r)=\{ y_1(x), y_2(x), \cdots, y_r(x),\} \\
&\dots \\
y_{s-1}(x_{l+1}) - y_{s-1}(x_l) &= \mathbf{F}_{s-1}(\mathbf{y}(x_l))(x_{l+1} - x_l) \\
\mathbf{y}_s(x_{l+1}) - \mathbf{y}_s(x_l) &= \left[\mathbf{F}_s(\mathbf{y}_s(x_l)) + \int_{x_l}^{x_{l+1}} G_s(x) dW(x) \right] (x_{l+1} - x_l) \qquad (12) \\
&\dots \\
\mathbf{y}_m(x_{l+1}) - \mathbf{y}_m(x_l) &= \left[\mathbf{F}_m(\mathbf{y}_m(x_l)) + \int_{x_l}^{x_{l+1}} G_m(x) dW(x) \right] (x_{l+1} - x_l)
\end{aligned}
$$

Here, W is a Wiener process, $y_1, \cdots, y_s$ are s components of the stochastic and $\mathbf{F}$ are the s non-linear relationships matrices. The s until m coefficients G are the gain factors for the noise n and x represents the pixel-point. The system of non-linear stochastic differential equations is solved by martingale theory, Liptser et al. (1977-78), Hetzheim (1993). Here, the effect of approximation is easy to understand because the calculation remains in the same space as the model given. For the non-linear function F for short distances can be used functions like exp(-ay(x)) or $y(x)^n$ with $n \geq 2$. For further distances can be used log(-ay(x)) or $y(x)^n$ with $n < 2$. The number of equations will be determined successively by searching the effect of change of the images after application of the non-linear filtering. If the change is less than a given threshold, than no more equations have to be generated. The solution of equations can be approximated by repetition of linear filtering to get the stochastic properties.

Estimation of texture properties coupled with areas of weak changing grey values

Texture properties are relationships between the grey value differences of neighbouring pixels and thus weakly influenced by change of the mean brightness. Such elementary properties are combined to isolate vehicles. A car can be described by a smooth texture and a similar brightness in a small area as consequence of the car's lacquer. A lorry has usual an inhomogeneous part near the wheels and the driving position and a homogeneous part of the payload cover. The shape of pedestrians is fuzzy. Interesting areas can be selected by an AND-operation of a very high and a very low bit of the image. The lower bit image describes the texture and the higher a bright vehicle if the higher bit is 1 or a dark vehicle if this bit is 0. The road homogeneity is to be analysed by texture. Structural properties, such as jumping of textures or uniform textures of a side view of vehicles are used to distinguish lorries from buses. By morphological edge detection, vehicle contours are isolated and connected with areas of uniform grey values of a car. The non-traffic objects, such as buildings, trees and road surfaces show another textural structure than vehicles do. The geometrical structure of buildings is simpler and can be characterised by edge enhancement s_{10} on the textures of image b_{tex} with the erode and dilate operation $s_{10} = f(dilate(b_{tex}) - erode(b_{tex})) > const$.

Another property is obtained from the flow of the vehicles during a time interval, generated by the difference of successive incoming images. By this difference of texture properties, the road is divided from moving vehicles. By the texture property, masts are isolated by textural edge detection, Hetzheim (2002), and following comparison with a line character represented by equal gradients at different points $x_1, y_1, \ldots, x_4, y_4$:

$$s_{11} = (s_{10}(x_1, y_1) - s_{10}(x_2, y_2)) \,/\, (s_{10}(x_3, y_3) - s_{10}(x_4, y_4)) \tag{13}$$

where s_{11} is nearly equal 1 along the edge-line of y_1 until y_4 and their extension. For the detection of trees and bushes or lawn, the texture properties are very effective for isolation.

Description of different kinds of properties by measure

The isolation of vehicles is very difficult, because of the high diversity of situations, vehicles and pedestrians. Only a combination of many properties can overcome this problem. To characterise vehicles often only parts of the properties (mathematical spoken a map of one property on another property) is needed. This problem demands a generalised property representation. For a better decision-making the combination of selected properties related to their importance is used. Considering a property's importance implicates the loss of additivity. Normalisation required by probability does not exist anymore. By the fuzzy measure the properties described by different kinds of relationships are mapped on the closed interval [0,1]. The fuzzy measure, first defined by Sugeno (1974), has, besides the additive terms as a probability measure, a term with the combination of all elementary fuzzy measures multiplied by a factor λ. The factor λ has an effect similar to a weight factor for the interaction between the properties. If $\lambda = 0$ then the fuzzy measure is equal to the probability measure. The coupling of the elementary fuzzy measures (densities) $g_1(q_1)$ over the elementary area q_1 with another elementary fuzzy measure $g(q_2)$ over the other elementary area q_2 is defined by Sugeno(1974):

$$g_\lambda(q_1 \cup q_2) = g_\lambda(q_1) + g_\lambda(q_2) + \lambda g_\lambda(q_1) g_\lambda(q_2) \qquad (14)$$

where $\lambda = (1+\lambda g(q_1))(1+\lambda g(q_2)) - 1$ is a coupling constant used as a substitution for the loss of additivity. For a set of elements $A = \{q_i\}$ the relationship above can be used recursively and gives:

$$g(A) = \sum_{i=1}^{n} g(q_i) + \lambda \sum_{i=1}^{n-1} \sum_{j=i+1}^{n} g(q_i) g(q_j) + \cdots + \lambda^{n-1} g(q_1) \cdots g(q_n) \qquad (15)$$

The measure $g(A)$ and the parameter λ for the coupled elementary properties for smaller areas are calculated by an iterated coupling with two properties, where one of them is the result before. Properties especially capable to be represented by a fuzzy measure are:

- logical function of some bitmaps of data within an interval,
- grey values of images higher/lower as a threshold
- areas given by a selected texture
- areas selected between two maxima of the histogram

For fusion texture properties are also represented by fuzzy functions which are mostly a collection of values over a number of single pixel points. The values of the neighbour pixels are of stochastic nature and often not directly correlated with this value. Normally, these fuzzy functions are described by a characterisation over a threshold. Outside of such a characteristic threshold the values have no effect. Inside the interval the values generate fuzzy properties for the adapted condition. Properties used for the fuzzy function are:

- differences of data values in different distances
- the weak change of the data values in related areas
- the difference of the stochastic in different directions
- the number of ranks related to different distances
- values obtained by the subtraction of the original and mean value

These fuzzy functions are also mapped on a closed interval [0,1]. The properties for the detection of a vehicle can be divided in a part describing an area and one describing values over an area. The first part is adapted by the fuzzy measure, describing areas with similar grey values, and the second part is described by a fuzzy function, characterising the texture within an area. So, the image can be divided in bitmaps. The highest and the two lowest bitmaps can be combined logically in a non-additive case or histogram values over small areas obtained within different intervals are combined. Roughly spoken, a more functional property is represented by a fuzzy function $h(x_{k,l})$ over an area of a fuzzy measure $g(x_{k,l})$.

Fusion of different kinds of properties for detection of vehicles

The properties defined by a fuzzy measure or fuzzy function are generalised properties, which can be fused to generate a more adapted property. Mostly the information in the properties exists only in a composed form and has to be decomposed. This decomposition is possible by

the fuzzy integral. In the fuzzy integral a part of the fuzzy measure and a part of the fuzzy function are liked together whereas other parts of properties in such a combination will be reduced. For the fusion is important, that a property less than an assumed threshold contradict the detection for a vehicle and neglect the decision for a vehicle. Fuzzy functions and fuzzy measures are used, because it is nearly impossible to know all contributions for estimation of a vehicle. Thus, the normalisation as the fundament for a probability cannot be used.
The fusion of fuzzy measures and fuzzy functions gives new properties, which are more characteristic for vehicles and not so dependent of ambient lighting as the properties alone. The fuzzy measure g and the fuzzy function $f(\xi)$ with values on the pixel points ξ within an area A are combined by the fuzzy integral of the form of Sugeno (1974):

$$\int_A h_\alpha(x) \oplus dg = \sup_{\alpha \in [0,1]} \{\min[\alpha, g(A \cap H_\alpha)]\} \quad (16)$$

with $H_\alpha = \{\xi \mid h_\alpha \geq \alpha\}$ where α is a cut value. By the fuzzy integral values over the possible area of a limited grey value, represented by a fuzzy measure $g(\xi_{k,l})$, are connected with the texture values $h(\xi_{k,l})$ defined over this area. Therefore, the fuzzy integral represents a functional relationship between the fuzzy measure and the fuzzy function. For $h_\alpha(x)$ the values over the area of selected pixel-points are used, representing a stochastic property. α is the threshold where the assumption is fulfilled, that the property is used in the minimal condition. The area A is given as the data area where surely a specific texture property is expected.
The benefit of a fuzzy measure and a fuzzy function is that different kinds of properties are mapped on the closed interval [0,1]. The value obtained by the fuzzy integral also is defined in the interval [0,1] and can be used as a new fuzzy measure g_2 or a new fuzzy function $h_{\alpha 2}$

$$g_2(A) = \int_A h_\alpha(\xi) \oplus dg_1 \qquad h_{\alpha 2} = \int_A h'_{\alpha'}(x) \oplus dg \quad (17)$$

The components of both sets are combined for selecting specific texture properties. That way, coincidences of the properties contained in the fuzzy measure and the fuzzy function are filtered with the fuzzy integral. The new fuzzy measures and fuzzy functions make sure that all possible properties in all combinations, which should be considered, are extracted. In such a way an image is obtained, where the grey values represent a measure of the membership to a texture. A threshold decides which pixel of the data belongs to a selected texture.
Roughly spoken, the fuzzy integral selects the largest value of all minimal fuzzy values mapped upon the fuzzy measure restricted by a defined area and cut by the constant α. Thus the importance of a property is included. This includes the neglecting of contributions if a value is less than an assumed threshold. The area may be described by the difference of two images taken at different time succeeding a given threshold. This area describes a changing area during time related to given parameters. This area is connected with stochastic values, produced by a combination of mathematical moments as variances combined with rank or sign statistic, representing selected texture properties. The obtained result is then used as a new area where e.g. bushes are detected.

Estimation of the type of vehicles and their velocity

For the determination of the type of the vehicle several properties are combined, such as contours of the vehicles or a jump in the texture and brightness. All possible properties have to be fused for a better decision making. Often the contours are occluded or not representative for a vehicle. Than the texture of the surface of the vehicle has to be analysed distinguish cars from the road. If the traffic light is red clusters of vehicles are observed. In this case the division is possible by shrinking the areas of the vehicles by application of morphological operations. By textural analysis it is possible to eliminate the shadow of a vehicle from the vehicle self what is important for the estimation of the size of the vehicle. Especially if the road is wet and the headlights by reflection enlarge the size of the vehicle in the image. This effect can be reduced by analysis the combination of brightness and texture. For the estimation larger vehicles, like lorries and buses a parameter is obtained by division of the quadrate of perimeter and the area of the vehicle in the image. This is a measure for the "fuzziness" of the vehicle. A lorry is more "fuzzy", and so this parameter is higher. On the other side, a lorry is longer as a car and so the parameter is here higher than for a smaller vehicle. For a smaller vehicle the relation is nearly a circle, where this parameter gives nearly $p \approx (2\pi r)^2 / (\pi r^2) \approx 4$. If this parameter is much greater than 4, then for a lorry is decided. The windows of the vehicles are all mad by glass, so that all give nearly the same grey value if not a special reflection exists. Such a property is also used in the decision. Normally, the wheels are very good visible in the image and are important for detection of a vehicle.

By the fusion of all properties, the decision in classes of vehicles, pedestrians and cyclists is realised. The velocity of the vehicles is determined by using differences of images, taken at 3 successive times. Because the shape of the vehicles in time different images is only slow changed if the time interval is less enough, the vehicle shape of the two difference images is similar. For the determination of the velocity characteristic points or areas (i.e. window or wheel) are used for the comparison. The normalisation is given by comparison of the area of the vehicle compared the shifted part produced in the difference image by the moving of the vehicle. This is possible, because the width of the vehicles is less than the width of the lanes on the road. Because the vehicles are normally longer than broad, the direction of the moving of vehicles can determined by the difference images by using endpoint of the moved vehicles and calculating the angel by the arcus-tangens-function.

TRAFFIC CHARACTERISTICS, TRAFFIC RECORDING, REPLAY AND RECORDING PROCESSING

Image processing results – shape, size, location, velocity, and direction of vehicles represent a microscopic traffic object feature set. Based on this raw data, a number of traffic characteristics are calculated. Traffic density is measured as the number of vehicles per kilometre. Estimates of density are often based on measurements of local flow and velocity and the continuity equation $q = \bar{v} * k$, where q is flow, k is density and $\bar{v}$ is assumed to be the current average velocity. To measure the velocity a big number of inductive loops should be used. With a camera based system the location of all vehicles is known and a measurement section comprising an adequate area of the road is described by its coordinates. Using these variables the traffic density can be measured as $k = N / l$, where N is the number of vehicles in the measurement section at a certain time and l is the length of this section. Direction dependent flow rates are important for road planning. Simple traffic flow measurements at the arrivals and drains of an intersection allow the calculation of origin-destination relations only in special cases. Using the microscopic traffic parameters, each vehicle passing the camera observed intersection can be tracked from image to image and be marked with a unique identifier. Following all vehicles from arrival to drain the optical information system (OIS) allows calculating direction dependent flow rates. Another parameter for traffic situation at an intersection is the queue length in front of traffic lights. Queue length describes the number of vehicles with a velocity equal or near to zero. If a measurement section describes the inflow area of an intersection all vehicles standing there can be found by the image processing by means of the vehicle's location and speed. The number of cars or the distance between the first and the last standing car can be used as the queue length measure. Determining the queue length of all inflow areas and its equalization as the optimisation criteria is successfully used by a traffic light control simulation developed in the Institute of Transport Research (Mikat, J. *et al. 2003*) to show intersection throughput optimisation. Time gap measurement is frequently used for green time extension. Inductive loops located at a distance to the stop line detect arrival and measure speed of vehicles. Knowing distance to the stop line a corresponding travel time is assumed and a green prolongation time is calculated. If the vehicle retards and arrives later, the additional green time is lost, because no further measurements were available. With a camera based sensor, the vehicle's distance and speed can be measured repeatedly. If for the remaining green time $t_g < v_{t+i} / l_{t+i}$, with v_{t+i} being the velocity and l_{t+i} the vehicle's distance at a subsequent sampling moment is valid, the green time can be finished and earlier be allocated to the conflicting traffic stream. Supervising the intersection's conflict area is useful to detect dangerous situations e.g. caused by stalled trucks or left turners which could not flow off. If a camera based traffic sensor is used, it is easy to observe this area. Long-lasting occupancy, determined by vehicles inside the conflict area with speed zero indicates to a significant problem. To evaluate the traffic dynamic in a road section, e.g. to study or improve road safety, the frequency of lane changes can be of interest. The basic traffic object parameter set allows to track all objects. Measurement sections are used to describe the geographical layout of the lanes. Following the track for all objects within a lane, until they move to another, permits to calculate the number and frequency of lane changes. Microscopic traffic object features provided by the image processing to accomplish. Other traffic flow characteristics like overtaking recognition, blockage detection, interaction of different traffic types, influence of road layout on vehicle behaviour etc. also require the availability of microscopic traffic object parameters and can directly be calculated. The continuous availability of microscopic information about all vehicles in an observed traffic area allows to implement a new approach of traffic recording. Instead of storing video images over a longer time, only the parameters list of all traffic objects provided by the image processing are stored for each image in a data base. Additionally, a background image of the observed traffic area has to be stored. To replay the traffic scene the image space coordinates of the traffic objects are used to map symbolic vehicles into the background image. Repeating this process periodically with all sets of vehicle coordinates describing their position for subsequent time steps the vehicle will virtually move along the road in the traffic area image. In comparison with video recording a high degree of data reduction is achieved. A set of traffic object parameters to be stored for one video image which contains 100 traffic objects requires about 10KByte whereas a high resolution jpeg compressed video image requires 100KBytes. Furthermore, the traffic object parameters can be used to directly execute post-processing algorithms to calculate traffic parameters. To do this no further time consuming image processing is necessary. These ways traffic parameters can be computed in just a fraction of the real duration of the traffic scene. Instead of to re-replay a video image record after the measurement section for a traffic parameter has been moved to a more appropriate location only a recalculation using the traffic object parameters is done. When the traffic scene is analysed by automatic traffic characteristics calculation interesting properties could be found. E.g. an accident situation could be found by recognising that the average speed in an inflow section decreases significantly. If one wants to see the original images for this situation and this set is available, the time stamp in the traffic object parameter set can be used to select the corresponding images.

Conclusion

By application of well adapted mathematical methods to image processing of video camera images, traffic scene characterisation, traffic object detection and derivation of microscopic parameters give good results. It can be shown, that by use of texture and the fusion of different kinds of properties and their generalisation as fuzzy measures, vehicles, pedestrians and cyclists are detected nearly complete in real-time over the entire image. Microscopic traffic parameter information stored in a data basis allows calculate a big variety of traffic flow characteristics in an innovative way. New traffic characteristics impossible to measure with conventional sensors can be derived. The approach was tested for many thousands of traffic images in an automatic working regime and has successfully demonstrated its suitability.
This paper contains results achieved in the project OIS which was supported by the BMBF.

References

Bickel, P. J., Klaassen, C. A. J., Ritov, Y. and Wellner, J. A. (1993) *Efficient and Adaptive Estimation for Semiparametric Models*, Baltimore and London, The Hopkins University Press.

Hetzheim, H. (1993) *Using Martingale Representation to Adapt Models for Non-Linear Filtering, Proceedings ICSP'9*, Beijing, pp. 32-35, Intern. Academic Publication.

Hetzheim, H. (1999) *Analysis of hidden stochastic properties in images or curves by fuzzymeasure and functions and their fusion by fuzzy or Choquet integrals, Proc.* SCI'99 and ISAS'99 , Orlando, Florida, vol. 3, pp. 501-508.

Hetzheim, H. (2002) *Algorithmen zum Auffinden von Fahrzeugen und Bestimmung deren Größe und Geschwindigkeit aus Infrarot-Bildern,* VDI-Berichte 1731, Optische Technologien in der Fahrzeugtechnik, pp. 307-314.

Hetzheim, H. (2002) *Mathematical description of the fine structure of edge areas for traffic-control ICSP'02 Proceedings*, Beijing, pp. 691-694. Beijing, pp. 691-694.

Hetzheim, H., Dooley, L. S. (1995) *Structured non-linear algorithms for image processing based upon non-parametric descriptions, Conference Proceedings ACVV'9),* Singapore, Dec.1995, Vol. III, pp. 402-406.

Leich, A., Fliess, T., Jentschel, H.J. (2001) *Bildverarbeitung im Straßenverkehr,* Technische Universität Dresden, Institut für Verkehrsinformationssysteme, Dresden.

Mikat, J. et.al. (2003) *Agent Based Traffic Signals on a Basic Grid.* Institute of Transport Research German Aerospace Center.

Transport Science and Technology
K.G. Goulias, editor

CHAPTER 22

DISTRIBUTED INTELLIGENT TRAFFIC SENSOR NETWORKS

Mashrur Chowdhury, Civil Engineering and K.-C. Wang, Electrical and Computer Engineering, Clemson University

ABSTRACT

For real-time traffic control, transportation communities have long depended on human operators to continuously monitor a large number of video cameras and other sensors installed along highways. This chapter presents a distributed intelligent traffic sensor network that will automatically detect traffic incidents and initiate automated response. This system is based on a new generation of transmission-based infrared sensors, which measure vehicle speed, classification and count. Compared to existing reflection-based infrared sensors, the transmission-based system detects vehicle presence when passing vehicles block the infrared signal transmitted by a sensor. The low cost sensor is an ideal candidate for saturated deployment in a distributed sensor network. To facilitate the design and implementation of a distributed traffic sensor network, we propose the hierarchical network architecture. A discussion of the distributed control methodologies and network protocols is given. The chapter also presents the integrated simulation solution developed for studying the distributed traffic sensor networks.

INTRODUCTION

Given anticipated increases in highway traffic, the problems of traffic management and control will continue to expand. Many cites around the world have been using technologies and systems to better manage and control their surface transportation network under the intelligent transportation systems (ITS) umbrella.

The future of real-time traffic management lies in an integrated sensor network where sensor nodes, controllers and a centralized management center collaborate in an integrated sensor system through hardware, software and communication modules to support reporting and response for safe and efficient operation of highways. There is a great need to find a low cost, easy to deploy and reliable sensor that is suitable for saturated deployment for collaborative decision-making in the integrated sensor network. These sensors must rely on a communication network that is flexible and inexpensive. In this network, sensors and controllers should communicate with each other without relying on existing communication infrastructures such as wired connections or wireless base stations.

Several technologies have been developed over the years to detect vehicles and assess traffic characteristics. At the present time, inductive loop detectors are the most widely deployed. Current traffic detector technologies include intrusive detectors, such as inductive loops, magnetometer and magnetic detectors, and non-intrusive detectors, such as microwave radar, infrared sensors, ultrasonic, acoustics and video image processing (Klein 2001). Public agencies are seeking a cost-effective easy-to-deploy alternative to inductive loops for accurate traffic flow estimates.

Although existing loop detectors have many problems, the U.S. transportation community is yet to adopt newer detectors as an industry standard. In the past few years, several novel non-intrusive sensing technologies have emerged (Klein, 2001). Some of them are promising, but costly, such as active infrared and acoustic array sensors; others are comparatively cheaper, but are weather sensitive, such as passive infrared and ultrasonic sensors. One of the most promising sensor technologies is microwave radar, which is insensitive to weather, multi-lane capable, and moderately priced. However, microwave radar cannot detect vehicles stopped or moving at speeds lower than 15mph (Zhang, 2004), which commonly happens during highway incidents.

A reflection-based laser sensor system was studied (Cheng *et al.*, 2005; Tropatz, 1999; Harlow *et al.*, 2001). In the existing infrared sensor concept, the distance of propagation of the laser is large, which may cause excessive attenuation, diffraction, and scattering of the signal. Existing sensor systems are unsuitable, because of cost and deployment infrastructure requirements, for saturated deployment that will be necessary for collaborative decision making.

The transmission-based sensor system described in this chapter is reliable, low cost, and easy to deploy, which makes it suitable for saturated deployment for collaborative decision-making. An integrated sensor network that distributes the management and control of highway operations can bring the surface transportation system in line with state-of-the-art technology and revolutionize real-time traffic management. To plan, deploy, and operate such a network effectively in a selected traffic management area, a new generation of transmission-based optical sensor system is needed together with a collaborative decision making algorithm. This hierarchical wireless ad hoc sensor network architecture will facilitate automated reporting and response for real-time traffic management.

SENSOR DEVELOPMENT

The proposed sensor system is based on the simple concept of vehicles passing a pair of laser transmitters (laser diodes, LD) and photo detectors (PD) block transmission of infrared signals. The sensor system's initial design included three LD's and three PD's that can work together to measure traffic speed, vehicle size, and traffic volume on multiple lanes. Equipped with a radio frequency (RF) transmitter, the sensor sends readings to a PC-based base station without processing. Figure 1 shows a schematic of this first-generation sensor design, which details how the sensor hardware transmits its raw measurements to a fixed base station. However, for wide-scale deployment, each sensor node will have a microcontroller responsible for the communication and processing. Figure 2 shows a photo of the sensor prototype in the laboratory.

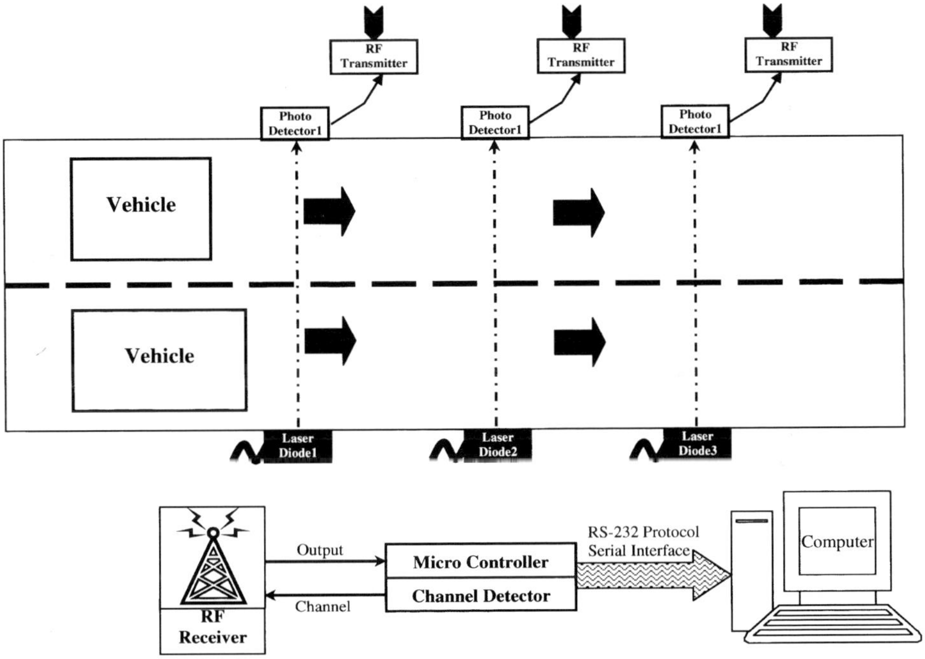

Figure 1. Schematic of the laboratory set-up of the sensor system.

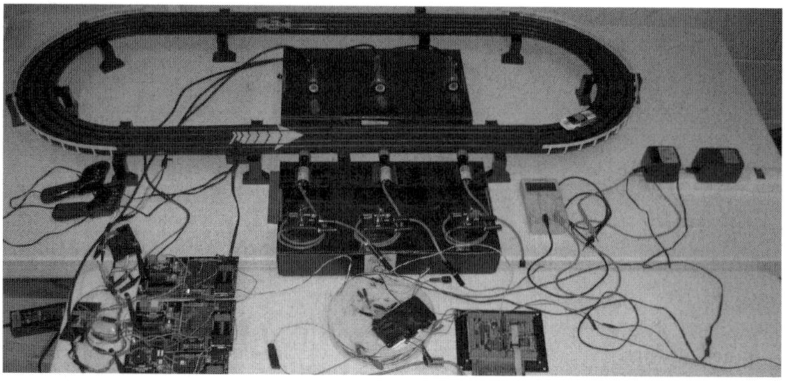

Figure 2. Transmission-based laser sensor system in the laboratory.

The communication platform will use wireless ad-hoc technologies, allowing each node to communicate to and through neighboring nodes, thus eliminating the need for communication infrastructure. The wireless traffic sensor network assumes that each node can independently sense, process, and communicate measurements to arbitrary devices over a wireless ad hoc network.

After the experimental prototype system was built and laboratory tested, the system was field tested on a north-bound section of US 123, a multilane divided highway in South Carolina, near Clemson (Atluri, 2005). In an earlier field test along a stretch of Ohio highway near Dayton, these sensors successfully counted the number of vehicles and accurately measured their velocities and sizes (Goodhue, 2004). Figure 3 shows a truck approaching a pair of sensors on US 123 in Clemson, South Carolina.

FIELD EVALUATION OF SENSORS

Statistical analysis was performed to compare the sensor collected data with data collected manually. Table 1 shows count data by vehicle types and mean speed and variance collected manually, and generated by the sensor system. A Chi-Square test of vehicle classification data, collected manually at a 95% confidence interval, revealed no significant difference between the classification of vehicles obtained from the sensor system and the data (Atluri, 2005). Similarly, the results of an independent t-test suggested similar differences of the mean speed of the manually collected data with the sensor generated data. This data also reflected a 95% confidence interval. A two-sample z-test was used to compare the count obtained by the sensor with the manual count. These results also indicated no significant difference between the manually collected vehicle count and sensor generated counts. A 95% confidence interval was again applied.

Figure 3. A truck approaching a LD/PD pairs on US 123.

Table 1. Manual recorded and sensor measured data.

Data Type	Passenger Cars	Single Unit Truck	Single and Multi-Trailers	Speed	
				Mean	Variance
Manual	337	11	11	58.033	7.882
Sensor	312	9	9	58.182	7.822

Although the tests provided encouraging results on the sensor system, US 123 is a low volume rural highway. Further tests should be conducted on highways with higher traffic volumes and consequently higher occlusion rates.

SENSOR INTELLIGENCE

Individual sensors provided with limited computing power and low level of intelligence will detect incidents using estimated traffic parameters. This detection is based on a lower level of abstraction of sensor intelligence and hence needs verification. For verification, multiple sensors will communicate with a cluster-controller, which has a higher level on intelligence and computing power. Multiple cluster controllers may also communicate with each other to verify an incident and initiate an action. Figure 4 shows the functional components of the proposed sensor. Computation algorithms will be embedded in each sensor to estimate traffic characteristics and initiate control functions. With the transmission-based laser technology, sensors can be effectively installed along highways and activated for independent sensing operations.

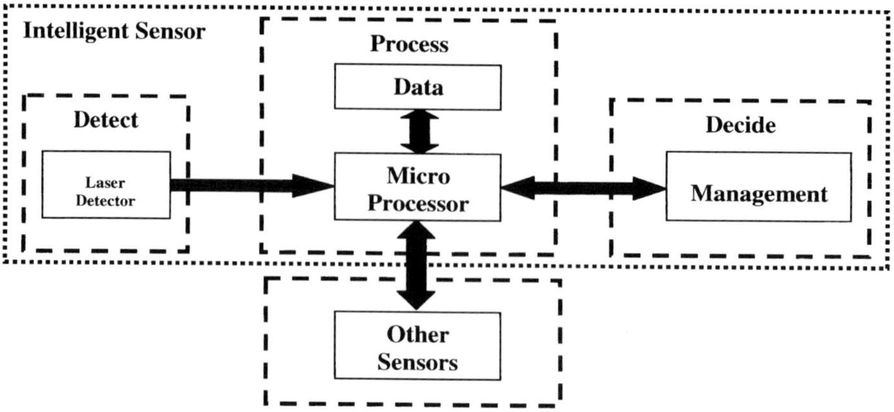

Figure 4. Sensor architecture for automated response.

Figure 4 also shows the process of highway incident management using the optical traffic sensors and ad hoc wireless network. Each sensor independently collects measurements in its monitored road segment at programmed intervals. Using on-board processing capabilities, each sensor processes its measurements into robust statistics. The sensor then uses these statistics to apply intelligent decision-making algorithms to detect any abnormal situations.

While these sensors are attractively simple, their inherently limited views render their decisions highly sensitive to short-term flow variations and may inaccurately represent global traffic conditions throughout the entire road. Therefore, it is essential that decisions made by sensors, leading to any form of control action, be both reliable and verifiable. In the traffic sensor network, individual sensors collaborate to produce reliable incident detections. Incident detection is always performed in two phases: local detection and clustered verification. When a sensor "decides" that an abnormal situation has occurred, it requests nearby sensors in the same cluster for verification of the incident. The verification can also be initiated by a cluster controller; a sensor that is also responsible for collecting detection alerts from multiple sensors in its cluster. Once verified, the incident is robustly detected and subsequent control actions may be initiated.

DISTRIBUTED DECISION MAKING

In the proposed architecture, the measurements are processed at each node, while retaining only statistics over specified periods. When traffic is steady, no communications are necessary for energy conservation. When abnormal measurements are observed, nodes exchange measurements with neighboring nodes for confirmation and alert local controllers. Based on its assigned task, a local controller can report the event to other controllers or control centers, issue commands to adapt the sensing operations (perhaps switching to a

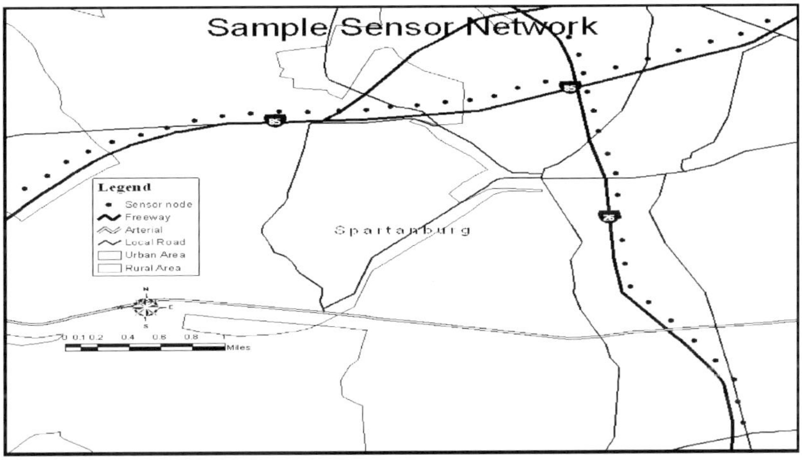

Figure 5. Sample wireless sensor network along the I-85 corridor in South Carolina.

higher resolution), or initiate any necessary automated responses. Local controllers and control centers are entities that can interface with other traffic control entities such as the traffic signaling systems.

Message communications in traffic monitoring sensor networks are highly structured in their temporal and spatial distributions, which are driven by road geometries and semantics of the supported monitoring and response operations. Figure 5 shows a sample network of the proposed physical network. A number of studies have exploited sensor networking methods for intelligent transportation systems (Goel *et al.*, 2003; Nadeem *et al.*, 2004; Sawant *et al.*, 2004), all of which have considered sensors mounted on moving vehicles that communicate opportunistically in a dynamic network. None have addressed reliable communication for fixed sensor networks deployed along roadways in extended ranges and massive quantities. The proposed project will identify the communication structure of fixed networks for which the hierarchical network architecture and associated networking methods are defined to facilitate operations with enhanced efficiency and assured quality of service.

WIRELESS NETWORK ARCHITECTURE AND PROTOCOLS

The intelligent traffic sensor network achieves its monitoring and control functions by transporting sensor data and control decisions over a wireless network. Because each sensor can only acquire measurements about its limited vicinity, information from multiple sensors are combined to form a coherent view of a road's traffic flow and to make robust control decisions accordingly. A limited number of studies have considered the use of sensor networks in intelligent transportation systems, which are limited to sensors placed on vehicles (Sawant *et al.*, 2004).

The problems of data transport and collaborative data processing in sensor networks have been studied extensively (Intanagonwiwat *et al.*, 2003; Wang and Ramanathan2005; Gridhar and Kumar, 2005). The majority of these studies have considered sensor networks for either monitoring a field or tracking targets intruding a closed area, both of which typically assume a *flat* network of numerous sensors deployed in the area. For monitoring, sensor data are transported towards a few *data sinks* that are placed among the sensors (Sankarasubramaniam *et al.*, 2003; Xu *et al.*, 2004). For target tracking, sensor data can be processed directly at sensor nodes given sufficient computing power, while the processed results (detection decisions) are sent to either a central server and/or their immediate neighbors (Wang and Ramanathan, 2005). In both scenarios, sensor data and decisions can be generated at arbitrary locations and then be transported towards any direction from a sensor. The incurred data pattern is not usually known before any event occurs.

On the contrary, traffic sensor networks are deployed along highways stretching long distances in regular geometries, and their communication patterns are inherently guided by the physical infrastructure. In a typical traffic sensor network, it is expected that upstream and downstream sensors in each road segment exchange their data in assessing the traffic flow, while traffic conditions may be exchanged among multiple road segments for making traffic load balancing decisions to facilitate dynamic traffic assignments in order to improve the overall network flow. By exploiting the inherent structures in their communications, the network can be more accurately provisioned and the sensor operations can be more systematically programmed. To this purpose, we define the hierarchical network architecture for data communication in traffic sensor networks.

The hierarchical network architecture consists of logical entities of sensors, clusters, controllers, and control centers. Each entity is defined with a specific set of operations. A physical device can serve one or more purposes according to its computation and communication capabilities and the system's operational needs. The entities' functions are summarized below:

- Sensors – Sensors produce raw measurements and process them into robust statistics and, optionally, perform local control decisions. Via their wireless network interface, measurements, statistics, and decisions can be communicated with other sensors and controllers for collaborative operations.
- Clusters – Clusters are sets of sensors grouped for collaborative operations. To facilitate operations of different scopes, clusters can be defined in multiple levels of hierarchy. A cluster can consists of any number of sensors, as well as any number of sub-clusters. A cluster provides the reference for programming data aggregation, information relay, and decision making within its defined scope.
- Controllers – Each cluster has a designated controller, while each controller can serve for one or multiple clusters. Controllers perform data aggregation and decision making for the cluster(s), and collaborative operations with controllers in other clusters. Based on collected information, a controller may detect abnormal traffic conditions, initiate response actions upon detected traffic incidents, and

potentially interface with actuation systems such as traffic signal control, reversible lane control, and freeway ramp metering for incident resolution.

- Control centers – Traffic control centers are anchors of the system administered by human operators. They are at the top of the hierarchy and oversee a traffic sensor network's operations over large geographic areas determined by highway geometry and administrative boundaries. Control centers obtain periodic status updates from controllers, audit automated operations, initiate or authorize incident response actions, and dictate specific operations to be done by the system.

The architecture allows transportation engineers to effectively program monitoring and control operations among logical entities within the hierarchy, rather than the individual physical devices. Once deployed and assigned its role in the hierarchy, each device carries out their designated operations and interacts with other entities according to their roles. The role assignment can be done in full or semi-automation with preloaded or adaptively discovered information. For example, each sensor may be loaded with minimal required information such as the identification of the highway they are servicing and location marker within that highway. Once deployed, the sensors can communicate with their neighbors, collaboratively determine their assignments, and start carrying out their corresponding tasks.

To facilitate efficient communication in the hierarchical network, networking protocols must be tailored to its unique structure. The following summarizes the unique challenges and potential solutions in various protocol layers:

Physical Layer (PHY)

Various predominant PHY protocols are used in wireless ad hoc networks with different bandwidth, fading, and power consumption characteristics, e.g., the IEEE 802.11, 802.15.4, 802.16 standards. The devices' transmission power most directly affects the choice of network topology. The larger the transmission power, the farther a device can communicate, and the shorter latency (less hops) needed for sensors to relay data or decisions towards controllers. Finally, a lesser number of sensors are needed along a road segment to maintain wireless connectivity. Wireless sensors are assumed to operate on batteries. A higher transmission power implies a shorter sensor lifetime before batteries must be replenished. On the other hand, one can provision selected devices with more resources (energy, processing power) such that they can take on additional responsibilities being controllers. It is expected that devices will have diverse communication capabilities, such as short range communication among sensors, moderate range communication among local controllers, and long range communication among higher level controllers and the control center.

Link Layer

Link layer protocols maintain connection between immediate neighbors and regulate their transmissions in the wireless medium to avoid collisions. While there is not a single protocol of choice for wireless ad hoc networks, industrial standards such as IEEE 802.11 and IEEE

802.15.4 are used in many experimental ad hoc networks. IEEE 802.11 and its multiple variants (IEEE 802.11 a/b/g) define a suite of Link and PHY layer protocols that provide an easily set up random access method with bandwidth ranging from 1 to 54 Mbps (higher bandwidth achieved at shorter distances). Originally designed for broadband local area networks, IEEE 802.11 is overkill for low data rate sensor applications and its power consumption is significant. IEEE 802.15.4, a low power, low data rate (20-250 kbps) link and PHY specification for ad hoc connection in personal area networks, is hence adopted in more recent sensor network applications. In addition to a random access protocol similar to that of IEEE 802.11, a contention free transmission mode is also available (IEEE 802.11 also specifies a contention free mode, which is not implemented in most devices today). While random access permits more efficient bandwidth utilization for on-demand communications due to spontaneous events, regular data exchanges are expected to benefit from predictable contention free accesses. The benefits of utilizing both scheduled and random access have been demonstrated; one such example can be seen in (El-Hoiydi, 2002).

Network Layer

Network layer serves the purpose of addressing devices in the network and routing data between any two devices. Because sensors are meant for producing location dependent measurements, it is the location of the device rather than device identities that matters in most sensing operations, which may request data from a certain geographic area rather than a named set of sensors. Sensors also respond to queries according to where they are. Such a location-centric concept is demonstrated in a target tracking application, where programmers define geographic regions (Wang, 2005). As such, they implicitly task all sensors within such regions to carry out the tracking algorithms. In this context, geographic routing schemes are needed to facilitate data transport (Ko and Vaidya, 2000; Karp and Kung, 2000; Rao *et al.*, 2003; Kim *et al.*, 2004). However, geographic location alone does not adequately reveal the structure in a traffic sensor network where sensors must transport data upstream and downstream a highway or among structurally related segments (for example, two segments branching from the same upstream segment.) Sensors and controllers may not know the location of all other controllers they may want to communicate with, and it may be extremely cumbersome to exploit and maintain such information.

Given the hierarchical architecture, a hierarchical addressing scheme can be effectively established. Utilizing hierarchical addresses, a routing protocol can readily identify paths from any device towards other hierarchically related devices of a specific hierarchical scope, such as "all upstream sensors within the same cluster", "controller of this cluster", or "all controllers within the same parent cluster." This protocol is adequate for defining an operation in a traffic sensor network, while maintaining the advantages of simplicity and generality.

A hierarchical addressing scheme naturally suggests the logical forwarding paths from any sensor in the network. As wireless connectivity among sensors do not necessarily align with the physical road structure, the addressing scheme may need to be bootstrapped with their associated road identity. Such configuration may be done at deployment, when each node is manually provided its road identity and mileage, with which a locally unique hierarchical addressing can be sought for the network nodes.

Transport Layer

User Datagram Protocol (UDP) and Transmission Control Protocol (TCP) are the two most widely utilized transport protocols; the former supporting connection-less unreliable datagram delivery and the latter supporting reliable data stream connection between two end points. Most sensing applications adopt UDP, as it is often unnecessary to assure the successful delivery of any particular packet, provided that data from many sensors in the same vicinity are highly correlated. The same principle may apply to data delivery in a traffic sensor network. Yet, delivery of safety critical control information must be explicitly assured. For example, control messages should either be sent with TCP connections, or reliable UDP transports with explicit acknowledgement and error recovery.

Application Layer

On top of the entire networking stack, application layer composes the logical operations of sensors and controllers. Typical sensor operations may include the following:

- periodic traffic measurements;
- periodic status reports;
- local detection of specified events (congestion, accidents, speeding, severe driving conditions;
- collaborative detection of specified events with upstream/downstream sensors, and;
- instantaneous notification of detected events.

Typical controller operations may include the following:

- aggregation of sensor reports;
- estimation and predicting traffic flow;
- local detection of specified events;
- collaborative detection of specified events with related cluster controllers;
- initiation of local control actions (alert display, traffic detour, metered ramp control, etc.), and;
- initiation of coordinated control actions with related cluster controllers.

In a hierarchical context, sensor and controller operations can be programmed without the knowledge of the actual network topology. Instead, all operations can be implemented by defining the behavior for each entity type and the interactions between any two entities. Additional support for robust computing can be provided by implementing standard distributed computing functionalities to address issues such as fault tolerance, synchronization, transaction consistency, and group agreement requirements.

NETWORK SIMULATION FOR DEPLOYMENT PLANNING

Research communities have been utilizing various academic or commercial simulation tool sets to study network protocols and applications. The most widely adopted simulator is the ns-2 (Network Simulator version 2) developed by joint projects at USC/ISI, LBL, Xerox PARC, and UCB (ISI, 1989). The ns-2 project commenced in 1989 and remains in active development today. The wide acceptance of ns-2 as a de-facto research platform has been due to its open source, extensible architecture, modular designs, and support of standard protocols ready for interfacing with new applications. The latest version (ns-2v2.28 on Feb 3, 2005) supports standard protocols for Internet, mobile, satellite and ad hoc wireless networks. Research groups also used this open platform to publish newly developed protocols and applications for community-wide evaluation. For example, the latest sensor network medium access protocol (standard IEEE 802.15.4) has been made available as an ns-2 module.

With ns-2, packet level simulations can be done by simulating all chosen protocols' processing logic with random network effects such as event occurrence, sensor noise, detection uncertainty, and communication latencies for application in the highway network. Figure 6 shows one example result obtained with ns-2, assessing the average communication latencies and packet loss probabilities of periodic UDP sensor reports between a sensor and a controller at different reporting rates and sensor-to-controller distances. Such decisions assist in determining the proper location of sensors and controllers in the highway network, the proper reporting rate to maintain reliable network operations, and the necessity of explicit delivery assurance mechanisms.

In addition to the wireless network oriented dynamics, random traffic patterns will also result in random network disturbances. To simulate dynamic vehicular traffic flow and their effects on a traffic sensor network, ns-2 can be extended to interface with available traffic simulators. The authors are currently integrating the traffic simulator PARAMICS with ns-2 to construct a comprehensive simulation platform of traffic sensor networks.

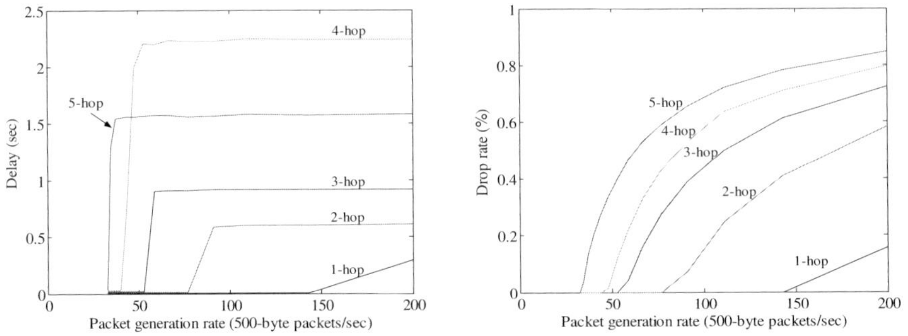

Figure 6. Sample results of average sensor report delay and drop rate simulated in ns-2.

CONCLUSIONS

This chapter presents a framework for an integrated sensor network that holds promise for real time traffic management in a large highway network. A new generation of low-cost laser sensor system that accurately measures vehicle speed, size and presence provides the foundation for saturated deployment necessary for the sensor network. Two field tests, one conducted in Dayton, Ohio and another in Clemson, South Carolina revealed that the sensor system was able to accurately provide vehicle speed, size and presence. However, further tests should be conducted under a wide range of traffic conditions to produce detailed information on system performance and modify the system for large scale deployments.

In addition to the low cost sensors, this chapter presented the ad hoc wireless network sustaining the sensors' distributed decision making. With wireless sensors and controllers communicating data and decisions, the network autonomously monitors traffic in a distributed and collaborative fashion, reporting and resolving any detected events. The strengths of the proposed sensor system include the ability of each node to independently sense and process information, and to initiate response with collaborations between the sensor nodes. The proposed hierarchical architecture facilitates effective design and implementation of a wide range of control operations over the highway network. The chapter ends with a discussion on the challenges and suggested strategies for designing the network protocols and an integrated simulation platform for studying the network performances.

REFERENCES

Atluri, M. (2005). *Design and Evaluation of Transmission-Based Optical Sensor System for Intelligent Transportation Systems Applications.* M.S. Thesis, Clemson University.

Cheng, H. H., B. D. Shaw, J. Palen, B. Lin, B. Chen, and Z. Wang (2005). Development and field test of a laser-based nonintrusive detection system for identification of vehicles on the highway. In: *IEEE Trans. on Intelligent Transportation Systems*, Vol. 6, pp. 147-155.

El-Hoiydi, A. (2002). Spatial TDMA and CSMA with preamble sampling for low power ad hoc wireless sensor networks. In: *Proc. ISCC 2002*, pp. 685-692.

Goel S., Imielinski T., Ozbay K., Nath B. (2003), Sensors on wheels – towards a zero infrastructure solution for intelligent transportation systems. In *Proceedings of the International Conference on Embedded Network Sensor Systems,* pp. 338-339.

Goodhue, P. (2004). *Development of Laser Based Sensor System.* M.S. Thesis, University of Dayton.

Gridhar, A. and P.R. Kumar (2005). Computing and communicating functions over sensor networks. In: *IEEE J. on Selected Areas in Communications*, Vol. 23, pp. 755-764.

Harlow, C. and S. Peng (2001), Automated Vehicle Classification System with Range Sensors. In: *Transportation Research Part C: Emerging Technologies*, Vol. 9, pp. 231-247.

Intanagonwiwat, C., R. Govindan, D. Estrin, J. Heidemann, and F. Silva (2003). Directed diffusion for wireless sensor networking. In: *IEEE/ACM Trans. on Networking*, Vol. 11, No. 1, pp. 2-16.

ISI (1989), The Network Simulator – ns-2. In: *http://www.isi.edu/nsnam/ns/.*

Karp, B. and H. T. Kung (2000). GPSR: greedy perimeter stateless routing for wireless networks. In: *Proc. MobiCom*, pp. 243-254.

Kim, Y., J.-J. Lee, and A. Helmy (2004). Modeling and analyzing the impact of location inconsistencies on geographic routing in wireless networks. In: *ACM SIGMOBILE Mobile Computing and Communications Review*, Vol. 8, No. 1, pp. 48-60.

Klein, L. A. (2001). *Sensor technologies and data requirements for ITS*. Artech House, Boston MA.

Ko, Y.-B. and N. H. Vaidya (2000). Location-aided routing (LAR) in mobile ad hoc networks. In: *Wireless Networks*, Vol. 6, No. 4, pp. 307-321.

Nadeem T., Dashtinezhad S., Liao C., Iftode L. (2004). Traffic view: traffic data dissemination using car-to-car communication, *ACM SIGMOBILE Mobile Computing and Communications Review,* vol. 8, no. 3, pp. 6-19.

Rao, A., C. Papadimitriou, S. Shenker, and I. Stoica (2003). Geographic routing without location information. In: *Proc. MobiCom*, pp. 96-108.

Sankarasubramaniam, Y., B. Akan, and I. F. Akyildiz (2003). ESRT: event-to-sink reliable transport in wireless sensor networks. In: *Proc. MobiHoc*, pp. 177-188.

Sawant, H., J. Tan, Q. Yang, and Q. Wang (2004). Using Bluetooth and sensor networks for intelligent transportation systems. In: *Proc. Int. Conf. Intelligent Transportation Systems*, pp. 767-772 .

Tropatz, S., E. Horber, and K. Gruner (1999). Experiences and results from vehicle classification using infrared overhead laser sensors at toll plazas in New York City. In *Proc. IEEE/IEEJ/JSAI Int. Conf. Intelligent Transportation Systems*, pp. 686-691.

Wang, K.-C. and P. Ramanathan (2005). Location-centric networking in distributed sensor networks. In: *Distributed Sensor Networks* (S. Iyengar and R. Brooks, eds.). CRC Press.

Xu, N., S. Rangwala, K. K. Chintalapudi, D. Ganesan, A. Broad, R. Govindan, and D. Estrin (2004). A wireless sensor network for structural monitoring. In: *Proc. Int. Conf. Embedded Networked Sensor System*, pp. 13-24.

Zhang, Y. and L. M. Bruce (2004). Automated Accident Detection at Intersections. In: *Federal Highway Administration and Mississippi Department of Transportation Project*, No. FHWA/MS-DOT-RD-04-150.

Transport Science and Technology
K.G. Goulias, editor

CHAPTER 23

USE OF TRAFFIC MANAGEMENT CENTER AND SENSOR DATA TO DEVELOP TRAVEL TIME ESTIMATION MODELS

Jiyoun Yeon and Lily Elefteriadou, University of Florida, U.S.A

ABSTRACT

Travel time is widely recognized as an important performance measure for assessing highway operating conditions. Traffic Management Centers (TMCs) however have typically collected and used speed and flow as performance measures. Also, previously developed travel time estimation models under congested conditions underestimate the travel time due to difficulties in calculating queue formations and dissipations. The purpose of this study is to develop a model that can estimate travel time on a freeway using Discrete Time Markov Chains (DTMC) with data typically available by TMCs. The methodology considers whether the link is congested or not based on the probability of breakdown. The estimated travel time for a given Origin-Destination (O-D) can be obtained for a given time period as a function of demand. In this paper, p.m. peak period data from a Philadelphia, PA, site are used to develop the model, and results are compared to field-measured travel time. It was concluded that the proposed model accurately predicted travel time along the route, at the 95% confidence level.

1. INTRODUCTION

Travel time is widely recognized as an important performance measure for assessing highway operating conditions. Turner *et al.* (2004) indicate that travel times are easily understood by practitioners and the public, and are applicable to both the users' and operators' perspective. Also, the California Department of Transportation (Caltrans) suggested that travel time is one of the possible performance measures in the PeMS (Performance Measurement System) which can provide information in various forms to assist managers, traffic engineers, planners, freeway users, and researchers (Choe, 2002). However, a number of empirical studies have demonstrated that travelers are interested not only in the time it usually takes to complete a trip, but also in the reliability of travel times (Turner *et al.*, 2004). There are two performance measures for assessing travel time uncertainties: travel time reliability and travel time variability. Travel time reliability has been defined as the probability that a trip can be made within a specific duration of time. Travel time variability represents the dispersion of travel time, i.e., it can be simply considered as a standard deviation of travel time.

A number of studies have attempted to develop algorithms for estimating travel time using advanced surveillance systems: video image processing, automatic vehicle identification, cellular phone tracking, probe vehicles, etc. However, most existing infrastructures managed by TMCs are loop detectors or microwave sensors which provide point detection data, and usually speeds are converted into travel time from those data. Oh *et al.* (2003) indicated that travel time estimates from point detection of speeds under congested conditions are underestimated.

The objective of this study is to estimate Origin (O) - Destination (D) travel time for a given route as a function of the demand at each segment of the route, using field data typically available by TMCs. This paper focuses on the usage and treatment of sensor data provided by TMCs to develop travel time estimation models. The methodology employs the concept of probabilistic breakdown for freeway segments, and is based on Discrete Time Markov Chains (DTMC). The concept of probabilistic breakdown is used to determine the probability of congestion occurrence at each freeway segment. Field data were obtained from the Philadelphia, PA, TMC for several weekdays without accidents/incidents or rain. The scope of the paper includes congestion due to heavy traffic, and not due to weather, accidents/incidents, or work zones. The next section of the paper briefly reviews the most relevant literature. After that, the data obtained for developing and validating the model are presented. The fourth section presents the methodology for developing the travel time estimation model using DTMC. To evaluate the model developed, estimated travel times are compared to field-measured travel time data, and the results are presented in the fourth section. The last section of the paper presents conclusions and recommendations.

2. LITERATURE REVIEW

During the past few years, many researchers have focused on estimating or predicting travel time with various methodologies such as loop detector data (Oh *et al.*, 2003; Chu, 2005), time series models (Shaw and Jackson, 2002), artificial neural network models (Shaw and Jackson, 2002), and regression models (Zhang, 2003). However, travel times are rather sensitive to the prevailing conditions, and the models developed do not work well for all conditions. It is particularly complicated to estimate travel time under congested conditions due to difficulties in calculating queue formations and dissipations. Thus, it is required to identify the impact of congestion in time and space in order to accurately estimate travel time. Stochastic processes are an appropriate method for solving the problem because they can analyze and predict conditions (states) in time and space, subject to probabilistic laws.

Yang *et al.* (1999) indicate that travel time is easily influenced by high demand which could result in a higher probability of breakdown. Breakdown can be defined as the time period during which speed drops sharply in a time series speed plot, and it can be considered as a transition state which can be used as an indicator to distinguish congested from non-congested conditions (Lorenz and Elefteriadou, 2000, and Persaud *et al.*, 2001). Once breakdown occurs, traffic typically becomes congested. However, even at the same location with the same demand levels, breakdown may or may not occur (Elefteriadou *et al.*, 1995). Thus, breakdown has a probabilistic nature, and in order to estimate travel times, it is essential to incorporate the breakdown probability occurrence.

Evans *et al.* (1998) developed an analytical model for the prediction of flow breakdown based on zonal merging probabilities. Firstly, the authors determined the arrival distribution of the merging vehicles, and then calculated the transition probabilities from state to state. The probability of breakdown was obtained using Markov chains and implemented in MATLAB. Castillo (2001) developed a model to reflect the propagation of speed drop when the traffic volume reaches capacity. The author considered a speed drop caused by lane changing or merging of other traffic streams as a complex stochastic phenomenon and developed a theoretical model that can provide normalization probability of speed drop behaviours. The proposed model sets a recurrence equation for the speed drop and its duration. Heidemann (2001) developed a model to describe the flow-speed-density relationship using concepts from queuing theory for non-stationary traffic flow which implies that input and output flow are not in statistical equilibrium. To illustrate how the queuing theory is applied, the author partitioned the road into pieces of length $1/k_{jam}$, where the processing time would be $1/(k_{jam} \cdot v_f)$ if the vehicles run at their desired speed. Thus, the flow rate is $\mu = k_{jam} \cdot v_f$ and the arrival rate is $\lambda = k \cdot v_f$. According to the developed model the space mean speed, denoted v(k), for traversing the distance $1/k_{jam}$ at density k with desired speed (v_f) is identified. Based on the derived space mean speed at density k, v(k), some traffic characteristics such as flow, density, queue length, travel time, etc. can be obtained. Kharoufeh and Gautam (2004) derived an analytical expression for the cumulative distribution function of travel time for a vehicle travelling on a freeway link. The authors assumed that vehicle speeds are a random environment and can be considered as a finite-state Markov process. The random environment process includes physical factors (e.g., roadway geometry, grades, visibility), traffic factors (e.g., density, presence of heavy vehicles, merging traffic), or environmental factors (e.g., weather conditions, speed limits, etc.). Based on the assumption that the environmental process is a Continuous Time Markov Chain (CTMC), an exact analytical expression is obtained for the Laplace transformation of the link travel-time cumulative distribution function.

In conclusion, stochastic processes have been used in the literature to analyze changes of traffic states over time and space, and thus they are an appropriate method for estimating travel time for congested conditions.

3. DATA COLLECTION

Model development requires the collection of appropriate data, which can be grouped into two categories; model development data and model validation data. Model development data are needed to build a model and apply the proposed methods, while model validation data are needed to verify statistically the accuracy of the model developed. In this paper, speed and flow data are used for model development, and link travel time or route travel time data are used to validate the developed model. To take into account the typical breakdown phenomena and day-to-day demand variations, the study area selected carries commuter traffic and has nearly ideal geometric conditions. i.e., level terrain, absence of severe curves, adequate sight distance, etc. The selected study site is a 9-mile long segment of US-202 Southbound which carries commuters in the Philadelphia, PA, area. The speed and volume data were obtained by Mobility Technologies. The data were collected by RTMS (Remote Traffic Microwave Sensors), owned by Mobility Technologies, for a 4-month period, from May to August, 2004. Specific sensor locations and distances between sensors are shown in Figure 1. RTMS provide 1 minute speed, volume, and occupancy data which were used for further analysis.

Travel time data were collected at the TMC located in Philadelphia, PA. In the TMC, there is a CCTV (Closed Circuit Television) monitoring system where images are displayed from cameras in the field. Each camera can be controlled manually by panning, tilting, and zoom-in, zoom-out from the TMC, and the image viewed can be recorded using a VCR (Videocassette Recorder). For the validation of this study, travel time data were collected between three pairs of cameras (Figure 1): C 206 & C 214, C 206 & C 212, and C 205 & C 213 during different dates and times. The cameras focused on the adjacent detector locations, for which speed and flow data were obtained from Mobility Technologies. Travel time data were collected on two weekdays: May 20th (Tue.), and May 25th (Tue.), 2004. Each camera had a time clock displayed on the screen. During the data collection, a vehicle would be identified in the first camera and its arrival time would be recorded for the first upstream camera observed. Then, when that same vehicle was identified in the downstream camera, its departure time would be recorded. Table 1 provides a summary of the travel time data collection.

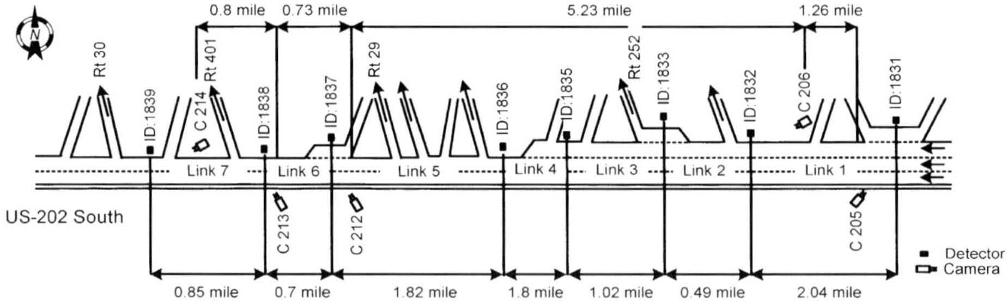

Figure 1. Camera and detector locations along US-202 SB (not to scale)

Table 1. O-D Travel time data summary

Camera O-D Pair	Detector ID Number	Date and Time	Recording Hours	Length[1] (mile)	Collected Sample Size
C 206 & C 214	1832 – 1839	5/20 15:15-16:20 5/25 11:05-13:00	1h 05m 1h 55m	6.68	120 199
C 206 & C 212	1832 – 1837	5/25 15:30-16:30	1h	5.13	102
C 205 & C 213	1831 – 1838	5/25 17:50-19:00	1h 10m	7.87	53

Note: 1) The length is based on the distance between detectors, not the distance between cameras.

3.1 Standardizing Analysis Intervals in Raw Data

Before using the raw data, it is necessary to standardize the time interval of each link data because the collected data are stored in slightly different time interval clocks at each detection location. The raw data collected by RTMS are not synchronized, and the data collection intervals by locations differ by more or less 1 minute. The starting time of data collection is not exactly the same at each location (Table 2), while the duration of each time interval may also not be exactly 1 minute. For example, the time between the start of intervals No. 10 and 11 is exactly 1 minute, while the time between the start of intervals No. 11 and 12 is less than 1 minute (i.e., 59 seconds). Thus, the data should be adjusted accordingly.

Table 2. Raw data set before standardizing time interval at detector ID 1836

Interval No.	Detector ID	Time	Speed (mph)	Volume (veh/min)	Occupancy (%)
:	:	:	:	:	:
10	1836	07:09:01	61.20	55	11
11	1836	07:10:01	60.27	64	11.5
12	1836	07:11:00	68.35	48	6.5
13	1836	07:12:00	59.65	58	12
14	1836	07:13:00	58.72	69	12.5
15	1836	07:14:00	59.03	72	12
16	1836	07:14:59	57.79	66	12
17	1836	07:15:59	65.24	43	6
18	1836	07:16:59	58.10	49	8.5
20	1836	07:17:59	56.86	83	17
21	1836	07:18:58	54.99	56	9.5
:	:	:	:	:	:

The main concept of the adjustment process is to apply the weighted average of time length of each interval. The methodology steps are as follows:

- *Step 1*: Find a time interval which starts with seconds as "00", for example interval No. 12 in Table 2.
- *Step 2*: Create standardized time intervals, such as 07:01:00, 07:02:00, etc. at the beginning of the corresponding time interval
- *Step 3*: Calculate the difference between the actual start time of interval i+1 and the standard time interval i, and call this time length A_i
- *Step 4*: Calculate the difference between the actual start time of time interval i+1 and the standard time interval i+1, and call this time length B_i
- *Step 5*: Calculate the ratios of time length A_i and B_i to 1 minute; call these ratios A_i' and B_i' respectively
- *Step 6*: Calculate the adjusted volume and speed (i.e., standardized time interval data)

Time lengths A, B, A', and B' can be expressed in the following equations:

$$A_i = | t_{Actual(i+1)} - t_{Standard(i)} | \qquad A'_i = \left(\frac{A_i}{1\,\text{min.}} \right)$$

$$B_i = | t_{Actual(i+1)} - t_{Standard(i+1)} | \qquad B'_i = \left(\frac{B_i}{1\,\text{min.}} \right) = 1 - A'_i$$

Figure 2 shows an example for calculating the standardized time for volume. In this example, the length of intervals between solid lines represent the standardizing time interval and the grey bars represent the actual data points which are more or less than 1 minute in duration. To obtain the standardized volume of time interval i, it is required to obtain the i-th and (i+1)th volume, and the converted ratio of time to 1 minute, A_i' and B_{i-1}'. Then, the standardized volume can be computed by $A_i' \times$volume (i) + $B_{i-1}' \times$volume (i+1). This is appropriate, because the i-th interval is closer to i-th standardized time interval. Speed can also be converted to standard 1- minute interval data using the same procedure.

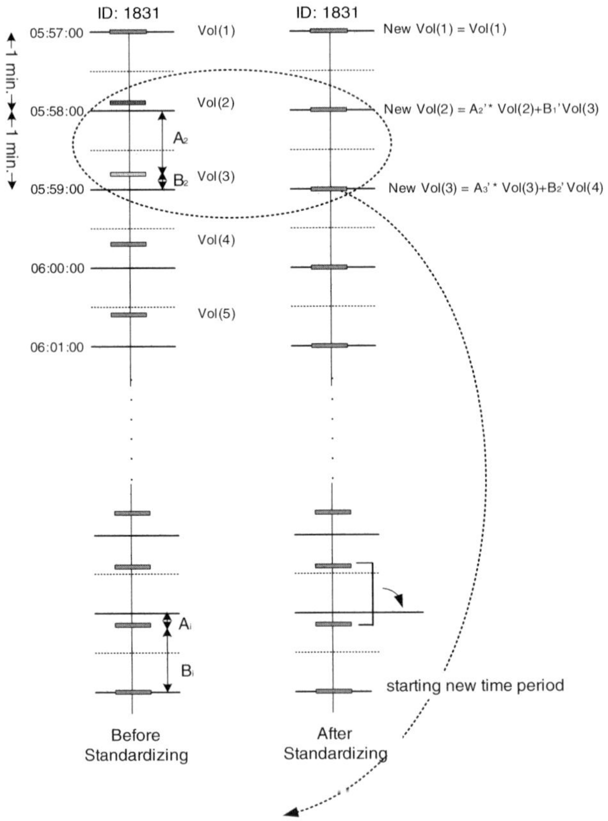

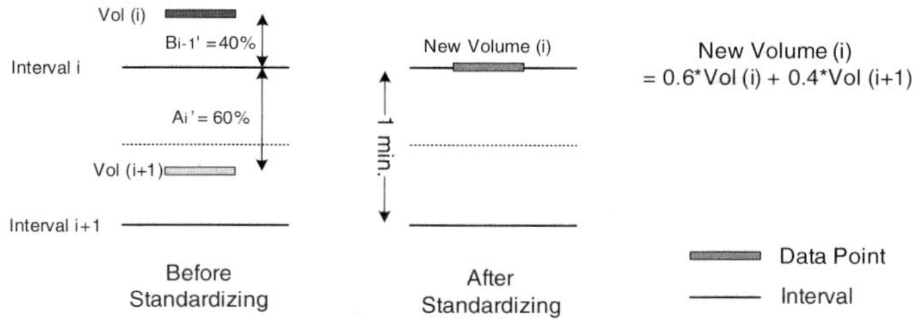

Figure 2. Example for obtaining standardized interval data

4. MODEL DEVELOPMENT AND MODEL VALIDATION

A stochastic process X = {X (t), t ∈ *T*} is a collection of random variables. Usually, t is interpreted as time and X (t) is the state of the process at time t (Ross, 1996). If the index set T is countable, the process is called a discrete-time stochastic process, while if T is a non-countable variable it is a called continuous-time stochastic process. This research focuses on how the system states change every time unit, so DTMC is applied. The procedures for model development are briefly presented here (additional information is provided in Yeon *et al.,* 2005).

4.1 Definition of States and Variables

A system is defined as an entire freeway route, and the system state variable can be defined as X_t at time t (t = 0, 1, 2, ...). The X_t can be described as a set of x_i (t), where x_i (t) is a link state variable. A link is defined by the area between detectors (Figure 1) and a route is composed of several links. The x_i (t) can be a binary variable: if the state of link i at time t is congested, x_i (t) is 1, otherwise it is 0.

4.2 Link Travel Time Estimation

Travel time estimated from input/output flows under non-congested conditions is regarded as constant due to their small variance, while under congested conditions, travel time increases with increasing demand, and the variance is high (Lint *et al.,* 2005). In this study, the travel time of each link is estimated for both non-congested and congested conditions.

4.3 Transition Matrix

To consider the impacts of congestion propagation in time and space, the effects of congestion to the upstream or downstream segments are described with a transition matrix. The transition is a change of state and the one-step transition matrix shows the changing rate from state i to j. As the number of steps increases, the system becomes more stable, and the probabilities when n goes to infinity are called long-run probabilities. When the Markov chain is *irreducible* (i.e., all states communicate with each other) and *ergodic* (i.e., a process will finally return to the starting state within a certain time period), there exists a unique stationary probability for all j.

4.4 Route Travel Time Estimation

The final task is to estimate route travel time using the output from the previous three tasks. This can be estimated as follows:

$$\bar{T} = \sum_{j=1}^{m} \sum_{i=1}^{k} \Pi_j \{(1 - x_i) \times NT_i + x_i \times CT_i(d)\}$$

where, Π_j is the long-run probability for state j

x_i is the state variable of link i for state j

NT_i is the non-congested travel time of link i

$CT_i(d)$ is the congested travel time of link i as a function of demand

m is the total number of possible states

k is the total number of links

4.5 Adjustment for Time of Day and Daily Variations

Demand varies daily, weekly, and monthly, thus the model has to distinguish between time periods, such as a.m. peak, off-peak, and p.m. peak, and P_{ij} values should be estimated for each time period. The peak time periods are determined based on traffic volume data in daily demand pattern graphs.

4.6 Model Validation

Two sets of O-D travel time data were used for model validation; C 205 & C 213 (e.g., link 1 to link 6) for 17:50-19:00 and C 206 & C 212 (e.g., link 2 to link 5) for 15:30-16:30 on May 25th, 2004. The estimated travel time for 17:50-19:00 is 7.87 min., while the average measured travel time is 7.92 min. and its variance is 1.28 min. In addition, the estimated travel time for 15:30-16:30 is 5.39 min., while the average measured travel time is 5.37 min. and its variance is 1 min. To compare the estimated travel time to measured travel time, the z-statistical test for means is conducted. The sample size for both time periods is greater than 30, thus, according to the central limit theorem, the measured travel time approximately follows the normal distribution. Table 3 shows the statistical test results, which show there is no statistical evidence that the estimated travel time differs from the measured travel time at the 95% confidence level for both time periods.

Table 3. Statistical results of the model validation (α = 0.05)

Time period	H_0	H_1	Test statistics	$z_{\alpha/2}$	p-value	Result
17:50-19:00	$\mu = 7.92$	$\mu \neq 7.92$	$z = \frac{\bar{y} - \mu_0}{\sigma / \sqrt{n}} = \frac{7.87 - 7.92}{1.13 / \sqrt{53}} = -0.322$	±1.645	$p(\lvert z \rvert \geq 0.322) = 0.747$	Do not reject H_0

15:30-16:30	$\mu = 5.37$	$\mu \neq 5.37$	$z = \frac{\bar{y} - \mu_0}{\sigma / \sqrt{n}} = \frac{5.39 - 5.37}{1/\sqrt{102}} = 0.202$	±1.645	$p(\|z\| \geq 0.202) = 0.840$	Do not reject H_0

5. CONCLUSIONS AND RECOMMENDATIONS

This study focuses on how to use traffic data typically provided by TMCs to develop travel time estimation models on freeways. The proposed model is based on DTMC and considers the probability of breakdown along each freeway link. It subsequently estimates the expected travel time for the entire route as a function of those probabilities of breakdown. The model developed was found to match field travel time estimates very well. According to statistical testing, there is no evidence that the estimated travel time differs from the measured travel time at the 95% confidence level.

Even though the model was found to be accurate when compared to field data, it was developed using data over two days, and validation data were collected during 2 hours (i.e., 15:30-16:30 & 17:50-19:00 on May 25th). Furthermore, the equations developed for estimating the congested travel time of each link were based on those limited data. Thus, to generalize the model, it is required to use additional data sets both for model development and validation. In addition, the breakdown over those days and for this particular freeway route was defined as occurring when speed drops below 50 mph for more than 5 minutes. Congestion occurrences during different times and at different sites may have different thresholds of breakdown. Additional research should be conducted to establish whether the probability of breakdown definition should be different for different sites.

The model developed provides expected O-D travel time for a given facility, which is constructed based on point detection data. Thus, the same methodology could be applied to other freeway segments if appropriate speed and flow data are available. TMCs can relatively easily develop similar models and calibrate them to their needs to obtain travel time estimates for various time periods. In doing this, the data used should be synchronized between detection locations such as every 20-second, 1-minute, 5-minute, etc. with the same starting time. The model development process would provide more accurate results in cases where the TMC has ramp data available over the study sites. Ramp information is needed for more detailed analysis of link travel time estimation, especially for congested link travel times. In addition, shorter distances between detectors would result in more accurate or reliable travel time estimation. Shorter distances would result in smaller variance in the estimation of link travel times, which would provide more reliable O-D travel times for traveller information systems.

6. ACKNOWLEDGEMENTS AND DISCLAIMER

This work was sponsored by the National Science Foundation (NSF), under project NSF-0338764. The contents of this paper reflect the views of the authors, who are responsible for the facts and the accuracy of the data presented herein. The contents do not necessarily reflect the official views of NSF. The authors wish to acknowledge Mobility Technologies for providing speed and flow data, and the Pennsylvania Department of Transportation for providing access to the Philadelphia Traffic Management Center.

7. REFERENCES

Choe T., Skabardonis A., and Varaiya P. (2002), Freeway Performance Measurement System (PeMS): An Operational Analysis Tool, *Transportation Research Board 81st Annual Meeting*

Chu, L. (2005), Adaptive Kalman Filter Based Freeway Travel time Estimation, *Transportation Research Board 84th Annual Meeting*

Del Castillo J. M. (2001), Propagation Perturbation in Dense Traffic Flow: A Model and its Implications, *Transportation Research Part B*, Vol.35, pp.367-389

Elefteriadou L., Roess R. P., and Mcshane W.R. (1995), Probabilistic Nature of Breakdown at Freeway Merge Junction, *Transportation Research Record 1484*, pp. 80-89

Heidemann D. (2001), A Queueing Theory Model of Nonstationary Traffic Flow, *Transportation Science*, Vol. 35, No.4, pp.405-412

Kharoufeh J. P., and Gautam N. (2004), Deriving Link Travel Time Distributions via Stochastic Speed Processes, *Transportation Science*, Vol.38, No.1, pp.97-106

Lorenz, M., L. Elefteriadou, "A Probabilistic Approach to Defining Capacity and Breakdown", *Transportation Research Circular E-C018*, Fourth International Symposium on Highway Capacity, Proceedings, June 27 – July 1, 2000, pp. 84-95.

Oh J., Jayakrishnan R., Recker W. (2003), Section Travel Time Estimation from Point Detection Data, *Transportation Research Board 81st Annual Meeting*

Evans, J., L. Elefteriadou, G. Natarajan, "Determination of the Probability of Breakdown on a Freeway Based on Zonal Merging Probabilities", *Transportation Research Part B: Methodological*, Vol. 35B, No. 3, March 2001, pp. 237-254.

Persaud B., S. Yagar, D. Tsui, and H. Look (2001), Breakdown-related Capacity for Freeway with Ramp Metering, *Transportation Research Record 1748*, pp. 110-115

Ross, S. (1996), Stochastic Processes, 2nd edition, John Wiley and Sons

Shaw T. and Jackson D. (2002), Reliability Measures for Highway Systems and Segments, TRANSYSTEMS Corporation

Turner S. M. (1996), Advanced Techniques for Travel Time Data Collection. *Transportation Research Board 75th Annual Meeting*

Turner S., Margiotta R., and Lomax T. (2004), Lessons Learned: Monitoring Highway Congestion and Reliability Using Archived Traffic Detector Data, FHWA-OP-05-003

Van Lint J.W.C. and van Zuylen H. J. (2005), Monitoring and predicting Freeway Travel Time Reliability Using Width and Skew of the day-to-day Travel time Distribution, *Transportation Research Board 84th Annual meeting*

Yang H., H. K. Lo, and W.H. Tang (1999), Travel time versus capacity reliability of a road network, In: Bell, MGH, Cassir, C, eds. Reliability of transport networks. Baldock: Research Studies Press, pp. 119-138

Yeon J., Elefteriadou L., and Lawphongpanich S. (2005), Travel Time Estimation On A Freeway Using Discrete Time Markov Chains, *Transportation Research Board 85th Annual meeting*

Zhang, X. and Rice, J. (2003), Short-Term Travel Time Prediction, *Transportation Research Part C*, Vol 11, pp.187-210

Transport Science and Technology
K.G. Goulias, editor

CHAPTER 24

AN EVALUATION OF THE EFFECTIVENESS OF PEDESTRIAN COUNTDOWN SIGNALS

Scott S. Washburn, Associate Professor, Department of Civil and Coastal Engineering, University of Florida, Gainesville, FL, swash@ce.ufl.edu

Deborah L. Leistner, Kimley-Horn and Associates, Ocala, FL, leistnerd@cox.net

Byungkon Ko, Graduate Research Assistant, Department of Civil and Coastal Engineering, University of Florida, Gainesville, FL, kobrain72@hotmail.com

ABSTRACT

A before-and-after study of pedestrian countdown signals was conducted at five intersections in Gainesville, Florida. All of the intersections had high pedestrian traffic volumes during certain times of the day. Additionally, several distinct pedestrian populations were present across the intersections. The data were collected over a period from October 2003 to April 2004. Overall, the countdown signals appear to have had an overall positive influence on crossing behavior. Proportions of pedestrians entering on the WALK indication increased, while proportions entering on the solid DON'T WALK indication decreased. The countdown signals have also had the effect of increasing the proportion of pedestrians exiting on the FDW interval as opposed to the DW interval; therefore decreasing the number of pedestrians that remain in the crosswalk at the release of conflicting traffic. Pedestrians appear to be adjusting their walking speed to finish crossing during the FDW interval. The countdown signals have not had the potentially negative effect of increasing the proportion of pedestrians entering the crosswalk during the FDW interval, for those pedestrians that also arrived at the crosswalk during the FDW. Overall, the countdown signals did not have a negative effect on pedestrian behavior such as running, hesitating and going back to point of start. In addition, the overall proportion of conflicts with vehicles decreased after the installation of the countdown signals.

INTRODUCTION

Pedestrian crossing behavior at signalized intersections has long been a safety concern in the United States. One of the main contributing factors to this issue is the fact that a large percentage of people do not completely understand the pedestrian indications, particularly the flashing 'DON'T WALK'. This often leads to pedestrians making an incorrect decision as to when it is acceptable to enter the crosswalk, which consequently often results in pedestrians being present in the crosswalk when a conflicting vehicular movement receives the green indication. Pedestrian countdown signals were developed in an effort to improve this situation.

Pedestrian countdown signals differ from traditional pedestrian signals in that a countdown timer is displayed during the flashing 'DON'T WALK' (FDW) period. This timer displays the time left until the solid 'DON'T WALK' (DW) period starts (see Figure 1). It has been purported that this design will lead to a higher level of pedestrian safety due to pedestrians making better crossing decisions with the added countdown information. Thus, an intended result is that a smaller percentage of pedestrians will be left in the crosswalk at the beginning of the DW period. It is has also been purported that compliance with the pedestrian indications will improve as a result of the countdown information. However, it has also been suggested that inclusion of the countdown timer will encourage more people to enter the intersection during the FDW period, rather than waiting for an ensuing WALK indication. This can be a risky decision as pedestrians may underestimate the time that it actually takes them to make a crossing movement, and thus be caught in the intersection at the start of a conflicting vehicular movement.

Due to considerable statewide interest from the traffic engineering community in Florida, the Florida Department of Transportation and the City of Gainesville sponsored a study of the pedestrian countdown signals with the Transportation Research Center of the University of Florida. This paper summarizes the results of this study. Additional details can be found in the project final report [1].

PREVIOUS STUDIES

The number of studies on pedestrian countdown signals has thus far been limited. This section gives a brief overview of a sampling of the studies that have been performed. Botha et al. [2] conducted a before-and-after study of pedestrian countdown signals at four intersections in the City of San Jose. They found that after the installation of the countdown signals, a significantly larger percentage of pedestrians that arrived during the FDW interval entered on that interval, while a smaller percentage waited for the ensuing WALK display. However, a smaller percentage of pedestrians exited the crosswalk during the DW interval in the after period. They also found that there was little difference in the number of pedestrians exhibiting erratic behavior (e.g., running, stopping/hesitating, turning-around) and the number of pedestrian-vehicle conflicts.

Huang and Zegeer [3] performed a study of pedestrian countdown signals at two intersections (one crosswalk at each) in Lake Buena Vista, Florida. These two intersections were "treatment" sites and were matched up with three "control" intersections which had conventional pedestrian signals. They found that compliance with the WALK indication (i.e., pedestrians who started crossing during the WALK interval) was significantly lower at the countdown locations. They did not find a significant difference in the percentage of pedestrians that were still left in the crosswalk when the signal indication turned to DW (after starting their crossing during WALK or FDW) between the treatment and control sites. They found that there were a significantly smaller percentage of pedestrians who started running when the FDW indication appeared at the treatment sites. Of course, with a treatment and control site study approach, it is more difficult to isolate the effects of the treatment (the pedestrian countdown signals in this case), as other intersection and pedestrian population specific characteristics may be influencing the results. It should also be noted that this study used a camera positioned at ground level on a sidewalk for data collection. It is possible that the visible presence of data collection devices influenced the behavior of some pedestrians.

Eccles et al. [4] conducted a study of pedestrian countdown signals at five intersections in Montgomery County, Maryland. They used a before-and-after study design. They concluded that the countdown signals had a generally positive effect on pedestrian behavior and did not have a negative effect on motorist behavior. Specifically, they found that 2 of the 20 crosswalks experienced a statistically significant decrease in the number of pedestrians who entered on WALK, but that 6 crosswalks experienced a significant increase. They also found that none of the intersections had a significant increase in the number of phases with pedestrians remaining in the intersection at the release of conflicting traffic. Additionally, they found a significant decrease in pedestrian-motor vehicle conflicts after the installation of the pedestrian countdown signals at four of the intersections where conflicts were observed.

Khasnabis et al. [5] found that pedestrians tend to ignore pedestrian signal indications under low traffic volume conditions, and that the compliance with flashing signals tend to be lower than for steady signals. Yauch and Davis [6] relate the problems of lack of compliance with pedestrian indications to the continuing changes in design and operation associated with the pedestrian signals, which generate confusion and distrust. A Dutch [7] review of pedestrian safety studies showed that on average only 35 percent of pedestrians cross during the WALK interval and that the type of destination had no impact on the probability of pedestrians crossing during the DW indication. Pedestrian behavior at the crossing and willingness to comply with the pedestrian signal indication is also influenced by pedestrian delay caused by the signal as a function of timing, by the volumes of pedestrian and vehicular traffic, and by roadway characteristics such as width. According to Zegeer et al. [8] "pedestrians that are willing to trust their own judgment of gaps in traffic incur less delay than those who comply with the signal".

RESEARCH APPROACH

A before-and-after study methodology was employed to evaluate the effectiveness of the countdown signals. The following sections provide the details about each component of the research approach.

Study Site Selection

Five intersections in the City of Gainesville were chosen as study sites for the countdown signals. These intersections are as follows: (a) W. University Ave / North-South Dr, (b) W. University Ave / 17th St, (c) W. University Ave / 2nd St, (d) E. University Ave / 1st St, and (e) Archer Rd / SW 16th St. All of the intersections have high pedestrian traffic volumes during certain times of the day. Additionally, several distinct pedestrian populations are present across the intersections. Intersections (a) and (b) are adjacent to, and provide access to, the University of Florida campus. The pedestrian population at these two intersections consists primarily of university students, faculty, and staff, with students being the large majority of users. Intersections (c) and (d) are in the downtown Gainesville area. Intersection (d) has a pedestrian population of professionals (e.g., lawyers, engineering and financial consultants, etc.), retail shop and restaurant employees, students (there is an adjacent bus station serving the university), and other miscellaneous users that visit the downtown area. Intersection (c) is three blocks west of intersection (d) and is located in the nightclub area. This intersection has a small amount of daytime pedestrian traffic, but has large volume of pedestrian traffic at night, particularly on Thursday, Friday, and Saturday nights. Intersection (e) is adjacent to, and provides access to, the Shands Hospital. This hospital is affiliated with the University of Florida, and is located on the southern edge of the campus. The pedestrian population at this intersection consists primarily of hospital employees (e.g., doctors, nurses, other staff), and some patients (this is not the primary access point for patients or visitors). Details about each site, including illustrations of surrounding land uses and geometry, can be found in the project report [1].

Data Collection

A data collection system developed at the University of Florida Transportation Research Center was used for this project. This system is capable of simultaneously capturing pedestrian and vehicular movements with a video camera and the corresponding signal indications. The result is a composite video scene with traffic movements and overlaid graphics indicating the current signal indications. A screen shot from one of these videos is shown in Figure 2. The system enables personnel to reduce the data accurately and efficiently. Another advantage to this system is that it is virtually undetectable, as all components but the camera are contained inside the signal controller cabinet, and the camera is mounted well above the ground. Additional details on this system can be found in a report by Washburn and Courage [9].

The before data were collected during the period from 9/30/03 – 11/1/03. The after data were collected during the period from 11/17/03 – 12/13/03 and 3/24/04 – 4/15/04. The collected data used for analysis in this paper are shown in Table 1. The countdown signals were installed between 10/28/03 and 11/04/03. Generally, a minimum of two weeks was allowed between the installation of the countdown signal at each site and the beginning of the 'after' data collection. A public education campaign did not accompany the activation of the countdown signals.

Data Reduction

For each intersection, data for one crosswalk crossing the major street and one crosswalk crossing the minor street were recorded. Peak pedestrian traffic periods were used for data reduction, typically 7:00 – 9:00 AM, 11:00 AM – 1:00 PM, and 4:00 – 6:00 PM. Also, for one intersection, a late night data collection period was utilized (10:00 PM – 2:00 AM) because of its vicinity to nightclubs.

The following events were manually recorded from each video: (a) time of pedestrian arrival at the curb, (b) whether pedestrian push button (if present) was pressed, (c) phase of pedestrian signal the pedestrian arrived and entered the crosswalk (i.e., WALK, FDW, DW), and during which signal cycle (e.g., current cycle or a following cycle), (d) phase during which pedestrian exited the crosswalk, (e) any erratic crossing behavior (e.g., running, turning around), and (f) any pedestrian-vehicle conflicts.

A number of performance measures were calculated for this study, including: (a) percentage of pedestrians entering the crosswalk during the WALK, FDW, and DW indications, (b) percentage of pedestrians exiting the crosswalk during the WALK, FDW, and DW indications, (c) compliance with the FDW indication, (d) percentage of pedestrians hesitating, running, or going back to the curb, and (e) percentage of pedestrian-vehicle conflicts.

These measures allow the following basic questions to be answered with respect to the implementation of the countdown signals: (a) are pedestrians more or less to comply with the WALK indication?, (b) are pedestrians more or less likely to comply with the FDW indication?, (c) are pedestrians more or less likely to still be making a crossing movement when the DW phase appears?, (d) is erratic pedestrian crossing behavior more or less frequent?, and (e) are pedestrian-vehicle conflicts more or less likely?

RESULTS

The results for the percentages of pedestrians entering the crosswalk during each of the signal intervals are shown in Table 2. The results for the percentages of pedestrians exiting the crosswalk during each of the signal intervals are shown in Table 3. The results for the percentages of pedestrians that arrive on the FDW indication and subsequently enter the crosswalk on the WALK, FDW, or DW signal intervals are shown in Table 4. The results for the percentages of pedestrians hesitating, running, or going back to the curb are shown in Table 5. The results for the percentages of pedestrian-vehicle conflicts are shown in Table 6.

Additionally, a test for differences in population proportions was conducted for each of the performance measurements. The intent of this test is to determine if there is a statistically significant difference between the before and after measurements, which would likely be a result of the countdown signal treatment.

The test statistic is given by the following formula [10].

$$z = \frac{\hat{p}_1 - \hat{p}_2}{\sqrt{\hat{p}\hat{q}(1/m + 1/n)}} \quad [1]$$

Where:

z = calculated statistic to be compared to critical z value from table of normal distribution probabilities for a given confidence level (1 - α),
$\hat{p}_1$ = estimate of before population proportion (for specified performance measure),
$\hat{p}_2$ = estimate of after population proportion (for specified performance measure),
$\hat{p} = \frac{m}{m+n}\hat{p}_1 + \frac{n}{m+n}\hat{p}_2$,
$\hat{q} = 1 - \hat{p}$,
m = before sample size, and
n = after sample size.

This test statistic is applicable when the sample sizes are large and when the size of the populations is much larger than the samples. Additionally, one assumption of this test is that the two population samples are independent of one another. This is a reasonable assumption as long as there was little duplication of the test subjects in both samples. It is unlikely that many of the pedestrians observed in the before period were also observed in the after period. Table 7 presents the results of the statistical tests.

The following sections summarize, for each intersection, the calculated performance measures, (a) – (e), previously described, and the corresponding tests for significance.

E. University Ave / 1st St

One of the most significant items to note for this intersection is the changes in entering and exiting statistics. After the installation of the countdown signals, the proportion of pedestrians entering on WALK and FDW decreased significantly, and the proportion of pedestrians entering on DW increased significantly. Likewise, the proportion of pedestrians exiting on

WALK[1] and FDW decreased significantly and the proportion of pedestrians exiting on DW increased significantly, after the installation of the countdown signal.

These results are primarily due to changes in signal timing and the resultant change in the number of crossing opportunities, and corresponding time allocated to the WALK interval. Due to equipment problems after the before data collection period, it was necessary for the City of Gainesville to replace the signal controller and update the signal timings at the same time. This ultimately led to less overall time allocated to the WALK interval during a given observation period at the observed crosswalks. Thus, it is difficult to directly compare the before and after results at this intersection due to this significant change at the site in between the before and after data collection periods.

In addition to the high proportion of pedestrians crossing the minor street on the DW indication (for the after condition) due to the timing issue and low vehicular demand on this approach, another observation was made regarding an adjacent bus transfer station. The Regional Transit System (RTS) downtown transfer station is in very close proximity to both the minor and major street crosswalks observed for this study. Often a high volume of pedestrians arrive at the intersection after departing from a bus. Pedestrians also cross from the other side of the major street to get to buses at the transfer station. There was little compliance with the pedestrian signal indication by pedestrians arriving at the intersection either coming from or going to the bus transfer station. This situation has certainly been exacerbated by the revised minor street crosswalk timing. Overall, a high level of disregard for the pedestrian signal indications was observed at this site (both before and after), particularly during lower vehicular volume conditions on the main street. Many pedestrians were willing to utilize gaps in the traffic stream to make their crossing, during the DW interval.

The other changes of significance at this intersection between the before and after conditions were related to hesitating and running in the crosswalk. A smaller proportion of pedestrians hesitated in the crosswalk after the installation of the countdown signals. A larger proportion of pedestrians ran in the crosswalk (not including joggers) during the after condition. This is likely a function of pedestrians monitoring the countdown timer and hurrying up to get out of the crosswalk when they see the timer getting low.

W. University Ave / 2nd St

At this intersection, the proportion of pedestrians exiting on the FDW indication increased significantly in the after condition. Again, this is probably an indication that the pedestrians are monitoring the countdown signal while crossing and adjusting their crossing speed as necessary. For those pedestrians that arrived during the FDW interval, the proportion that

---

[1] The proportion of pedestrians exiting on the WALK interval is not normally a very useful measure, which is why it is not included in the statistical tests of Table 7, but due to significant timing changes at this intersection it is discussed here.

waited for the ensuing WALK indication to cross increased significantly in the after condition. Correspondingly, the proportion that entered on the FDW (when arriving on FDW) decreased significantly in the after condition. Another significant change was with respect to erratic behavior—the proportion of pedestrians hesitating increased significantly in the after condition. Finally, the proportions of pedestrians running or stopping in the crosswalk to avoid a potential vehicle conflict increased significantly in the after condition.

This intersection has a large amount of pedestrian activity on weekend nights (for which most of the reduced data correspond), with two nightclubs located on the corners of this intersection. Thus, it is important to keep in mind that these observations are predominately of college students that have probably been drinking alcohol. It was also observed that between 1:30 – 2:00 AM (last call and closing time for the clubs), there was general disregard for the pedestrian signals (both non-countdown and countdown). Often large groups (20-30) of people would cross the street on the DW indication while cars would yield the right-of-way.

W. University Ave / 17th St

This intersection has a very high volume of pedestrian traffic, and pedestrians are continually present at all corners of the intersection during the peak time periods of the day. At this intersection, the proportion of pedestrians entering on the WALK indication increased significantly in the after condition. Likewise, the proportion of pedestrians entering on the DW indication decreased significantly. Pedestrians appeared to be more willing to use the push buttons to activate the countdown signals and wait for the WALK indication before making their crossing maneuver than they were with the previous signals.

Also, the proportion of pedestrians exiting on the FDW indication increased significantly in the after condition, with a related significant decrease in the proportion of pedestrians exiting on the DW indication. Overall, it appears the countdown signal is having the desired effect at this intersection of clearing pedestrians out of the crosswalk before a conflicting vehicular movement is initiated.

Lastly, there was a significant decrease in the proportion of pedestrians hesitating in the crosswalk, as well as stopping to avoid a vehicle conflict. There was, however, a significant increase in the proportion of evasive maneuvers to avoid a vehicle conflict.

W. University Ave / North-South Drive

At this intersection, the proportion of pedestrians entering on the WALK indication increased significantly in the after condition. Likewise, the proportion of pedestrians entering on the DW indication decreased significantly. Also, the proportion of pedestrians exiting on the FDW indication increased significantly in the after condition, with a related significant decrease in the proportion of pedestrians exiting on the DW indication. It appears the countdown signals are having the desired effect at this intersection of clearing pedestrians out of the crosswalk before a conflicting vehicular movement is initiated. The proportion of pedestrians stopping to avoid a vehicle conflict decreased significantly in the after condition.

Archer Rd. / SW 16th St

The results at this intersection are consistent with those of the previous two, where the proportion of pedestrians entering on WALK increased significantly, the proportion entering on DW decreased significantly, the proportion exiting on FDW increased significantly, and the proportion exiting on DW decreased significantly in the after condition. One difference, however, is that the proportion of pedestrians entering on FDW increased significantly in the after condition. At this intersection, the eastern side of the intersection has a six-lane cross section, whereas the western side has a five-lane cross section. The eastern side has an additional right-turn lane for the westbound approach. The countdown time for the major crosswalks at this intersection is timed for the six-lane cross section, which is 18 seconds. The countdown time for the major crosswalks at the other intersections range from 13-15 seconds, with all of them having 5-lane cross sections for the major street. Thus, for pedestrians crossing the major crosswalk on the west side of this intersection (the one recorded for this study), they essentially have three extra seconds of crossing time. So it is not surprising that the proportion of pedestrians entering on FDW increased in the after condition, as the pedestrians probably learned that they had more than enough time to clear the intersection. Thus, pedestrians were probably comfortable entering the crosswalk a few seconds later after the start of the FDW interval than at other locations. Additionally, the proportion of pedestrians running or stopping to avoid a vehicle conflict decreased significantly in the after condition.

SUMMARY

For the after condition, the proportion of pedestrians entering on the WALK indication increased at three of the five observed intersections. A complimentary reduction in the proportion of pedestrians entering on the DW interval was also present at the same three intersections. At these intersections, it appears that the countdown signals are having the desirable effect of increasing the proportion of pedestrians entering on the WALK interval versus the other intervals. At the intersection of E University Ave/1st St, the results were in the other direction, largely due to changes in signal timing. At the intersection of W University Ave/2nd St, there was no significant change in these measures.

The countdown signals have not had the potentially negative effect of increasing the proportion of pedestrians entering the crosswalk during the FDW interval, except at one intersection where there is extra time for crossing on the observed crosswalk. Pedestrians generally appear to be using the additional information provided by the countdown timer to adjust their walking speeds and complete the crossing prior to the release of conflicting traffic. Except for E

University Ave/1st St, the countdown signals have had the desirable effect of increasing the proportion of pedestrians exiting on the FDW interval as opposed to the DW interval. This may result in a decrease in pedestrian injuries and fatalities at signalized intersection crossings due to reduced pedestrian exposure to conflicting vehicle movements.

The results for erratic crossing behavior and pedestrian-vehicle conflicts are somewhat mixed. In addition, the overall frequency of these types of events is much smaller compared to the normal crossing maneuvers; thus, it is difficult to draw any general conclusions from these particular data.

Despite the differing populations of pedestrians (e.g., students, business professionals, health care professionals, etc.), the countdown signals appear to have had an overall positive influence on crossing behavior. However, two segments of pedestrian population were absent to any degree: school-age children and the elderly. While Florida does have a very high percentage of older persons, this is a very small percentage of the population in Gainesville, as would be expected in a small university city. Obviously, older pedestrians may have significantly different crossing behavior from younger pedestrians. Thus, these results cannot necessarily be extended to areas with significantly different pedestrian populations. Further study is also needed to assess the long-term effect of countdown signals; that is, will pedestrian compliance with the countdown indications decrease once pedestrians grow accustomed to the new devices.

ACKNOWLEDGEMENT

The authors would like to thank Conrad Renshaw, Kris McCoy, Phil Mann, and Brian Kanely with the City of Gainesville; Ken Courage, Christian Gyle, Jessica Morriss, and Jennifer Webster with the University of Florida.

REFERENCES

1. Washburn, Scott S. and Leistner, Deborah (June 2005). "Evaluation of the Effectiveness of Pedestrian Countdown Signals". Final Report. Prepared by the University of Florida Transportation Research Center. Prepared for the Florida Department of Transportation. Tallahassee, FL.

2. Botha, J.L., Zabyshny, A.A., and Day, J.E. (May 2002). "Pedestrian Countdown Signals: An Experimental Evaluation". Final Report. Prepared by the Department of Civil and Environmental Engineering, San Jose State University. Prepared for the City of San Jose Department of Transportation.

3. Huang, H., and Zegeer, C. (November 2000) "The Effects of Pedestrian Countdown Signals in Lake Buena Vista". Final Report. Prepared by the University of North Carolina Highway Safety Research Center. Prepared for the Florida Department of Transportation Safety Office.

4. Eccles, K.A., Tao, R., and Mangum, B.C. (January 2004). "Evaluation of Pedestrian Countdown Signals in Montgomery County, Maryland". Presented at the 83rd Annual Meeting of the Transportation Research Board.

5. Khasnabis, S., Zegeer, C.V., and Cynecki, M.J. (1982) "Effects of Pedestrian Signals on Safety, Operations, and Pedestrian Behavior – Literature Review". *Transportation Research Record 847*, pp. 78-86.

6. Yauch, P.J. and Davis, R.E. III. (April 2001). "Pedestrian Signals – A Call to Action". *ITE Journal*, pp. 32-35.

7. U.S. Department of Transportation (December 1999). *Dutch Pedestrian Safety Review*. Publication No. FHWA-RD-99-092. Online at: http://www.fhwa.dot.gov/tfhrc/safety/pubs/99092/99092.pdf Accessed April 14, 2005.

8. Zegeer, C., Opiela, K.S., and Cynecki, M.J. (1982) "The Effect of Pedestrian Signals and Signal Timing on Pedestrian Accidents." *Transportation Research Record 847*, pp. 62-72.

9. Washburn, Scott S. and Kenneth G. Courage (May 2001). "Development and Testing of a Red Light Violation Data Collection Tool". Southeast Transportation Center. University of Tennessee, Knoxville. Final Report. http://stc.utk.edu/htm/researchcom.htm

10. Devore, Jay L. (1991). "Probability and Statistics for Engineering and the Sciences, Third Edition". Brooks/Cole Publishing Company. Pacific Grove, California.

Table 1. Collected data used for analysis.

SITE	Data Collection: BEFORE		Data Collection: AFTER	
E University Ave & E 1st St	10/03/03 Friday	7:00 AM - 9:00 AM 11:00 AM - 2:00 PM 3:00 PM - 6:00 PM	4/02/04 Friday	7:30 AM - 8:45 AM 11:45 AM - 1:00 PM 4:30 PM - 6:00 PM
	10/07/03 Tuesday	7:00 AM - 9:00 AM 11:00 AM - 2:00 PM 3:00 PM - 6:00 PM	4/13/04 Tuesday	7:30 AM - 8:45 AM 11:45 AM - 1:00 PM 4:30 PM - 6:00 PM
	10/13/03 Monday	7:00 AM - 9:00 AM 11:00 AM - 2:00 PM 3:00 PM - 6:00 PM	4/14/04 Wednesday	7:30 AM - 8:45 AM 11:45 AM - 1:00 PM 4:30 PM - 6:00 PM
			4/19/04 Monday	7:30 AM - 9:00 AM 11:30 AM - 12:50 PM 4:40 PM - 6:00 PM
	TOTAL HOURS: 24		TOTAL HOURS: 16	
W University Ave & W 2nd St	10/02/03 Thursday	7:00 AM - 9:00 AM 11:00 AM - 2:00 PM 3:00 PM - 6:00 PM	12/05/03 Friday	10:00 PM - 2:00 AM
	10/07/03 Tuesday	7:00 AM - 9:00 AM 11:00 AM - 2:00 PM 3:00 PM - 6:00 PM	12/06/03 Saturday	10:00 PM - 2:00 AM
	10/17/03 Friday	10:00 PM - 2:00 AM	4/03/04 Saturday	11:30 AM - 12:50 PM 4:40 AM - 6:00 PM
	10/18/03 Saturday	10:00 PM - 2:00 AM	4/05/04 Monday	12:30 PM - 1:50 PM 5:40 PM - 7:00 PM
			4/13/04 Tuesday	7:40 AM - 9:00 AM 11:30 AM - 12:50 PM 4:40 AM - 6:00 PM
	TOTAL HOURS: 24		TOTAL HOURS: 24	
W University Ave & W 17th St	10/9/03 Thursday	2:00 PM - 6:00 PM	11/17/03 Monday	7:00 AM - 9:00 AM 11:00 AM - 2:00 PM 3:00 PM - 6:00 PM
	10/10/03 Friday	3:00 PM - 6:00 PM	11/18/04 Tuesday	11:00 AM - 2:00 PM 3:00 PM - 6:00 PM
	10/13/03 Monday	7:00 AM - 9:00 AM 11:00 AM - 2:00 PM 3:00 PM - 6:00 PM	11/19/04 Wednesday	7:00 AM - 9:00 AM
	10/14/04 Tuesday	7:00 AM - 9:00 AM 11:00 AM - 2:00 PM	3/26/04 Friday	11:00 AM - 2:00 PM 3:00 PM - 6:00 PM
	TOTAL HOURS: 21		TOTAL HOURS: 16	
W University Ave & W N/S Dr	10/01/03 Wednesday	10:30 AM - 6:30 PM	12/09/03 - Tuesday	4:30 PM - 6:00 PM
	10/07/03 Tuesday	7:30 AM - 8:45 AM 11:45 AM - 1:00 PM 4:30 PM - 6:00 PM	12/10/03 Wednesday	7:30 AM - 8:45 AM 11:45 AM - 1:00 PM
	10/08/03 Wednesday	11:45 AM - 1:00 PM 4:30 PM - 6:00 PM	3/24/04 Wednesday	7:30 AM - 8:45 AM 11:45 AM - 1:00 PM 4:30 PM - 6:00 PM
	10/09/03 Thursday	7:30 AM - 8:45 AM	3/25/04 Thursday	7:30 AM - 8:45 AM 11:45 AM - 1:00 PM 4:30 PM - 6:00 PM
	TOTAL HOURS: 16		TOTAL HOURS: 12	
Archer Rd & SW 16th St	10/07/03 Tuesday	6:45 AM - 8:00 AM 11:45 AM - 1:00 PM 4:30 PM - 6:00 PM	3/25/04 Thursday	6:45 AM - 8:00 AM 11:45 AM - 1:00 PM 4:30 PM - 6:00 PM
	10/09/03 Thursday	6:45 AM - 8:00 AM 11:45 AM - 1:00 PM 4:30 PM - 6:00 PM	3/26/04 Friday	6:45 AM - 8:00 AM 11:45 AM - 1:00 PM 4:30 PM - 6:00 PM
	10/10/03 Friday	6:45 AM - 8:00 AM 11:45 AM - 1:00 PM 4:30 PM - 6:00 PM	4/06/04 Tuesday	6:45 AM - 8:00 AM 11:45 AM - 1:00 PM 4:30 PM - 6:00 PM
	10/13/03 Monday	6:45 AM - 8:00 AM 11:45 AM - 1:00 PM 4:30 PM - 6:00 PM		
	TOTAL HOURS: 16		TOTAL HOURS: 12	

Table 2. Pedestrians entering crosswalk.

SITE	TOTAL		ENTERING AT "W"			ENTERING AT "FDW"			ENTERING AT "DW"		
	Before	After	Before	After	Diff.	Before	After	Diff.	Before	After	Diff.
EUA & E 1st ST	808	501	51.73%	34.73%	-17.00%	13.99%	6.39%	-7.60%	34.28%	58.48%	24.20%
WUA & W 2nd ST	1434	1076	32.43%	33.74%	1.31%	9.41%	11.34%	1.93%	58.16%	54.93%	-3.23%
WUA & W 17th ST	3378	3225	27.03%	40.74%	13.71%	8.14%	9.27%	1.13%	64.83%	49.98%	-14.85%
WUA & N/S DR	409	259	75.79%	85.71%	9.92%	8.56%	6.56%	-2.00%	15.65%	7.72%	-7.93%
ARCHER & SW 16th AVE	1610	1278	53.66%	69.41%	15.75%	13.35%	17.37%	4.02%	32.98%	13.22%	-19.76%

Table 3. Pedestrians exiting crosswalk.

SITE	TOTAL		EXITING AT "W"			EXITING AT "FDW"			EXITING AT "DW"		
	Before	After	Before	After	Diff.	Before	After	Diff.	Before	After	Diff.
EUA & E 1st ST	808	501	6.56%	2.79%	-3.77%	61.26%	35.93%	-25.33%	32.18%	61.28%	29.10%
WUA & W 2nd ST	1434	1076	14.09%	6.88%	-7.21%	26.92%	36.71%	9.79%	59.00%	56.41%	-2.59%
WUA & W 17th ST	3378	3225	4.77%	4.40%	-0.37%	43.64%	53.61%	9.97%	51.60%	41.98%	-9.62%
WUA & N/S DR	409	259	0.98%	1.16%	0.18%	77.75%	88.42%	10.67%	21.27%	10.42%	-10.85%
ARCHER & SW 16th AVE	1610	1278	8.01%	8.45%	0.44%	58.32%	72.85%	14.53%	33.66%	18.70%	-14.96%

Table 4. Compliance with FDW indication.

SITE	ARRIVALS AT "FDW"		COMPLIANCE			NON-COMPLIANCE					
			WAITING FOR *"W"*			ENTERING AT *"FDW"*			ENTERING AT *"DW"*		
	Before	After	Before	After	Diff.	Before	After	Diff.	Before	After	Diff.
EUA & E 1st ST	104	26	4.81%	3.85%	-0.96%	87.50%	88.46%	0.96%	7.69%	7.69%	0.00%
WUA & W 2nd ST	81	98	0.00%	12.24%	12.24%	98.77%	86.73%	-12.04%	1.23%	1.02%	-0.21%
WUA & W 17th ST	241	263	2.07%	1.90%	-0.17%	91.70%	90.87%	-0.83%	6.22%	7.22%	1.00%
WUA & N/S DR	27	16	3.70%	6.25%	2.55%	92.59%	93.75%	1.16%	3.70%	0.00%	-3.70%
ARCHER & SW 16th AVE	202	195	18.81%	14.87%	-3.94%	71.78%	76.92%	5.14%	9.41%	8.21%	-1.20%

Table 5. Erratic pedestrian behavior.

SITE	TOTAL		HESITATING			RUNNING			GOING BACK		
	Before	After	Before	After	Diff.	Before	After	Diff.	Before	After	Diff.
EUA & E 1st ST	808	501	2.60%	0.80%	-1.80%	5.32%	10.98%	5.66%	0.12%	0.20%	0.08%
WUA & W 2nd ST	1434	1076	0.14%	1.21%	1.07%	6.00%	7.90%	1.90%	0.07%	0.00%	-0.07%
WUA & W 17th ST	3378	3225	0.47%	0.09%	-0.38%	3.37%	3.50%	0.13%	0.24%	0.09%	-0.15%
WUA & N/S DR	409	259	1.47%	1.16%	-0.31%	2.20%	5.02%	2.82%	0.49%	0.00%	-0.49%
ARCHER & SW 16th AVE	1610	1278	0.37%	0.08%	-0.29%	6.52%	5.79%	-0.73%	0.25%	0.23%	-0.02%

Table 6. Pedestrian-vehicle conflicts.

SITE	TOTAL		RUN			STOP			EVADE		
	Before	After	Before	After	Diff.	Before	After	Diff.	Before	After	Diff.
EUA & E 1st ST	808	501	0.37%	1.40%	1.03%	3.22%	3.19%	-0.03%	0.00%	0.00%	0.00%
WUA & W 2nd ST	1434	1076	0.07%	5.95%	5.88%	0.42%	2.51%	2.09%	0.14%	0.09%	-0.05%
WUA & W 17th ST	3378	3225	0.15%	0.03%	-0.12%	3.46%	1.80%	-1.66%	0.03%	0.25%	0.22%
WUA & N/S DR	409	259	0.24%	0.00%	-0.24%	2.44%	0.38%	-2.06%	0.00%	0.00%	0.00%
ARCHER & SW 16th AVE	1610	1278	0.02%	0.01%	-0.01%	0.04%	0.02%	-0.02%	0.01%	0.01%	0.00%

Table 7. Calculated Test Statistic (*z* value) by Performance Measure.

Event	Intersection[1,2]					
	1	2	3	4	5	Overall
Exit-FDW	**9.231**	**-5.211**	**-8.148**	**-3.729**	**-8.307**	**-8.693**
Exit-DW	**-10.670**	1.296	**7.864**	**3.909**	**9.322**	**6.624**
Enter-W	**6.161**	-0.690	**-11.882**	**-3.268**	**-8.791**	**-3.829**
Enter-FDW	**4.639**	-1.556	-1.628	0.964	**-2.960**	0.479
Enter-DW	**-8.759**	1.617	**12.329**	**3.242**	**13.111**	**15.804**
Compliance FDW – Wait for WALK	0.223	**-3.698**	0.139	-0.361	1.051	0.083
Non-Compliance – Enter at FDW	-0.136	**3.306**	0.329	-0.147	-1.175	**4.525**
Non-Compliance – Enter at DW	0.000	0.131	-0.119	1.019	0.122	1.017
Hesitating	**2.623**	**-3.077**	**2.933**	0.346	1.724	**3.015**
Running	**-3.525**	-1.840	-0.288	-1.832	0.815	**-3.482**
Going Back	-0.323	1.000	1.447	1.418	0.075	1.480
Conflict - Run	-1.811	**-8.115**	1.602	1.001	**2.839**	**4.065**
Conflict - Stop	0.024	**-4.129**	**4.247**	**2.407**	**3.792**	**14.289**
Conflict - Evade	0.000	0.344	**-2.363**	0.000	0.465	1.133

[1] Intersections are: 1) E University Ave and 1st St; 2) W University Ave and 2nd St; 3) W University Ave and 17th St; 4) W University Ave and N/S Dr; and, 5) Archer Rd and SW 16th St.

[2] Bolded values are significant at the 95% confidence level ($z_{0.05}$ = 1.96 for two-tailed test).
A negative *z* value means the before proportion is lower than the after proportion.
A positive *z* value means the before proportion is higher than the after proportion.

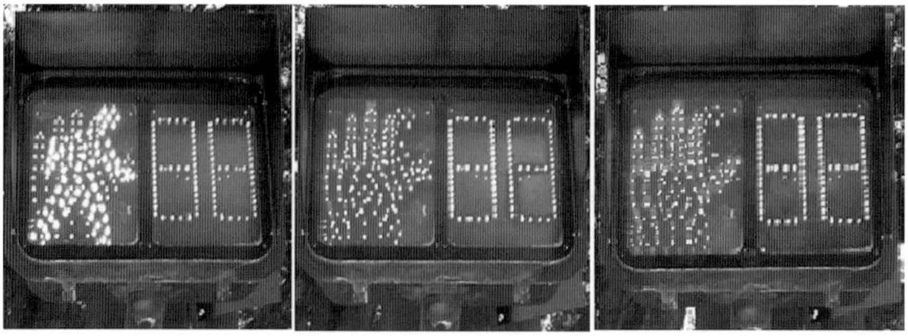

(a) (b) (c)

Figure 1. Pedestrian countdown signal indications; (a) WALK, (b) flashing DON'T WALK with countdown timer, and (c) solid DON'T WALK.

Figure 2. Composite video scene used for data reduction (University & 17th)

Transport Science and Technology
K.G. Goulias, editor

CHAPTER 25

AN ANALYSIS OF THE CHARACTERISTICS OF EMERGENCY VEHICLE OPERATIONS

Konstantina Gkritza, School of Civil Engineering,

Purdue University, West Lafayette, IN 47907

Phone: (765) 494-2206; Fax: (765) 496-7996; Email: kgkritza@purdue.edu

John Collura, P.E., Department of Civil and Environmental Engineering,

University of Massachusetts, Amherst, MA 01003-9293,

Phone: (413) 545-5404; Email: collura@ecs.umass.edu

Samuel C. Tignor, P.E., Department of Civil and Environmental Engineering,

Virginia Tech, Northern Virginia Center, Falls Church, VA 22043

Phone: (703) 538-8456; Email: stignor@aol.com

Dusan Teodorovic, P.E., Department of Civil and Environmental Engineering,

Virginia Tech, Northern Virginia Center, Falls Church, VA 22043,

Phone: (703) 538-8436; Email: dteodorovic@vt.edu

ABSTRACT

Concerns about increased emergency vehicle response times in the greater Washington D.C. Region, especially during peak periods, have led to the implementation of signal preemption systems to facilitate the efficient and safe movement of emergency vehicles. However, to date only limited research has been carried out on the travel characteristics of emergency vehicles. This paper presents an analysis of emergency vehicle operations, travel patterns and associated impacts to enhance our understanding and assist public agencies and other stakeholders in the planning and deployment of emergency vehicle preemption systems. Emergency vehicle characteristics that are examined include temporal and spatial distribution of emergency vehicle travel; frequency and duration of preemption requests; platoon responses; and crashes involving emergency vehicles. Data on major corridors in Fairfax County, Virginia and Montgomery County, Maryland are used in the analysis. The analysis indicates that such data are useful to assess the need for a preemption system along major arterials. Moreover, the analysis demonstrates the importance of considering emergency vehicle preemption impacts regarding delay to other vehicles. It was also found that there is some variability in the emergency vehicle characteristics depending on the proximity of a firehouse to an intersection and other factors. It is proposed that future efforts build upon this research to develop warrants to be used in determining the appropriateness of installing preemption systems at signalized intersections.

INTRODUCTION

Response time is a prime measure of emergency vehicle operational efficiency. Emergency response times in many States are threatened by a growing population, outdated technology and tight budgets. This is especially important in the National Capital Region where heavy traffic is considered a thorn in the side of firefighters and paramedics. The heavy traffic levels experienced during peak periods have a negative impact on emergency vehicle response times. Concerns about increased emergency vehicle response times in the region have led to the implementation of traffic signal control strategies, such as signal preemption systems, to facilitate the efficient and safe movement of emergency vehicles, as well as to resolve the challenges that gridlock situations present to drivers of emergency vehicles.

While there is a great deal of information available about the travel characteristics of individuals traveling for all kinds of purposes on a day to day basis, little is known about the operations, travel patterns and associated characteristics of emergency vehicles. Emergency vehicle characteristics that merit special attention include temporal and spatial distribution of emergency vehicle travel; frequency of emergency vehicle responses by time of day; the extent to which such responses include two or more emergency vehicles; and the impacts of emergency vehicle preemption regarding delay to other vehicles.

With the concern for providing "first responders" with efficient transportation resources coupled with the increase in emergency vehicle preemption system deployment, a study on the traffic flow characteristics of emergency vehicles is of great interest. In reviewing the current state of the art, it was found that adequate guidelines or criteria have not been developed for the placement of emergency vehicle preemption systems at existing signalized intersections. An improved understanding of the travel characteristics of emergency vehicles may assist transportation planners and engineers in identifying emergency vehicle preemption candidate intersections based on traffic operations and safety objectives.

This research study presents the results of the analysis of emergency vehicle operations in the greater Washington D.C. Region The analysis uses data on emergency vehicle responses in Fairfax County, Virginia, and data on the characteristics of emergency vehicles with respect to the preemption system deployed on U.S.1 in Fairfax County, Virginia and, in Montgomery County, Maryland. Emergency vehicle involvement in crashes on U.S.1 in Fairfax County, Virginia is also discussed. The remainder of this paper is organized as follows: the next section reviews major issues concerning the improvement of emergency service delivery and provides an overview of the factors affecting the need for emergency vehicle preemption. The third section describes the data used for analysis and presents the major findings. Finally, the last section offers some concluding remarks and recommendations for future research efforts.

BACKGROUND

Fairfax County includes 395 square miles of land area. There are 35 fire stations providing fire protection to the County residents and businesses, and 470 Firebox areas. The County's Rescue Squad Committee has defined effective response times within 6 minutes of dispatch for providing advanced life support and within 5 minutes of dispatch for providing fire suppression. Statistics presenting the operating performance indicators show that the response time criteria are not often met (Table 1). Furthermore, the unit arrival rates that satisfy the above response time criteria have deteriorated between the years 2000 and 2002, while the number of incidents has increased during that period (Fairfax County, 2003).

Table 1 Response Statistics in Fairfax County, Virginia

Unit Arrival Rates (within the response time criteria)	*2000*	*2001*	*2002*
Advanced Life Support (ALS)	81.31%	78.24%	78.63%
Suppression	57.93%	54.57%	56.28%
Incidents by Category	*2000*	*2001*	*2002*
Public Service	4,432	4,642	4,982
Fire	21,872	22,677	23,579
Emergency Medical Services (EMS)	55,552	57,800	60,685

To this end, Emergency Medical Services and Fire and Rescue administrators in the region seek methods to enhance system performance. One component scrutinized is the response time interval between call receipt and arrival on the scene, because of the heavy traffic levels experienced in the region during peak periods. Emergency Medical Services and Fire and Rescue Authorities in different regions have considered various methods so as to improve emergency response time; such as, strategic positioning of the new stations, adopting systems of the latest technology, adding more staff, and prioritizing urgent emergency medical runs versus non-urgent (Langbein, 2003; Ludwig, 2002). Another reaction to improve emergency response time involves the implementation of emergency vehicle preemption systems that are a part of a fast growing ITS interest area (Gifford et al., 2001). There is some evidence that the implementation of emergency vehicle preemption systems may reduce travel times for emergency vehicles with a relatively minor impact on the network (Traffic Engineering, Inc., 1991; BRW Inc., 1997; Bullock et al., 1999; Nelson and Bullock, 2000; McHale and Collura, 2002; Collura and McHale, 2003*)*. In addition, emergency vehicle preemption may decrease the number and severity of accidents involving emergency vehicles at signalized intersections (Louisell, 2003; Louisell et al., 2003). Empirical evidence from an emergency vehicle accident study conducted in St. Paul, Minnesota indicated an accident rate reduction of greater than 70% between 1969 and 1976 after the installation of 285 signal preemption systems on 308 signalized intersections (St Paul, 1977).

Review of the available literature led to the identification of the main factors that need to be reviewed when considering the deployment of emergency vehicle preemption (EVP) systems. Some of these factors are as follows (Straub, 2000):

1. *Significant congestion and queuing at intersection approaches.* It has been observed that the need of EVP is most needed when the level of service (LOS) is poor and becomes even worse during the peak hours. Thus, traffic volumes and the time of the day are two of the main factors that contribute to delays.

2. *Number of accidents involving emergency vehicles* is a clear indication of the need for EVP, but the lack of accidents does not indicate that EVP should not be provided.

3. *Number of emergency runs* indicates the likelihood of delays to emergency vehicles and the need for EVP.

4. *Large sizes of some types of emergency vehicles* (i.e. fire trucks, paramedic engines etc.) normally have low acceleration rate, have more difficulty in navigating through congested intersections and are likely to have a greater impact to the other traffic since they require more road space. In such cases, providing EVP may help.

5. *Geometrics of the intersection and roadway* may indicate the need for EVP; lack of shoulders and auxiliary lanes; inadequate corner sight distance; and/or complex or

unusual intersections with severe skewness may make the safe movement of the emergency vehicle difficult.

The aforementioned factors alone are not sufficient to justify the deployment of an EVP system. The implications that the deployment of signal preemption systems could have on the general-purpose traffic is a major concern for transportation planners, stakeholders, and general public that also needs to be evaluated. Preempting a traffic signal will unconditionally interrupt the normal timing plan by inserting a special plan or phase to accommodate a request from an emergency vehicle that has the potential to affect negatively the flow of traffic (Obenberger and Collura, 2001). The impact of emergency vehicle preemption on the general-purpose traffic will be related to several factors including:

1. *Frequency of preemption requests*; the lower the number of preemption requests the less the impact on the other traffic.

2. *Platoon responses*; the smaller the size of vehicle platoons the shorter the duration of the preemption phase.

3. *Average duration of preemption phase*; the shorter the duration of the preemption phase the less the disruption to the traffic signal timing.

4. *Transition strategy selected*; the shorter the time required to serve a preemption control plan and transition back to the coordinated operation of the normal signal timing plan, the less the impact on the general-purpose traffic.

5. *Side street volume*.

The next section describes the data collection and collation efforts.

DATA DESCRIPTION AND ANALYSIS

The main factors that are examined in this analysis of emergency vehicle characteristics include: emergency vehicle responses; frequency of emergency runs by time of day and platoon responses; frequency and average duration of preemption requests; and crashes involving emergency vehicles. The analysis uses data obtained from: 1) the emergency response log data from three fire stations in Fairfax County, Virginia, and 2) the Virginia Department of Transportation (VDOT) crash database, to assess whether an EVP system is warranted in the study area according to the factors listed in the previous section. In addition, data on the characteristics of emergency vehicles with respect to the preemption system deployed obtained from 3) the 3M™Opticom™ Priority Control System deployed in Fairfax

County, Virginia, and 4) the preemption system deployed in Montgomery County, Maryland, are used to estimate the potential impact of emergency vehicle preemption on the general-purpose traffic according to the factors identified at the end of the previous section. Several methods and techniques are applied for summarizing and interpreting data, including graphical representations and statistical tests.

Analysis of Emergency Vehicle Responses on U.S.1 in Fairfax County, Virginia

U.S.1 is one of the major arterials in Northern Virginia connecting Prince William County to the Capital Beltway (I-495) which in turn acts as a connector to Washington D.C. The corridor considered in this study is between Fort Belvoir and Capital Beltway and is approximately 8 miles in length encompassing 28 signalized intersections. There are three major fire stations (fire station 9, Mt Vernon; fire station 11, Penn Daw; and fire station 24, Woodlawn) and two hospitals in the field of interest, which are considered the major sources of emergency vehicle travel. The data collected include emergency vehicle responses for the three fire stations for the year 2000, maintained by Fairfax County Fire and Rescue Department. The data provide the following information: firebox number; incident number; event type (fire, basic or advanced life support); dispatch hour; day of the week; location of the incident; month; and unit ID (ambulance, rescue engine, fire truck, paramedic engine, medic, or other).

The results indicate that the characteristics of emergency vehicle trip generation in terms of temporal distribution of emergency vehicle travel (by time of day, day of week and month of year) exhibit significant variability among the three fire stations at the 95% confidence interval. Fire stations 9 and 11 received on average per day nearly the same number of calls (25.5 and 25.8 calls, respectively); twice as many as fire station 24 received (on average, 13.3 calls per day).

The pattern of variation of emergency calls over the course of the day is very useful in an analysis of projected traffic conditions for both emergency vehicles and other vehicles. Different months of the year, different days of the week and different time periods of the day (AM and PM peak periods, midday and night) are compared to assess the variability in the frequency of emergency calls. Statistical tests are then applied to assess whether the observed differences in the temporal distribution of emergency responses among the three fire stations can be explained by the natural sampling variability or are attributed to other factors (Washington et al., 2003). It was found that the frequency of emergency calls is higher during daytime (8:00AM -8:00PM) than it is during the nighttime. Figure 1a provides supplemental information and shows that the frequency of emergency calls is higher during the PM peak period and lower during the AM peak period for all fire stations. A higher frequency of emergency calls during the daytime is likely to result in greater implications to the other traffic since traffic is more during daytime. In turn, because of higher levels of traffic during daytime, emergency vehicle response times are anticipated to be higher in comparison to nighttime. Thus, it becomes more difficult for the fire and emergency medical services personnel to meet the response time criteria set by the Fire and Rescue Department during daytime. Further analysis indicates that there is not significant variability in the frequency of emergency calls by day of week (Figure 1b) or month of year.

Each fire station under study has heavy rescue vehicles and ladder mounted trucks with heavy axle weights, large turning radii and low acceleration rates. It was found that a large vehicle is involved in most cases of an emergency response; a paramedic engine is dispatched along with a medic vehicle (44%) or a fire truck (19%) to serve the majority of incidents (53%) that require provide advanced life support. Under these circumstances, a preemption system might be beneficial.

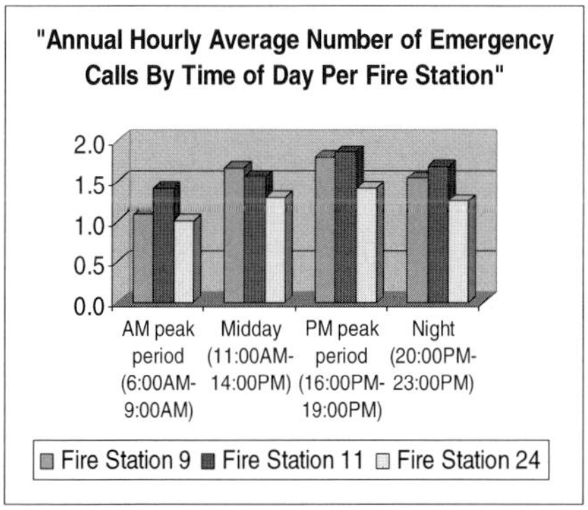

Figure 1a. Annual Hourly Average Number of Emergency Calls By Time of Day per Fire Station

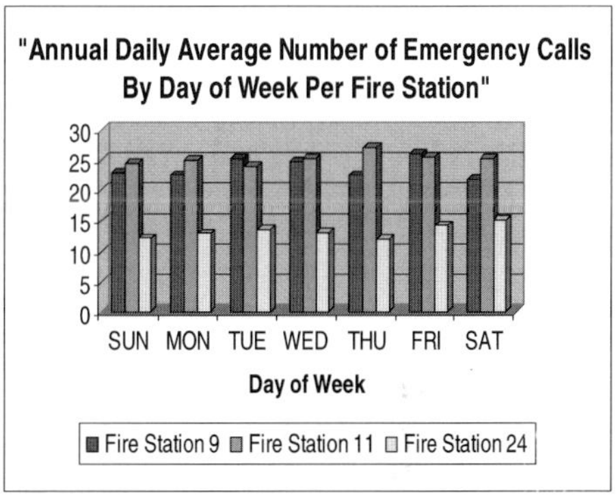

Figure 1b. Annual Hourly Average Number of Emergency Calls By Day of Week per Fire Station

Analysis of Emergency Vehicle Preemption Requests on U.S.1 in Fairfax County, Virginia

The corridor considered for this part of analysis is a segment of U.S.1 approximately 1.4 miles in length encompassing 7 signalized intersections. The signalized intersections are spaced randomly with a significant variability in the distance between any two intersections. Two of the 7 intersections are very closely spaced (distance less than 200 ft); as such, a total of 6 intersections are considered in the analysis. This segment of U.S.1 is under the service area of Fire station 11. All 6 intersections are equipped with the 3M™Opticom™ Priority Control System that provides preferential treatment to emergency services (fire and medical) and other vehicles such as transit, as needed. Emergency vehicles have first priority, thus eliminating any confusion. The whole procedure is achieved in three steps that occur within seconds , as follows (Mittal, 2002): i) an emitter mounted on the emergency vehicle (or bus) is activated to send encoded infrared communication; ii) a detector located near the intersection receives the signal and converts it into electronic communication and; iii) a phase selector, housed in the controller cabinet, discriminates and authorizes the user, logs management information and requests priority advantage for the controller to extend a green light or truncate a red (only in the case of emergency vehicles), thus giving the vehicle an efficient, natural appearing right of way.

The emergency vehicle preemption request data were obtained after the deployment of the priority control system at the 6 intersections along U.S.1 and represent a 53-day period from July 16, 2002 to September 6, 2002. During this period preferential treatment was provided only for emergency vehicles. The preemption data provide information regarding the number of emergency vehicle preemption requests granted and denied; their associated frequency and duration; and the size of vehicle platoons per preemption request.

The results indicate that the daily occurrence of preemption requests ranges from 0 to 21 requests with the average value fluctuating from 6 to 12 requests, exhibiting significant variability by intersection, as shown in Table 2. The frequency of preemption requests varies also by time of day. It is lower during the AM peak period at all intersections (up to one request in three hours); during the other three time periods of the day the frequency is a little higher and ranges between one and two requests in three hours (Figure 2a). In addition, there seems to be significant variability in the frequency of preemption requests by day of week; however, there does not seem to be a clear pattern between weekdays and weekends (Figure 2b).

Table 2 Daily Frequency of Emergency Vehicle Preemption Requests per Intersection

Intersection	Number of Emergency Vehicle Requests/day			
	Mean	Standard Deviation	Min	Max
RT.1 & Popkins Lane	5.7	3.7	0	18
RT.1 & Memorial St.	6.9	4.0	1	21
RT.1 & Beacon Hill Rd.	6.6	3.4	1	18
RT.1 & Southgate Dr.	11.6	4.4	1	21
RT.1 & South Kings Hwy	8.5	4.1	0	21
RT.1 & North Kings Hwy	8.4	4.1	0	21
Summary Statistics of One–Way ANOVA Test				
F		14.5		
P-value		<0.0001		

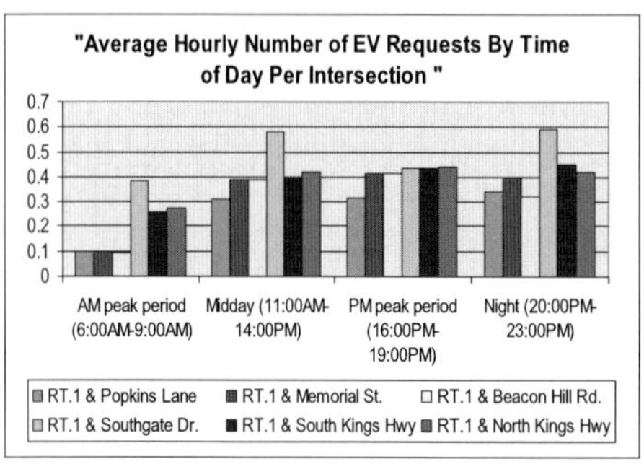

Figure 2a. Frequency of Emergency Vehicle Preemption Requests by Time of Day

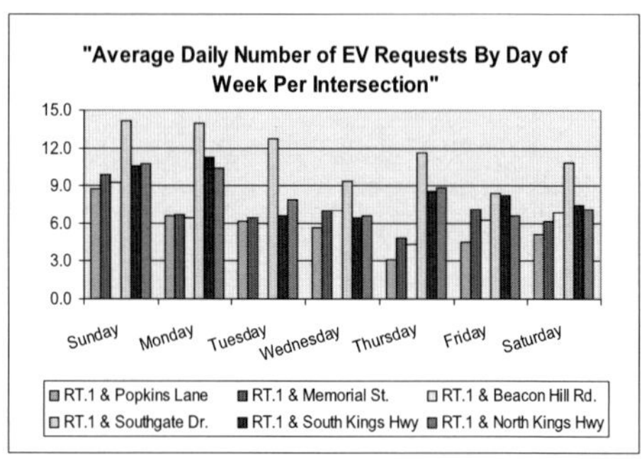

Figure 2b. Frequency of Emergency Vehicle Preemption Requests by Day of Week

Regarding the preemption system functionality, it was found that very few emergency vehicle preemption requests were denied. The number of preemption requests denied ranges from 1 to 2% of the total number of requests. In most cases, it appears that the reason for a request being denied is when having two or more simultaneous preemption requests. Another reason that was identified is a low measured intensity of an emitter's signal during the whole time of the call; a threshold of intensity equal to 200 has been set below which, the preemption request might be denied. In a few cases, a request was probably denied when made within the pedestrian phase.

Another important consideration when deploying a signal preemption system is the average duration of the preemption phase. As the length of time required to serve a preemption control plan and transition back to the coordinated operation of the normal signal timing plan increases, the impacts to the traveling public typically increase (Obenberger and Collura, 2001). The analysis indicates that the average duration of preemptions is relatively low; on average, it ranges from 16 to 26 sec, exhibiting significant variability by intersection, as shown in Table 3. However, there appears to be no significant variability in the average duration of preemptions by time of day, day of week or direction of travel.

Table 3 Average Duration of Emergency Vehicle Preemptions per Intersection

Intersection	Average Duration of Emergency Vehicle Preemptions (sec)			
	Mean	Standard Deviation	Min	Max
RT.1 & Popkins Lane	20.9	7.4	6	106
RT.1 & Memorial St.	26.4	10.4	6	131
RT.1 & Beacon Hill Rd.	18.7	9.8	6	131
RT.1 & Southgate Dr.	15.7	6.0	6	47
RT.1 & South Kings Hwy	22.7	6.8	6	86
RT.1 & North Kings Hwy	17.1	6.9	6	55
Summary Statistics of One–Way ANOVA Test				
F		112.3		
P-value		<0.0001		

The duration of preemption is also proportional to the number of emergency vehicles in a platoon responding to an emergency call; the higher the number of vehicles in a platoon the longer the duration of the preemption resulting in more disruption to the traffic signal timing and in turn, in a more severe impact on the genral-purpose traffic. The analysis suggests that the size of vehicle platoons per preemption request seems to be relatively small; in 73% of the cases under study, each platoon included only one emergency vehicle. This finding can be considered a positive sign for the traffic engineers engaged with signal preemption systems, as it indicates that in most cases the duration of preemption would be relatively low resulting in less disruption to the traffic signal timing and the general-purpose traffic, especially on the side streets. In addition, it appears that the likelihood of having some preemption requests denied because of interference with another request is anticipated to be relatively low.

Analysis of Emergency Vehicle Preemption Requests in Montgomery County, Maryland

Complementary to the previous analysis, it was of interest to assess the variability of the frequency and duration of emergency vehicle preemption requests by geographic location. Twenty-five major signalized intersections in Montgomery County, Maryland where a preemption system is installed were considered to serve the purpose of this comparative analysis. Montgomery County is located adjacent to the nation's capital, Washington, D.C. and includes 497 square miles of land area. There are 19 fire stations providing fire protection to the County residents and businesses, and 535 Firebox areas. The County's Rescue Squad Committee has defined effective rescue squad coverage as reaching 90% of the population within a 10-minute response time after placing a 911 call (Montgomery County, 2003).

The data were obtained from Department of Public Works and Transportation (DPWT) of Montgomery County, Maryland. The preemption data represent a 5-weekday period from April 20, 2000 to April 26, 2000 and offer the following information: the date of the preemption; the start and end time of the preemption mode; the preemption status (on/off); and the intersection number.

The analysis indicates that the average number of preemption requests per day per intersection is 12.6 requests (s.d. =10.3), and the average duration of the signal preemption time per intersection is 47.5 sec (s.d. =25.3). At any particular intersection, the number of preemption requests as well as the average duration of the signal preemption time within any weekday seems to be similar. However, the frequency of emergency preemption requests is statistically different for different hours of the day and different time periods of the day among the 25 intersections, with higher occurrence during the daytime.

The results of the comparative analysis indicate that the observed difference in the daily frequencies of preemption requests between the two Counties appears to be rather marginal at the 95% confidence interval. However, there seems to be a significant difference in the duration of preemptions between the two Counties. It appears that in Montgomery County, Maryland, the average duration of preemptions is higher. This fact could be attributed to several factors including the proximity of the firehouse to the intersection, roadway geometrics and traffic and operating characteristics of the intersection at which the preemption system is deployed. The longer duration of the preemption phases in Montgomery County could be possibly also attributed to larger platoons of emergency vehicles responding to an emergency call.

Study of Emergency Vehicle Crash History on U.S.1 in Fairfax County, Virginia

The main goal of the analysis of the emergency vehicle crash data is to provide information on the crash situation involving emergency vehicles before the preemption system was installed on a major arterial (U.S.1) in Fairfax County, Virginia. The data maintained by the Virginia Department of Transportation (VDOT) for a five-year period 1997-2001 provide information in terms of the accident date and time; description of the location (type of intersection, type of facility, type of traffic control, and number of lanes); collision type and severity; number of fatalities, injuries and amount of property damage; number of vehicles involved; environmental conditions; and other contributing circumstances. The findings of this analysis could be useful in the future-when post crash data become available- to conduct a before-after study to document the overall effect of the devices on crashes. This information could be useful to traffic engineers considering a preemption system at signalized intersections for safety purposes.

In total, 22 crashes involving emergency vehicles were reported during the analysis period. Figure 3 illustrates that from 1998 to 2000 the number of emergency vehicles involved in a crash on U.S.1 in Fairfax County, Virginia has increased, suggesting that the safe movement of emergency vehicles along this corridor is an issue that needs to be addressed. The crash statistics indicate that in total, there were no fatalities reported but there were 6 injury crashes that resulted in injuries to nine individuals. In 86% of all crashes, two vehicles were involved. Of the 22 crashes involving emergency vehicles on U.S.1 during the study period, a total damage cost of $124,570 was estimated; on average, this cost is about $5,700 per crash. There were crashes of cost as low as $250 to as high as $21,000. This finding reinforces the notion that providing a safer movement of emergency vehicles can save money to the Fire and Rescue Community; money that could possibly be allocated in improving emergency vehicle operations.

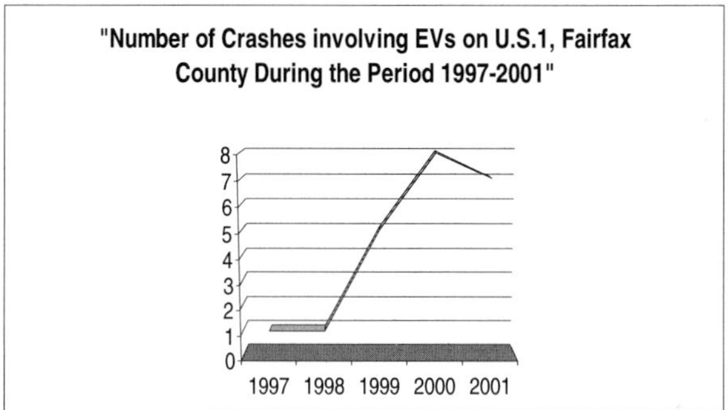

	1997	1998	1999	2000	2001
Number of Crashes	1	1	5	8	7

Figure 3. Number of Crashes involving Emergency Vehicles on U.S.1 in Fairfax County during the period 1997-2001.

Further analysis indicated that the majority of crashes on U.S.1 (64%) occurred at intersections, most of which (79%) were signalized. Furthermore, more crashes (41%) were identified as angle type in comparison to other types of collisions such as rear end (32%) or sideswipe (18%). This suggests that more crashes occurred when the emergency vehicle was maneuvering trying to pass vehicles or was making a left turn, and could be also indicative that an appropriate warning sign was absent.

CONCLUSIONS AND RECOMMENDATIONS

The analysis of the emergency vehicle characteristics in the greater Washington D.C. Region revealed the following:

1. The characteristics of emergency vehicle trip generation in terms of temporal distribution of emergency vehicle travel vary by fire station.

2. The frequency of emergency calls is higher during the daytime, with higher frequency during the PM peak period (on average, two calls per hour) and lower during the AM peak period (on average, one call per hour).

3. Heavy emergency vehicles are present at the fire stations; at least one is involved in each response.

4. The crash situation involving emergency vehicles has grown worse from 1998 to 2000 on U.S.1 in Fairfax County, Virginia. Most crashes occurred at intersections (64%), most of which were signalized (79%); in addition, more crashes were identified as angle type (41%) in comparison to other collision types such as rear end or sideswipe.

5. The frequency of emergency vehicle preemption requests on U.S.1 is lower during the AM peak period at all intersections (up to one request in three hours); during the other three time periods of the day the frequency ranges between one and two requests in three hours.

6. Very few emergency vehicle preemption requests on U.S.1 were denied (1 to 2%).

7. The average duration of emergency vehicle preemptions on U.S.1 is relatively low; on average, it ranges from 16 to 26 sec with no significant variability by time of day.

8. The size of vehicle platoons per preemption request on U.S.1 is relatively small; in 73% of the cases, each platoon included only one emergency vehicle.

9. The characteristics of emergency vehicle preemption requests are dependent on the conditions specific to each intersection at which the preemption system is installed. There is also some variability of the frequency as well as the average duration of preemption requests by geographic location. This could be attributed to several factors including the proximity of the firehouse to the intersection, roadway geometrics,

traffic characteristics, traffic control capabilities as well as to the size of the emergency vehicle platoons responding to an incident.

Considering the critical factors affecting the need for preemption including emergency runs and time of day; emergency vehicle crash history; and the presence of heavy emergency vehicles, it can be concluded that the need for a preemption system on U.S.1 does exist to enhance the performance of emergency vehicle operations and provide a better environment for just in-time delivery. The empirical results reported in this study regarding the frequency and average duration of preemption requests on U.S.1 suggest that the disruption to the other traffic is anticipated to be low or even negligible. Field results from earlier preemption studies (Mittal, 2002; Louisell, 2003) on U.S.1 regarding delay to other vehicles and queue lengths on the side streets reinforce the above notion. While the results may suggest that an EVP investment is warranted, consideration must also be given to the investment requirements associated with EVP installation and operation. Such an installation needs to identify the directions of flow to be provided EVP and the corresponding initial costs of detectors, phase selectors, emitters, warning lights (if, desired), software, and other necessary equipment and anticipated operating and maintenance costs. These costs will vary depending on the type of EVP system selected and the vendor.

Finally, while this research has attempted to lay the groundwork for understanding the characteristics of emergency vehicle travel and operations in the greater Washington D.C. Metropolitan area, it is proposed that future research efforts build upon this research in order to:

1. Positively conclude the overall effect of the preemption devices on emergency vehicle response times and crash rates. An ex-post analysis of emergency responses and crashes would be useful to evaluate the effectiveness of the preemption strategy on the study corridor.

2. Develop warrants to be used in determining the appropriateness of installing preemption systems at signalized intersections, taking into account institutional challenges, traffic characteristics, traffic signal control capabilities, operational limitations, and roadway geometric constraints.

ACKNOWLEDGEMENT

The authors would like to thank the research team involved in this project with special thanks to Chuck Louisell and Houng Soo for their cooperation and expert guidance. In addition, the authors want to acknowledge the funding support provided by the Washington Metropolitan Council of Governments, the Virginia Department of Transportation, and the Maryland Department of Transportation.

REFERENCES

BRW Inc. (1997). An Evaluation of Emergency Vehicle Preemption Systems.

Bullock, D., J. Morales and B. Sanderson (1999). Impact of Signal Preemption on the Operation of the Virginian Route 7 Corridor. Paper presented at the 1999 ITS America Annual Meeting.

Collura J. and G. McHale (2003). Improving Emergency Vehicle Traffic Signal Priority System Assessment Methodologies. Paper presented at the 82nd Transportation Research Board Annual Meeting.

Fairfax County, Virginia, Statistics & Annual Progress Reports, 1999-2002. http://www.fairfax.va.us/ps/fr/general/stats.htm http://www.fairfax.va.us/ps/fr/general/anlrpt.htm Accessed February 21, 2003.

Gifford, J., D. Pelletiere and J. Collura (2001). Stakeholder Requirements for Traffic Signal Preemption and Priority in Washington, D.C., Region. *Transportation Research Record* **1748**, TRB, pp. 1-7. National Academy Press, Washington D.C.

Langbein S. (2003). Emergency Response Caught Between Growth and Urgency, Fort Collins, Colorado Website:http://www.coloradoan.com/news/stories/20030119/news/807868.html Accessed March 3, 2003.

Louisell, C. (2003). A Proposed Method to Evaluate Emergency Vehicle Preemption and the Impacts on Safety – A Field Study in Northern Virginia. Paper presented at the 2003 ITS America Annual Meeting and Exposition in Minneapolis, MN.

Louisell, C., J. Collura, D. Teodorovic and S. Tignor (2003). Assessing the Safety Benefit of Emergency Vehicle Preemption at Signalized Intersections. Paper presented at the 10th Annual ITS World Congress in Madrid, Spain.

Ludwig G. (2002). Emergency Medical Runs: Urgent Vs. Non Urgent, EMS@Firehouse.com Website: http://www.firehouse.com/ems/ludwig/2002/june.html Accessed March 7, 2003.

McHale, G. and J.Collura (2002). An Assessment Methodology for Emergency Vehicle Traffic Signal Priority Systems. Proceedings of ITS Congress in Sidney, Australia.

Mittal M. (2002). Assessing the Performance of Emergency Vehicle Preemption System: A Case Study on U.S.1 in Fairfax County, Virginia. M.S. thesis, Virginia Polytechnic Institute and State University, Virginia.

Montgomery County, Maryland – Department of Fire and Rescue Website: http://www.montgomerycounty.gov/mc/dfrs/index.asp Accessed May 30th, 2003.

Nelson, E. and D. Bullock (2000). Impact Evaluation of Emergency Vehicle Preemption on Signalized Corridor Operation. Paper submitted for Publication to the 80th Transportation Research Board Annual Meeting.

Obenberger, J. and J. Collura (2001). Transition Strategies to Exit Preemption Control: State-of-the-Practice Assessments. *Transportation Research Record* **1748**, pp. 72-79, National Academy Press, Washington, D.C.

St Paul (1977). *Emergency Vehicle Accident Study*. Department of Fire and Safety Services, St. Paul, MN.

Straub, G. (2000) *Emergency Vehicle Preemption and Emergency Traffic Signals – Suggested Guidelines*, Maryland State Highway Administration.

Traffic Engineering. Inc. (1991). *Emergency Response Management System Study*. Prepared for the City of Houston and Metropolitan Transit Authority.

Washington S., M.G. Karlaftis and F.L. Mannering (2003). *Statistical and Econometric Methods for Transportation Data Analysis*. Chapman & Hall/CRC.

Transport Science and Technology
K.G. Goulias, editor

CHAPTER 26

U.S. TRANSPORTATION POLICY AND SUPPLY CHAIN MANAGEMENT ISSUES: PERCEPTIONS OF BUSINESS AND GOVERNMENT

Evelyn Thomchick, The Pennsylvania State University

INTRODUCTION

While the availability of a wide variety of transportation services may be taken for granted in a modern society, the establishment of transportation policy and the provision of the transportation services stems from a highly complex process. Because transportation has elements of a public utility (Spychalski, 2004), governments at all levels are generally involved in establishing the political environment in which transportation systems can operate. Within this framework, a wide variety of private and public enterprises establish and conduct business. There are several different modes of transportation, each comprising their own systems, and in many cases, interfacing with each other. For most of the modes, there are commercial and personal users and the "cargo" may be freight or people. It is not surprising that transportation policy, if it exists at all in a society, has developed in a fragmentary manner.

The principal participants in transportation may be classified as follows: 1) government agencies at different levels, 2) the transportation service providers, and 3) the users of transportation, which would include shippers of cargo and people who use transportation for personal mobility. Because transportation is associated with politics, there are also special interest groups that can affect the direction of transportation policy. Transportation is a broad and diverse subject, and transportation policy makers have a wide range of issues requiring their attention. Likewise, transportation research spans a variety of disciplines, ranging from the social sciences to engineering.

The main focus of this article is on the role of transportation in business, specifically supply chain management. In concise terms, supply chain management is the management of the flow of the materials, information, and funds involved in the production of goods or services. Although formal hypothesis testing was not within the scope of this research, a research question may be posed as follows: Do corporate supply chain managers, corporate transportation managers, and government transportation planners differ in their perception of the importance of transportation and supply chain issues? The original intent was to administer a mail survey to individuals who are employed in the above positions to assess their importance of transportation and supply chain management issues. However, construction of a survey that would be unbiased and span the wide range of transportation expertise of the business and government positions was a challenge. Thus, as a first step, transportation research was analysed on the justification that the research reflects the priorities of the different transportation participants in their respective publication outlets.

The objective of this article, then, is to present the results of an analysis of transportation research from selected business and government publication outlets. Comparisons of topics are made among carrier-, shipper-, and policy-oriented research in the different publications and research gaps are identified. The analysis of the transportation research publications is used as a proxy for assessing the importance of transportation issues among different business and government transportation professionals.

ROLES AND ISSUES OF TRANSPORTATION PARTICIPANTS

In most developed countries, the establishment of transportation policy that helps to determine what types of transportation systems exist in those countries stem from the decisions and actions of government agencies. Transportation government or government-related agencies exist at several different levels: international, national, regional, and local. These levels are defined by the political jurisdictions of specific countries. The roles of the government agencies include planning, policy development, and public promotion of transport programs and generally extend to the actual construction, operation, and maintenance of large portions of the transport infrastructure, and sometimes even the provision of transportation services. Government transportation agencies are also concerned with social and economic impacts of transportation, including safety, environmental impacts, and economic development. A major issue of most government transportation agencies is the financing of transportation systems.

Transportation service providers are organizations who are in the business of providing transportation services. They may be public or private organizations. In either case, the main roles are to manage the business of providing transportation services, be responsive to customers' demands, and comply with government policy and regulations. Transport service providers on the other hand, driven by the demand requirements of their customers, are concerned with how they will organize labor, material, and financial resources to meet customer requirements while earning attractive enough profits to remain in business over the long term. Thus, they concern themselves with issues such as operating management systems; availability, cost and organization of qualified labor; selecting, financing, operating, and maintaining equipment and terminals; procuring critical supplies, such as fuel; organizing strategic partnerships and alliances; and costing and pricing of services. They also are concerned with the compliance and cost of environmental and other regulatory requirements governing the ownership and operation of transport equipment and access to the transportation infrastructure.

Finally, there are the users of the transport systems, whose roles are to procure transportation services and use the transportation systems responsibly. The users may be shippers of freight or personal users. In the most general sense their issues deal with the reliability of transport, including frequency of service, geographic coverage, timeliness, special services, security, safety, and visibility. Other issues include the cost-effectiveness of transport and the ability of the service provider to harmonize their processes with the users' requirements.

The relationships among the above transportation participants are dynamic and the efficacy of the overall transportation system can be affected by the actions and decisions of any one of the participants. This is illustrated in Figure 1. An example of such an interrelationship was

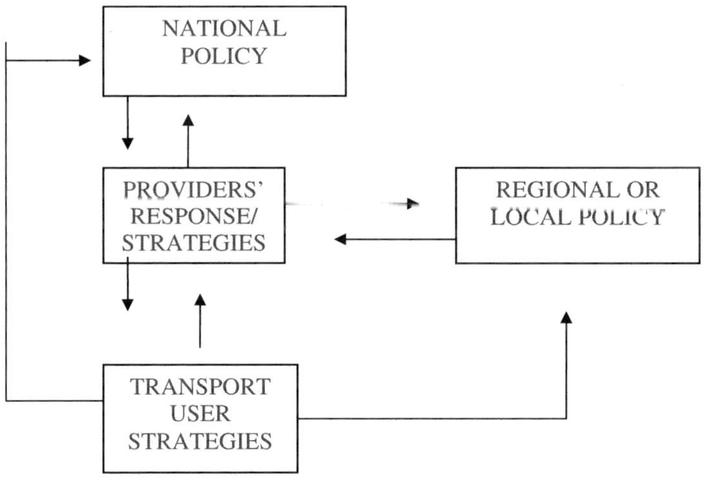

Figure 1. Interrelationships Among Government, Transportation Providers, and Transportation Users.

found by Gittings and Thomchick (1987) in their analysis of rail freight line abandonment following transportation deregulation in the U.S. A change in national policy (transportation deregulation) led to abandonment of unprofitable rail lines by rail carriers (provider response strategy). Shippers using these rail lines had to switch to trucking services (transportation user strategy). Trucking companies were offering very favourable rates following deregulation (provider response strategy). so shippers initially were satisfied with using trucking. However, some shippers, particularly of oversized cargo, couldn't use trucking and required rail service. After a certain period had elapsed, other shippers became dissatisfied with trucking service and wanted to switch back to rail. The state transportation agency found it necessary to provide and subsidize rail service on certain lines. The chain of events that occurred resulted from a change in national policy. However, a change by any of the participants can change the dynamics of the interrelationsips. The advances in the practice of supply chain management along with transportation deregulation in the U.S. have affected how shippers use transportation in their businesses. In the 1980s, certain manufacturers began using just-in-time (JIT) material management and production practices. JIT required small, more frequent shipments, which contributed to traffic congestion problems during business hours in urban areas (Rao and Grenoble, 1991). In this case a shift in user strategy, brought about partly by a change in national transportation policy and partly by business and technological developments, changed the interrelationship among the transportation participants and created issues for the transportation agencies. Thus it is important that the participants are aware of events and decisions occurring in each of the other participants' domains and understand the impacts of said events and decisions on their own domains. This is the main issue of this research.

METHODOLOGY

The methodology employed was a review of literature of selected U.S. based transportation journals. The following journals were selected: the *Transportation Journal,* the *Journal of Business Logistics,* and publications from the Planning area of the Transportation Research Board. It is acknowledged that these publications represent a small percentage of available transportation publication outlets because of the wide range of transportation research areas. Some journals are specific to a research area, such as, *The Journal of Public Transportation.* The three publications sources were selected because the focus of this analysis is on business freight transportation, and the three publication outlets represent a breadth of topics related, but not limited, to business freight transportation. The lengths of time all three publication outlets were in existence are comparable.

The *Transportation Journal* is affiliated with the American Society of Transportation and Logistics, the professional organization that grants certification in transportation and logistics (www.astl.org, 2005). The *Journal of Business Logistics* is affiliated with the Council for Supply Chain Professionals (www.clm.org, 2005), one of the largest associations of logistics and supply chain professionals. These two journals can be said to represent the shipper and the carrier perspectives. The Transportation Research Board is a division of the National Academy of the Sciences and Engineering (www.trb.org, 2005). The Planning area is the area that was judged to best represent policy-making in freight transportation. Article abstracts dating from 1987 to 2004 were reviewed and were classified as shown in Table 1. A representative article title is listed along with the classification definition for illustrative purposes.

Table 1. Classification of Transportation Articles

Government-centered – having to do mainly with policy-making and government concerns (Transportation, International Trade, and Economic Competitiveness (Transportation Research Board, 2003))

Carrier-centered – having to do with the operations of a transportation company (Truck Driver Recruitment: Some Workable Strategies (Lemay and Stephen, 1988))

Shipper-centered – having to do with the business of arranging for freight transportation for a corporate entity (Strategic Logistics Capabilities for Competitive Advantage and Firm Success, (Morash *et al.*, 1996))

Carrier-shipper-centered – combined interest of the carriers and shippers (A Longitudinal Assessment of Motor Carrier-Shipper Relationship Trends (Crum and Allen, 1997))

Government-carrier-centered – combined interest of government policy and carrier operations or strategy (Airline Financing Policies in a Deregulated Environment (Chow *et al.,* 1988))

Government-shipper-centered – combined interest of government policy and shipper operations or strategy (Paying Our Way: Estimating Marginal Social Costs of Freight Transportation, (Transportation Research Board, 1996))

Government-carrier-shipper-centered – combined interest of government policy, carrier operations and strategy, and shipper operations and strategy (Motor Carrier Selection in a Deregulated Environment, (Bardi *et al.* 1989))

The number of articles falling into each category was tabulated. Since the publications' schedules were not the same, the numbers were converted to percentages. The results are presented in the next section. It should be noted that articles on passenger transportation and transportation education and professional development appeared in all of the publication outlets but were not included in the tabulation of the articles.

RESULTS AND CONCLUSIONS

The tabulated results are shown in Table 2. A statistical analysis was not conducted. A chi-square test would be the most likely statistical instrument. However, the large number of cells with low frequencies or zero undermine the validity of the test. A visual examination of the data suggests statistical significance, and much can be gleaned from an examination of the tabulations, particularly when analyzed against expected results. While both the *Transportation Journal* and the *Journal of Business Logistics* are logistics and supply chain journals, their evolution has been from different origins. The American Society of Transportation and Logistics has more of a carrier heritage, while the Council of Supply Chain Management Professionals (originally the National Council of Physical Distribution Management) has more of a shipper heritage. Of course, the Transportation Research Board's focus is on government policy.

Given these orientations, one might expect the *Transportation Journal* to contain more carrier-centered articles, the *Journal of Business Logistics* more shipper-centered articles, and the Transportation Research Board more government policy-centered articles. This is, in fact, reflected in the results. However, it is interesting to note that the *Transportation Journal* covers the broadest range of topics, the *Journal of Business Logistics* is almost exclusively shipper-oriented, and the Transportation Research Board is almost exclusively government-policy-oriented. If one were to identify gaps, they would be: the government-shipper category in all three publication outlets, and the government-centered and the government-carrier categories in the *Journal of Business Logistics*. It is interesting to note that the Transportation Research Board contains the most articles, both in absolute numbers and percentage, of the "integrated" area of government-shipper-carrier.

The intent of these comments is not to criticize these publication outlets for publishing articles in what is considered their primary domain. However, to avoid a fragmented transportation policy, all of the transportation participants must be considered in an integrated manner, so it is desirable to have research that spans the interests of as many transportation participants as possible. (This includes passengers, which were not included in the tabulation of these articles because of the focus on freight). Sophisticated research in any discipline becomes focused and in-depth, and this does not always allow for breadth in scope, and, of course, not all research can be boundary-spanning. However, in order to solve the increasing transportation problems of today, a broader research framework should be adopted. There should be more research that falls into the boundary-spanning categories, particularly, the government-carrier-shipper category, which should reflect both passenger and freight transportation. This study focused on selected business logistics-related and policy publication outlets that were assumed to be the most likely to contain these types of articles. Perhaps an examination of transportation-related journals in other disciplines would produce different results.

CLASSIFICATION OF JOURNAL ARTICLES GAPS						
	Transportation Journal		Journal Of Business Logistics		Transportation Research Board	
Classification of article	Number of articles	Percentage	Number of articles	Percentage	Number of articles	Percentage
Government centered	16	5%	1	0.3%	96	60.4%
Carrier centered	111	36%	9	2.8%	0	0.0%
Shipper centered	28	9%	236	72.6%	0	0.0%
Carrier-shipper	59	19%	59	18.2%	0	0.0%
Government-carrier	71	23%	1	0.3%	29	18.2%
Government-shipper	4	1%	6	1.8%	0	0.0%
Government-carrier-shipper	17	6%	13	4.0%	34	21.4%
Total	306		325		159	100.0%

Table 2. Classification of Journal Articles

REFERENCES

American Society of Transportation and Logistics, *Transportation Journal,* issues from 1987-2004, www.astl.org, accessed August 2004.

Bardi, E.J., P. K. Bagchi, and T. S. Raghunathan (1989). Motor Carrier Selection in a Deregulated Environment. *Transportation Journal*, Vol 29, Iss. 1, pp 38-44.

Chow, G., R. D. Gritta, and R. Hockstein (1988). Airline Financing Policies in a Deregulated Environment. *Transportation Journal,* Vol. 27, Iss. 3, pp. 37-43.

Council for Supply Chain Management Professionals, *Journal of Business Logistics,* issues from 1987-2004, www.cscmp.org, accessed August 2004.

Crum, M. R. and B. J. Allen (1997). A Longitudinal Assessment of Motor Carrier-Shipper Relationship Trends. *Transportation Journal,* Vol. 37, Iss. 1, pp. 5-18.

Gittings, G., and E. A.Thomchick, (1987). Some Logistics Implications of Rail Freight Abandonment. *Transportation Journal,* Vol. 26, No. 4, pp. 16-25.

Lemay, S. A, and G. S. Taylor (1988). Truck Driver Recruitment: Some Workable Strategies. *Transportation Journal.* Vol. 28, Iss. 1, pp. 15-23.

Morash, E. A. and L.M. Cornelia, and S. K. Vickery (1996). Strategic Logistics Capabilities for Competitive Advantage and Firm Success. *Journal of Business Logistics*, Vol. 17, No. 1, pp. 1-22.

Rao, K., and W.L. Grenoble, IV, (1991) Modelling the Effects of Traffic Congestion on JIT. *International Journal of Physical Distribution and Materials Management,* Vol. 21, Iss. 2, pp. 3-10

Spychalski, J.C., P.F. Swan, (2004). U.S. Rail Freight Performance Under Downsized Regulation. *Utilities Policy,* **12,** 165-179.

Transportation Research Board (1996). *Paying Our Way: Estimating Marginal Social Costs of Freight Transportation,* TRB Special Report 246.

Transportation Research Board (2003). *Transportation, International Trade, and Economic Competitiveness.*

Transportation Research Board, publications in the Planning category, www.trb.org, accessed August 2004.

Transport Science and Technology
K.G. Goulias, editor

CHAPTER 27

USING PERFORMANCE MEASURES TO DETERMINE WORK NEEDS: AN OPERATOR'S PERSPECTIVE

Katherine D. Jefferson

ABSTRACT

Transportation agencies are adopting performance measurement as an attribute of their organizational cultures. These measures evaluate effectiveness, operating efficiency, productivity, service quality, customer satisfaction and cost-effectiveness (Poister 2003). Performance measurement evaluates progress toward specifically defined organizational objectives and includes both evidence of actual fact and measurement of customer perception (FHWA 2001). Performance measurement is not confined to the upper echelons of organizations. Rather, it is diffused throughout all organizational levels as government agencies account for their stewardship of public funds, and reporting outcomes as proof of efficiency becomes a fixture within normal operating procedures. Beyond illustrating good stewardship and efficiency, performance measures are used to determine work needs and to influence decision-making with respect to staffing levels, funding needs, or the choice between direct service delivery and outsourcing.

Within this context a case study is offered to trace the evolution of a work unit from the level of reporting work activity completion to its use of performance measures to demonstrate cost-effectiveness and to determine the percentage of work that could be outsourced. The case study shows how performance measurement was used to: determine the classification and compensation levels of staff members; support the use of alternative work schedules; validate an equipment procurement proposal; and change work practices and the general utilization of staff.

INTRODUCTION

Transportation agencies are adopting performance measurement as an attribute of their organizational cultures. Performance measures are multidimensional objective quantitative indicators of the various aspects of a public program or agency operations. These measures evaluate organizational or work unit effectiveness, operating efficiency, productivity, service quality, customer satisfaction and cost-effectiveness (Poister 2003). Performance measurement evaluates progress toward specifically defined organizational objectives and includes both evidence of actual fact and measurement of customer perception (FHWA 2001).

Performance measurement is not confined to the upper echelons of organizations. Rather, it is diffused throughout all organizational levels as government agencies account for their stewardship of public funds, and reporting outcomes as proof of efficiency becomes a fixture within normal operating procedures. Work units are increasingly challenged to move beyond enumerating outputs to measuring their effectiveness and the quality of the work they perform. When performance measurement is undertaken with a focus on effectiveness and quality, work needs can be determined and decisions such as staffing levels, funding needs, or the choice between direct service delivery and outsourcing are possible.

Within this context a case study is offered to trace the evolution of a work unit from the level of reporting work activity completion to its use of performance measures to demonstrate cost-effectiveness, the overall operational efficiency of the work unit, and to determine the percentage of work that could be outsourced. The case study highlights the activities of the Pavement Markings Unit in the Northern Virginia District of the Virginia Department of Transportation (VDOT) between 1995 and 2001. During the seven-year observation period, the unit advanced from assuming that they operated efficiently (because funding allocations were exhausted and materials and supplies were used) to using performance measures to facilitate a wide range of policy decisions. The adopted measures are not overly elaborate, which may explain their acceptance by unit members. However, they allowed unit members to gain an understanding of the relationship between their field activities and the broader VDOT organizational objectives of safety and mobility.

Before presenting the case study, a general overview of performance measurement approaches that have been applied within the private and public sectors is provided. This is followed by a review of measurement methods used in transportation agencies at large and those that have been used exclusively for pavement marking operations. Next the case study is presented, in which the application of performance measurement at the VDOT work unit level is discussed. Following this is a discussion of the types of decisions that have been made the performance measures. Finally, the paper concludes with an exploration of the policy implications, future opportunities and challenges that the work unit faces as it broadens its use of performance measures.

Private Sector Approaches

Performance measurement in private sector organizations dates back to the early 1900s. Frederick W. Taylor, a mechanical engineer and the founder of Systems Engineering, developed the scientific method, which entails measurement of work activities to facilitate efficiency (Taylor 1911). In the mid-1940s, Dr. W. Edward Deming developed the philosophy and methodology for Total Quality Management (TQM). TQM comprises fourteen principles, tools and procedures that provide guidance for managing organizations. The methodology entails monitoring critical variables and outputs and charting analytical findings (Mead 1996). Two essential concepts of TQM are the notion of continuous learning and improvement, and the idea that customer needs and expectations define quality, as opposed to agency-established standards (Maas 2004). In the mid-1980s, Motorola, Inc. pioneered the Six Sigma Methodology, a business improvement process that focuses on customer requirements, process alignment, analytical rigor and timely execution (Motorola 2004). Six Sigma is a measure of quality that uses data and meticulous statistical analysis to: identify defects in a process or product, reduce variability, and achieve as close to zero defects as possible (Motorola 2004). Although the profitability imperative that exists in private sector organizations is absent in the public sector, the guiding principles of private sector performance measurement may be adapted and subsequently applied within the public domain.

Public Sector Approaches

Interest in performance measurement within the United States government can be traced to the Kennedy Administration when system analysis processes were brought into the Department of Defense (Poister 2003). Performance measures associated with the budgeting process within the Planning Programming and Budgeting (PPB) system were introduced in the Johnson Administration (Poister 2003). In the 1990s, taxpayer revolts, pressure for privatization of public services, legislative initiatives aimed at controlling "runaway" spending, and the devolution of many responsibilities to lower levels of government generated increased demands to hold governmental agencies accountable to legislators and the public in terms of what they spend and the results they produce (Poister 2003). In response to this call for greater governmental accountability, performance measurement was stipulated in the Government Performance and Results Act of 1993 (GPRA), a centerpiece of Vice President Al Gore's effort to reinvent the federal government. GPRA (1993) defines performance measurement as a process of assessing progress toward achieving predetermined goals. The legislation argues that performance measurement can encompass measures of efficiency, effectiveness and quality.

Measurement in Transportation Agencies and Pavement Marking Operations

Basilica et al. (2000) argue that a growing number of transportation agencies are embracing performance measurement as a management tool, but few agencies use performance measurement across the full range of their activities. The authors also observe that existing measurement systems have not been externally validated. Kassoff (2000) argues that most of the state Departments of Transportation (DOTs) have initiated or experimented with performance measures to some degree. However, in no two cases have state DOTs undertaken performance measures for identical reasons and implemented them in the same way (Kassoff 2000). Many states measure the progress of capital improvement projects with respect to timeliness and budgetary constraints. In other state DOTs, measures of reductions in pollutant emissions, vehicle stops, delay, fuel consumption and travel time are used to assess how effectively the DOT manages its arterial network.

The body of literature relative to performance measurement for pavement marking operations contains studies of data that may be used for developing performance measurement systems. However, there is significant latitude in how these data may be used to gauge efficiency, effectiveness and quality in pavement marking operations. In some instances, evaluations of material durability, cost and service life are offered as the basis for developing performance measures. In others cases, efforts have been undertaken to correlate accident events with the condition of pavement markings. When this approach is used, measures of effectiveness or quality are related to how the condition and type of installed pavement markings contribute to an increase in (or aids in the reduction in) the number of accidents that occur. Researchers have also attempted to establish minimum acceptable brightness levels (retroreflectivity) as the centerpiece of pavement markings performance measurement. Additionally, qualitative analyses have been used to determine the level of motorist satisfaction as an indication of pavement marking effectiveness and/or quality. Opportunities abound to develop performance measures for pavement marking operations that are uniquely suited to local circumstances and individual organizational goals and objectives.

Having explored performance measurement and its various applications (within the private sector, public agencies, transportation departments, and pavement marking operations), a case study is presented to illustrate the use of performance measurement at the work unit level.

CASE STUDY

Background

The Pavement Markings unit, a group within VDOT's Traffic Field Operations section, is responsible for installing new and maintaining existing pavement markings and messages (i.e., stop bars, crosswalks, arrows, symbols, gore area markings, etc.). Its territorial responsibility is a four-county area in Northern Virginia – Arlington, Fairfax, Loudoun and Prince William counties. The unit manages over 30,000,000 linear feet of pavement markings (of the various material types) and over 4500 pavement messages. When fully staffed, three crews are deployed to install and maintain longitudinal lines and two crews install and maintain pavement messages. The inventory of installed markings increases annually as new streets are added to the state system, when roadways are widened, or when asphalt is applied to previously unpaved roadways.

The unit measures its operational efficiency by complying with established federal, departmental and internal guidelines. With respect to maintenance of longitudinal lines (the focus of the case study), the unit assumes a one-year service life for makings installed with water-borne or latex paint. One hundred percent of markings installed with this material are retraced annually to assure an acceptable level of retroreflectivity. The effectiveness of the annual cycle is measured by qualitative and quantitative measures. The qualitative measure is a de facto indicator of customer satisfaction, which is presumed to be at an acceptable level when there is an absence of complaints regarding unmarked pavement. (The unit does not conduct customer satisfaction surveys.) The quantitative measure is related to cost-effectiveness; the unit strives toward achieving the goal of installing markings at or below $0.05 per linear foot.

Performance Measurement

Osborne and Gaebler (1992) argue that organizations that measure the results of their work – even if they do not link funding or rewards to those results – find that the information transforms them. People begin to ask the right questions, to redefine the problem they are trying to solve, and to diagnose that problem anew (Osborne and Gaebler 1992). The authors conclude that what gets measured gets done. In 1995, hand-written daily work reports were used to document that employees performed pavement marking activities, and utilized materials and supplies. However, there was no annual report of accomplishments for the 1995 painting season, and there was no compilation of work report information to develop an overall assessment of performance with respect to goals of operational efficiency or cost-effectiveness. The unit believed that most roads were retraced during the painting season, and that the costs associated with their activities were acceptable. There were no measures of effectiveness or quality, other than a general perception that markings were adequate (based on visibility).

In 1995, one-third of the Pavement Markings unit's employees, including the unit manager, departed state service in accordance with the early retirement provisions contained within Virginia Workforce Transition Act (WTA). The next year, an existing staff member was promoted to the vacated position of unit manager. One of his initial decisions was to replace hand-drawn sketches of installed markings with automated data files that correlated with records of daily work activities. As the annual inventory of installed markings increased, sketches were redrawn to reflect modifications to roadways and the types of markings that were installed on them. Staff members completed this task during winter months and confirmed the accuracy of sketch modifications by conducting site visits, since weather conditions precluded pavement-marking installation.

The unit manager's motivation for automating sketch data was not based upon a premeditated intention to create a performance measurement system. However, his actions served as the catalyst for the unit's efforts in this regard. Since sketches were created and modified in accordance with records of completed work, it became possible for the unit to begin to understand the magnitude of its responsibility for the first time in its existence. This knowledge enabled the manager to concretely assess the operational efficiency of his group, as opposed to assuming that the unit operated efficiently (i.e., because funding was allocated and expended, work activities were undertaken when weather permitted, and materials and supplies were used).

In 1996, the daily routines of unit members were altered. Staff members were tasked with collecting data about the roadways on which work was performed, and with entering these data into spreadsheets. Most staff members were not computer literate and the unit had access to a limited amount of computer hardware and software. Consequently, the automation of daily work reports and the ability to demonstrate compliance with performance measures began slowly. Records from the 1996 painting season were limited to rudimentary data, such as lists of route numbers, roadway names, and retracing dates.

By the end of the 1997 season, the unit was able to compile an annual report showing accomplishments by month, though data were not disaggregated by material type. The annual data allowed the manager to begin to understand how much work had not been accomplished, and to see the consequences of the lack of coordination between pavement marking operations and highway maintenance activities. It was not uncommon to have newly retraced markings eradicated by roadway resurfacing activities, or for citizens to complaint about unmarked (newly resurfaced) roadways. The annual report informed the decision to incorporate pavement markings installation within the annual highway resurfacing contract. An outcome of this decision was that pavement markings personnel were invited to participate in the highway resurfacing schedule planning and implementation processes. During the planning stage, when the listing of roadways to be resurfaced is developed, markings personnel were asked to provide data on the linear footage and type of markings that exist on the roadway. These data were included in the solicitation for bidders and were subsequently used by inspectors who monitor contractor performance. Implementation of this policy change would not have been possible without the initial quantification of the inventory of installed markings.

During the 1998 and 1999 painting seasons, data files were enhanced to include: references to roadways identified by their locations within map grids, start and end points of routes, average daily traffic volumes on individual roadways, total linear footage of installed markings by route, and total linear footage of installed markings on each route by material type. Using these data, it became possible to create and use monthly updates to support requests for additional staffing (e.g., temporary employees or high school students during summer months) and the use of alternate work schedules (including overtime). Data were used to illustrate the connection between the level of inputs, outputs/productivity and cost-effectiveness.

During the 2000 painting season, a standardized format for reporting monthly and annual accomplishments was developed. Daily summaries of the number of linear feet of markings retraced were compiled. In addition, daily calculations of materials used, man-hours expended to complete the work, labor costs per man-hour, equipment costs and non-productive labor costs (e.g., when employees are on leave or perform duties that are not directly related to pavement marking operations) were performed. Aside from capturing the costs for retracing operations, data were collected on work performed by contractors (e.g., markings eradication) and work charged to alternate funding sources (e.g., capital improvements or special projects).

In the annual report from the 2000 painting season, managers were able to show that: 100% of water-borne paint markings (over 14,000,000 linear feet) were retraced at an overall unit cost of $0.04 per linear foot; the cost-effectiveness goal ($0.05 per linear foot) was met or exceeded seven of the eleven-month painting season; and the unit exceeded its cost-effectiveness goal for the year.

During the 2001 painting season, the unit suffered significant personnel losses. Four employees whose combined work experience exceeded 130 years retired. Two other long-term employees transferred to a different work unit, and another employee was forced to retire due to disability. As one would expect, the annual report for the 2001 painting season reflected a decrease in productivity. Nevertheless, the unit was able to meet its overall cost-effectiveness goal by installing markings at $0.05 per linear foot. However, only 84% of water-borne paint markings (nearly 13,000,000 linear feet) were retraced. During nine of the eleven months when pavement markings were applied, the cost-effectiveness goal was met or exceeded, and the cost-effectiveness goal was met for the year.

Work Decisions

The performance measurement efforts that began in 1996 were used to inform various decisions relative to work unit operations. An initial action involved increasing the use of durable marking materials (thermoplastic, extended life tapes) to reduce pavement-marking deficiencies in high traffic areas. Water-borne markings have been found to exhibit a one-year service life and the life cycle of durable markings ranges from 3-7 years (Abboud and Bowman 2002). Considering this, the Department adopted a policy for installing extended life tapes on interstate routes and for the use of thermoplastic markings on designated primary and high-volume secondary routes (ADT>10,000 vehicles) within the Northern Virginia district. In addition, as mentioned above, installation of extended life tapes and thermoplastic markings was incorporated into annual resurfacing contracts. These actions, which improved the visibility of installed markings and facilitated the achievement of organizational goals of safety and customer satisfaction, were precipitated by the development and retention of measures of the installed markings inventory and to quantification of the unit's productivity.

Performance measurement also affected decisions regarding the general utilization of staff. Due to agency-wide staffing reductions, it was not possible to replace the employees who departed during the 2001 painting season. However, senior managers provided authorization to restructure the work unit and to develop contracts to outsource thermoplastic pavement marking installation. To create the organizational capacity to perform contractor oversight, two existing crewmember positions were reclassified to the Team Leader level. Doing so created the opportunity increased compensation, and personal/vocational development.

Relative to influencing changes in work unit operations, performance measures were used as the foundation for a proposal to procure a vehicle to install longitudinal thermoplastic markings. Two of the underlying premises of the procurement proposal were that obtaining a thermoplastic vehicle would reduce the unit's dependence on contractors to provide this service and that the vehicle would create the organizational capability to install durable markings at safety- or politically-sensitive locations. Although initial approval was granted to procure the vehicle, budgetary constraints subsequently derailed its final delivery.

At the conclusion of the 2001 season, the unit manager determined that additional organizational changes were necessary. In addition to contracts for installation of thermoplastic markings, service contracts were developed and solicitations for bidders were disseminated to outsource 50% of water-borne paint retracing responsibilities in two counties. In addition, a provision for application of markings was included in all contracts for new traffic signal construction (which is performed by developers or traffic signal contractors). Members of the work unit absorbed oversight responsibilities associated with these changes. Additionally, based on their confidence in their performance measures, standards with respect to cost-effectiveness and material service life were incorporated into the contracts for outsourced activities.

POLICY IMPLICATIONS

Osborne and Gaebler (1992) argue that adoption of crude performance measures, followed by protest and pressure to improve the measures, followed by the development of more sophisticated measures is common wherever performance is measured. The authors emphasize that this practice explains why so many public organizations have discovered that even crude measures are better than no measures (Osborne and Gaebler 1992). However, in light of this, how can public organizations assure that adequate consideration is given to the interests of stakeholders as measurement systems increase in complexity? In addition, how does one assure that performance measures will ultimately provide a means to justify rewarding successful endeavors instead of practices that support maintaining the status quo?

Poister (2003) cautions that performance measurement is not a panacea for all the problems and challenges that confront effective organizations and programs. Decisions regarding strategies, priorities, goals and objectives are often made in heavily politicized contexts, which are characterized by competing interests at different levels, forceful personalities and the abandonment of principle in favor of compromise (Poister 2003). Considering these observations, what happens when an organization changes its focus? What are the consequences of changing measures to align with a new organizational vision?

Poister (2003) also argues that performance measures, while descriptive, may not provide a clear indication of cause and effect. Measures may describe deficiencies; however, once deficiencies are found, is the agency not required to make changes in its operational practices and/or organizational structures to adequately address areas where they are lacking? In addition, what are the consequences if known deficiencies are ignored?

Performance measures that are designed to illustrate the achievement of tangible goals (especially within the transportation arena) may overlook the intangible or undesirable consequences of achieving those goals. For example, one of the implications of adopting the policy of outsourcing involves the loss of the organizational capacity to perform the outsourced activities. If contractors are responsible for service delivery, it arguably ceases to be cost effective for the agency to retain the equipment, supplies and personnel formerly required to provide the services. In some cases, it is possible to assign personnel to other tasks. In other instances this may not be possible. In the latter case, what are the associated human resource costs? What are the short and long-term consequences associated with the loss of institutional memory and technical expertise the outsourced personnel once provided?

One of the specific outcomes of the adopted performance measures is an increase in outsourcing. However, from an agency liability perspective, if a contractor is reluctant, refuses or demands additional compensation to perform a particular activity, the agency may be at a disadvantage. Although pavement markings unit employees were reclassified to reflect their responsibility for performing contractor oversight duties and, while these employees possess the technical expertise necessary to engage in pavement marking operations, skills in negotiation and persuasion are required to assure that work performed by contractors is accomplished in accordance with established guidelines and within specified timeframes. Possession of these non-technical skills is not generally emphasized within operator qualifications and courses to develop and inculcate these skills in maintenance employees have only recently been added to the VDOT training curriculum. The agency is challenged to develop its personnel to discharge these duties.

In light of these policy implications as their use of performance measures expands, the challenge for the pavement markings unit is to align its work activities and performance measures with the overall agency mission of promoting safety and mobility.

CONCLUSION

The VDOT Pavement Markings unit uses performance measures to define work needs and influence decision-making. Measures of operational efficiency and cost-effectiveness have been used to inform decisions about: retracing designated roadways via contract, the percentage of work to be outsourced, classification and compensation levels of staff to represent changes in roles and responsibilities, utilization of alternate work schedules, the types of markings that are used on designated roadways (and the associated phasing out of ineffective marking materials), equipment procurement proposals, work practices, and the general utilization of staff.

Documentation of accomplishments and the use of performance measures have also facilitated cultural changes within the unit. Overall, there is an increased sense of accountability among staff members with respect to expenditure of public funds. Success and failure to achieve goals are routinely communicated to senior managers. The desire to develop proficiency in computer software usage is prevalent. In addition, the drive to improve over time has intensified. Efforts to improve the quality of installed markings and thereby reduce citizen complaints about unmarked pavements have been undertaken.

An unanticipated outcome of the unit's performance measurement activities has been a reduction in the need for the unit manager to offer expert testimony during court cases in which the functionality of pavement markings has been questioned. This success has been fuelled by the extant knowledge of the installed inventory, practices to assure that retroreflective pavement markings exist on high-traffic roadways and the retention of detailed automated records of work activities performed.

In the future, the unit is challenged to further measure the quality of its work. Research of equipment to measure retroreflectivity and ways to incorporate its measurement into daily work activities has commenced. While their use of performance measures is expected to increase in the future, the unit has obstacles to overcome to achieve this goal. There continues to be a loss of personnel, and a consequent loss of technical expertise. The prevailing solution to this problem has been to resort to outsourcing. Although outsourcing enhances the chances of achieving performance targets, the possibility exists that this practice may eventually cause the unit to become obsolete.

The unit has progressed significantly over the seven years under review. Their use of performance measurement has helped to determine work needs, facilitates fact-based decision-making and informs their outsourcing strategy. By adopting a culture of performance measurement, the Northern Virginia Pavement Markings unit has solidified their contribution to the overall VDOT organizational objectives of safety and mobility.

REFERENCES

Abboud, N. and B. L. Bowman (2002). Cost- and Longevity-Based Scheduling of Paint and Thermoplastic Striping. In: *TRB 2002 Annual Meeting CD-ROM,* Transportation Research Board, National Research Council, Washington, D.C.

Basilica, J., D. Bremmer, P. Plumeau, K. Sadekas, R. Winick, E. Wittwer, M. Wolfgram, and D. Zimmerman (2000). Workshop Summary: Linking Performance Measures and Decision Making. In: *Performance Measures to Improve Transportation Systems and Agency Operations*, Transportation Research Board Conference Proceedings 26, National Research Council, Washington, D.C.

Federal Highway Administration (2001). Serving the American Public: Best Practices in Performance Measurement. http://govinfo.library.unt.edu/npr/library/papers/benchmrk/nprbook.html.

Kassoff, H. (2000). Resource Paper: Implementing Performance Measurement in Transportation Agencies. In: *Performance Measures to Improve Transportation System and Agency Operations.* Transportation Research Board Conference Proceedings 26, National Research Council, Washington, D.C.

Maas, K.F. (2004). Total Quality Management and Reinventing Government. http://home.t-online.de/home/kfmaas/q_tqm.html.

Mead, A. (1996). Deming's Principles of Total Quality Management (TQM). In: Deming Distilled – Essential Principles of TQM. http://wwwwell.com/user/vamead/demingdist.htm.

Motorola, Incorporated. (2004) What is Six Sigma? https://mu.motorola.com/sixsigma.shtml.

Osborne, D. and T. Gaebler. (1992). Reinventing Government. Addison-Wesley Publishing Company. Massachusetts.

Performance Plans and Reports. Government Performance and Results Act of 1993, 103rd United States Congress, Section 1115 (f).

Poister, T.H. (2003). Measuring Performance in Public and Nonprofit Organizations. John Wiley & Sons, Inc. California.

Taylor, F.W. (1911). The Principles of Scientific Management. Harper Brothers. New York.

Transport Science and Technology
K.G. Goulias, editor

CHAPTER 28

PLANNING HUCKEPACK TECHNOLOGY – ADVANCED TRANSPORT TECHNOLOGIES IN EU

M. Sc. Nikolina Brnjac, Prof. d. Sc. Dragan Badanjak, M. Sc. Vinko Jenič; Faculty of Transport and Traffic Engineering, Vukelićeva 4, 10000 Zagreb, Croatia

ABSTRACT

In the last thirty years the West European area has noted a permanent economic growth that has been reflected also on the growth of the overall transport operations. The transport sector in the European Union has recorded an estimated 4 per cent gross national product of the Union and employed more than 6 million people. Road transport of goods is constantly increasing and occupies mainly the dominant position in cargo transport. At the same time the share of railways has been reduced over the last thirty years. Therefore, there is an increasing need for the data on intermodal transport. "European transport policy for 2010: time to decide" forecasts a transport demand growth in European Union of 38% for freight transport. The rail/road system combines the advantages of both rail and road and can make a considerable contribution towards resolving intra-European traffic problems. **The success of huckepack transport must be based on consistent policies with respect to standards for both loading units and the vehicles carrying them.** The carriage of goods, especially in international transport requires co-operation between several carriers from all the traffic branches. One carrier cannot deal with the whole transport task due to the diversity of transport routes. Depending on the organisation and the technical and technological co-ordination and co-operation of the participating transport branches, by simultaneous usage of transport and manipulation means in various combinations of transport, different forms have developed, that is, different techniques of rationalisation in the exploitation of the intermodal transport. The objective of this work is to resolve the problem of congestion, to improve safety and efficiency of huckepack transport, to minimise transport impact on the environment, problems of bottlenecks and impact of new technologies in combined transport as the basis for the development of a sustainable European transport system and to support

EU industry's competitiveness in the production and operation of transport means and systems.

1. INTRODUCTION

Combined transport in Europe is a fast growing part of the transport market, but it is under constant pressure by the competition of road transport. Freight transport is of major importance to society, especially for the development of the European Union. Because of the large number of trucks (and cars) the congestion of the road network is constantly increasing. In the future the transport sector will continue to grow strongly.

The policy of EU is to reduce the negative effects of transport. There are several strategies to reach this goal. One of the strategies is to replace pure road transport by intermodal transport. In many European countries, the need for an intermodal transport policy is increasing because of environmental concerns, reasons of overall efficiency and the benefits of co-ordination of modes to cope with the growing transport flows. In the United States, intermodal transport options are being developed because they are considered to be cost-efficient in the North American context. Eastern countries are still developing the unimodal form and they are not occupied with intermodal transport. Unless the transport sector considers mode-independent service requirements and utilizes spare capacities in other modes, road transport is likely to continue to increase its present market share.

Implementing the European intermodal transport system requires coordinated development of transport policy on the European, national and regional levels.

The objective and the main aim of this work is to develop the methods suitable for the development of intermodal transport technologies and to indicate the essential problems of intermodal transport in the EU countries and to present the problem itself on the example of Croatia - central eastern country (future EU member), for an optimal integration of different modes of transportation customer-oriented door-to-door services without favouring one mode of transport over the others.

What is the present and future role of huckepack transport? This paper attempts to address these issues by:

explaining the necessity of transport policy development as a means of solving obstacles in intermodal transport,

analysing huckepack transport in EU and Croatia with the suggested development measures and huckepack transportation routes in Croatia

2. TRANSPORT POLICY AS A MEANS OF EU POLICY DEVELOPMENT

The basic methods for the EU transport system development are reflected in the following: the development of a transport system that will be capable of meeting the transport needs efficiently and at a very high level of quality with minimum negative impact on the environment. This means that EU favours combined transport exclusively for environmental reasons, prevention of bottlenecks, destruction of road infrastructure, as well as for safety reasons in order to prevent catastrophes on road routes.

There are three attributes which are the basis of the new EU transport system:

- interconnectivity (connections among different traffic networks),
- intermodality (connections of services between branches),
- interoperability (inter-branch and intra-branch connections of services) which is the basis of the idea of sustainable development in the area of transport.

The combination of road and rail transport represents a logical step towards a more efficient transport over greater distances. The main problem of combined transport represents the organisation of the transport process. The combined transport development is most influenced by the government policy in which there is conflict between different interest groups. Joint transport policy as the means of transport policy development of EU is considered to be the great solution.

2.1. OBSTACLES OF INTERMODAL TRANSPORT

The market share of intermodal transport in the total European transport is limited: 8% of all intra EU transport (in tkm) takes place via intermodal transport. Nevertheless, all forms of European intermodal transport have shown a considerable growth over the last decade. Between 1990 and 1996 the average annual growth in tkm amounted to 9.3% for all the forms of intermodal transport[1].

In fact, the performance of intermodal transport varies considerably with the mode used for the main haulage phase, with intermodal traffic representing as much as 36% of total international traffic for rail, but only 13% for Short Sea Shipping and as little as 4% for Inland Waterways[2].

In the modal-oriented transport system, any change of mode within a journey involves the change of system rather than just technical transshipment. This creates *friction costs* which can make intermodal transport uncompetitive in comparison with unimodal transport. Friction costs are the measurement of the inefficiency of transport operation.

They are expressed in the form of higher prices, longer journeys, more delays, or less reliability on time, lower availability of quality services, limitations on the type of goods, higher risk of damage to the cargo, more complex administrative procedures. To make intermodal transport attractive for the user, friction costs must be reduced.

One of the problems of intermodal transportation is infrastructure. Inadequate access by rail, road or waterborne transport to the existing transfer points can hamper the integration of these modes and transfer between modes. The lack of interoperability within some modes poses significant problems. The obstacles are for example, the different railway signalling systems

---

[1] Source : EU Energy and Transport in figures, DG TREN website.

[2] *IMPRINT-EUROPE,* Andrea Ricci; Pricing of intermodal transport: lessons learned from RECORDIT

and loading gauges, and different bridge heights along Europe's inland waterways. Technical specifications for transport means are often regulated differently by country and by mode, which also raises issues of interoperability.

In theory it may be regarded as ideal but in practice, combined transport is accompanied by certain problems that need to be solved. The requirements of the industrial market are changing and with it also the intermodality requirements.

In order to achieve the goals of combined transport development, it is necessary to harmonize integration among the transport modes, infrastructure, hardware (cargo units, vehicles, telecommunications), operations and services as well as the legislation.

Table 1 shows the situation of the international combined transport in 2002 studied on 18 European corridors. The overall traffic amounted to 4,741,653 TEU or 54.4 million tonnes, out of which 44.1 million tonnes (81%) were transported unaccompanied and 10.4 million tonnes accompanied[3].

Form	**TEU (millions)**		**Net tonnes (million tonnes)**		
	2002	**2015**	**2002**	**2015**	**2015/2002**
unaccompanied	3.48	8.70	44.10	103.60	+135%
accompanied	1.26	1.50	10.40	12.40	+19%
Total	**4.74**	**10.20**	**54.50**	**116.00**	**+113%**

Table 1: International combined transport 2002/2015

Data on international accompanied transport in 2002 include the results of all the 17 existing "rolling highways"[4]. Transport included 547,000 trucks, out of which 1/3 used the services on the Brenner corridor and 20% the Tauern. The total volume of unaccompanied transport was about 3.5 million TEU. The market research showed the following:

In 2002 forty (40) companies offered unaccompanied transport services on the above mentioned corridors. Forty-nine (49) percent of the total was assigned to intermodal operators joined in UIRR, 19% Intercontainer-Interfrigo (ICF), and 32% to other operators. Some fifteen years ago, the intermodal services were provided only by UIRR companies and ICF.

In 2002, 60% of the overall European unaccompanied CT was carried out by land transport and 40% by water. Between the CEEC and EU countries, the container transport on inland waterways amounted to 80% of the total volume, whereas the land transport accounted for 20%.

[3] Study on Infrastructure Capacity Reserves for Combined Transport by 2015, May 2004.
[4] Transport of entire road vehicles by railway.

3. HUCKEPACK TRANSPORT IN EU AND CROATIA (DEVELOPMENT AND POSSIBILITIES)

The problem of huckepack technology and the combined transport in this section has been analysed by taking as example Croatia, as a typical Central-Eastern European country. The example of huckepack technology in Croatia could be implemented on the example of other European developing countries.

As already analysed in the paper, the huckepack technology considered from the aspect of economy facilitates efficient and rational division of work between the road and the rail transport.
Depending on the achieved organisational and technical and technological co-ordination and co-operation of the participating transport branches by simultaneous usage of transport and manipulation means in various transport combinations (road-railway, road-sea, road-river, railway-sea, railway-river and sea-river), various forms i.e. types of rationalisation techniques of transport and manipulation work in exploitation of the intermodal transport have been developed and are implemented in practice.
This technology, however, is little used in Croatia. The reasons for poor usage can be found primarily in the lack of equipment, first of all regarding low-floor wagons and loading-unloading technique. Low-floor wagons are necessary because of the structure gauge, and the advanced manipulators (portal cranes) for fast loading and unloading. Recently, a low-floor wagon of the Saadkms-z (498) series has been produced in Croatia. The wagons are 8-axle low-floor special wagons with 4-axle bogies, intended for transport of tractor trailers and trucks with trailers. The conditions of running of these wagons regarding the total weight, axle load as well as the total length of the truck trailer have been regulated by the directives of the EU member countries.

Figure 1 – Existing and suggested relations for huckepack transport from Zagreb

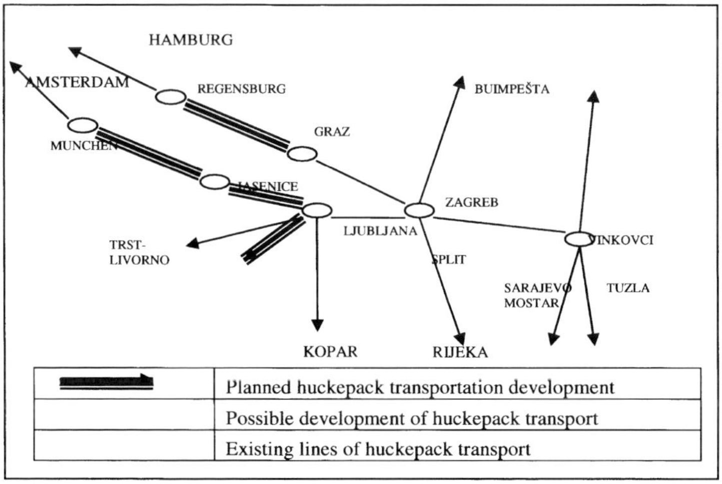

Analysing it may be concluded that Croatia needs to give priority to the construction and modernisation of the traffic routes in order to be able to follow the development of roads. It is necessary to establish container and huckepack trains (Zagreb-Rijeka, Hamburg-Jasenice, Graz-Zagreb-Rijeka). By investing into traffic routes, Croatia, as well as other East European countries can become part of the connected traffic network and the EU traffic market. For the development of huckepack technology, the construction and modernisation of terminals and cargo-transport centres is absolutely necessary. Being a maritime country Croatia has all the preconditions to develop combined transport. The objectives necessary for the development of combined transport include provision of conditions for facilitating traffic infrastructure, devices and vehicles that would be close to the European quality thus removing the barriers for equal participation of Croatia on the European traffic market.

According to AGTC, there are certain conditions that have to be fulfilled in order to achieve efficiency of the international combined transport, such as e.g. train performance parameters, minimal standards for railway lines, minimal standards for terminals and minimal standards for stations[5].

a) Train performance parameters (Table 2)
Trains used for international combined transport have to comply with the following minimal standards:

Table 2 – Train performance parameters

Minimal standards	Current situation	Targeted values
Nominal minimal speed	100 km/h	120 km/h
Train length	600 m	750 m
Train weight	1200 t	1500 t
Axle load (of wagons)	20 t	20 t (22.5 t at speed of 100km/h)

b) Minimal standards for railway lines

All the railway lines of rail transport have to have adequate daily capacity, avoiding the stay and delay of the trains.

c) Minimal standards for terminals

The main task of the terminal is to reduce the time of waiting and the manipulation time, which means that the time necessary to perform activities in the terminal and the waiting time

[5] PHD, R. Sabolovic: Multimodal and combined transport in a function of railway development, Zagreb, 1999.

of road vehicles is reduced to a minimum. The location of terminals has to have good road and railway network, and simple access to the terminal by roads.

d) Minimal standards for stations

En-route stations have to have adequate daily capacity of the tracks in order to be able to avoid delays in combined transport. The tracks have to contain the cargo profile as well as railway lines and the length sufficient to accommodate the entire train of combined transport.

Taking as model the western countries it is necessary to introduce "direct" container trains from origin to destination stations. The forming of such "direct" container trains would be possible at the Zagreb Shunting Yard. The Zagreb Shunting Yard meets the requirments since it is located on the international railway corridor X and is intersected by the Corridor Vb (Budpest-Zagreb-Rijeka) which has great predispositions for usage.

The main driving force of the development of combined transport in Croatia is urgent organisation of such transport modes in Croatia and the transit through Croatia, including all the traffic branches. One of the positive advantages of Croatia is that it is a maritime country and combined transport can include also the sea and the river transport. However, one of the most significant items for promoting combined transport is the development of railway and rail traffic as the main carrier of huckepack technology.

The integration of Croatian railways as well as other railways of Central and Eastern Europe into the EU railways needs to be realised in two basic fields: through legislative and standard conditions and through solving of technical drawbacks.

The liberalisation in the approach to railway infrastructure is the last in the series of liberalisation approaches to the infrastructures of traffic branches, and in the majority of countries this process is underway. In Croatia the legal framework has been created by adopting the new Act on Railways, and it will start to be implemented on 1 Jan. 2005 with the Act coming into force.

Since there is very little time until the new Act starts to be implemented, the restructuring of the company is necessary in order to be ready for the competition. This is possible by respecting all the mentioned advantages of the society and by using them as the basis for building the future, as well as by fast repair of those characteristics of the society which have been listed as comparative drawbacks.

4. CENTRAL-SOUTHEAST EUROPEAN HUCKEPACK LINES

The total distance for goods transport between Central and Southeastern European countries is about 2000 km. The transport uses the routes shown in Figure 3.

Figure 2. European main lines from Northwest to Southeast

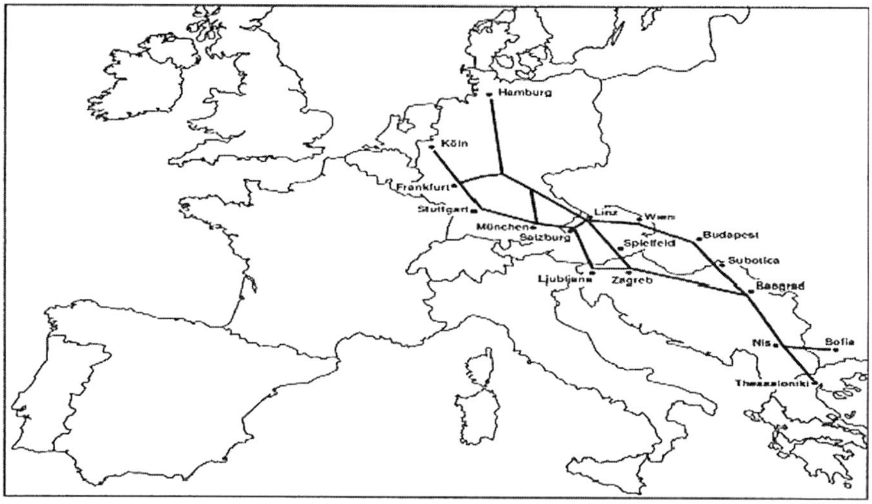

The course of the rail and the road infrastructure is almost identical. Huckepack transport is carried out as unaccompanied traffic- between NL/B/D and Austria, including Sopron at the Hungarian/Austrian border (semi-trailers and swap bodies); between NL/B/D/A and Greece (containers and swap bodies), and as accompanied traffic – between D and Y; between D and Austria. The weakest points of huckepack transport in Central-Southeast European lines are: long periods between the lates cargo delivery time and the departure of trains as well as between arrival of trains and the beginning of unloading, long overall transport times (long delays at the borders).

At present it is not possible to make full use of the shoulder height of 4m permitted on a large part of the route since limitations still exist on the section south of Ljubljana and Zagreb. Consignments are therefore restricted to a height of 3.80 m for pocket wagons and 3.70 m for low-loader wagons.

The analysis of transport times and delays shows that weak points are caused by: operational and administrative obstacles, transport volume which is still low on individual sections of the line. Since the railways are liable to organize transport at the lowest possible cost, huckepack transport must accept long stops[6].

5. CONCLUSION

The leading European countries have recognised the significance of the development of traffic branches which allow more economic transport of goods with respect to their advantages. In

[6] ECMT

order to provide the high-quality and efficient role of traffic, the traffic has to follow the development of science and technology. In satisfying the services for faster, safer and more cost-effective traffic, the development and implementation of huckepack transport technology is of great significance. The main objective of research in this work is the proof of the need and possibility of developing the combined transport and advanced transport technologies in the countries of Central and Eastern Europe for overall further traffic development of the European Union countries. The main research results show the necessity to reorganise the railways as one of the essential factors of the huckepack technology development. The process of globalisation requires from traffic not only increased quantity but also higher quality of transport and the harmonisation process with the developed and advanced world market.

According to the past experiences of the European countries, the necessary incentive measures for the development of combined transport are:

investment policy of combined transport,
reduction and complete tax exemption when purchasing the means for combined transport,
provision of favourable loans,
determining of the number of truck certificates.

The problems of advanced transport technologies in the developed EU countries are present regardless of their level of development but there is clear evidence of the orientation towards further development of the combined transport.

6. REFERENCES

Brnjac, N., Jolić, N and Božičević. D (2003). *Analysis of the possibilities for introducing piggyback technology in Croatia*, Annals of DAAAM for 2003&Proceedings of the 14th International DAAAM Symposium, Vienna

Kreutzberger, E., Macharis, C., Vereecken, L and Woxenius.J (2003): *Is intermodal freight transport more environmentally friendly than all-road freight transport? A review*, NECTAR Conference No 7, Umeå, Sweden June.

Ricci, A. (2002) *IMPRINT-EUROPE,*; Pricing of intermodal transport: lessons learned from RECORDIT

Sabolovic, R. PHD, (1999).: Multimodal and combined transport in a function of railway development, Zagreb

Vukić, D., Badanjak, D and Brnjac, N.(2004).: *Liberalisation of the Approach to the Railway Infrastructure: Possibilities and Dangers*, 3. EUROPSKI PROMETNI KONGRES, 22.-23.travnja, Opatija

EU Energy and Transport in figures (2003)., DG TREN

Freight intermodality (2001)., The EXTRA project, Transport RTD program

External costs: Research results on socio-environmental damages due to electricity and transport (2003).

RECORDIT: Final Report (2001).: Actions to Promote Intermodal transport

SPIN (Scanning the potential of intermodal Transport) (2002)., Deliverable 1

Transport Science and Technology
K.G. Goulias, editor

CHAPTER 29

A DYNAMIC PROCEDURE FOR REAL-TIME DELIVERY OPERATIONS AT AN URBAN FREIGHT TERMINAL

Gaetano Fusco, Università degli studi di Roma "La Sapienza", Italy
Maria Pia Valentini, ENEA – Ene/Tec, Centro Ricerche Casaccia – Roma, Italy

ABSTRACT

The paper describes a procedure for real-time management of an urban logistic centre, where deliveries addressed to the inner city are consolidated and distributed by a specialized fleet of vehicles. Optimisation of delivery operations is formulated as a dynamic vehicle routing problem with time windows and solved by a two-level genetic algorithm, which runs in real-time and applies a clustering algorithm both to generate the initial population and to update it when some unexpected event modifies the current conditions. In addition, to enable the optimisation of non-planned items, which arrive at the logistic centre at random during the service time, a specific operational scheme is devised for the quick coding of these items. Simulation tests show that the algorithm reacts to new events quickly and finds out a quite effective solution, so that it is suitable for real-time applications.

INTRODUCTION

Easy and straightforward distribution of goods is vital for economic activities of urban centres. On the other hand, town livableness needs mitigate the heavy impact of goods transportation on urban traffic, noise and air quality. Recent surveys highlighted that in many cities freight traffic is about 15% of total traffic (Ambrosini and Routier, 2000), but it contributes as 30% to the road occupancy (Patier-Marque, 2000) and as 50% to pollutant emissions of particulate. It has also a huge impact on traffic congestion, as unloading and delivering operations are often carried out in double-parking. Usual practices to manage urban goods traffic include a set of restrictions to access to the inner city, which may depend on vehicle type, load, fuel used and concern authorized hours, parking spaces and specific urban areas. Such measures can only postpone major decisions facing the problem of a better organization of the process. Surveys conducted in many European Countries outlined the large inefficiency of the urban delivery system: small shops are restocked by a number of different suppliers, who very often use their own vehicles or apply to small carriers. This leads to at least two inefficiency factors: increase the number of trips and use old and inefficient vehicles, which have often a very low load factor. In principle, this situation could be improved by introducing platforms where goods addressed to city centres are collected, rearranged and then dispatched to final destination using suitable, clean and well-loaded vehicles. Several pilot experiences have been conducted in last few years in several European cities regarding the implementation of Urban Logistic Centres (ULC), many of which co-founded by EU within research projects (eDrul, CityPorts, Merope,....). Such experiences demonstrate that ULCs management has to face many administrative and operational problems, due -on one hand- to a low tendency by transport operators to delegate others to perform their last-mile and -on the other hand- to a strong need of reducing time and costs of delivering service to attract potential customers. Optimisation of available resources is then a success feature. However, as most goods often arrive at ULC without any previous notice and are not identified by a standard code, deliveries are usually processed by following a simple First In-First Out rule and no optimisation of vehicle tours is carried out.

FUNCTIONAL SCHEME OF THE URBAN LOGISTIC CENTRE

The role of the Urban Logistic Centre (ULC) is to collect goods to be dispatched to a certain urban area in order to allow a reorganization of vehicle loads and delivery tours. ULC operations consist of goods receiving and stocking, vehicles tours scheduling, vehicles loading, transport and delivering. In the operational scheme here assumed, items of different kinds of goods can arrive at the logistic centre continuously during the service time. Part of arriving goods is announced in advance by telematic means, while others arrive without notice. We assume that each vehicle may operate more than one delivery tour and that vehicles of different classes are available. Each tour is defined by the starting time and, differently from the static case, can collect only deliveries that have arrived and can be loaded before the starting time. Parcels are characterized by weight, volume, desired time windows and possible specific requirements, as temperature control or urgency. Capacity of docks and warehouse is supposed to be always sufficient to ensure constant loading and unloading times. Fig.1 shows the specific operational scheme that has been devised to associate a barcode (numerical or alphanumerical) to each parcel arriving at the logistic centre, as it is required by optimisation programs. If a parcel arriving at the base has been notified in advance through EDI transmission, the parcel code has been already assigned when the logistic centre has accepted the delivery request from the supplier; the supplier himself provides for labelling as well as for administrative and operational data, so that the optimisation program can pre-process all notified deliveries. On the contrary, not pre-notified parcels require a first identification activity (1st ID) at the logistic centre, during the downloading phase. Respective delivery notes are then conveyed to the billing office and all associated information is input into the computer. After receiving 1st ID, all parcels are conveyed to the warehouse area, where they are sorted accordingly to the last number of their code. The deliveries database is updated and taken by the optimisation procedure, which runs continuously and computes the tour list. Each tour is defined by its starting time and the ordered list of parcels, whit related destination addresses. On the basis of this list, items are picked at the warehouse and conveyed to the weighing and labelling system, where a barcode label (*routing label*) containing all information regarding the delivery (address, time window, number of items, tour code, loading plan and additional information) is attached to each item.

As the optimal departure time of a tour computed by the optimisation system has come, all items belonging to that tour are loaded into the vehicle which the tour has been assigned to and tour starts. The operational scheme here assumed ensures necessary time sequence for running optimisation routines without limiting the ULC potential customers.

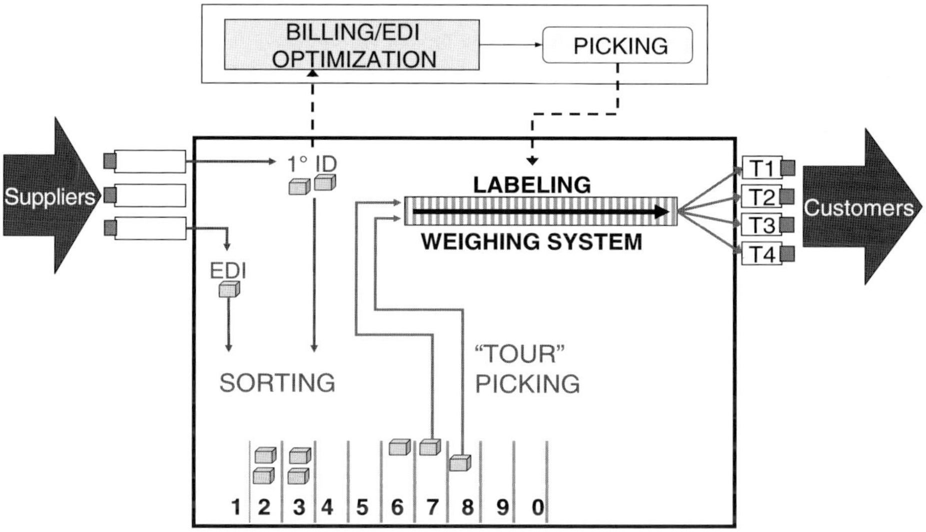

Fig.1. Layout and operations at the urban logistic centre, with flows of goods (continuous lines) and information flows (dotted lines).

DYNAMIC FORMULATION OF VEHICLE ROUTING PROBLEM

The dynamic formulation introduced here generalizes the usual static formulation of the vehicle routing problem with hard time window and capacity constraints on an asymmetric graph (Bodin *et al.*, 1999) to take into account the arrival time of each delivery at the logistic centre. As usual the road graph is composed by a set D of nodes, which include the depot at the origin 0 and the subset D' of nodes $\{i\}$ where deliveries have to be dispatched, and a set E of links $\{i,j\}$, which represent the best routes connecting destinations i and j. Corresponding least travel times and operative costs at time interval τ are denoted by $c_{ij}(\tau)$ and $t_{ij}(\tau)$,

respectively. Each node i has a service time σ_i required to unload the quantity q_i from a vehicle. A vehicle is permitted to arrive at destination i before the beginning of the time window u_i and wait at no cost until u_i. Delayed deliveries are allowed but involve an additional costs. Each vehicle k has a capacity p_k and is time-constrained at the depot, which it must leave and where it must return back within the time window $[a_0,b_0]$.
Two 0-1 binary and one continuous decision variables are introduced: x_{ijp} is 1 if and only if in the tour p the visiting point i is immediately before the delivery point j and 0 otherwise; y_{kp} is 1 if and only if tour p is assigned to vehicle k and 0 otherwise; s_{0p} is the departure time of tour p from the depot. The objective function has to be minimized with respect to the aforementioned variables y, x, z, s_{0p}:

$$\min_{y,x,z,n,s_0} \sum_{i\in D(\tau)} \sum_{j\in D(\tau)} \sum_{k=1}^{m} \sum_{p=1}^{n} c_{ijk}^{\tau} y_{kp} x_{ijp} + \sum_{i\in D(\tau)'} \sum_{h=1}^{H} b_{1,h} \left[\delta\left(s_{0p}^{\tau}\right)\right]_i^2 + b_2 n \tag{1}$$

The first term represents the operative costs of the delivery process, the second is a penalty function associated to delay δ of deliveries and the third is a fixed cost associated to any tour. Coefficients $b_{1,h}$ and $b_{2,h}$ are weights depending on the category h of the delivery. Time dependency implies that a generic solution found at time τ can be changed during the process, because travel times vary, new deliveries are taken at the logistic centre or the departure time of a tour is reached. As each vehicle can operate more than one tour during the day, the usual static formulation that assigns deliveries to vehicles is here re-formulated in terms of tours, each of them has its own starting time. Constraints of the problem are explained in the following.
Equation (2) states that all destinations j must be visited exactly once within the same tour p:

$$\sum_{j\in D(\tau)} x_{ijp} = 1, \quad \forall i \in D'(\tau),\ p = 1,2,\ldots,n \tag{2}$$

where n is the total number of tours carried out from τ to the end of the service time. Constraints (3) state that a destination i visited during a tour p is also the departure point for the successive delivery j.

$$\sum_{i\in D(\tau)} x_{ijp} - \sum_{i\in D(\tau)} x_{jip} = 0, \quad j \in D(\tau);\ p = 1,\ldots,n \tag{3}$$

Constraint (4) states that each tour must start to the depot. It is worth noting that this formulation allows a vehicle to pass to depot even more than once.

$$\sum_{i\in D'(\tau)} x_{i0p} > 0, \quad p = 1,\ldots,n \tag{4}$$

Constraint (5) requires that all tours must be assigned to vehicles, provided that each tour can be assigned at most to one vehicle, as stated in constraint (16).

$$\sum_{k=1}^{m} \sum_{p=1}^{n} y_{kp} = n \tag{5}$$

Constraint (6) imposes that all the $r(\tau)$ deliveries remaining at time τ must be released before the end of the service time.

$$\sum_{i\in D'(\tau)} \sum_{j\in D'(\tau)} \sum_{p=1}^{n} x_{ijp} = r(\tau) \tag{6}$$

Constraints (7) and (8) impose that the total weight and volume of each tour do not exceed the load and volume capacity of the vehicle (Q_k and V_k), which the tour is assigned to.

$$\sum_{i \in D'(\tau)} q_i \sum_{j \in D'(\tau)} x_{ijp} y_{pk} \le Q_k, \; k = 1,2,\ldots,m; \; p = 1,2,\ldots,n \tag{7}$$

$$\sum_{i \in D'(\tau)} v_i \sum_{j \in D'(\tau)} x_{ijp} y_{pk} \le V_k, \; k = 1,2,\ldots,m; \; p = 1,2,\ldots,n \tag{8}$$

Constraints (9) require that each tour p cannot start before the last delivery arrived at time τ has been loaded.

$$s_{0p} \le \max_j \left\{ \alpha_j^\tau \sum_{i \in D'(\tau)} x_{ijp} + \sum_{j \in D'(\tau)} \rho_j \sum_{i \in D'(\tau)} x_{ijp} \right\}, \; p = 1,\ldots,n \tag{9}$$

where s_{0p} is the starting time of tour p, α_j^τ is the actual (or forecasted) check-in time of delivery $j \in D'(\tau)$ at the depot and ρ_j is its loading time.

Constraints (10) define the arrival time ω_{jp}^τ at destination j in the tour p as a function of the travel time t_{ij}^τ of route ij forecasted at the time τ and the unloading time σ_i at the preceding destination i:

$$\omega_{jp}^\tau = s_{0p} + \sum_{i \in D(\tau)} t_{ij}^\tau x_{ijp} + \sum_{i \in D(\tau)} \sigma_j \sum_{i \in D'(\tau)} x_{ijp} \qquad \forall j \in D(\tau); \forall p = 1,2,\ldots n \tag{10}$$

Constraints (11) express that the service of a delivery s_j can not start before the beginning of the time window, u_j:

$$s_{jp} = \max \{\omega_{jp}^\tau, u_j\} \; j \in D(\tau); \; p = 1,2,\ldots n \tag{11}$$

Constraints (12) require that tour p' cannot start before vehicle k which it is assigned to has come back to the depot from its previous tour $p'-1$. The new notation of index p' denotes that the tours are ordered by their starting time.

$$y_{k,p} s_{0p'} \ge y_{k,p'-1} \omega_{0p'-1}^\tau + \sum_{j \in D(\tau)} \rho_j \sum_{i \in D'(\tau)} x_{ijp'} \qquad \forall p' = 2,3,\ldots n; \; k = 1,2,\ldots m \tag{12}$$

Equations (13) define the delay of a delivery i arrived after the end of the window, w_j:

$$\delta_{jp} = \omega_{jp}^\tau - w_j \ge 0, \; j \in D'(\tau); \; p = 1,2,\ldots n \tag{13}$$

Finally, expressions (14)-(16) are definitional constraints.

$$x_{ijp} \in \{0,1\} \; \forall i, j \in D(\tau); p = 1,\ldots,n \tag{14}$$

$$y_{kp} \in \{0,1\} \; k = 1,\ldots,m; \; p = 1,\ldots,n \tag{15}$$

$$c_{ijk}^\tau \ge 0; \; b_{1,h} \ge 0; \; b_{2,h} \ge 0 \tag{16}$$

REAL-TIME SOLUTION PROCEDURE

The capacitated vehicle routing problem with time windows has been shown to be a non convex NP-hard problem. Several effective exact algorithms have been proposed to solve static problems with up to 100 destinations (Mingozzi *et al.*, 1998). Anyway, many authors prefer heuristic approach to deal with realistic applications. Laporte (1992) provides a survey of both exact and approximate solving methods. Among several various heuristics proposed in the last years, one of the most effective resulted to be the tabu search technique (Rochat and Taillard, 1995; Gendrau *et al.*, 1994). Genetic algorithms have been widely implemented to solve this problem, also by the same authors (Fusco *et al.*, 2003), and revealed to be a quite effective and efficient tool (Baker and Ayechew, 2003). Additional difficulties arise when the problem includes capacity constraints and time windows (Desrosiers *et al.*, 1986) or time dependency (Ichoua *et al.*, 2003).

In a dynamic context, the problem is further complicated because of the constraint (13) on the starting time of the tours. Moreover, both travel times and the demand to serve can change continuously during the day and this requires re-calculating the solution. Thus, the solution procedure of the dynamic problem has to be so efficient to provide a good feasible solution for large scale problems of several hundreds or thousands of deliveries in no more than about 30 minutes. As genetic algorithms simulate an evolutionary process, they are flexible enough to be applied in real-time by continuously updating the givens and adjusting the solution.

On the basis of this assumption, the solution procedure is based on a bi-level genetic algorithm, whose chromosomes are defined by following a path representation (Choi *et al.*, 2003) and provide, at a given time instant, the delivery list of each vehicle. The population is updated every time the givens change because a tour starts or a delivery arrives. As stochastic algorithms are heavily affected by the goodness of the initial solutions, a specific algorithm has been devised for this task. The structure of the general solution procedure (Fig.2) takes planned deliveries as initial givens, applies a clustering algorithm to generate the first population and then runs the optimisation algorithm to compute sub-optimal delivery tours.

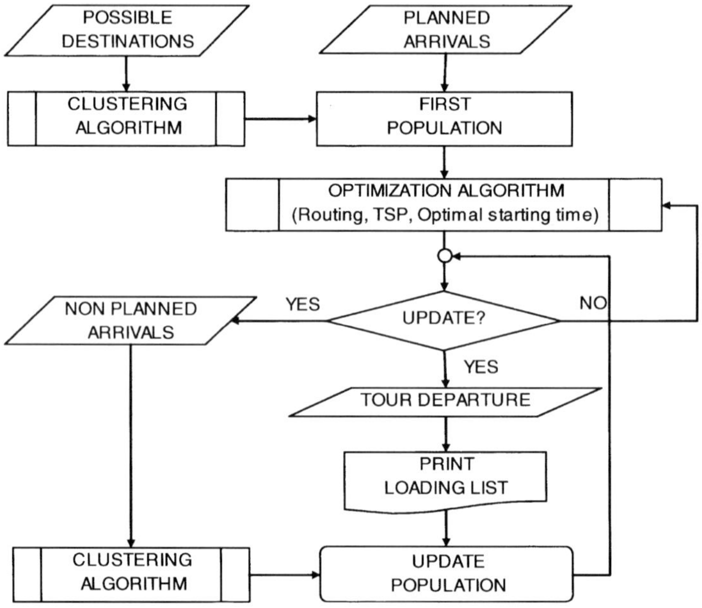

Fig.2. Structure of the solution algorithm.

Initialising algorithm and generation of first population

The initialising algorithm first applies a clustering algorithm and then combines a random and a serial selection scheme to generate the initial population. The clustering algorithm is derived from AGNES (Kaufmann and Rousseeuw, 1990), which splits all the possible destinations both in the time and in the space, by minimizing the average distance between destinations of the same cluster. As each destination is so assigned to one cluster, it is possible also to define the cluster of a tour, as the prevailing cluster of destinations belonging to it. This information is exploited by the solution procedure, by assuming that a rational solution is to assign a destination to a tour of the same cluster. The procedure of serial selection assigns a delivery by random to each given vehicle and generates a new tour when a given quota of the vehicle capacity is exceeded. The procedure of random selection assigns a generic delivery to one tour, which is taken with probability P_1 among those belonging to the same cluster of that delivery and with probability $(1-P_1)$ to the other clusters. In any case, only tours having available capacity are considered and a new tour is generated if necessary.

Optimisation algorithm

The optimisation algorithm computes sub-optimal delivery tours and to this goal applies a first genetic algorithm to solve the classic Vehicle Routing Problem (VRP), a second genetic algorithm that solves a Travel Salesman Problem (TSP) for each tour of the solution of VRP and then computes its optimal starting time. Both genetic algorithms have the usual framework composed by: cross-over, mutation, constraints verification, fitness computation, reproduction and population update. Peculiarities of the dynamic VRP problem required introducing some modifications, with respect to usual algorithms, namely:

Cross-over. As different individuals of a solution of the VRP (i.e., different tours) may have a different number of chromosomes, to avoid repeating or missing some customers, a matrix incidence has been introduced, which associates tours and customers univocally, and then makes it possible tie-break crossover on a randomly selected column.

Mutation. Mutation alters the structure of chromosomes to explore some new portion of the search space. Information regarding clusters of tours is here exploited to guide the process toward presumably better solutions. For each individual, a tour selected at random is assigned to another tour, which is taken with probability P_2 among those belonging to the same cluster of that delivery and with probability $(1-P_2)$ to the others.

Computation of tour starting time. Since early arrivals always would require waiting for the beginning of the time window, the starting time is computed as the earliest time instant that complies at least one of time window constraints. To do so, equations (10) and (17) are computed recursively from the first to the last destination.

$$s_{0p} = min\left\{ \max_j \{u_j, \omega_{jp}\}, \min_j \{w_j, \omega_{jp}\} \right\}, \quad p = 1, \ldots, n \tag{17}$$

Fitness computation and reproduction. Fitness of all individuals is computed by applying eqn. (1). The bi-level structure implies that fitness is firstly computed for the inner problem (TSP) and then for the outer (VRP). Finally, all members of the current population are ranked by their fitness value and best individuals are selected for reproduction by stochastic sampling.

Population update

The current solution is continuously updated by checking whether the starting time of a tour has been reached or a new arrival has come at the logistic centre. In the first case, provided that a given minimum number of generations has been processed, the program prints the loading list of the tour and removes it from the current population. In the second case, it codes the new deliveries in terms of chromosomes, computes an initial solution for them and adds such a solution to the current population.

SIMULATION

Preliminary numerical tests have been carried out by simulating the dynamic delivery procedure is applied to an Urban Logistic Centre planned in Cosenza, an Italian town of 75,000 inhabitants. The road network is modeled by a graph of 491 nodes and 1,029 links. The fleet is composed by 9 vehicles: 4 vans having capacity of 3,000 kg and $7.5m^3$ and 5 small electric vehicles of 1,500kg and $3.5m^3$. In Cosenza there are 1,391 retailer shops that attract 65,500 kg and receive 511 deliveries every day, on average. By assuming that ULC deals with 200 deliveries in a generic day, we generated 100 planned and 100 non planned deliveries, each of them is composed by 1 to 10 parcels and has two time windows: in the morning and at the afternoon, respectively. Both the optimisation algorithm and the simulation of the delivery process have been implemented in a software package coded in C++ (Dylog, 2005). Numerical tests were carried out on a Pentium 4-1800MHz computer.
An example of simulation results is shown in Fig.3, which depicts the number of deliveries arrived, released and waiting at ULC, the number of vehicles in tour and the value of the fitness normalised with respect to the number of deliveries at ULC. Planned deliveries are available at ULC at the beginning of the service time (6:00). Non planned deliveries arrive at ULC in 4 different shipments: at 8:40, 11:20, 13:10 and 15:10. At the beginning of operations, the fitness has very low improvements (1.4% after 170 generations) and this highlights the effectiveness of the clustering algorithm applied to find out the initial solution.
At each new arrival, the fitness function increases, but returns back to similar values after only few minutes. Specifically, the first two unpredicted arrivals at 8:40 and 11:20 are very well absorbed by the algorithm and do not affect fitness values. Further unexpected arrivals at 13:10 occur at a very unfavourable time instant, because just few minutes before two vehicles had left the logistic centre, so that there are fewer available resources. The fitness value has a sharp increase up to 4 times the previous values. In such a condition, the optimisation algorithm reveals to be both efficient and effective, as it takes only 7 minutes (in simulation) to reduce the fitness as 70%. It is worth noting that the speed of the simulation was about 2 times more than the real time. In real applications the algorithm could process 2 times more computations and so would achieve the same result in a 2 times shorter time, that is in only 3.5 minutes. After that critical condition, several tours start and, each time a vehicle leaves the logistic centre or a group of items is delivered, the value of the average fitness changes. At 15:10, a new group of unexpected deliveries arrives at the terminal, when only 1 vehicle is available. Improvements of the fitness are quite small and so its unitary value (fitness per

delivery) is slightly increasing, due to the decreasing number of waiting deliveries. As the number of deliveries to release diminishes, the fitness function decreases and the process ends when the last item is delivered.

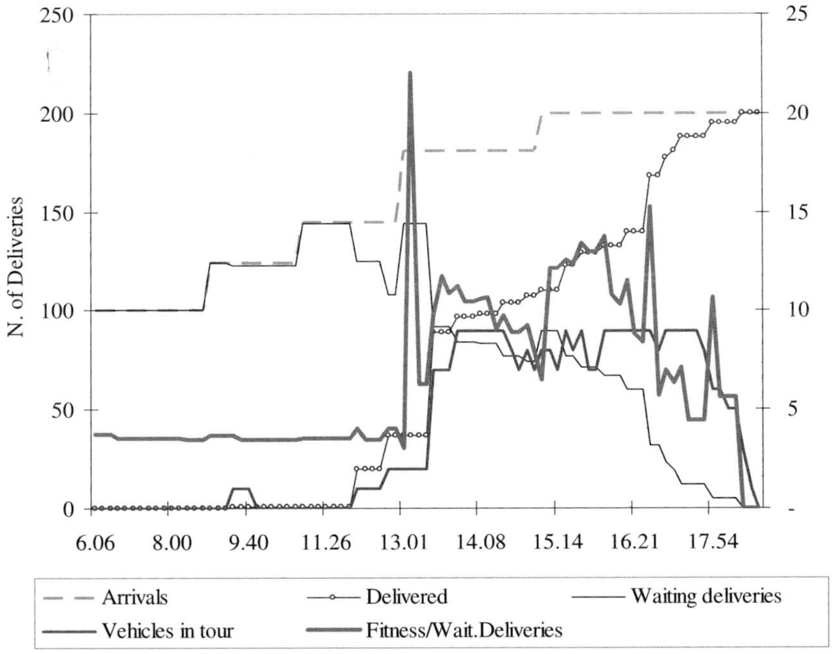

Fig.3. Simulation results of the real-time optimisation of deliveries at the logistic centre.

Sensitivity analysis on algorithm parameters underlines that the algorithm is robust with respect to the probabilities of mutation and cross-over, for both routing and TSP genetic algorithms (values of about 0.80 being sub-optimal). With respect to usual static applications of genetic algorithms, a real-time application requires much smaller values of those parameters that would improve algorithm effectiveness but reduce its efficiency. In our tests we experienced better solutions by using small values for the population size (less than 100 individuals) as well as by reducing the number of generations of the TSP algorithm (less than 50). In fact, the peculiar framework of the solution procedure exploits the clustering algorithm to find a good initial solution and benefits from genetic algorithms almost to update the current solution after a heavy modification of initial data.

CONCLUSIONS

In this paper we have described a procedure for real-time management of deliveries at an urban logistic centre, where items can arrive continuously, even without notice. In these conditions, critical issues are the check-in operations of items that arrive at the logistic centre without a standard identification code and the promptness of real-time procedure. As for the first issue, a specific operational scheme has been devised to ensure the application of the optimisation algorithm to unexpected items. As for the second one, many numerical tests have been conducted by simulating the operations at a logistic centre during a typical day. Results have shown that the algorithm has succeeded in improving the current solution in very few minutes, even when strong perturbations occurred.

REFERENCES

Ambrosini, C. and G.L. Routhier (2000). Objectives, methods and results of surveys carried out in the field of urban freight transport: an international comparison. *World Conference on Transportation Research*, Seoul, Paper Number 2504.

Baker, B.M. and M.A. Ayechew (2003). A genetic algorithm for the vehicle routing problem. Computers & Operations Research, 30, 787-800.

Bodin L., V. Maniezzo and A. Mingozzi (1999). Street Routing and Scheduling Problems. In: *Handbook of Transportation Science*, (R. W. Hall ed.), Kluwer Academics Publ.

Choi, I., S. Kim and H. Kim (2003). A genetic algorithm with a mixed region search for the asymmetric traveling salesman problem. *Computers & Operations. Research*, **30**, 773-786.

Desrosiers, J., F.Soumis, M. Desrochers and M. Sauve (1986). Methods for routing with time windows. *European Journal of Operations Research*, **23**, 236-245.

Dylog (2005). Dylog 1.0 User's Manual. ENEA, Ene/Tec-Università di Roma "La Sapienza", Dipartimento Idraulica, Trasporti e Strade. Technical Report.

Fusco, G., L. Tatarelli, and M.P. Valentini (2003). Last-Mile, a Procedure to Set-Up an Optimized Delivery Scheme. In: *Logistics Systems for Sustainable Cities* (E.Taniguchi and R.G. Thompson eds.), Elsevier.

Gendrau, M., A.Hertz, and G. Laporte (1994). A tabu search heuristic fort he vehicle routing problem. *Management science*, **40**, 1276-1290.

Ichoua, S., M. Gendrau, J.Y. Potvin (2003). Vehicle dispatching with time-dependent travel times. *European Journal of Operations Research*, **144**, 379-396.

Kaufmann, L. and P.J. Rousseeuw (1990). *Finding Groups in Data: An Introduction to Cluster Analysis*, John Wiley & Sons.

Laporte, G. (1992). The vehicle routing problem: an overview of exact and approximate algorithms. *European Journal of Operations Research*, **59**, 345-358.

Mingozzi A., R. Baldacci and D. Palumbo (1998). An Exact Algorithm for the Vehicle Routing Problem with Time Windows and Precedence Constraints. *Working Paper*. Department of Mathematics, University of Bologna, Italy.

Patier-Marque, D. (2000). Which tools to improve urban logistics? *World Conference on Transportation Research*, Seoul, Paper Number 2508.

Rochat, Y. and É. Taillard (1995). Probabilistic diversification and intensification in local search for vehicle routing, *Journal of Heuristics*, **1**, 147-167.

Transport Science and Technology
K.G. Goulias, editor

CHAPTER 30

THE LOGISTIC SERVICES IN A HIERARCHICAL DISTRIBUTION SYSTEM

Tomasz Ambroziak, Marianna Jacyna, Mariusz Wasiak
Warsaw University of Technology, Faculty of Transport, 75 Koszykowa Str, 00-662 Warsaw

INTRODUCTION

Constantly growing expectations of the customers of the logistic service providers make them search for new organisational solutions. Any organisational improvements are to increase the customer service standards and at the same time reduce the costs incurred by the company.
A major problem in multi-tiered distribution systems is to determine their best organisation, taking into account the existing capacities of the logistic service providers and the expectations of their customers. It is equally important to identify the optimum system development priorities on the basis of the expected volume and structure of the demand for logistic services. Thus, it is reasonable to carry out a multiple criteria evaluation (that is evaluation accounting for the interests of both service providers and customers) of the existing and designed distribution systems (Jacyna, 2001; Jacyna et al., 2003(1); Mindur, 2000).

For the purpose of this article, logistic services mean comprehensive services relating to the carriage, loading, storage, product packing and labelling, consulting and customs clearance, as well as a range of auxiliary services. Naturally, the establishment and development of logistic centres contributes to logistic service quality and efficiency (Węgrzyn, 2003).
The logistic centre is defined as an independent business entity which is located close to connections representing two or more different modes of transport; which has a separate area connected to the transport environment (mainly to the road network), infrastructure (roads, storage areas, parking lots, buildings and facilities), equipment for the change of the means of conveyance, as well as personnel and organisation; and which provides logistic services pursuant to once-off assignments or long-term contracts with external companies (Jacyna et al., 2003(2); Wasiak, 2004).

SUBJECT OF ANALYSIS

The subject of our analysis is the hierarchical distribution system organised to provide logistic services to the companies within the given area. In a hierarchical distribution system, the distribution of goods from suppliers (producers, manufacturers or importers) to customers (manufacturers, exporters or wholesalers) is indirect, i.e. carried out by logistic service providers, i.e. carriers, forwarding and shipping companies, distribution or logistic centres. For the purpose of this article, we assume that it is the logistic centres, of various range, which are involved in the flow of goods from suppliers to customers.

Such a system is referred to as the logistic service system for the area (LSSA). And its main features are as follows, (Jacyna et al., 2003, Wasiak, 2004):

- structure, which represents the connections between LSSA clients and the potential logistic centre locations as well as between the potential locations of logistic centres themselves (at different distribution tiers);
- characteristics of the structure components, which reflect the (major) features of the real-life structure components;
- volumes of logistic tasks, reflecting the existing or expected demand for LSSA services; and
- organisation, which describes the potential locations of logistic centres and transport connections used.

As noted above, to develop a model of LSSA it is necessary to know its structure, the characteristics define thereon, the logistic tasks within the analysed area and the system organisation adopted.

Let us assume that $\boldsymbol{G}$ denotes the LSSA structure mapping, $\boldsymbol{F_G}$ denotes the set of the characteristics defined on the LSSA structure, **ZL** denotes the volume of logistic tasks and **O** denotes the organisation, then the model of the logistic service system for the area, ***MLSSA***, is defined as the following ordered quadruple:

$$\boldsymbol{MLSSA = \langle G, F_G, \mathbf{ZL}, O \rangle}$$

Consequently, in order to develop a model of LSSA, one has to know its structure $\boldsymbol{G}$, the characteristics defined on the structure elements $\boldsymbol{F_G}$, the logistic tasks within the analysed area **ZL**, and the adopted system organisation $\boldsymbol{O}$.

STRUCTURE OF A MULTI-TIERED DISTRIBUTION SYSTEM

We assume that the set of the logistic service system elements is composed of the system clients (suppliers and customers using the services of LSSA) and the logistic centres within the area. In the analysed model, LSSA clients are divided into internal clients and external clients, that is the logistic centres located in the nearest proximity of the system and linked to the system. LSSA clients form the first tier, while logistic centres form the higher tiers of the distribution system.

For the sake of unambiguity, let s be the index of subsequent distribution tiers and S be the total number of the distribution tiers in the system. Consequently, let ***S*** be the set of numbers of the distribution tiers defined as follows:

$$\boldsymbol{S} = \{1, ..., s, ..., S\}$$

Let us further assume that ***W*** is the set of numbers of the elements of a multi-tiered distribution system, i.e.:

$$\boldsymbol{W} = \{1, ..., w, ..., w', ..., W\}$$

Let us assume that ***KZ*** denotes the set of the external clients of the distribution system, ***KW*** is the set of its internal clients and $\boldsymbol{W}^s$ is the set of the locations of the logistic centres of the tier $s \in \boldsymbol{S} \setminus \{1\}$. These are disjoint sets, which means that:

$$\boldsymbol{W} = \boldsymbol{KZ} \cup \boldsymbol{KW} \cup \boldsymbol{W}^2 \cup ... \cup \boldsymbol{W}^S$$

In any real-life distribution system, the connections between its elements are the transport connections operated. We assume that LSSA operates the following direct transport connections between: the external logistic centres linked to the system and the logistic centre locations within the system; the internal clients of the system and the logistic centre locations; and the logistic centre locations themselves.

We assume that the set of the transport connections identified within a multi-tiered distribution system is defined as follows:

$$\boldsymbol{L} = \{(w,w'): \xi_{w,w'} = 1 \text{ for } w, w' \in \boldsymbol{W} \text{ and } w \neq w'\}$$

where ξ denotes a mapping such that $\xi(w,w') \equiv \xi_{w,w'} = 1$ if and only if there is a direct operational transport connection between the nodes w and w'; otherwise, $\xi(w,w') = 0$.
When it is thus viewed, the LSSA structure is defined as the set of its elements and the relations between them. Consequently, the structure of a multi-tiered distribution system may be represented by the following graph ***G***:

$$\boldsymbol{G} = \langle \boldsymbol{W}, \boldsymbol{L} \rangle$$

where: ***W*** – the set of the graph nodes (numbers of the elements of a multi-tiered LSSA);
L – the set of the graph edges (interpreted as the existing transport connections between the relevant elements of LSSA).

Characteristics of the structure elements

The set of the characteristics defined on the LSSA structure, $\boldsymbol{F_G}$, includes the characteristics $\boldsymbol{F_W}^{\mathbf{P}}$ defined on the set of the potential logistic centre locations and the characteristics $\boldsymbol{F_L}$ defined on the set of the transport connections. Consequently, the logistic service network, ***LN***, is defined as the following ordered triple:

$$\boldsymbol{LN} = \langle \boldsymbol{G}, \boldsymbol{F_W}^{\mathbf{P}}, \boldsymbol{F_L} \rangle$$

In the model discussed, we assume that for the set of the potential logistic centre locations, one should identify the characteristics of the following (Ambroziak et al., 2003): gravity radius; maximum goods stream load; minimum goods stream load; local fees and taxes; average labour cost per employee of pre-defined skills; proximity of supraregional roads; connectibility to the existing supraregional road network; environmental and municipal restrictions; plot size; number of the owners of the land; land purchase opportunities; attitude of local authorities to investment projects; development limitations; unit cost of the land; quantity of adaptable facilities; standard of adaptable facilities.

As for the transport connections, one should identify their length and the unit transport costs involved.

Identification of logistic tasks

For the purpose of the identification of logistic tasks, we assume that the main services provided by the logistic centre are related to carriage and storage, while all the other services, such as complementation and decomplementation, banking, postal and financial services, customs clearance services, etc. are of auxiliary nature. Consequently, it is sufficient to account for the primary services, i.e. carriage and storage, while mapping the demand for the services of the logistic centres within the LSSA under design. The demand for such services depends directly on the demand for/supply of goods from different cargo categories, as identified for the potential clients of LSSA.

Let $\boldsymbol{R}$ be the set of cargo groups, i.e.:

$$\boldsymbol{R} = \{1, ..., r, ..., R\}$$

Let the volume of logistic tasks identified for the analysed area be represented by the matrix **ZL**, in which the first R columns are interpreted as the supply m_n^r of the goods of different cargo categories ($r \in \boldsymbol{R}$), as declared by particular clients ($n \in \boldsymbol{W}^1$), while the next R columns are interpreted as the demand a_n^r for the goods from different cargo categories, as declared by particular clients:

$$\mathbf{ZL} = \begin{bmatrix} m_1^1, & ..., & m_1^r, & ..., & m_1^R, & a_1^1, & ..., & a_1^r, & ..., & a_1^R \\ m_2^1, & ..., & m_2^r, & ..., & m_2^R, & a_2^1, & ..., & a_2^r, & ..., & a_2^R \\ & & & & ... & & & & & \\ m_n^1, & ..., & m_n^r, & ..., & m_n^R, & a_n^1, & ..., & a_n^r, & ..., & a_n^R \\ & & & & ... & & & & & \\ m_{N^k}^1, & ..., & m_{N^k}^r, & ..., & m_{N^k}^R, & a_{N^k}^1, & ..., & a_{N^k}^r, & ..., & a_{N^k}^R \end{bmatrix}$$

Organisation of logistic services within LSSA

In the model discussed, the organisation of the logistic services provided within LSSA is determined by the logistic centre locations, the links between the LSSA clients and logistic centres as well as the links between the logistic centres themselves. Thus, in order to define such organisation, it is necessary to introduce three types of decision variables, namely:

– matrix of binary decision variables, $\mathbf{X} = [x_{w'}]_{W^P}$, of the following elements:

$$x_{w'} = \begin{cases} 1, & \text{if there is a logistic centre at the potential location } w' \in \boldsymbol{W}^{\mathbf{P}} \\ 0, & \text{otherwise} \end{cases}$$

– matrix of binary decision variables, $\mathbf{Y} = [y_{w,w'}]_{KW \times W^P}$, of the following elements:

$$y_{w,w'} = \begin{cases} 1, & \text{if the transport connection } (w,w') \in \boldsymbol{L}^{\mathbf{KW,P}} \text{ is used} \\ 0, & \text{otherwise} \end{cases}$$

– matrix of non-negative decision variables, $\mathbf{Z} = \left[z^{r}_{w,w'}\right]_{(KZ+W^{P})\times(KZ+W^{P})\times R}$, of the elements

$z^{r}_{w,w'} \in \mathcal{R}^{+}$ interpreted as the cargo loads, in the relevant direction, on the transport connection $(w,w') \in \boldsymbol{L}$, for the cargo category $r \in \boldsymbol{R}$.

Let $\boldsymbol{O}$ be the set of the numbers of alternative organisations of LSSA, i.e.:

$$\boldsymbol{O} = \{1, ..., o, ..., O\}$$

in which the *o*th organisation of LSSA is defined as follows:

$$o = < \mathbf{X}(o), \mathbf{Y}(o), \mathbf{Z}(o) >$$

The purpose of the development of the model of LSSA is to determine the optimum organisation of the latter. In the process, one must consider numerous factors affecting the economic efficiency of future logistic centres. Needless to say, the indicated optimum organisation of LSSA should be a feasible one, that is it should meet all the problem constraints. And the most important constraints incorporated into the LSSA model are related to:

- performance of the logistic tasks requested by the internal clients of LSSA and the task assignment to the potential logistic centre locations;
- performance of the logistic tasks requested by the external clients of LSSA;
- actual assignment of the internal clients of LSSA to the potential logistic centre locations;
- cargo flows, particularly: prevention of impossible flows, flow non-negativeness, flow conservation, prevention of flow cycles.
- potential logistic centre locations, particularly ensuring: minimum load, and capacity.

MULTI-TIERED DISTRIBUTION SYSTEM OPTIMISATION CRITERIA

The previous discussion indicates that the process of determining the optimum organisation of LSSA should be based on the optimisation with respect to multiple criteria, reflecting different points of view on the evaluation of the solution and constituting the partial criteria for the optimisation problem. The most important partial criteria which should be considered in the evaluation of the feasible solutions of the LSSA organisation include: transport costs; amount of transport work; volume of logistic tasks; estimated costs of logistic centre operation; transport infrastructure and its accessibility from the logistic centre locations; convenience of the logistic centre locations; cost of the acquisition of the land assigned for the logistic centres; and proximity of existing facilities which could be adapted for LSSA purposes.

Let k be the number assigned to the partial criterion considered in the evaluation of a multi-tiered distribution system. Thus, the partial criteria set $\boldsymbol{K}$ is defined as follows:

$$\boldsymbol{K} = \{1, ..., k, ..., K\}$$

And let $f^{k}(o)$ be the value of the partial criterion $k \in \boldsymbol{K}$ for the *o*-th organisation of LSSA. Consequently, the global objective function $\mathbf{F}(\boldsymbol{O})$ may be defined as follows:

$$\mathbf{F}(\boldsymbol{O}) = [f^{1}(\boldsymbol{O}), f^{2}(\boldsymbol{O}), f^{3}(\boldsymbol{O}), f^{4}(\boldsymbol{O}), f^{5}(\boldsymbol{O}), f^{6}(\boldsymbol{O}), f^{7}(\boldsymbol{O}), f^{8}(\boldsymbol{O})]$$

where $\boldsymbol{O}$ is the set of the alternative organisations (designs) of LSSA.

The transport cost criterion ensures that the actual locations of logistic centres in a multi-tiered distribution system are determined in such a manner as to minimise the costs of performance of logistic tasks. It depend mainly on the length of the operated connections and

the transport technology applied. Needless to say, a different set of logistic centre locations will result in different transport connections used and thus different costs of transport.
If we assume that $m1^r_w$ ($a1^r_w$) denotes the demand (supply) declared by the internal clients of LSSA, $c^r_{w,w'}$ denotes the unit cost of the transport of the cargo group r, and $d^r_{w,w'}$ denotes the length of the transport connection (w,w'), then the transport cost criterion $f^1(\mathbf{Y},\mathbf{Z})$ may be expressed as follows:

$$f^1(\mathbf{Y},\mathbf{Z}) = \sum_{r\in R}\sum_{w'\in W^P}\Bigg[\sum_{w\in KW}\left(\left(m1^r_w + a1^r_w\right)\cdot y_{w,w'}\cdot c^r_{w,w'}\cdot d^r_{w,w'}\right) + \\ + \sum_{w\in KZ}\left(\left(z1^r_{w,w'} + z2^r_{w',w}\right)\cdot c^r_{w,w'}\cdot d^r_{w,w'}\right) + \sum_{w\in W^P}\left(\left(z3^r_{w,w'} + z4^r_{w',w}\right)\cdot c^r_{w,w'}\cdot d^r_{w,w'}\right)\Bigg] \longrightarrow \min \tag{1}$$

The amount of the transport work generated within the system is another factor which is important for the evaluation of logistic centre locations. Within the given logistic service area, one can transport huge loads for short distances or vice versa.
When used in addition to the transport cost criterion, the criterion of the transport work amount makes the analysed distribution system well-adapted to both bulk and highly processed goods. Formally, this second partial criterion, $f^2(\mathbf{Y},\mathbf{Z})$, may be expressed as follows:

$$f^2(\mathbf{Y},\mathbf{Z}) = \sum_{r\in R}\sum_{w'\in W^P}\Bigg[\sum_{w\in KW}\left(\left(m1^r_w + a1^r_w\right)\cdot y_{w,w'}\cdot d^r_{w,w'}\right) + \\ + \sum_{w\in KZ}\left(\left(z1^r_{w,w'} + z2^r_{w',w}\right)\cdot d^r_{w,w'}\right) + \sum_{w\in W^P}\left(\left(z3^r_{w,w'} + z4^r_{w',w}\right)\cdot d^r_{w,w'}\right)\Bigg] \longrightarrow \min. \tag{2}$$

The volume of logistic tasks depends on the total number of clients served by the given logistic service system for the area and the structure of the demand for logistic services declared by particular clients. The volume of the potential cargo stream determines the number and size of the logistic centres to be established.
The criterion of the volume of logistic tasks, $f^3(\mathbf{Y}, \mathbf{Z})$, is defined as follows:

$$f^3(\mathbf{Y},\mathbf{Z}) = \min_{w'\in W^P}\left\{\sum_{r\in R}\left[\sum_{w\in KW} m1^r_w\cdot y_{w,w'} + \sum_{w\in KZ} z1^r_{w,w'} + \sum_{w\in W^P}\left(z3^r_{w,w'} + z4^r_{w,w'}\right)\right]\right\} \longrightarrow \max \tag{3}$$

By incorporating the criterion of the logistic centre operating costs, one can account for a portion of the unit costs of the services provided by such centres. We assume that the operating costs are a function of the index of local fees and taxes, h^1_w, and labour costs, h^2_w. Consequently, the criterion of the logistic centre operating costs, $f^4(\mathbf{X})$, may be defined as follows:

$$f^4(\mathbf{X}) = \sum_{w\in W^P}(h^1_w + h^2_w)\cdot x_w \longrightarrow \min. \tag{4}$$

The proximity of major transport routes featuring high cargo flows means that it would be possible to take over some logistic services related to the long-distance cargo transport. This conclusion leads to another partial criterion, which should be taken into account in the evaluation of the design of multi-tiered distribution systems. Formally, the criterion related to the transport infrastructure accessibility, $f^5(\mathbf{X})$, is defined as follows:

$$f^5(\mathbf{X}) = \frac{\sum_{w \in \boldsymbol{W}^P} \left[(h_w^3 + h_w^4) \cdot x_w \right]}{\sum_{w \in \boldsymbol{W}^P} x_w} \longrightarrow \max. \tag{5}$$

where: h_w^3 –parameter indicating the proximity of supraregional roads for the with potential logistic centre location; h_w^4 – parameter indicating the accessibility of supraregional roads for the *w*th potential logistic centre location.

Apart from ensuring the smooth access from the given logistic centre location to the existing transport network, one should also consider a number of other factors affecting the location suitability, such as the plot size (h_w^6), existing environmental and municipal restrictions (h_w^5), development limitations (h_w^{10}), the number of the owners of the land (h_w^7), the land purchase opportunities (h_w^8) and the general attitude of local authorities to investment projects (h_w^9). Therefore, we add the criterion of the convenience of the logistic centre locations, $f^6(\mathbf{X})$, which is defined as follows:

$$f^6(\mathbf{X}) = \frac{\sum_{w \in \boldsymbol{W}^P} \left(h_w^5 + \frac{h_w^6}{\max_{w \in \boldsymbol{W}^P} (h_w^6)} + h_w^7 + h_w^8 + h_w^9 + h_w^{10} \right) \cdot x_w}{\sum_{w \in \boldsymbol{W}^P} x_w} \longrightarrow \max. \tag{6}$$

While choosing the logistic centre location, one should also consider the land acquisition cost, which depends on the plot size (h_w^6) and the unit cost of the land (h_w^{11}). It leads to the criterion of the cost of the acquisition of the land assigned for the logistic centres, $f^7(\mathbf{X})$, which is defined as follows:

$$f^7(\mathbf{X}) = \sum_{w \in \boldsymbol{W}^P} (h_w^6 \cdot h_w^{11}) \cdot x_w \longrightarrow \min. \tag{7}$$

Finally, another important criterion to be applied is the adaptability of the existing infrastructure facilities to logistic purposes. We assume that this eighth criterion, i.e. the criterion of the proximity of existing facilities which could be adapted for LSSA purposes, $f^8(\mathbf{X})$, is a function of the quantity (h_w^{12}) and the standard (h_w^{13}) of the existing facilities which can be adapted for the project purposes. It is defined as follows:

$$f^8(\mathbf{X}) = \frac{\sum_{w \in \boldsymbol{W}^P} (h_w^{12} \cdot h_w^{13}) \cdot x_w}{\sum_{w \in \boldsymbol{W}^P} x_w} \longrightarrow \max. \tag{8}$$

The formal definition of the partial criteria discussed above was a precondition for the development of the method for the multicriteria evaluation of the design of multi-tiered distribution systems.

ALGORITHM OF THE MULTICRITERIA EVALUATION

The optimum organisation of the logistic service system for the given area is expected to maximise the benefits for both the clients of the system and the companies involved in the establishment of future logistic centres. Thus, the goal is to find a compromise solution which will be Pareto-optimal, that is a solution for which all the criteria considered will have non-inferior values (but not necesserily maximal or minimal). The presented method for the multicriteria evaluation of the LSSA organisation enables determining the best organisation of a multi-tiered distribution system with respect to the criteria adopted and the pre-defined preferrences of the decision maker. The algorithm of the multicriteria evaluation of the LSSA organisation consists of seven computational stages which we identifed on fig.1, (Jacyna et al., 2003; Wasiak, 2004):

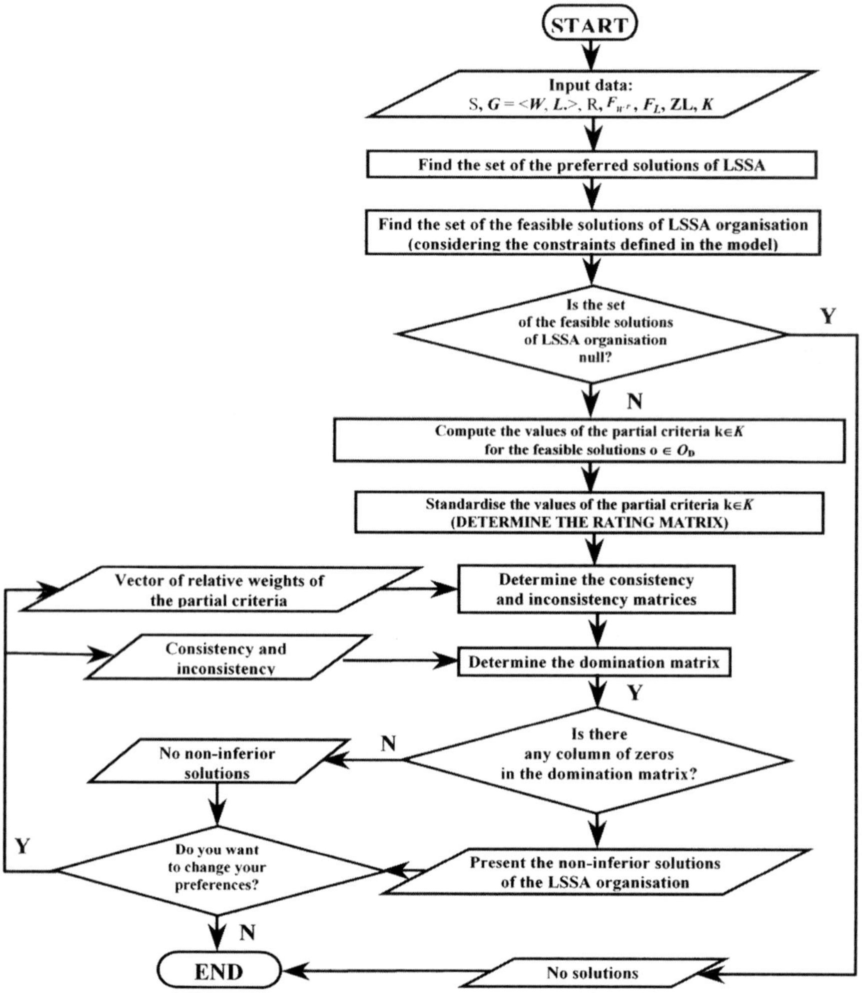

Fig. 1. Algorithm for solving the multi-objective optimisation problem of the LSSA organisation evaluation

Computational example

The presented approach has been verified for the test case of the distribution system operated by FRAMEX. It is a three-tier system. The first tier comprises the suppliers of goods from various cargo categories and the consignees. The second tier is the company's regional warehouses, while the third one is the central warehouse, which is currently located at Teresin. The regional warehouses are located at Gdynia, Poznan, Wroclaw, Gliwice and Cracow.

In the existing distribution system (Fig. 2a), the goods are carried from the suppliers to the company's central warehouse in Teresin and then to the regional warehouses in Gdynia, Poznan, Wroclaw, Gliwice and Cracow. Each regional warehouse has its own clients assigned to it. Currently, FRAMEX has over 300 clients.

Prior to computations, FRAMEX' clients were aggregated. Then, the volumes of the logistic tasks requested by the clients of the analysed distribution system as well as the characteristics of the potential distribution centre locations and transport connections were identified.

The purpose of the multicriteria evaluation of the analysed distribution system was to find the best organisation, considering Szczecin, Mszczonów and Teresin as potential locations of additional regional warehouses and Gliwice, Mszczonów and Gdynia as potential locations of additional central warehouses. As a result of the relevant computations, 136 feasible solutions of the distribution system organisation were selected. The solutions differed with respect to the distribution centre locations, the clients assigned to particular centres and the cargo flows along particular connections.

In order to determine the optimum organisation of FRAMEX' distribution system, all the feasible solutions were evaluated with respect to multiple criteria.Initially, in line with the adopted multicriteria evaluation procedure, the rating matrix was computed and the relative weights of the partial criteria were assumed. In the next step of the method, the consistency matrix and the inconsistency matrix, for FRAMEX' distribution system organisation ratings, were computed on the basis of the feasible organisation rating matrix and the adopted relative weights of the partial evaluation criteria.

Next, the solution rating consistency threshold, $\alpha=0.55$, and the inconsistency threshold, $\beta=0.45$, were determined by a process of trial and error. Then, they were used to find the domination matrix, which in turn constituted the basis for determining the optimum (or rather non-inferior or non-dominated) solution.

The non-dominated solution proved to be the solution #66 (Fig. 2b). It calls for five regional distribution centres (in Gdynia, Poznan, Mszczonów, Wroclaw and Cracow) and one central warehouse in Gliwice. Should the distribution system organisation indicated as the optimal be implemented, FRAMEX will achieve a significant reduction in both transport costs and the transport work involved.

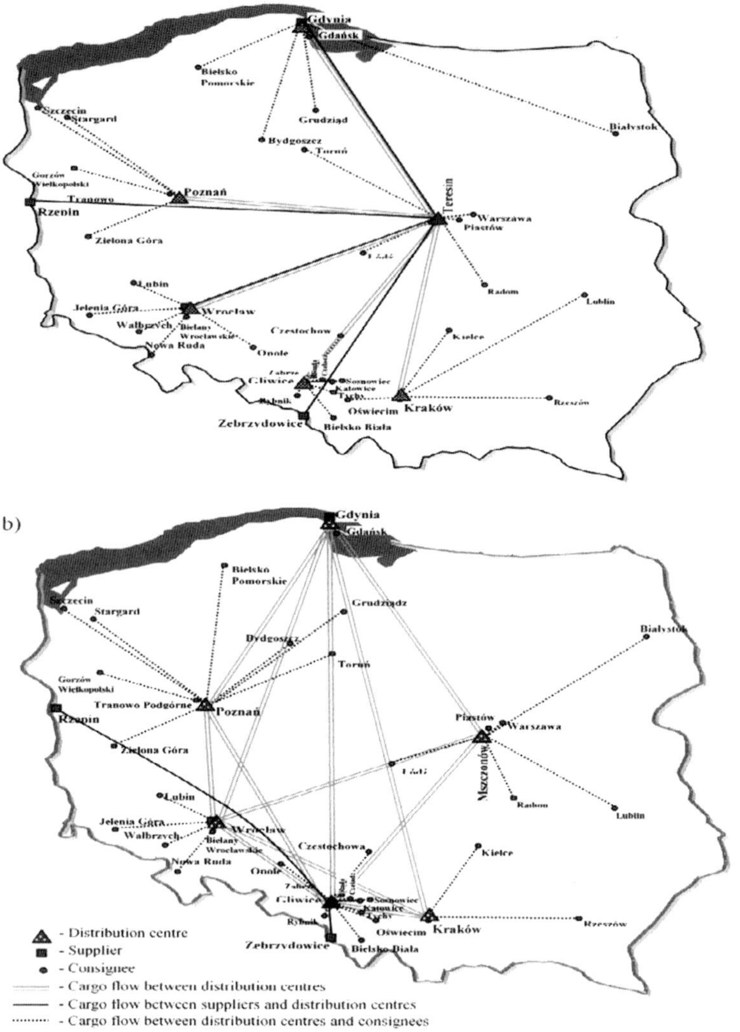

Fig. 2. FRAMEX' distribution system: (a) existing, (b) optimum

CONCLUSIONS

A major advantage of the presented approach to determining the optimum organisation of multi-tiered distribution systems is the fact that multiple criteria can be considered in the evaluation of alternative solutions. As a result, the outcome is a compromise solution, maximising the benefits for both the companies involved in the logistic centre operation and their clients. The final selection of the optimum organisation of the distribution system is based on the values of the partial evaluation criteria, taking into account the preferences of the decision maker. The latter are expressed through the relative weights of the partial objectives as well as the requirements (thresholds) relating to the so called consistency and inconsistency ratios.

Another important advantage of the presented approach is the fact that it can be applied in the evaluation of the suggested locations of distribution centres and the assignment of the clients to such centres as well as in the evaluation of the locations of the existing distribution centres and the logistic tasks performed by them in multi-tiered distribution systems.

References

Ambroziak T. Jacyna M.(2003).Chosen aspects of logistic centre organisation, Publishers Scientific Works "TRANSPORT' vol 1(17)2003, Radom University of Technology, Polish Acad. of Sc. – Transport Com., Szczyrk

Jacyna M.(2001). Application of multicriteria modelling for evaluation of transport systems. Scientific Papers Warsaw University of Technology, Transport, vol.47, Warsaw.

Jacyna M. Wasiak M.(2003). Multicriteria Evaluation of Logistic Centres Configuration in a Hierarchical Distribution System. 11th International Scientific Conference – Science, Education and Society, University of Žilina, Žilina, September.

Jacyna M. Wasiak M.(2003). Multiaspects evaluation of logistic distribution centres in hierarchical distribution system. National Sc. Conf. – Transport in Logistics. Logistic Chain, Gdynia Naval Academy.

Mindur L.(2000). Methodology of localisation and forming of logistic centres in Poland. Railway Publisher, Warsaw

Wasiak M.(2004). Multicriteria evaluation of logistic service of chosen area in three-level distribution system. Intern. Sc. Conf. – Transport in the 21st Century, Warsaw

Węgrzyn B.(2003). Logistic distribution centres as a factor of economic development. National Sc. Conf. – Transport in Logistics. Logistic Chain, Gdynia Naval Academy

Transport Science and Technology
K.G. Goulias, editor

CHAPTER 31

THE MULTIOBJECTIVE OPTIMISATION TO EVALUATION OF THE INFRASTRUCTURE ADJUSTMENT TO TRANSPORT NEEDS

Marianna *Jacyna*
Warsaw University of Technology, Faculty of Transport, 75 Koszykowa Str 00-662 Warsaw

PROBLEM IDENTIFICATION

Any general discussion relating to transport revolves around the fact that it satisfies one of the basic human needs, namely the need to overcome distances. It is always necessary to determine the relations between the transport and its environment, including the demand for the transport tasks and the impact of the performance of these tasks on the natural environment and socio-economic processes (Basiewicz et al., 2000; Jacyna, 2001(1)). Thus, the works on the subject often focus on finding the optimal solutions for transport functioning (Steenbrink, 1978).

One of the major problems relating to modern transport systems is the accurate evaluation of the suitability of the transport network infrastructure components for the actual transport tasks performed. Such evaluation proves difficult owing to the problem complexity, which results from both technological constraints, financial limitations, ecological considerations determined by the public interest, etc., and the conflicting interests of the different participants of the transport process, who try to maximise their own profits. Hence, it is necessary to develop multiple criteria methods which would support the process of making decisions on the transport network modernisation and development, taking into account such conflicting interests.

Thus, the modern theory of transport calls for the development of the methods for the review and evaluation of various options of the traffic organisation in terms of the suitability of the infrastructure components for the transport tasks performed. For example, the development of the transport system in Poland and other Central and Eastern European countries is determined by the expected demand for transport services and the need to adapt the system infrastructure to the EU's standards and requirements.

For the purpose of the international cooperation it is necessary to develop a common transport network which could satisfy growing transport needs. Therefore, the transport system development should be based on the analysis of the relations between the expected tasks, required system facilities and the cost of the effective performance of such tasks. Nevertheless, the modernisation and development should not be limited to establishing an integrated transport network and increasing its capacity, however important it is, but should also promote safety and environmental protection improvements (Jacyna, 1997).

As discussed in (Jacyna, 2001(2)), only the solutions of multiobjective optimisation problems constitute the proper basis for the multicriteria evaluation of any transport system. However, if more than one criterion is considered in the model, it is likely that no all-best solution will be found, as the solution that is optimal according to one criterion is not necessarily the best in terms of another one.

Therefore, it is necessary to develop methods accounting for the multicriteria nature of the problems to be solved (Ackoff, 1968; Jacyna, 2001; Korhonen et al., 1986; Roy, 1990(2), Zeleny, 1992) multicriteria methods that will support making optimum decisions on the scope of the transport system development, e.g. transport network modernisation or extension, taking into consideration the interests of different user groups.

COMPONENTS OF THE TRANSPORT SYSTEM MODEL

Transportation as a whole is usually analysed in system terms. The purpose of any transport system is to ensure the flow of passengers and cargoes, which is described by the type, quantity and characteristics of the carried objects as well as movement relations and quality parameters. The analysis of any transport system aims at the identification of the processes involved. As real-life transport systems are far too complex for a direct analysis, one has to develop their models. Any good model of a transport system should reflect the complexity of the processes within the system and their interdependencies as well as the system relations with its environment.

Generally speaking, the main purpose of the model is to understand the real world better. Through models we can verify the laws governing phenomena and objects, that is pieces of the reality we want to investigate. The data from the testing of the model may be used at a later stage of the research. Thus, the model is a tool enabling us to achieve the actual objectives of the research (Magnanti, et al., 1984).

The way the real things are reflected in the model depends on many factors, particularly the skills and abilities of the researcher as well as the purpose of the research. As the model is only a substitute for the reality, it reflects only that part there of which is required to achieve the purpose of the modelling, while it is deprived of many unimportant details and attributes.

While developing a model of any transport system, it is necessary to reflect the four basic features of the system, namely its structure, characteristics of the structure components, traffic stream and organisation (i.e. the description how the structure components are used to meet the transport needs). Hence, a transport system model (TSM) is defined as the following ordered quadruple:

$$\mathbf{TSM} = \langle \mathbf{G}, \boldsymbol{F}, \mathbf{T}, \mathbf{O} \rangle$$

where: **G** – structure graph, ***F*** – set of functions defined on the structure graph nodes and/or edges, **T** – traffic flow, **O** – organisation.

The system structure and the characteristics of its components can be represented by a network within the meaning of the graph theory (Korzan, 1978). The network approach determines the language to be used for describing the transport system model, which is accurate in mathematical terms and easily understandable for transport experts.
Thus, we assume that the network **S** is the following ordered triple:

$$\mathbf{S} = \langle \mathbf{G}, \boldsymbol{F_W}, \boldsymbol{F_L} \rangle$$

where: $\mathbf{G} = \langle \boldsymbol{W}, \boldsymbol{L} \rangle$ – transport network structure graph, in which $\boldsymbol{W}$ is the transport node set, i.e.: $\boldsymbol{W} = \{1, ..., i,j, ..., W\}$, while $\boldsymbol{L}$ is the edge set of the graph **G**, i.e. a set composed of the elements defined as follows: $\boldsymbol{L} = \{(i,j): \eta_{ij}=1,\ i \neq j,\ i,j \in \boldsymbol{W}\}$ corresponding to direct connections (road legs) between transport nodes i and j; $\boldsymbol{F_W}$ – set of functions defined on the structure graph nodes; $\boldsymbol{F_L}$ – set of functions defined on the structure graph edges.
Moreover, let $\boldsymbol{E}$ be the set of transport relation pairs:

$$\boldsymbol{E} = \{(a,b):\ (a,b) \in \boldsymbol{A} \times \boldsymbol{B},\ a \in \boldsymbol{A},\ b \in \boldsymbol{B}\}$$

i.e. a set of relation pairs of the transport needs specified within the given transport network; where $\boldsymbol{A}$ is a set of names (numbers) of traffic flow sources, while $\boldsymbol{B}$ is a set of names (numbers) of traffic flow destinations.

We further assume that there is a mapping x defined over the transport relations set $\boldsymbol{E}$, which associates any element of the set $\boldsymbol{E}$ with a positive real number, i.e.:

$$x: \boldsymbol{E} \longrightarrow R^+$$

while $x(a,b) \equiv x^{ab} \in R^+$ is interpreted as the volume of transport needs for the relation (a,b).
Let $\boldsymbol{XE}$ be a set composed of the elements defined as follows:

$$\boldsymbol{XE} = \{x(a,b) \equiv x^{ab}: \rightarrow (a,b) \in \boldsymbol{E}\}$$

while $x(a,b) \equiv x^{ab}$ is interpreted as the volume of the traffic within the transport network, between the pair of the relevant nodes, i.e. between the source a and the destination b.
Let us further assume that for any relation $(a,b) \in \boldsymbol{E}$ there is a set $\boldsymbol{P}^{ab}$ of the paths which connect the relevant nodes. Consequently, we assume that there is a mapping o defined over the Cartesian product $\boldsymbol{P}^{ab} \times \boldsymbol{XE}$, which associates any element of that product with a positive real number, i.e.:

$$o: \boldsymbol{P}^{ab} \times \boldsymbol{XE} \longrightarrow R^+$$

while $o(p,x^{ab}) \in R^+$ is interpreted as the organisation of the traffic stream x^{ab} on the path $p \in \boldsymbol{P}^{ab}$ for the relation (a,b). We assume that the mapping o is given for any relation (a,b), which means that we have defined the traffic organisation for the whole transport network.

Thus, the mapping o determines how the performance of the transport tasks is organised within the analysed transport network, i.e. the network of the structure represented by the graph **G** and constraints determined by the function sets $\boldsymbol{F_W}$ and $\boldsymbol{F_L}$, which are defined on the graph nodes and edges respectively. Naturally:

$$\boldsymbol{O} = \{o:\ o(p,x^{ab}) \in R^+,\ p \in \boldsymbol{P}^{ab},\ x^{ab} \in \boldsymbol{XE},\ (a,b) \in \boldsymbol{E}\}$$

MULTIOBJECTIVE OPTIMISATION PROBLEM OF TRANSPORT TASK ALLOCATION TO ROADS

The problem of finding the best solution for the transport task performance organisation, i.e. the problem of selecting the traffic organisation ensuring optimum achievement of conflicting partial objectives (e.g. maximisation of carriers' profits, maximisation of the value for transport service buyers and minimisation of losses for the society) is in fact a multiobjective optimisation problem.

We assume that partial objective functions are measurable, i.e. that their values are real numbers. The value of the partial objective function for any X (not necessarily the optimum one) is interpreted as the level of achievement of the relevant partial objective. It means that the measurability of the level of *achievement* of a partial objective is linked to the concept of a partial criterion. Consequently, the level of achievement of the global objective is measured by a global criterion, which is a function which associates the levels of achievement of partial objectives with real numbers, which form a scale with a defined unit of measurement. In a multiobjective optimisation problem, the necessary condition for the partial objectives to be linked to one another is that the set of feasible solutions ***D*** is common for all the criteria.

For the purpose of the multiobjective optimisation of the task performance organisation within a transport network, let us consider an objective function of three components:

$$< F_1(X(r)), F_2(X(r)), F_3(X(r)) >$$

where:

- $F_1(X(r)) = c^{p;ab}(X(r)) = c^{p';ab}(X(r))$ is interpreted as the mean cost;
- $F_2(X(r)) = m^{p;ab}(X) = m^{p';ab}(X)$ is interpreted as the marginal cost;
- $F_3(X(r)) = z^{p;ab}(X) = z^{p';ab}(X)$ is interpreted as the external cost.

Thus, we can formulate a tri-objective optimisation problem, which will be subject to minimisation, i.e.:

$$\min_{\{[X(r)]\}} < F_1(X(r)), F_2(X(r)), F_3(X(r)) >$$

The solution of the problem with the aforementioned objective function and the given set of constraints is the task performance matrix $X(r)=[x_{ij}^{p;ab}(r)]$.

Solving the optimisation problem with the aforementioned objective function produces the optimum task performance matrix $X^*(r) =[x_{ij}^{*p;ab}(r)]$, which defines traffic streams for the individual edges of the structure graph.

Naturally, the minimisation is carried out for the following set of constraints (which must be satisfied for any option of the task performance organisation $r \in R$):

(1) condition that the transport task $[x^{ab}(r)]$ should be performed;

(2) constraints resulting from the predefined capacity of road connections (legs);

(3) constraints on the traffic flow: traffic flow non-negativeness (NP); traffic flow additiveness (AD); and traffic flow behaviour (ZP).

METHOD FOR MULTIPLE CRITERIA EVALUATION OF THE INFRASTRUCTURE SUITABILITY FOR TRANSPORT TASKS

General assumptions for the evaluation method MAJA99

Any well-defined objective function used for the evaluation of the transport infrastructure components should reflect the modern approach to solving traffic problems and 'represent' the interests of all the parties involved, directly or indirectly, in such evaluation, i.e. transport service providers, transport service buyers and the general public opting for the environment protection. It is worth noting that each party participating in the decision-making process represents its own interest, which makes it a multiple criteria problem. Because of the diverse nature of the partial objective functions representing all the participants of the decision-making process, it is unlikely to find a solution which would satisfy everybody. Consequently, it is not possible to adopt just a single criterion.

The multicriteria evaluation method discussed in this paper is based on the concept of using the domination relation defined by means of the so called weight consistency/rating inconsistency ratios in order to determine the best traffic stream distribution, meeting the pre-defined technical constraints imposed on the transport network infrastructure.

Choosing the right option of the transport corridor infrastructure modernisation or construction is a difficult problem. First, one should consider several alternatives of the transport task allocation and then, evaluate them, thus making a definite choice of one of the options according to a multicriteria objective function.
Solving the formulated single-objective optimisation problems for different objective functions and different sets of constraints produces the input data required for a multiple criteria evaluation of the transport task allocation. And such multicriteria evaluation is conducted using the MAJA method (originally programmed in Delphi environment) (Jacyna, 1998).

The results of the multiple criteria evaluation of the transport task allocation within the given portion of a transport network, are then submitted for approval by experts (decision makers). If the solution is approved, it ends the problem solving procedure. Otherwise, the optimisation problems are reformulated, the suggestions of the experts being accounted for through incorporation of additional constraints or modification of the objective function. Then, the software package MAJA'99 is used to carry out the multicriteria evaluation.
The iterative procedure for solving the problem of the transport task allocation within the given portion of a transport network ends when the experts raise no objections as to the results of the multicriteria evaluation of the task allocation problem.

General description of the method

The MAJA method for multiple criteria evaluation of the transport task allocation is based on the detailed rating of alternative traffic stream distributions, considering the relative weights of the partial criteria. The method enables the selection of the best option of the traffic stream distribution. In numerical terms, the MAJA procedure computes the so called consistency/inconsistency ratios and uses the domination relation to determine a non-inferior option of the traffic stream distribution.

Let $\boldsymbol{X}$ be a finite set of alternative transport task allocations within a transport network, i.e.:

$$\boldsymbol{X} = \{X(1), ..., X(r), ..., (R)\}$$

where R is the total number of the analysed options.

Let as assume that there is a pre-defined set of the partial criteria which will be used for the evaluation of the alternative transport task allocations within a multimodal transport corridor:

$$\boldsymbol{F} = \langle f_1(X), f_2(X), ..., f_k(X), ..., f_K(X)\rangle$$

The value of each partial criterion $f_k(X)$ is expressed in measurable (numerical) units, e.g. local currency units. It means that any components affecting the value of the criterion are also expressed numerically. In the method under discussion, we compare R options of the transport task allocation with respect to K partial criteria.

We assume that there is a mapping w defined over the Cartesian product $\boldsymbol{X}\times\boldsymbol{F}$, i.e.:

$$w: \boldsymbol{X}\times\boldsymbol{F} \longrightarrow R^+$$

for which $w(X_r, F_k) \equiv w(r,k) \in R^+$ is interpreted as the rating of the *r*-th option of the transport task allocation X(r) according to the *k*th partial criterion F(k). In a special case, the mapping w may be given by a matrix W of the elements w(r,k).

Let $\boldsymbol{FX}$ be a set composed of the elements defined as follows:

$$\boldsymbol{FX} = \{f(X(r),k): r\in \boldsymbol{R},\ k\in \boldsymbol{K}\}$$

and interpreted as the value of the criterion *k* for the transport task allocation option *r*.

And let $\boldsymbol{WR}$ be a set composed of the elements defined as follows:

$$\boldsymbol{WR} = \{w(r,k): r\in \boldsymbol{R},\ k\in \boldsymbol{K}\}$$

and interpreted as the rating of the option *r* with respect to the criterion *k*.

We assume that there is a mapping θ defined over the set $\boldsymbol{FX}$, which transforms the latter onto the set $\boldsymbol{WR}$, i.e.:

$$\theta: \boldsymbol{FX} \longrightarrow \boldsymbol{WR}$$

for which $\theta(f(X(r),k))=w(r,k)$ is interpreted as the rating of the option *r* with respect to the criterion *k*.

Thus, for each option of the transport task allocation and each partial criterion, there are the following ratings:

- $f(X(1),1)=w(1,1)$, $f(X(2),1)=w(2,1)$, ..., $f(X(r),1)=w(r,1)$, ..., $f(X(R),1)=w(R,1)$
- ....
- $f(X(1),K)=w(1,K)$, $f(X(2),K)=w(2,K)$, ..., $f(X(r),K)=w(r,K)$, ..., $f(X(R),K)=w(R,K)$

In the next step of the MAJA method, the partial criteria from the set $\boldsymbol{F}$ must be assigned numerical values c(k), k=1, ..., K, which are interpreted as their relative weights.

The previous discussion indicates that the determination of the coefficients c(k) is not easy. One of the methods which may be applied is the procedure proposed by R.W. Ackoff (Ackoff, 1968), in which the criteria are assigned certain numbers corresponding to their relative weights. It is further assumed that the relative weight c(k) of each partial criterion is a number from the interval $\langle 1, 10\rangle$ and that the bigger c(k), the more important the criterion k.

The next step of the MAJA method for the multicriteria evaluation of the transport task allocation within the given part of a transport network is to develop the consistency matrix. Comparing any two alternative transport task allocations, X(r), X(r')∈***X***, i.e. any pair (X(r), X(r')), we determine for which criteria f(k), k=1, ..., K, the option X(r) is rated higher than the option X(r'). This formally is defined as follows:

$$z(X(r), X(r')) = \frac{1}{c} \sum_{f(k)\in \boldsymbol{F}:\, w(r,k)>w(r',k)} c(k) \qquad \text{where} \qquad c = \sum_{k=1}^{K} c(k)$$

The consistency ratio has the following properties: its range is the interval <0, 1>; it is 1 if and only if the option X(r) has the consistent rating across all the criteria f(k).
The consistency ratio reflects the relative importance of the set ***F*** of the criteria for which the option X(r) is rated higher than the option X(r'), z(X(r), X(r'))≡z(r,r'). It is convenient to present the criteria consistency ratios for all the option pairs in the consistency matrix Z
As already mentioned, the consistency ratio reflects the relative importance of the transport task allocation option X(r) versus the alternative option X(r'). However, for the purpose of solving the multicriteria optimisation problem of the transport task allocation, it is also important to determine how much the rating of the option X(r) is worse than the rating of the alternative option X(r'), or for how many criteria from the criteria set ***F*** the option X(r) is rated lower than the option X(r'). This is achieved by means of the rating inconsistency ratio.
Thus, the ratio of the inconsistency of the ratings of any pair of options (X(r),X(r')), which is denoted n(X(r),X(r')), is determined by comparing for which criteria f(k) the transport task allocation option X(r) is rated lower than the alternative option X(r'). It is convenient to present the inconsistency ratios in the inconsistency matrix N.
The inconsistency ratio n(X(r),X(r'))≡n(r,r') is equal to the ratio of the maximum difference between the option ratings according to the given criterion and the maximum difference between the option ratings in the rating matrix W. Formally, the inconsistency ratio n(X(r),X(r')) is defined as follows:

$$n(X(r), X(r')) = \frac{1}{d} \max_{(r,k)\in \boldsymbol{X}\times\boldsymbol{F}:\, w(r',k)>w(r,k)} \{w(r',k) - w(r,k)\}$$

where: d – the maximum difference between the least and the biggest element in the matrix W of the option ratings, i.e.:

$$d = \max_{(r,k)\in \boldsymbol{X}\times\boldsymbol{F}} \{w(r,k)\} - \min_{(r,k)\in \boldsymbol{X}\times\boldsymbol{F}} \{w(r,k)\}$$

It is easy to notice that the range of the rating inconsistency ratio is the interval <0, 1> (like for the consistency ratio). The inconsistency ratio of one, i.e. n(X(r),X(r'))=1, indicates the maximum inconsistency between the ratings of the options X(r) and X(r'). The inconsistency ratio is zero, if the transport task allocation options X(r) and X(r') are rated the same across all the criteria f(k).

The interesting point to be emphasised is that the consistency matrix reflects the differences between the relative weights of the criteria, while the inconsistency matrix reflects the differences between the ratings of the alternative options of the transport task allocation within the given part of a transport network. This is an important advantage of this method over other multicriteria evaluation methods. It should be noted that the maximum consistency of the options X(r) and X(r'), i.e. z(X(r), X(r'))=1, implies that n(X(r), X(r'))=0, and vice versa.

Once the criteria consistency matrix X and the rating inconsistency matrix N are created, the profitability thresholds v and q, which are referred to as the "consistency threshold" and "inconsistency threshold" respectively, must be determined. The purpose of these thresholds is to select the effective options of the traffic stream distribution from the set ***X***. The range of both thresholds is the interval (0, 1). Normally, the consistency threshold v tends towards one, while the inconsistency threshold q tends towards zero. The thresholds v and q act as a 'sieve', letting through only those options X(r) from the set ***X*** that meet both the consistency and inconsistency thresholds.

For any pair of two different options X(r), X(r') of the transport task allocation within the given part of a transport network from the feasible option set ***X***, i.e. $X(r), X(r') \in \boldsymbol{X}$, which are compared using K partial criteria and the thresholds v and q, the option X(r) is said to be superior to the alternative option X(r') if and only if the pair (X(r), X(r')) meets the following condition:

$$z(X(r), X(r')) \geq v \wedge n(X(r), X(r')) \leq q$$

Graphical illustration of the above condition, for each pair (X(r),X(r')), is the graph ***Gf***, which is defined as follows:

$$\boldsymbol{Gf} = (\boldsymbol{X}, \boldsymbol{U})$$

for which: ***X*** – the set of the graph nodes representing all the analysed options of the transport task allocation; ***U*** – the set of the graph edges (X(r), X(r')), such that:

$$\{(X(r), X(r')) \in \boldsymbol{U}\} \equiv \{[z(X(r), X(r')) \geq v] \wedge [n(X(r), X(r')) \leq q]\}$$

For different thresholds v and q, one can generate a family of subgraphs of the graph ***Gf***. Such subgraphs are generated by relaxing both the consistency condition (through a decrease of the threshold v) and the inconsistency condition (through an increase of the threshold q).
The final selection, which is based on the graph ***Gf***, of non-dominated (non-inferior) solutions is performed by separating those nodes of the graph ***Gf*** that have no edges (arcs) coming to them. Thus, if we identify a range of options of the transport task performance, we can select an option which will be optimal in terms of some pre-defined criteria.

Method verification for the corridor Gdansk-Warsaw-Katowice-Żilina

The methodology for the multicriteria evaluation of the infrastructure suitability for the transport tasks was verified using the corridor Gdańsk-Warsaw-Zilina (road and railway traffic) as an example. For the structure of the analysed part of the transport corridor, for the base option, see the graph shown in Fig.1. For the purpose of the analysis, we considered two cases, namely: the structure of the analysed part of the transport network remains unchanged; and the network infrastructure is modified by adding motorway A-1.

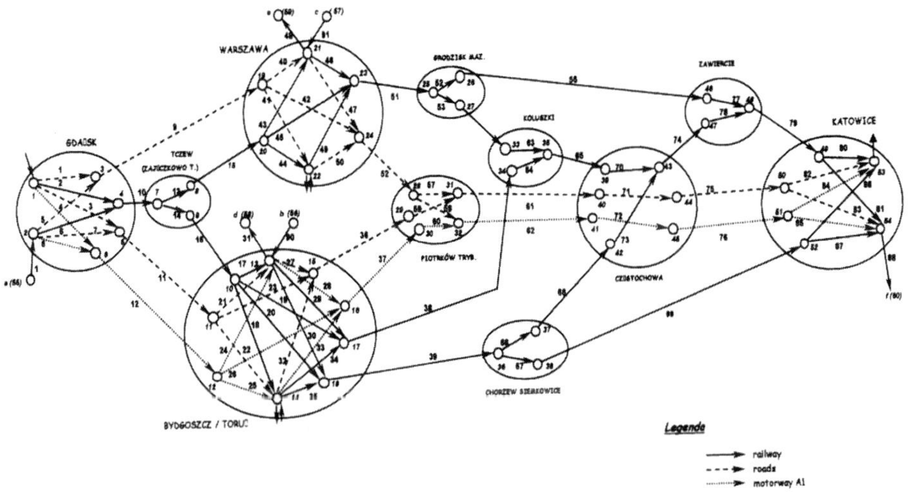

Fig.1.The Graph structure of the corridor Gdansk-Warsaw-Katowice

The multicriteria evaluation of the suitability of the analysed transport infrastructure for the demand for transport services (existing or expected) was performed for three different options. The global objective function F(X(r)), which consists of three partial criteria: F1(X(r)) – mean cost, F2(X(r)) – marginal cost and F3(X(r)) – external cost, used for the evaluation of each option X(r) of the organisation of the task performance by the analysed transport system, is the following vector:

$$F(X(r)) = \langle f1(X(r)), f2(X(r)), f3(X(r)) \rangle$$

Naturally, each option of the task allocation X(r) meets the pre-defined conditions and constraints on the traffic flow. The total costs corresponding to the partial criteria for all the options of the transport task allocation, for the cargo traffic, are shown in Table 1.

Table 1. Total costs corresponding to the partial criteria for all the options of the cargo traffic

Cargo Traffic			
Criterion	$F_1(X(r))$	$F_2(X(r))$	$F_3(X(r))$
	Mean Cost	Marginal Cost	External Cost
Option	*PLN billion*	*PLN billion*	*PLN billion*
W1	120,263.94	120,181.38	155,468.55
W2	534,494.14	533,452.17	668,946.40
W3	231,691.78	230,331.61	448,436.71

Source: Results of the analysis conducted.

The results of the analysis of the transport task allocation, in consideration of the partial criteria, for the options 1 and 2, for the cargo traffic, are shown in Table 2. The review of the results shows that some legs are overloaded. It means that these components of the transport network should be modernised or additional connections should be built.

Table 2. Comparison of the traffic flow on edges before and after the internalisation of external transport costs, for the options 1 and 2.

Edge (Arc)	Type of Connection	Cargo Traffic – Option 1			Cargo Traffic – Option 2		
		Mean Cost	Marginal Cost	External Cost	Mean Cost	Marginal Cost	External Cost
No.	-	'000 tonnes	'000 tonnes	'000 tonnes	'000 tonnes	'000 tonnes	'000 tonnes
7	road	6970.58	7049.44	6000.86	13738.01	13277.82	12171.17
8	railway	16438.87	16314.95	17040.80	32853.82	33037.10	33728.27
9	road	1312.72	1357.79	1680.52	2852.79	3129.70	3545.17
12	railway	27.31	28.00	29.17	57.92	61.39	61.88
13	railway	16411.56	16286.95	17011.63	32795.90	32975.70	33666.39
28	road	4023.87	4157.12	4037.03	8556.99	8932.81	8532.78
29	railway	8.01	8.36	9.58	17.86	19.71	20.88
30	railway	7827.36	7614.22	8780.50	15343.70	15422.75	16927.76
42	railway	1357.98	1452.37	2152.98	3163.59	3584.01	4532.16
43	road	11777.13	11762.29	10014.27	22906.83	22029.69	19975.39
46	railway	4.26	4.71	8.37	10.94	13.92	19.81
47	railway	1353.72	1447.66	2144.61	3152.65	3570.09	4512.35
51	road	1782.48	1856.08	3348.44	4240.45	4992.33	6876.32
52	road	14018.52	14063.33	10702.85	27223.37	25970.17	21631.85
55	railway	1361.73	1456.02	2154.19	3170.51	3589.80	4533.23
58	railway	2777.67	2645.00	3091.63	5263.71	5180.18	5775.67
59	railway	5049.69	4969.22	5688.87	10079.99	10242.57	11152.09
64	railway	4139.39	4101.02	5245.82	8434.21	8769.98	10308.89
65	road	15801.00	15919.41	14051.30	31463.82	30962.50	28508.17
68	railway	4143.66	4105.73	5254.19	8445.15	8783.89	10328.70

Source: Results of the analysis conducted.

We successively analysed the options corresponding to the fixed traffic volume and modified infrastructure in order to select the optimum one. The simulations show that decreasing the consistency threshold and increasing the inconsistency threshold leads to elimination of less desirable options.

CONCLUSIONS

The analysis of the results leads to the following conclusions:

1° If we assume an increase in the traffic volume on individual connections, the optimum option in terms of the suitability of the infrastructure of the analysed part of the transport network for the transport needs is the option 3, i.e. the option with the motorway A-1. It means that the construction of a new road connection, namely the aforementioned motorway, is necessary and justified.

2° Taking into account the cost internalisation, for all task allocation options there is a significant decrease in the traffic stream on road connections:

For option 1: up to 3,315.67 thousand tonnes p.a. on the arc Piotrków Trybunalski – Częstochowa;

For option 2: up to 5591.52 thousand tonnes p.a. on the arc Piotrków Trybunalski – Częstochowa (Road #1);

For option 3: up to 6078.88 thousand tonnes p.a. on the arcs Piotrków Trybunalski – Częstochowa – Katowice (Motorway A1).

As a result, that portion of the cargo traffic is taken over by the railway transport.

3° The proposed method for the multicriteria evaluation of the transport system development facilitates selection of the optimum option of the transport network infrastructure development with respect to its suitability for the traffic volume.

REFERENCES

Ackoff R. W.(1968) Optimal decision in applied researches, Polish Sc. Publisher, Warsaw
Basiewicz T., Jacyna M., Ambroziak T.(2000). The Logistical Point of View on Transport Infrastructure Needs Assessment/Corridors: the Baltic Sea – the Adriatic/International Scientific Symposium Traffic connection between the Baltic and the Adriatic/ Mediterranean, Croatian Academy of Sc.and Arts, Zagreb, November 22 and 23.
Jacyna M.(1997). The Structure and characteristics of the elements of a multimodal transport corridor in respect to traffic distribution modelling. Archives of Transport, vol.9 iss.1-2, Warsaw, 5-18
Jacyna M.(1998). Some aspects of multicriteria evaluation of traffic flow distribution in a multimodal transport corridor. Archives of Transport, Polish Academy of Sciences, Com. Of Transport, vol.10 iss.1-2, Warsaw, 37-52.
Jacyna M.(2001). Multicriteria Evaluation of Traffic Flow Distribution in a Transport Corridor. Railway Engineering, London 30.04-1.05.
Jacyna M.(2001). Application of multicriteria modelling for evaluation of transport systems. Scientific Papers Warsaw University of Technology, Transport, vol.47, Warsaw.
Korhonen P., Laakso J.(1986). A visual interactive method for solving the multicriteria problem. European Journal of Operational Research, vol. 24, No.2, 277-287
Korzan B.(1978). Elements of theory of graph and webs. Methodology and applications. Scientific and Technical Publishers, Warsaw
Magnanti T.L. and Wong R.T.(1984). Network Design and Transportation Planning: Model and Algorithm. Transportation Science, 18.
Roy B.(1990). Decision-aid and decision making. Europen Journal of Operational Research, vol. 45, 3 4-331.
Roy B.(1990). Multicriteria assistance of decision making. Sc. and Tech. Publishers, Warsaw,
Steenbrink A.(1978). Optimalisation of transport networks. Transport and Com.Pub. Warsaw.
Zeleny M.(1982). Multiple criteria decision making. McGraw-Hill, New York.

K.G. Goulias, editor

CHAPTER 32

ANALYSIS OF THE GREEK COASTAL SHIPPING COMPANIES WITH A MULTI-CRITERIA EVALUATION MODEL

Dr. Orestis D. Schinas
Professor Harilaos N. Psaraftis
Maritime Transport Laboratory
National Technical University of Athens

INTRODUCTION

This paper deals with the application of Multi-Criteria Decision Making (MCDM) techniques for the problem of the overall evaluation of the Greek Coastal Shipping (GCS) companies, and it focuses on the needs of lenders and investors. The importance of GCS is very high for the Greek society, as it demands regulation consistent also with European practices. It also demands efficient transport network operations for social and economic reasons and close financial monitoring as the important actors are listed at the Athens Stock Exchange (ASE). Lately, these companies have experienced growth, as a result of the partial deregulation of the industry, of the equity inflow from the ASE and the introduction of new vessels in service. The analysis period is extended from the fiscal year of 1997 up to 2002; this is a result of the availability of the data and their integrity check.

As revealed from the literature, there is no prior work on this issue, as well as there is no MCDM approach reported for the niche market of the GCS. Shipping finance literature is basically focused on time-series analyses and the lending (risk assessment) criteria are drafted in textbooks for many decades already.

The selected MCDM method is the Analytic Hierarchy Process (AHP). This method was preferred from other established methodologies as it does not demand prior knowledge of the utility function, it is based on a hierarchy of criteria and attributes reflecting the understanding of the problem, and finally, because it allows relative and absolute comparisons, thus making this method a very robust tool. AHP is adequately discussed and reviewed; the method allows combinations with other techniques, as well as, scenario analysis and simulation exercises. Last but not least, AHP allows group decision-making and is convenient in numerical handling.

The issue of the operational risk and of the risk assessment is critical, as most of the capital gearing this industry is coming from lending financial institutions (FI); the capital inflow from ASE amounts only to less than 10% of the total liabilities of these companies. The main issue of evaluating the overall ranking of GCS companies is addressed with the construction of a rather expanded hierarchy. A full justification for the selection as well as for the relative weighting is provided below. In most cases the selection of the criteria was instructed from the availability of data. For the upper levels of the criteria both scenario analyses and simulation exercises have been considered. The resulting ranking and indices provide a clear track of the course of the companies over the period of analysis (1997-2002). It is possible to monitor their course over time (overall index), as well as over partial attributes, such as the external and the internal criteria. A deeper degree of analysis is also possible but it cannot be visualized in planar graphs.

The model can be validated according to the shift-share calculations. In 60% of the cases, the shift-share analysis of the turnover data may explain the differences of the indices and in the rest cases there is a consistency with the shift-share calculations of the traffic. Traffic and turnover are intertwined yet the issue is their annual difference. Dependencies may be explained if only the criteria-weights are thoroughly analyzed. Apart from the overall ranking and monitoring over the years it is possible to use the same hierarchy or elements of the structure for practical problems, such as the corporate planning and mergers. In corporate planning it is possible to estimate the final position of the company if some tactical movements occur or to foresee the result of these movements, by using the same hierarchy and weights. In cases, where new actors or new parameters shall be taken into account, elements from the hierarchy may assist in the planning. This is also the case for the selection of the optimum merger-alternative; mergers and acquisitions are in the current business agenda. The model aims to have capabilities for practical use; nevertheless, it is consistent with the basic theory as described in the textbooks.

In all cases the consistency of the judgments remained under the limit of 10% as demanded from the theory, so there is confidence in the final numerical results. Different opinions or approximations may stem out of different sets of criteria. It is possible to include more alternatives and more criteria in the hierarchy as is but there is always the question over the availability and the integrity of the data for all companies in the set and for all years of the analysis.

Last but not least, the work presented here is based on the author's Doctoral Thesis available at www.martrans.org in English (Schinas, 2005). As a result, many details especially information related to the analysis of the criteria as well as of the calculations will be omitted due to space limitations.

GREEK COASTAL SYSTEM

The Greek Coastal Shipping (GCS) system is a very interesting case for research, not only within the Greek business pattern but also from a European perspective. The problem of the system is not solely of financial or transportation efficiency but embraces also a variety of interests of the State and of the society. The State and rest actors, such as carriers, local communities, and port authorities have to find an equilibrium that all partial interests can be merely satisfied and operate within a compromise pattern.

From a carrier perspective, the fleet has to serve necessarily ports with both adequate and inadequate traffic volumes. Furthermore, as the traffic is seasonal, a rational carrier would operate the fleet only during those months with adequate volumes or serve destinations of his interest. From a local community perspective, vessels have to connect the island with many other destinations for tourist-related purposes and to the mainland for commercial and social cohesion purposes. Local port authorities are the new players in the game and increase their significance steadily. Local ports have not only to serve the traffic but also to ensure funding from the State for local investments. Despite the fact that not all ports have been converted according to the Law 2932/2001 to corporations, local interests participating in the management of the port (in most cases these are just harbors) can seriously affect the operation of any fleet. The quality of port infrastructure and facilities is critical for safety, operational, security and economic grounds. Although most of the facilities are treated as 'ports' by the Law, they only offer harboring and in some cases of a very low quality. Last but most important actor is the State expressed through the Minister of Mercantile Marine (MMM), the Coast Guard and various other agencies, authorities, etc. that regulate or operate the system.

The implementation of EU Regulation 3577/92 on maritime cabotage aimed to liberalize the maritime services to the benefit of shipowners who have their ships registered and flying a flag of a Member State provided that their ships comply with the conditions for carrying out cabotage in the Aegean. It is interesting to note that Greece and other Mediterranean States have been granted temporal exception by means of derogation. Thus in 2004 the GCS should have been liberalized and the institutional framework harmonized with EU legislation. During these twelve years of grant-period too many things have been changed in the Greek economy and the GCS respectively. The changes and the new Law 2932/2001 did not really resolve any problems of the past: the State has still to find ways to ensure proper connection of the islands to the mainland and the carriers have to operate profitably in total, even when servicing destinations of no commercial interest.

This paper is focused on the listed coastal shipping companies, due to the lack of adequate data for the non-listed companies. The period of the analysis is the fiscal years 1997 – 2002. The companies under evaluation are ANEK, MINOAN Lines, NEL, Strintzis Lines (Blue Star Ferries) and EPATT (Attica Enterprises). It is interesting to note for the reader who is not familiar with the GCS, that ANEK and MINOAN are operators based on the island of Crete, enjoying dominant position in the lucrative Piraeus – Chania and Piraeus – Heraklion lines respectively. ANEK and MINOAN used to enjoy monopolistic status in these respective lines up to 2002. Furthermore, ANEK and MINOAN are powerful players in the Adriatic routes, as well as expand their interests to other sub-systems of the GCS. For simplicity reasons, one may distinct the GCS into the following sub-systems (see also the map):

- Piraeus – Crete
- Piraeus – Cycladic Islands
- Piraeus – Dodecanese Islands
- Piraeus – Eastern Aegean Islands
- Piraeus – Northern Aegean Islands
- Piraeus – Nearby Islands (Saronic Gulf)
- Thessaloniki (Macedonia) – various destinations
- Ionian Islands

Figure 1: The map of Greece

In the current study, all data concern the total market, and not a specific sub-system; only the routes between Piraeus and Crete are not as seasonal as the other entire are. Even the routes linking Piraeus with the main islands of Rhodes (Dodecanese) and Lesvos (North Aegean) and Samos (East Aegean) are seasonal (Psaraftis et al., 1994 as well as data provided by the operators). In that sense companies having the license to serve Crete have had an advantage over the others. The high seasonal attributes of the traffic forced also major players to pursue ventures in the Adriatic Corridor. The traffic over Adriatic is experiencing continuous growth and is fully deregulated as it is considered as international trade. The simultaneous deployment of vessels in the Adriatic and the Aegean is a difficult task but it was necessary for the companies aiming at the dominance of the market.

MODEL STRUCTURE AND APPLICATION

In most cases the evaluation of any company is solely based on financial ratios as well as on the 'story' behind the company. Although financial ratios offer a quantitative yet incomplete approach, this 'story' notion encompasses a rather vague and duly qualitative set of attributes of the company under scrutiny. These attributes are usually some of the following, although the list is not exhaustive:

1. Capabilities and track-record of the top management team
2. Sector perspectives
3. Perspectives of the company within this sector

4. Business strategy
5. Market perception - image
6. Familiarity with the investors

In the shipping market these attributes are commonly discussed among investors, lenders and borrowers, as both keep a very close track of the respective market. In cases of a public offering these are the critical points for an underwriter to communicate with the investors, who are not really aware or do not necessarily monitor this sector closely.

In the modern business environment, financial ratios and relevant analyses are not reflecting the whole picture; statements, structures and financial instruments are very complicated as companies expand operations to several institutional, legal and tax settings. Furthermore, the use of options and other off-balance-sheet risk tools may easily hide the actual risk exposure and financial status of the company. Nevertheless, financial ratios are important elements for the analyst to take into account but not the sole ones.

The 'story' set of attributes is extremely vague and is in no case uniform. 'Good' and 'bad' are very subjective notions. Furthermore, it is very difficult to quantify the above mentioned attributes, although there are some ideas and approaches for some of them.

As shipping is a cyclical market, financiers are monitoring the cycles and trying to predict future prices of indices or time-charter rates. The market is classified into several niches and modeling based on time-series is performed. This approach may provide a basis for further discussion on the demand as well as on the supply side, yet it can highlight only some aspects of the problem. Financiers translate the market and the company data into a risk figure that will feed a risk management system and yield a result suggesting the acceptance or the rejection of the project or provide the basis for a sensitivity analysis and guidance to the client.

The evaluation of a company is of pure multi-criteria analysis (MA) nature. Pardalos stresses the importance of development of multi-criteria decision support systems (MCDSS) as a future direction. A MCDSS is defined as a decision support system (DSS) that helps to implement MCDM methods (1995, p. xvi). In that sense what Chou and Liang have theoretically presented is an MCDSS (2001). The same applies for the evaluation model developed here. It is interesting to note that apart from any theoretical classification, MCDSS aim to include as many as possible parameters bearing information, while the traditional approaches disaggregate the problem into a set of quantitative elements, (financial and risk related ones) as well as, of qualitative attributes ('good' management team, perspectives, strategy). Any MCDSS quantifies the problem and therefore it harmonizes the approach and draws attention to other elements not included in the analysis yet. The added-value from any MCDSS is the foundation of knowledge, as the criteria and the hierarchies used underline our understanding; redundant elements are eliminated and new elements are entered in order to explain deviations and gray zones.

The reasons for the selection of AHP were the following:

1. in absolute comparison mechanisms it is not possible to experience rank-preservation problems;

2. the set of data is very large and the relative comparison of every alternative for every year available would increase the numerical and decisional burden exponentially;

3. it is easy to add alternatives (existing or dummy ones), to experiment with the sensitivity of parameters, or to estimate the outcome of an action (element sensitivity);

4. the focal attention lies on the hierarchy, i.e. on the insights and on the parameters determining the phenomenon

5. there is no need to estimate a utility function (or marginal utilities)

6. AHP can be combined with other methods

7. AHP-required hierarchies can be further developed to networks and systems with dependencies and influences (commonly addressed by ANP)

Hierarchy of the Criteria

The most creative part of decision making and the most critical one is the structuring of the hierarchy. This process has a significant effect on the outcome of the model as well as on explaining the phenomenon under scrutiny. Saaty (2001, p. 58) proposes a basic principle in the form of a question: *can I compare elements on a lower level meaningfully in terms of some or all of the elements of the next higher level?* (2001, p.59):

It is interesting to note that the issue of building hierarchies is not largely discussed in the literature; in a sense the decision-maker is 'uncontrolled' from any expert or academic guidance in structuring a hierarchy. In the book of Keeney and Raiffa there is a brief discussion on this issue (1993, p. 41). They base their comments on an essay of Manheim and Hall (1969), who prompted the issue of specification and means. By *specification*, one understands the subdivision of an objective into lower-level objectives or more details, thus clarifying the intended meaning of the general objective (goal). These lower-level objectives are considered as *means to the end*, thus by identifying them to very precise objectives, one can build the whole hierarchy up to the highest level. Keeney and Raiffa highlight also the importance of setting a sound and achievable overall goal, so to point the upper end of the hierarchy. A vague goal such as 'the good life' is not a very successful example for further elaboration. However they suggest going down the hierarchy as long as the advantages of doing so are more than the disadvantages. Complexity and data availability are the most common disadvantages. Regarding the formalization of the problem, Keeney and Raiffa discuss also the 'test of importance' of Ellis (1970). Ellis suggests before adding any objective (sub-criterion according to Saaty's terminology) into the hierarchy to consider whether this would change the course of action. This approach offers a basis for eliminating or including criteria into the hierarchy, yet it is not necessarily correct. It is possible to include criteria that fail the test of importance separately but collectively they are important. Last but not least, hierarchies for a specific problem are not unique. Their differences may be attributed to the degree of formalization as well as the point of view (subjectivity) of the decision-maker. In the case of the GCS it is possible (if not certain) that investors and financiers have a totally different structure of criteria from the users of the GCS services; even between them there would be a different structure between the residents of the islands and the tourists. In this context the interest lies currently with the financiers and the investors.

For the purposes of evaluating the coastal shipping companies the following hierarchy is considered. The overall index is estimated on the basis of two distinct sets of criteria: the internal and the external forces. As internal forces are considered attributes that are determined by the management of the company and more specifically in this case, all attributes related to the fundamental, the logistics services offered and the management. The criteria are presented below (see Figure 2, Figure 3, and Figure 4 respectively) and discussed thoroughly in the given literature (Schinas, 2005). The criteria-sets that are not directly controlled by the management are considered as external forces. The stock-performance, the market environment as well as the competition fall into that category. The subcriteria are presented below (see Figure 6, Figure 7 and Figure 8).

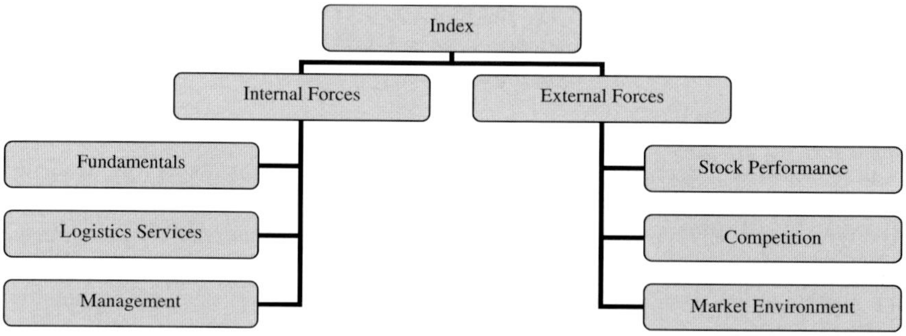

Figure 2: *Levels I, II and III of the hierarchy*

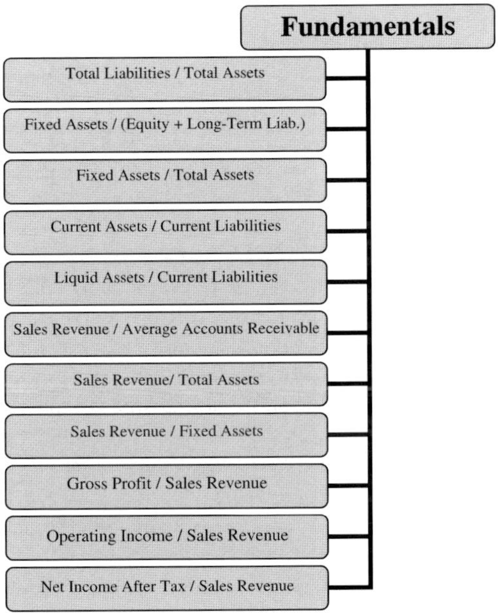

Figure 3: *Level IV – Fundamental Data*

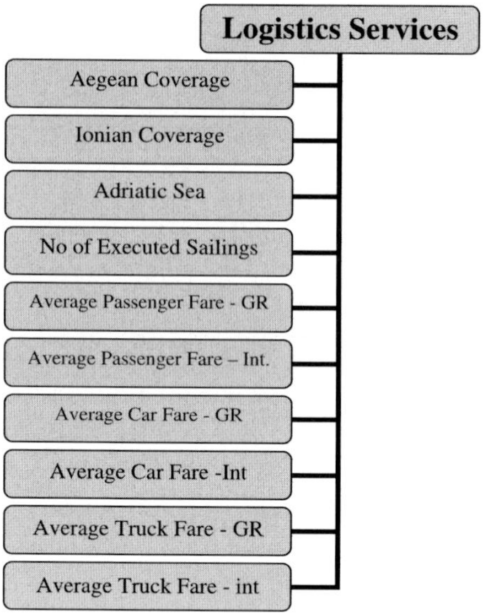

Figure 4: *Level IV – Logistics Service Data*

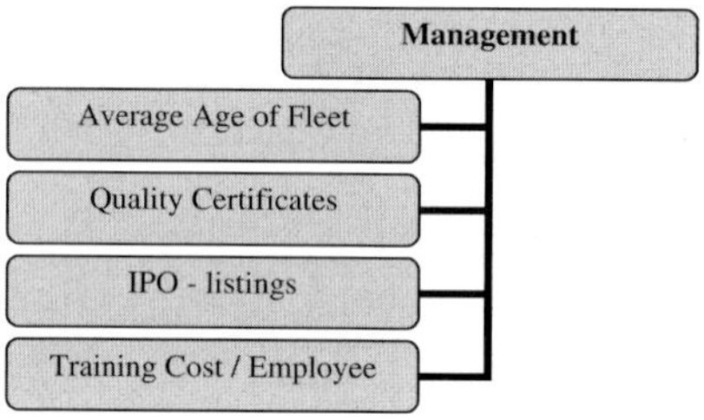

Figure 5: *Level IV – Management Data*

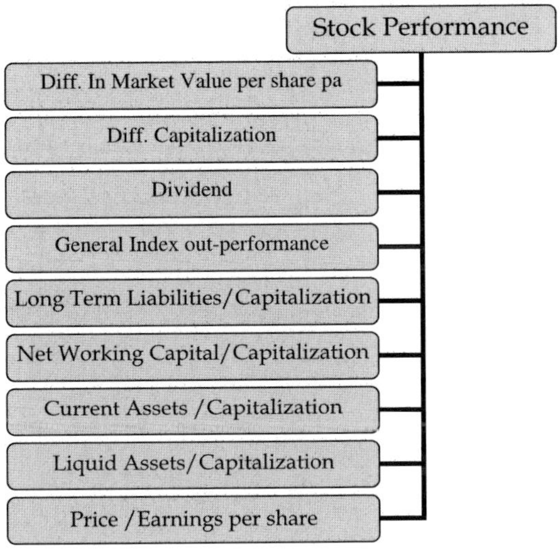

Figure 6: *Level IV – Stock Performance Data*

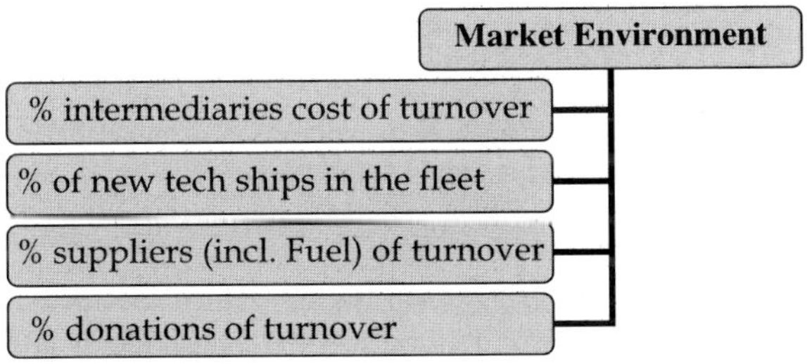

Figure 7: *Level IV – Market Environment Data*

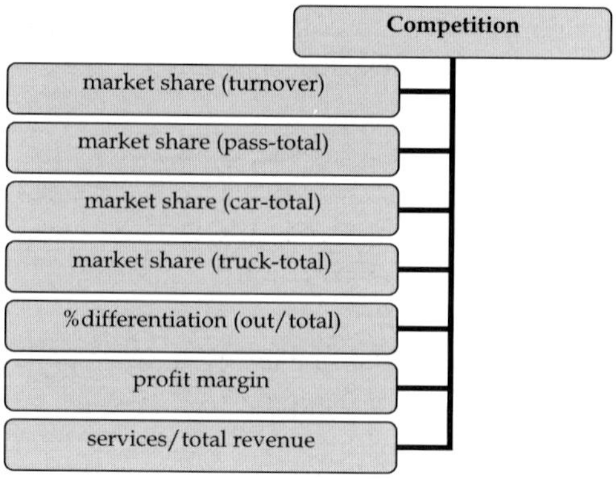

Figure 8: ***Level IV – Competition Data***

Detailed Analysis of the criteria as well as of the rational of selecting them is available in the original thesis.

APPLICATION OF THE MODEL AND RESULTS

From the basic theory of AHP it is necessary to pair-wise compare the attributes of various alternatives and then pair-wise compare criteria. This leads to a relative comparison approach, i.e. if the set of alternatives changes, then the whole procedure shall be executed again. Furthermore, what is really necessary is the absolute comparison of the alternatives per year. In order to find the attributes a_{ij} of a company per criterion (and per year) given the real figures the following procedure is followed. All relevant figures for all companies and for all years are estimated and then they are categorized into distinct spaces according to the quartile statistical function. So it is possible to assign letters A, B, C, D, or E for the attributes that fall into the respective quartile. Then, according to the technique used by Liberatore (1987 and 1992), as well as proposed in many books of Saaty (e.g. 1994, p.17), these A, B, C, D, and E are evaluated and their vector is extracted. Then this vector is idealized (i.e. all elements of the vector are divided by the largest one). The product of the idealized vector and of the vector of the criterion is the one that contributes to the overall index. For example let's use the following data:

	1997	1998	1999	2000	2001	2002
TA/TL	1,775	2,569	2,076	1,584	1,446	1,522

Table 1: Sample data – ANEK / Fundamental Data

Along with all other data (from all the companies and all the years) referring to TA/TL ratio, these data are gathered and classified according to the quartile statistical function. Quartiles often are used in sales and survey data to divide populations into groups. For example, one can use the quartile function to find the top 25 percent of incomes in a population. The result of the quartile function is:

1,903	2,194	2,263	3,060

That means that 1,903 is the bound for E, 2,194 for D and so on, considering the highest value as the best. The above figures are 'translated' as:

1997	1998	1999	2000	2001	2002
E	B	D	E	E	E

Then these spaces (letters) are evaluated to each other:

	A	B	C	D	E
A	1	3	5	7	9
B	1/3	1	3	5	7
C	1/5	1/3	1	3	5
D	1/7	1/5	1/3	1	3
E	1/9	1/7	1/5	1/3	1

(A is considered as the best set of values-options)

and that yields the following vectors:

	vector	Idealized Vector
A	0,510	1,000
B	0,264	0,517
C	0,130	0,254
D	0,064	0,125
E	0,033	0,065

The table with letters is now a vector with the following elements:

0,065 0,517 0,125 0,065 0,065 0,065

From the criteria evaluation procedure, based on biased preferences of the author, it is known that the TA/TL criterion is 0,041 (Schinas, 2005, Table 14), so by multiplying the 0,041*0,065 = 0,00264 the contribution of the TA/TL criterion in the overall index of ANEK in 1997 is extracted. This measurement is absolute. The classification of attributes to spaces (A, B, ..., E) is helpful for many reasons, although it would also be easy to compare directly the attributes of the companies per criterion and then to 'normalize' them by using the fundamental scale. Saaty considers this an approach that exploits accumulated experience (1994, p. 18). The assignment of letters for a given set of data (per criterion) enables the decision-maker to have a clear picture of the values and the averages in the sector (i.e. of all companies under analysis). Extreme values that occurred, because of many uncontrolled reasons, are only assigned a letter A or E and can therefore be evaluated with the rest. Furthermore the use of absolute measurements leads to a better understanding of the evaluation problem. It is like having a button with five options; extremes and inaccuracies are allayed in that sense. Last, but not least, is that by using the absolute measurements in such a large model, it is possible to keep the overall consistency as low as possible. The table above has a consistency ratio of 5.29%, which is acceptable and well below the limit of 10%. In the relative measurement approach the modeler could never be sure of the consistency as the numerical burden would be considerably higher. So the elements a_{ij} of every alternative (and per year) are the elements of the idealized vectors. It is reminded that the index will be derived as the sum of the product $w_j a_{ij}$, where w_j the criteria weights. Up to this point, the weights of the Level III have been estimated. The focus now shifts to the estimation of Level II and Level I (index), so the w_j vector will be estimated.

The first approach is descriptivist as some evaluations reflecting the biases and the beliefs of the decision maker will be presented and analyzed through a sensitivity analysis. Then for the same values of the first approach (lower levels) a simulation procedure will yield the possible response of a group of decision makers.

Scenario-Based Outcome

As the criteria weights have been set for the lower levels III and IV, the decision-maker has to come to a decision on the level II and level III criteria. At these levels, decisions are critical and reveal biases as well as 'stimuli' towards the final outcome. More specifically, the decision-maker has to make the following comparisons:

1. Internal (INT) vs External (EXT)
2. Fundamentals (F) vs Logistics Services (LS)
3. Fundamentals vs Management Related Criteria (M)
4. Logistics Services vs Management Related Data
5. Stock Performance (SP) vs Market Environment (ME)

6. Stock Performance vs Competition (C)
7. Market Environment vs Competition

The above seven comparisons are the ones missing in the two decision tables of level II and the one of level I. It is obvious that any judgment would be very subjective; commonly groups of decision makers focus on the criteria they understand better. Most probably an accountant would consider fundamentals as the most important set. A customer (client of the system) would consider the LS set as the most important. In order to come to a conclusion and to expose the capabilities of the model the following scenarios (values) will be discussed:

1. INT / EXT $\in$ [1/5, 1/3, 1, 3, 5]
2. F / LS $\in$ [3, 5]
3. F / M $\in$ [3, 5]
4. SP / ME $\in$ [3, 5]
5. SP / C $\in$ [3, 5]

The values considered (3 and 5) aim to highlight pure and clear preferences as well as to avoid extreme values, such as 1, 7 and 9. Obviously there is a bias towards financial criteria, as the fundamental accounting data (F) and the stock performance (SP) is considered as more important than the level of the service offered (LS), of the management (M), of the competition (C) and of the market environment (ME). The results are available however some highlights are presented below. It has to be mentioned that no extreme values have been taken into account. Considering the biased case with the following values:

1. INT/EXT = 3
2. F/LS = F/M = 5
3. SP/ME = 3
4. SP/C = 1/3

the yielded result is presented in the coming figures and the justification of this judgment is as follows:

- A company can control the internal forces. Although external ones are not controllable, they are still important. Therefore, the INT are slightly more important than the EXT.
- In all cases the fundamental accounting data are more important from the services and the managerial indicators for financiers and investors. As the other factors cannot be neglected, the value of 5 indicates their balance.
- Stock performance is slightly more important than the market environment as such. It is possible for a company to boom despite a recession and vice versa. Financiers and investors appreciate generally a good stock performance, therefore the ratio SP/ME >1 is considered. In contrast stock performance data are not as important as competition figures; market shares are generally more

important than stock performance for long-term placements, so SP/C<1. The values of SP/ME=3 and SP/C=1/3 reveal at least a slight preference.

The results reflect widely the perception of market experts. The data are better presented below (see Figure 9 and Figure 10):

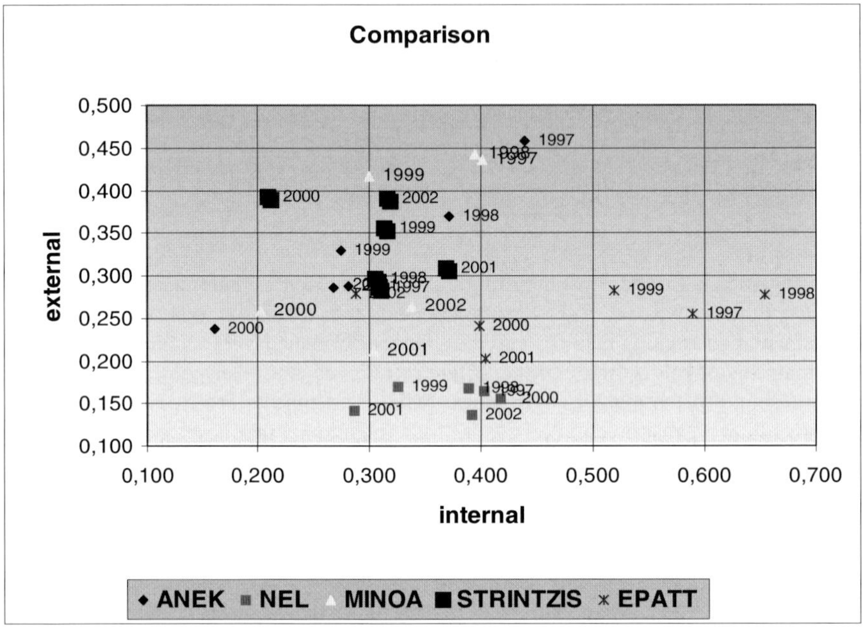

Figure 9 : Relative Position of Every GCS company on an annual base

In this typical representation of such results, the upper right part of the chart contains the companies with the highest grade. It is easy to see that EPATT gets a relatively high grade in INT and remains practically stable in EXT. Companies closer to the axes are underperforming in relation to others. Furthermore one can also monitor the 'track' of a company throughout the period of consideration. As the information contained above is adequate to support various conclusions, decision makers usually need only a ranking value. The next figure consolidates much of the above information:

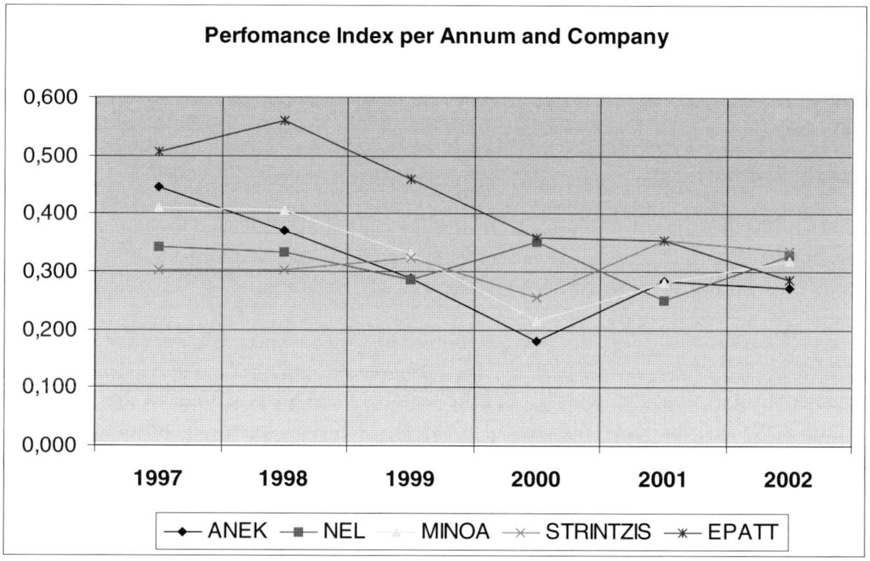

Figure 10: Total Performance Index of every GCS per annum

The usability of the above charts is obvious; one can easily track the performance of a company per se as well as per set of criteria. Furthermore, one can proceed in various analyses that will be discussed thoroughly in coming paragraphs. The analysis here would not be over unless the most critical criteria and elements would be identified. By applying the methodology described by Triantaphyllou (1996, 1997, 1998) for a sample year (say 1997) and for the criteria weights assigned previously it is found that the F is the most sensitive criterion ($sensC_F = 0{,}0392$) followed by C, F, ME, M, SP. The corresponding $\delta'_{k,i,j}$ quantity is $\delta'_{LS,2,4} = -4{,}65\%$. In other words, $\delta_{F,2,4} = 0{,}136618$ and $W^*_F = [0{,}536 - (0{,}136618)] = 0{,}399$. By normalizing the weights the new weights that will reverse the ranking are:

W_{SP}	WM_E	W_C	W_F	W_{LS}	W_M
0,067	0,022	0,200	0,462	0,124	0,124

With the new weights the previous ranking $A_5>A_1>A_3>A_2>A_4$ (where '>' stands for 'better than') becomes $A_5>A_1>A_3>A_4>A_2$. It is reminded that A_1≡ANEK, A_2≡NEL, A_3≡MINOAN, A_4≡STRINTZIS, A_5≡EPATT. Triantaphyllou and Sanchez (1997) expanded the above technique in order to estimate the most critical element. In a similar way definitions were given and similar theorem proven. By applying their methodology, it is estimated that the most critical element is a_{24}, i.e. the element of NEL for the fundamental criterion. Originally, its value was 0,456 and it was estimated that a reduction to the value of 0,400 would reverse the rank of A_2 and A_4. So the original rank of A_5>A_1>A_3>A_2>A_4 (where '>' stands for 'better than') becomes again A_5>A_1>A_3>A_4>A_2

Simulation-based outcome

The scenario-based solution is biased from the decision-maker. That does not necessarily scrap the applicability of the model. However, some times it is desired to get a more objective perspective. By looking back into the AHP the basic formula is:

$$Index_i = \sum w_j a_{ij}$$

that implements 'objective' elements a_{ij} and the weights w_j. This is similar to the 'weighted sum method' (WSM); the only difference is that in WSM is based on absolute weights (weights assigned by the decision maker directly) while in AHP come from a relative comparison procedure. The method is based on the assumption of the additive utility. That means, according to Keeney and Raiffa (1993, p. 231) that the attributes Y and Z are independent (two-attributes case that can easily be considered for N attributes). This assumption is the basis for almost all MCDM methods used in practice and in the academia. Furthermore, the issue of weights is dealt in most (if not all) cases by direct solicitation from the decision-maker. This is a means to understand the utility function of the decision-maker; in some cases the weights stem out a utility-function determination procedure, which is mostly derived by using questionnaires and structured questions.

In many cases it is necessary to combine the judgments of many persons or decision makers in order to get a better understanding of the problem as well as to achieve criteria-weighting that is considered as 'generally' acceptable. In the literature, such issues are called 'group decision making'. Saaty is suggesting the use of geometric mean in order to synthesize the judgments of individuals in to a group property. He bases this argument on the work of Kenneth Arrow (Saaty, 2001, p. 62). Identical approach is adopted by Keeney and Raiffa (1993, p. 523). The following is an illustrative example: say that three decision makers are assigning the following values according to the fundamental scale 2, 3, 7, then the group decision is estimated as $\sqrt{3;2*3*7} = 3{,}476$ and the outcome is rounded to 3. The above approach has been used in many cases and applications, yet more interesting and practical applications have been developed lately. A very interesting approach is the one of fuzzy sets. As a decision maker is asked for his judgment, this can happen in three different ways:

1. Each individual judgment is modeled with a probability distribution. For example, a decision maker may use the triangular distribution, where only the high, the low and the most likely value is asked. Then the mean of all judgments is estimated and therefore the group value is obtained.

2. All decision-makers are asked for a point-estimation. This assumes that all decision-makers have an equal probability of being correct, as the group decision is derived as average of all point-judgments.
3. All decision makers are asked to estimate high and low values and their judgment is assumed to be with equal probability within that range. The group decision judgment is derived by using the minimum lower and the maximum upper bounds of each individual judgment.

The following figure is explanatory:

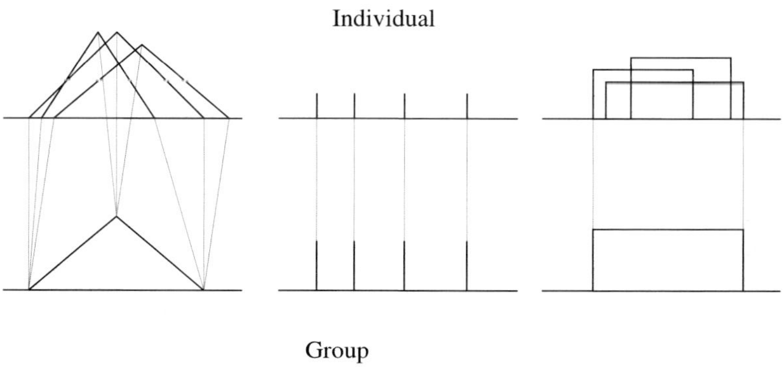

Figure 11: Individual and Group Interval Judgments

In the above figure, one can see the way a group decision is extracted out of individual responses. At the left side, individual triangular responses are combined into a new consolidated one. The same applies for the right side, where range-responses are combined. In the middle, individual point-estimations are the same with the group one before the averaging procedure. Generally the averaging procedure attracts the interest of researchers as weights and approaches influence heavily the final result. In the literature simple fuzzy averaging is widely used. The most 'complicated' case from the above group decision making option is the triangular one.

When the triangular distribution is used, it is better to consider the use of fuzzy sets and of their numerical operations. A triangular fuzzy number (TFN) A has the following membership function $\mu_A(x)$:

$$A \equiv \mu_A(x) = \begin{cases} \dfrac{x - a_1}{a_M - a_1} & a_1 \le x \le a_M \\ \dfrac{x - a_2}{a_M - a_2} & a_M \le x \le a_2 \\ 0 & otherwise \end{cases}$$

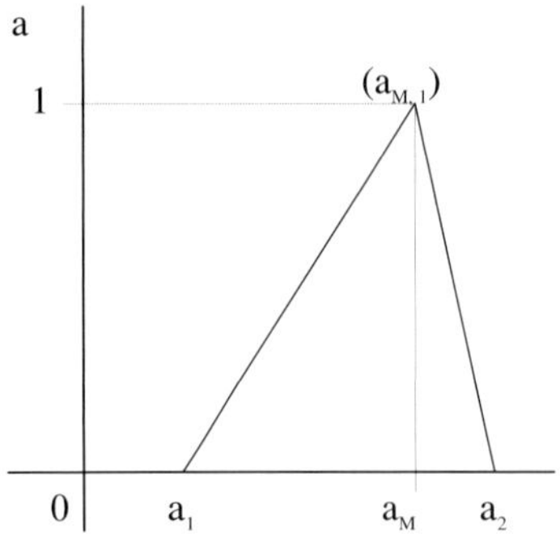

Figure 12: Graphical Representation of a TFN

Commonly in the literature of fuzzy applications, a triangular fuzzy number (TFN) is referred as M = (m, L, R), where m = a_M, L = a_1 and R=a_2. According to the theory the addition of TFN is defined as:

$M_1 + M_2 = (m_1+m_2, L_1+L_2, R_1+R_2)$

and the fuzzy averaging as:

$$M_{average} = M_1 + M_2 + \ldots + M_n = \left(\frac{1}{n}\sum_{i=1}^{n} a_m^i, \ \frac{1}{n}\sum_{i=1}^{n} a_L^i, \ \frac{1}{n}\sum_{i=1}^{n} a_R^i \right)$$

This fuzzy averaging is used in various forecasting methodologies. In the literature there are three ranking methods available:

1. Weighted method
2. The method of Chang, and
3. Kaufmann – Gupta method

All of them are thoroughly described in the literature. Although there are no significant numerical differences between them, the method of Chang is preferred because the dominant alternative is the one with the largest mathematical expectation E_j:

$$E_j = \frac{(R_j - L_j)(m_j + R_j - L_j)}{6}$$

Triantaphyllou and Lin (1996) proceeded into the fuzzification of crisp MCDM methods, such as WSM, WPM, AHP and TOPSIS and used the fuzzy ranking approach for determining the best alternative. They run the models with TFN and the outcome was a new TFN for every alternative. Then the ranking of TFNs yielded the outcome. In other applications 'defuzzification' techniques have been employed that yielded the final ranking (e.g. in Bojadziev et al., 1997, p.147). The entire above are very useful in structuring the simulation procedure, as they provide options and alternatives to the modeler. More specifically by using the above tools the model may select one (or a combination) of the below options:

1. To select a number of decision makers (say S), then simulate their response and finally to enter the geometric mean as input to the final AHP decision matrix. The distribution of the response can be:
   1.1. discrete values (1/9, 1.7, ..., 7, 9)
   1.2. uniform (1/9, 9)
   1.3. triangular (1/9, random value, 9)
2. To select a number of decision makers (say S), then to simulate their response as a TFN. Then a fuzzy AHP procedure (identical with the normal one but with numerical hurdles) can follow. The result can be ranked according to:
   2.1. The method of Chang
   2.2. Defuzzification procedures

A simulation procedure consists of basically five steps, namely *development*, *building*, *verification and validation*, *design of experiments*, and *analysis of the results*. The first one is the development of the conceptual model. In this specific case there are practically two options. The first is to simulate the responses first and consider their outcome as input to the model. The second is to run the model for every response considered as appropriate and then to consider the results of the model as basis for further elaboration. Both are achievable, although the second one is accompanied by numerical problems due to the large number of estimations. For consistency reasons, it is preferred to simulate the responses of a group of decision-makers to the very same ratios (preferences) used for the scenario-analysis before.

The second step is building of the simulation model. Apart from the various relations between the data, which are given in this case, it is important to estimate the distributions of the uncertain variables. This step is critical as it introduces a great deal of subjectivity in the model. Say that ten decision-makers are selected and their responses are feeding the formula of geometric mean. A sample of no less than 300 trials (no numerical justification) indicates that the mean of a discrete distribution is close to 2,75. The distribution contains all possible values according to the fundamental scale (1/9, 1/7, ..., 7, 9) with an even probability of occurrence. By using the uniform distribution for the same minimum and maximum (1/9 and 9) the mean is close to 4.53. The mean standard error is in these respective cases 0,37 and 0,22, thus favoring the uniform distribution. This was expected as the discrete distribution is feeding the geometric mean with values with 'concentration' close to or smaller than 1 (1/9, 1/7, 1/5, 1/3, 1).

In order to achieve a large sample of trials without extreme numerical burden, a group of 20 decision makers are selected. For the five ratios that were selected in the scenario-based analysis a simultaneous procedure is designed and executed. For every ratio a distribution is selected and the outcome of this (mean value) is then taken as input in the decision matrix. Specifically, the following distributions were selected:

1. For the INT/EXT ratio a uniform distribution is selected and the 'responses' of the decision maker are filtered through the non-linear geometric mean function. It is expected that this distribution will favor values over 1 (indifference point) reflecting also a bias.
2. For the F/LS ratio a triangular distributions are selected as responses of the decision makers; then the responses are filtered through fuzzy average function. It is also expected to feed the model with values over 1; the mean is expected close to the mean of the triangular distribution.
3. For the F/M ratio a triangular distribution is also selected but filtered through geometrical mean. The result is expected to slightly vary from the previous one. (However, different seeds are used, but the outcomes are expected to have similar distribution attributes.)
4. For the SP/ME and SP/C ratios discrete distributions {1/9, 1/7, ..., 7, 9} are selected and filtered through geometric mean functions. It is expected that the outcomes should be close to indifference point (1).

By running the model for 10,000 trials the yielded outcome is conforming to the biases expressed above. The distributions for the INT/EXT, F/LS and F/M are considered as normal ones. The mean is close to the median (if not identical), the skewness is within the range of (-0.5, 0,5), the kurtosis is around the value of 3 and the Kolmogorov-Smirnov test yield values of less than 0,03 for the normal distribution, indicating a very good fit of the data. The distributions of the SP/M and SP/C ratios are very close to the lognormal one, thus indicating very small probability of occurrence of extreme values. The fit data are also very good. The third step is devoted to verification and validation. By the term verification it is understood a procedure that ensures a free from logical errors model. By the term validation is understood that the model can adequately represent reality. In this specific case, there is no consideration of the verification. On the validation issue it has to be highlighted the subjective point of examination. A usual way is to ask experts; another way is to compare the outcome with historical trends or values. It is, however, noted that these ways are not applicable in cases without previous experiences and one should only seek for extreme values; in that case an extreme rank reversal would be the case. The fourth step can be omitted in this specific case; fifth step is analyzed thoroughly below.

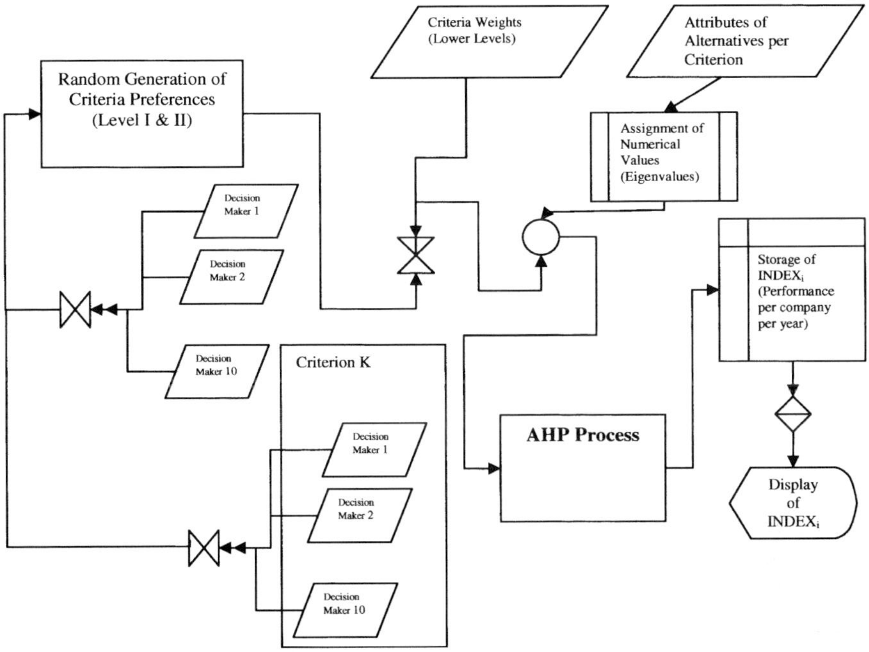

Figure 13: Simulation Flowchart

The outcome of the simulation procedure was yielded after few minutes, thus ensuring that similar exercises in a future application will also demand little time. The simulated responses also yielded results that could easily be fitted with normal and lognormal distributions. All distributions have been tested by using Kolmogorov-Smirnov criterion. By comparing the ranking of these results with the ranking from the scenario based on the following values:

1. INT/EXT = 3
2. F/LS = F/M = 5
3. SP/ME = 3
4. SP/C = 1/3

it is easily observed that there are no real ranking reversals for the 'best' performer. There are some reversals for some other elements as expected:

	1997	1998	1999	2000	2001	2002
ANEK			1		2	
NEL			-2		-1	

MINOAN			2		-1	
STRINTZIS			-1			
EPATT						

Table 2: Rank comparison between the results of simulation and scenario analysis

In the last table, the minus "-"sign indicates better position in relevance to the scenario-based results. Finally, it is interesting to note that such simulations are addressed as *decision making with uncertain judgments* in the literature. The reader may find interesting the contribution of Hahn (2003) with practical examples of the use of logit modeling, Markov chains and Monte Carlo simulations and adequate theoretical justification.

PRACTICAL APPLICATIONS AND FURTHER DEVELOPMENT

The practical application of the model can be expanded according to the needs of the modeler and the focus of the hierarchy. The modeling developed in the previous chapter can easily accommodate more companies and fiscal years for further analysis. One can input data of the new company or of the new fiscal year and estimate its relative position. That is a straightforward procedure and is not sophisticated. The model can lead to a comparison of 'non-accrued' operations or even fictitious data. Nevertheless, one can also proceed in exploratory analysis. Say that a company may consider the renewal of its fleet.

For presentation purposes let's consider ANEK and the fiscal year 2002. Say that ANEK wishes to reduce the age of fleet from 18 to 15 years. That would affect the internal set of criteria, as an element of the management-related data set. The original value was 0,268 and the new one is 0,312. Taking as basis previous criteria-weighting the ranking of the company was improved. From the 5th position it shifted to the 4th (total index 0,273 $\rightarrow$ 0,306). This is attributed to the fact that the original ANEK-2002 value was B and new one is A. As this a criterion with the highest importance in the analysis (6.84% of the global hierarchy), this shift was significant. However, this is a rather naïve approach. The ship costs money that is reflected in the balance sheet. Suppose the cost of the new ship is around 20bnGrD (~60m€). Then the values of the total assets, total liabilities, fixed assets and the long-term liabilities are affected.

By elaborating this exercise further one would say that the impact of a new vessel is not only financial but also affects other factors. For example, one would estimate (or assume) that the addition of the new vessel in the fleet affects the logistics services offered and the competition pattern. The new vessel may call four (4) new ports in the Aegean (say in the Cycladic complex of destinations) and improve the market shares of the company by 3%. The new values would alter the ranking, from the 4th position to the 1st! Although the addition of the new ports of call does not alter the absolute value (it remains C), so the internal value is not altered, the external one is significantly affected. The new values in the competition set of criteria shifts the absolute values from the original set (D, C, C, B, D, C, D) to the new one (C, B, B, A, D, C, D) that alters the final value of the external criteria from to 0,303 to 0,446. Thus the final index value equals 0,446*0,25+0,312*0,75=0,345 (0,273 originally).

Figure 14: Relative Improvement due to the new ship

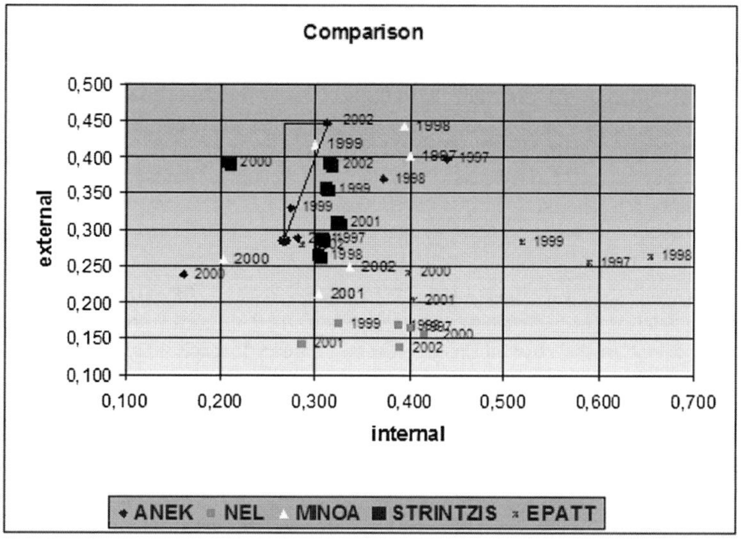

Apparently, this is an exploratory scenario that deals with the effect of a trend or a decision in the model. From an anticipatory and normative point of view, the process would demand a different course of action. Say that the management of NEL (fiscal year 2002, total index 0,328) desires to consider what should change so NEL could get the pole position instead the second place. It is reminded that STRINTZIS got the best ranking (total index 0,335). Following the sensitivity methodology selected, it is possible to evaluate the most critical criterion and the most critical element. Obviously, it is erroneous to alter the criteria weights at this stage! It is necessary to find the most sensitive element that would alter the ranking between NEL and STRINTZIS. After performing the necessary calculations according to Triantaphyllou and Sanchez (1997, p. 178) it is possible to identify the appropriate element. The original value per fundamental accounting criteria-set is 0,371 (NEL 2002) and it has to change to 0,396 in order to reverse the ranking. With the new ranking NEL surpasses STRINTZIS and all the other rankings remain stable.

It is clear that something has to change in the fundamental data-set so the management can proceed into further elaboration. As the fundamental data are filtered through ratios that feed absolute values, it is unlikely to match the 'desired' value (0,396) with a specific new input value. Normally a value close to the desired one is assigned. For example, the management of NEL would like to approximate this value by 'perturbing' the sales-revenues or the net income after taxes. The result would be the following: the sales revenue should increase by 11% (from €13,7m to €15,2m) in order to achieve the value of 0,399 (it is numerically impossible to reach the exact value of 0,396) or the net income after tax should increase by 911% (from €351k to €3,552k). Obviously, the increase of the net income after taxes by that percentage is out of question, while the increase of the sales revenue is not impossible. In a more elaborated accounting exercise a set of values could slightly alter in order to achieve the desired total index value.

Apart from the above application, one can use parts of the hierarchy as qualitative characteristics of the companies under evaluation are adequately quantified, or even reduce the hierarchy to basic features according to simplified analysis of needs, as in the following cases:

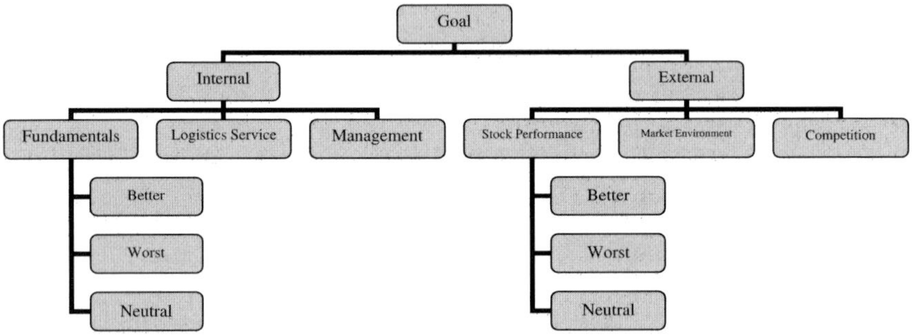

Figure 15: Determining the future position of a company – Forward Planning

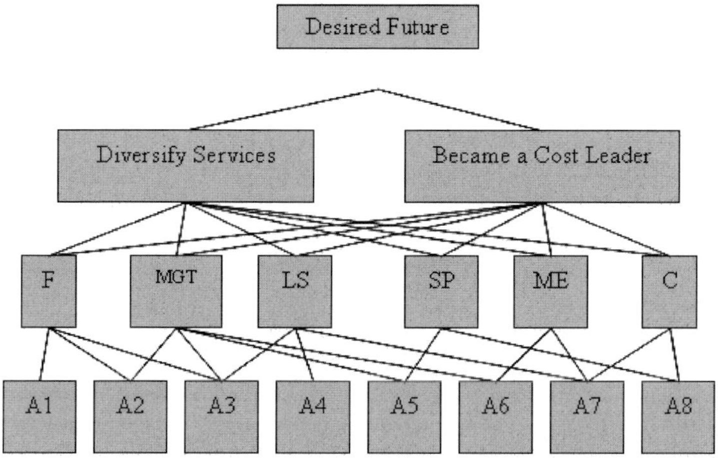

Figure 16: Determining Policies and Actions - Backward Planning

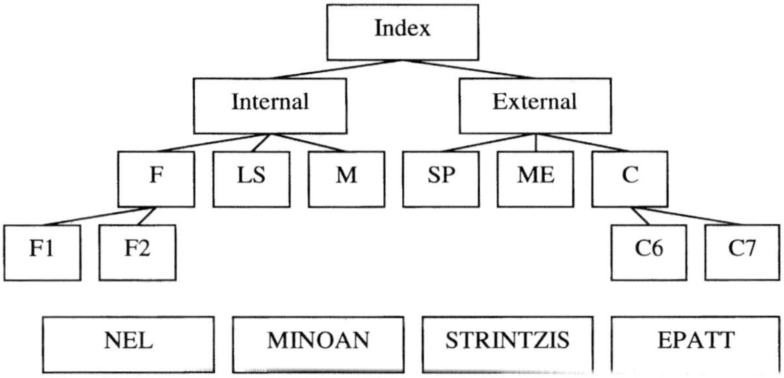

Figure 17: Choosing among merger candidates – sample hierarchy

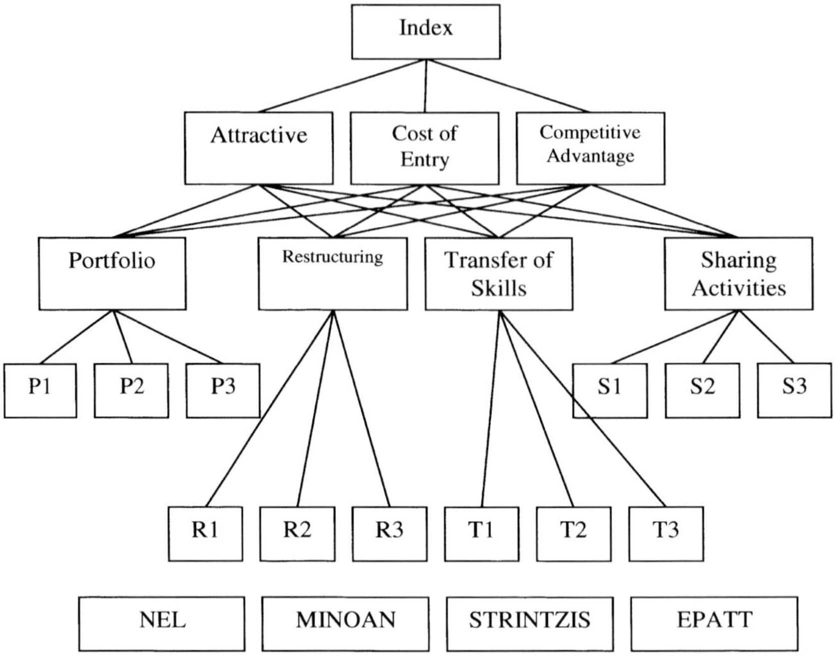

Figure 18: Selecting a target company based on Porter's theory

However, many issues have either to be reconsidered or resolved. The model is data-driven as well as goal-oriented. This is a conflicting compromise, as data lack for the accomplishment of mere goals. For example, the issue of Level of Service (LOS) provided by a carrier is here not adequately considered. No data over customer's satisfaction exist nor is it easy to extract an aggregate picture. In the year-round service different categories of customers are served, with different needs and attributes. Another issue that remains to be adequately handled in the model is the consideration of sub-systems. The performance of a carrier in a sub-system might be of interest; in the current format it is not possible to draw conclusions. Apart from data-related problems, the modeler shall also consider basic MCDM issues. Is AHP the appropriate methodology, or a combination would provide better results? In all cases the method shall not demand prior knowledge of the utilization function.

The validation of the results for the period 1997-2002 permits the use of the quantified attributes for further research. So, one can further add the attributes for the 2003 and 2004 fiscal years, as well as to draft scenarios. In the current turbulent business environment of the GCS a solid model is valuable for the State regulating the market still through licensing and some subsidies as well as for the carriers, who will experience fierce competition and business opportunities.

REFERENCES

Pardalos, P. (1995) eds

Chou, T.-Y., Liang, G.-S. (2001). Application of a fuzzy multi-criteria decision-making model for shipping company performance evaluation. Maritime Policy & Management, 28, (4), 375-392

Hahn, E. (2003) *Decision Making with Uncertain Judgments: A Stochastic Formulation of the Analytic Hierarchy Process.* Decision Sciences, Vol. 34, No 3

Keeney, R. L., Raiffa, H. (1993). *Decisions with Multiple Objectives. Preferences and Value Tradeoffs.* Cambridge University Press (reprinted 1999)

Liberatore, M. J., (1987). An Extension of the Analytical Hierarchy Process for Industrial R&D Project Selection and Resource Allocation. *IEEE Transactions on Engineering Management*, EM-34 (1), pp. 12-18.

Liberatore, M. J., Nydick, R. L. and Sanchez, P.M., (1992). The Evaluation of Research Papers (Or How to Get an Academic Committee Agree on Something). *Interfaces*, 22(2), pp.92-100.

Psaraftis, H.N., Magirou, V.F., Nassos, G.C., Nellas, G.J., Panagakos, G., Papanikolaou, A.D. (1994). Modal Split in Greek Shortsea Passenger Car Transport. European Short Sea Shipping, 2nd Roundtable, Delft University Press and Lloyd's of London Press

Saaty, T.L, Vargas, L.G. (1994). *Decision Making in Economic, Political Social and Technological Environments.* The Analytic Hierarchy Process Series, Vol. VII. RWS Publications, Pittsburgh, PA

Saaty, T. L. (2001) *Decision Making with Dependence and Feedback: The Analytic Network Process.* RWS Publications, Pittsburgh, PA

Schinas, O. (2005). Application of Multi Criteria Decision Making Techniques in Finance: the Case of the Greek Coastal Shipping Companies. Doctoral Thesis. National Technical University of Athens

Triantaphyllou, E. (1999). Reduction of Pairwise Comparisons in Decision Making Via a Duality Approach. Journal of Multi-Criteria Decision Analysis, 8, pp299-310

Triantaphyllou, E., Lin, CT (1996). Development and Evaluation of Five Fuzzy Multi-attribute Decision-Making Methods. International Journal of Approximate Reasoning, 14, pp281-310

Triantaphyllou, E., Sánchez, A., (1997). *A Sensitivity Analysis Approach for Some Deterministic Multi-Criteria Decision Making Methods* Decision Sciences, Vol. 28, No 1, pp. 151-194

Triantaphyllou, E., Shu, B., Sanchez, S.N., Ray, T. (1998). *Multi-Criteria Decision Making: An Operations Research Approach.* Encyclopedia of Electrical and Electronics Engineering (J.G. Webster, Ed.), John Wiley & Sons, NY, Vol.15, pp. 175-186

Transport Science and Technology
K.G. Goulias, editor

CHAPTER 33

AN EVALUATION MODEL FOR FORECASTING METHODOLOGIES USED BY PORTS

Orestis Schinas, Dr-Eng

Harilaos. N. Psaraftis, Professor, National Technical University of Athens

INTRODUCTION

The deregulation of the port industry as well as the increased significance of ports in the transportation chains has deliberated hidden potential. Policy- and business decision-makers have shifted their interest to port operations and mainly to the collaboration of transportation means at the nodal point of a port, seeking primarily for a more active role of the ports along the logistics chains. This paper is based on experiences gained from various research and development projects, mainly from the Tools and Routines to Assist Ports and Improve Shipping (TRAPIST) funded by the European Commission. Therefore much of the information provided is referred to European cases and pattern. However it is expected that similar problems have also been identified and have to be addressed in other regions as well.

At a European level, the role of ports is increasing and there is an issue of promoting the significance of small and medium ports (SMP) along the logistics chains. Most commonly these ports do not possess enough resources and expertise for the drafting of marketing plans or for the launching of aggressive strategies. Furthermore there is a gap in the management culture, as most SMP were and still are operating in sheltered business environment due to the regulation or due to the limited options users enjoyed in the past. However this is not the case in the coming years, as users enjoy more transport options because of the launching or the completion of various infrastructure projects. Consequently regardless of the institutional pattern SMP will face competition in their region.

A finding from the research was that most SMP do not base their forecasts on a specific methodology but a combination of historical data and conferring with the major users. Given also the results of the European Seaports Association (ESPO) survey, many ports base their forecasts on national macroeconomic data (ESPO, 2001), but this is not a very sound method as many trades are regional and not national or pass transit through the port. In many cases there are no adequate regional data or the correlation of commonly used macroeconomic parameters with the transit movements is weak (Psaraftis et al, 2003a).

The purpose of this paper is to present a methodology assisting SMPs to select a forecasting methodology. As there are practically three main categories of forecasting methodologies, a port has to focus on a specific one and then to customize it to the specific needs (or available data). The outcome of the methodology has to be an answer to the problem what methodology to choose, and therefore to prepare the management for the trade-off between quality and resource-utilization. The projection methodologies are strongly depended on the available data, their integrity and quality. Data collection is a rather complicated task as it is not only time-consuming and effort-demanding but it stipulates planning months and years ago, in order to achieve a specific goal. In many cases it is also necessary to combine data from various sources. This issue melts down to the trade-off between resources and accuracy. Last but not least the whole reengineering of the forecasting process in a SMP demands also a change in the management culture as well as the commitment of the senior management towards the new direction.

In the literature there is no previous work found to the best of the authors' knowledge. As long as the procedure is concerned, the Analytic Hierarchy Process (AHP) has not been used yet for the valuation of forecasting methodologies, also to the best of the authors' knowledge. There is no relevant example in the Dictionary of Hierarchies (Saaty and Forman, 2003), as well as in the related literature.

The issue of traffic or volume flow projection has been an old one and has been thoroughly treated in the literature. There are various demand- and supply-side approaches, replying to several related problems at a different level of accuracy. Nevertheless, the most of the methodologies are used by research institutions or in cases demanding sophistication and advanced analytical procedures. Most of them are not used by ports on frequent basis. In a recent survey released by the ESPO it has been made known to the wider academic and business community that ports collect data and proceed to forecasting for reasons of internal organization and planning as well as to justify infrastructure requests or financial support from the local or national authorities. It seems that forecasting is mainly an internal operation of the port, which is practically seldom outsourced to specialized agencies or consultants who work along with the port. Another result is that projections are rather short-term: few ports plan ahead as the business horizon may be completely different within a decade. Ports planning for the next 10 to 25 years are concerned on investing in costly infrastructure therefore they have to justify the involved risk. But, commonly, ports seem to concentrate their efforts in short periods, no more than five years. This complies also with the necessity to report and plan with the national and regional authorities. Maybe this also reveals the inability of ports to estimate future traffic relatively accurately. Research and experiences support that ports use mainly past statistics and economic indicators. The use of historical data cannot accurately assist in predicting figures for a long period of time, although it can predict quite satisfactorily the figures for the next fiscal or operational period. It is however interesting to note that ports tend to get data from their clients (shipping operators), their sub-contractors (terminal operators) and the competition. This is a good amount of information, which is however not easily quantifiable and harmonized (ESPO, 2001).

Moreover, it became evident that the port does not employ sophisticated methods for the forecasting of its future business. Most of such ports collect data on aggregated form per commodity category or even worst per tariff structure class. These data reflect flows of cargo handled in the specific port. The management of the port usually gets the picture on the logistics chains servicing these cargoes mainly from information provided by clients or agents, but this picture often is not reflected in any statistics (Psaraftis et al, 2003b)

It has been admitted that most of the ports collect data on an aggregate form, which are usually elaborated in an unsophisticated way; such as extrapolation of differences between years, basic regressions and correlations with macroeconomic data, such as the GNP or the production index. In most cases the results of these techniques are not tested from a statistical point of view and therefore their forecasting capability is rather limited. Also, large ports with advanced collection capabilities extend their analyses to two-digit NSTR classifications and check their forecasts with agents or clients of the specific niche of the market. Lately some large ports have acquired advanced software and systems for the forecasting of flows. Most of them are based on time-series analysis (Psaraftis et al, 2003a).

In both cases, the result is practically the same: ports usually do not base their forecasts on sophisticated methodologies. Additionally, they do not base their projections on logistics-related data but on historically handled volumes. The difference between large and small ports is that large ports have a data-warehouse with a capability to provide information at various levels of detail, and the smaller ports do not. In general, the forecasting methodologies have to offer a sound decision-support mechanism to the port management. In the contemporary business pattern this means that logistics aspects, such as the time and the cost to and from the target market of the port have to be taken into account, so the port management can set targets for internal operations and compete with other ports, routes and modal sequences.

The valuation methodology followed is based on the Analytic Hierarchy Process (AHP). This method is widely applied and has been extensively discussed in academic journals. AHP is a powerful tool when the sample of alternatives (in this case the number of methodologies under evaluation) as well as the number of criteria is relatively small. Furthermore the use of relative comparisons suggests and highlights the subjective point of view a decision-maker may have, given the operational framework and personal experiences. Last but not least is that the method can easily come to a conclusion if only consistency ratios are kept under given limits, avoiding calibration or other relevant procedures. In any case the hierarchical structure of the problem is essential for any methodology used and reveals the level of understanding and sophistication of the decision-maker (Saaty, 1994, pp.95-98). Three groups of methodologies will be evaluated in this paper:

1. the time-series analyses,
2. the transport-supply methods based mainly on the idea of generalized cost, and
3. the simulation techniques.

Of course the model can incorporate other methodologies but necessary understanding of the AHP method is necessary to accommodate more alternatives (forecasting methodologies).

This document is structured as follows: in the next section the hierarchy of evaluation of forecasting methodology is presented. Then this problem is approached with the assistance of AHP and discussed accordingly. As AHP is a rather well known technique, even the basic aspects of the methodology are omitted. The results of the analysis are presented in the next section and their meaning as well as their limitations is taken into further consideration. The paper concludes with a brief concluding summary.

THE FORECASTING METHODOLOGY EVALUATION HIERARCHY

The hierarchy constructed for the specific port forecasting problem is presented in the figure below (Chart 1). It is a hierarchy having as a focus the validation of three specific methodologies, labeled M_1, M_2, and M_3. Therefore the numerical outcome of the mechanism will be a number ranking every alternative, i.e. every methodology. The top index, i.e. the ranking of the methodologies, is typically called Level I. In the lower levels, criteria, sub-criteria and the alternatives are provided. In the Level II, three desirable main attributes of the methods are used as main criteria sets:

1. Usability [U],
2. Validity [V], and
3. Ability to support decisions [A].

These are considered the three basic desirable properties for the validation of the methodologies. At the next level of the hierarchy, Level III, criteria, attributes, aspects and characteristics of these three basic properties are presented. As it will be shown in the coming paragraphs, six criteria fall under the 'usability' set (U_1, U_2, ..., U_6), five under the 'validity' (V_1, V_2, ..., V_5) and four under the 'ability' (A_1, ..., A_4). Then the alternatives, i.e. the methods M_1 to M_3, are evaluated for every single one of them.

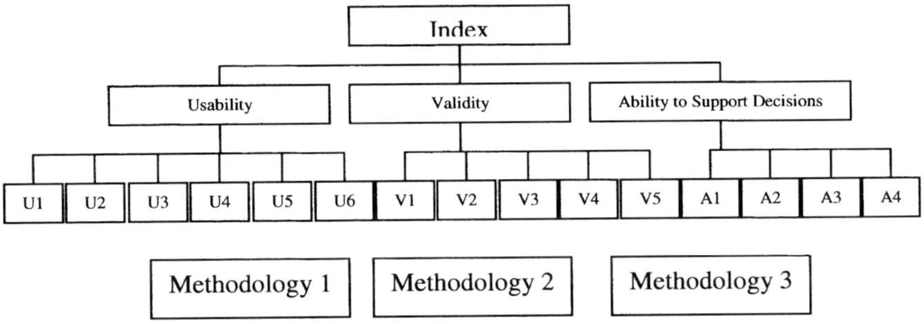

Chart 1: The Hierarchy of the Problem

By *usability* we mean a set of characteristics and capabilities of a method that determine the extent to which this method can be used in practice. These characteristics and capabilities form also the criteria at the next level. These criteria are namely the following:

U1:	easiness of data collection	The degree of difficulty in data collection limits the ability of the management for trade-offs among resources. As cargoes pass through the gate, data are collected by electronic clearance devices for the commodities entering or leaving the port zone. A methodology that can make use of such data maximizes the positive effects for the organization.
U2:	use of modern technology (or automation capability)	It deals with the exploitation of current technological facilities in ports. It is very important that a port can make maximum use of the installed technology in the yard. Usually, the port uses wireless technology, while shippers and carriers use mobile telecommunications, satellite applications and advanced internet-based systems. Evidently these systems may collect every necessary data for the needs of the statistics department. The higher the degree of compatibility of the data structure of the methodology and of the collection system the better rate will the methodology get finally.
U3:	cost (initial or maintenance cost)	As cost element in the hierarchy is considered the initial or the operating cost, that is necessary for the continuation of the application of a methodology.

U4:	time and effort devoted towards a result	The time and effort element completes the cost-related attributes of the hierarchy. By time and effort it is understood an expense of resources for the execution of the methodology from the starting point of collecting the data up to the end. a.
U5:	resources (people and level of experience)	Further to U3 and U4, the 'quality' of the resources, mainly the quality and the expertise level of the people, who were assigned the forecasting task, affects the outcome of any methodology. Generally, complicated and advanced methodologies demand a higher level of expertise.
U6:	necessity to cooperate with other parties in order to get a result	If the required data for a methodology can be found from the port archives, then this methodology has an inherent advantage for the port. Data collected from various sources have to be compiled and properly combined together; this task is not always easy as various sources collect data for different purposes. Therefore the decision-maker has to evaluate the need stemming out of a methodology to use data from various sources.

Validity stands for another set of criteria related to the methodology. The aim of this set of criteria is to explore the soundness of the methodology, highlighting attributes related to the data and the output. These criteria are:

V1:	revealed or stated preference	Revealed preference techniques are based on historical data, which reveal a trend but not necessary the reasons generating the trend. As sophisticated the model becomes, i.e. as many parameters there are involved the researcher may get into the essence of the problem, provided that there are enough data supporting the modeling and the samples are manageable. The stated preference techniques use sampling statistics and may focus to the heart of the problem. A decision-maker has to use this criterion of revealed vs. stated preference in order to determine the methodology that leaves as fewer gray zones as possible in the final result. That depends heavily on the nature of the port and

		the trade its serves..
V2:	data input procedure	This criterion applies obviously to the way data are entered in the system. This criterion is complementary to the criteria U2, U3, U4 and U5. The data input procedure is critical and in many cases, raw data entering automatically the system without any previous filtering may lead to flawed results. On the other hand keyed-in data may incorporate too many human errors, especially when humans do not have a clear understanding of or control over the data (Schinas et al, 2002).
V3:	data manipulation	In many methodologies, it is possible to manipulate data in such a way so to lead to desired or specified outcome. This is not necessary decayed but it may also be obligatory bringing the model back to the target-track. However, there are other methodologies not so sensitive in data manipulation and even if the user intervenes to the data base, then the outcome is not practically altered. Generally the wider and the deeper the data base is, the less sensitive is to data-manipulation. Data manipulation may also permit the extraction of other results, such as what-if scenarios or sensitivity checking.
V4:	Output	The outcome of any methodology shall be reviewed under two sub-criteria: the soundness of the results according to market practices, common sense and the rationality governing normal operations, as well as the sensitivity of the outcome over small changes of data. The sensitivity analysis provides also proof for the soundness or the robustness of the outcome.
V5:	self-control loops	The ability of the method to check the integrity of the procedure by filtering the data at various stages is a very important feature, as large amounts of data are commonly processed.

Finally the last set of attributes is the ability to support decisions; this set deals with the practical use of the outcome.

A1:	Product vs Aggregate results	In the modern logistic chains it is necessary for a port not only to focus on aggregated flows: unitized – bulk, import-export, but also on product specific ones. By the term 'product specific' it is not necessarily understood a specific product but a product range, a notion similar to NSTR or SITC 3-digit classification or in some cases of 2- or 4-digits. Evidently as ports become integral parts of logistics chains, which in most cases are product-based for consumption, packaging, marketing and value of time grounds, ports shall focus more on the characteristics of the transported cargoes.
A2:	reliability	The issue of reliability is as critical as the *data-manipulation* and the *data-input-procedure*. Obviously the reliability of the methodology depends heavily on the reliability of the data especially for deterministic systems, such as the time-series analysis and the transport-supply methodologies. For the stochastic systems, the issue of the reliability is critical as well but the methods incorporate also a degree of vagueness for the used data. In any case the reliability of the data is critical, as well as the reliability of the method as such. A problem of stability of the solution is usually dealt with sensitivity analysis but not always, and is in most cases an issue for an academic institution or research center.
A3:	Endurance	The endurance of the results against time is an essential criterion for the selection of costly forecasting method. The longer the results stand to time the better for the methodology
A4:	strategic vs tactical decisions	A strategic decision is taken by the upper management of the port and has also longer effect in the organization. A tactical decision is taken by the middle management, most commonly, and aims to fulfill the needs or the requirements of a strategic decision. In the case of port-potential quantification, a strategic decision is related to an issue of market penetration or expansion, while a tactical one is related to issues of generalized cost

		or other relevant data, by changing them cargo is attracted to pass through the port. Both decision levels are important but in most cases it boils down to an issue of available resources, management sophistication and structure of the market.

This hierarchy will be used for the evaluation of the forecasting methodologies. For the purposes of this paper and the accuracy of the results, it cannot be altered so to accommodate more or less criteria. A hierarchy reflects the understanding of the problem by the decision-maker and this level of accuracy is considered as adequate for the needs of SMP. Obviously another decision maker with a different perspective could use a different hierarchy.

APPLICATION OF THE MODEL

The methodologies under evaluation are presented thoroughly in the literature and are practically known to everybody with basic knowledge of transportation engineering and planning. The AHP methodology is very well known in the academic community as well as to practitioners. Given the hierarchy one can proceed to relative comparisons of the alternatives per attribute. The relative comparisons are based on the fundamental scale:

Verbal Value	**Numerical Values**
Equally important, likely or preferred	1
Moderately more important, likely or preferred	3
Strongly more important, likely or preferred	5
Very strongly more important, likely or preferred	7
Extremely more important, likely or preferred	9
Intermediate values to reflect compromise	2,4,6,8

Table 1: The fundamental scale (Saaty, 1994)

An illustrative example is the following: say that three alternatives (say the methodologies) have to be evaluated according to an attribute (say sub-criterion U2), then the evaluation in a tabular format is the following:

	A	B	C
A	1	5	7
B	1/5	1	3
C	1/7	1/3	1

The elements of the table are commonly symbolized as a_{ij}. The meaning of unit value along the diagonal of the table reflects the idea that the result is the same when comparing alternative i with itself. The upper triangle is the one a decision-maker fills with data; the elements of the lower triangle have to comply with the idea of reciprocal values, i.e. $a_{ji}=1/a_{ij}$ The meaning of this condition stems out of rationality; when comparing A to B by a_{ij} then when comparing B to A the element a_{ji} shall be equal to $1/a_{ij}$

When the tables with the judgments are set the priorities can be extracted by various methods. The priority vector expresses the relative importances implied by the previous comparisons. Saaty asserts that one has to estimate the right principal eigenvector of the matrix. As there are various eigenvalue approaches, for the scope and the needs of the current study the revised method is selected: the geometric mean of a row elements is calculated and then the numbers are normalized by dividing them with their sum. The consistency of a judgment table has to comply with the simple rule $a_{ij} = a_{ik}\, a_{kj}$. As in very few cases all comparisons are consistent, Saaty has developed a coefficient ratio (CR) that has to be less than 10% for a table.

The same approach applies also for tables that consist of criteria as in the case of level I and level II. After the alternatives are compared with each other in terms of each one of the decision criteria and the individual priority vectors are derived, then the next step is to involve the criteria in the calculations. The priority vectors become the columns of a decision matrix. The multiplications of the weights of criteria and of the priority vectors yield the final priorities:

$$A^{i}_{AHP} = \sum_{j=1}^{N} a_{ij} w_j$$

for i = 1, 2, …, M

If a problem has M alternatives and N criteria then the decision maker has to construct N judgment matrices (one for each criterion) of M×M order and one judgment matrix of N×N order (for the N criteria). Finally all A^i are calculated as above.

The judgment matrices are not presented due to space limitations. The weighting of the criteria at level III is provided in the following paragraphs, while the weighting of the level II is used for scenario analysis. The final results will be presented in the following section. The relative comparisons of the sub-criteria at level III consist an interesting task that reveals the subjective-perspective of the decision maker. As long as the ‘usability’ set of criteria the judgment matrix is the following:

	U1	U2	U3	U4	U5	U6	priorities
U1	1	5	1	1	3	5	29%
U2	1/5	1	1/3	1/3	½	1	7%
U3	1	3	1	1	3	3	24%
U4	1	3	1	1	3	3	24%
U5	1/3	2	1/3	1/3	1	1	9%
U6	1/5	1	1/3	1/3	1	1	7%
			CR	1.24%			

The elements of cost (U3 and U4) as well as the attribute of easy data-collection are considered as very important, therefore the priorities vector yield almost the same importance for these three sub-criteria. This is an assumption and a subjective point of view simultaneously yet it is not considered that it contradicts common practices. For the validity-related set of criteria the following judgment table is produced:

	V1	V2	V3	V4	V5	priorities
V1	1	1/3	1/3	1/5	1	7%
V2	3	1	3	1	3	30%
V3	3	1/3	1	1/5	3	14%
V4	5	1	5	1	5	41%
V5	1	1/3	1/3	1/5	1	7%
			CR	5.09%		

Similarly to the previous judgment matrix, the result that V2 (data-input) and V4 (output) sub-criteria are considered as most important, does not contradict common practices. However it is important to note that V3 (data-manipulation) is also an important feature, contrary to V1 and V5 that are not so important from a practical point of view. Finally, the judgment matrix for the 'ability to support decisions' sub-criteria is presented:

	A1	A2	A3	A4	priorities
A1	1	2	2	1/3	23%
A2	1/2	1	2	1/3	16%
A3	1/2	1/2	1	1/3	12%
A4	3	3	3	1	49%
			CR	4.49%	

The sub-criterion A4 (strategic vs tactical) is considered as the most important, while all others are also important but not dominant.

By applying scenarios in the judgment table at level II it is possible to highlight the importance of a specific criterion over the others and make some reasonable remarks over the meaning of this analysis. The last judgment matrix is the following:

	U	**V**	**A**
U	1	w_{12}	W_{13}
V	$1/w_{12}$	1	W_{23}
A	$1/w_{13}$	$1/w_{23}$	1

By setting w_{ij} as in the following table and by executing the necessary numerical calculations the following result table per scenario is extracted:

	enhanced usability	moderate usability	enhanced ability	moderate ability	enhanced validity	moderate validity	neutral
w_{12}	7	3	1	1	1/7	1/3	1
w_{13}	7	3	1/7	1/3	1	1	1
w_{23}	1	1	1/7	1/3	7	1/3	1
Result Cells							
CR	2.33%	2.43%	2.67%	2.64%	2.78%	7.03%	2.59%
A (time series)	**56%**	**51%**	30%	35%	**45%**	39%	**43%**
B (supply-side)	30%	34%	**52%**	**48%**	39%	**45%**	41%
C (simulations)	14%	15%	18%	17%	16%	17%	16%

Table 2: Result Table

First of all it has to be noted that the judgment matrices are consistent because the total consistency ratio of the hierarchy is less than 10%. In case that a scenario had a CR>10% then it would be necessary either to revaluate the sets of criteria used or rethink the attributed judgments per criterion and alternative.

From the above table is easy to understand that ports considering 'usability' as the most important criterion shall use time-series techniques for the necessary forecasting. Evidently this is consistent also with the experience from ports of various sizes and not only SMP. Furthermore ports with a neutral attitude towards these criteria would also select time-series as forecasting method. On the contrary ports highlighting the importance to base decisions on the results of a methodology would prefer the supply-side forecasting techniques. As expected also from the judgment tables, supply-side techniques can offer a better understanding over the logistic chains and insights for product and aggregate flows. For ports seeking validity in the final result, the above scenario analysis suggests that the selected methodology would be either time-series or supply-side techniques. The compliance of the results with current practices at ports is also a measurement for the effectiveness of pairwise comparisons and of the deriving priorities (Saaty, 2001, p.p. 84-85).

Simulation techniques seem to attract limited interest for port applications. As in no priority vector gets no more than 20%, they would easily be eliminated from further analysis. A more detailed analysis would include only the other two alternatives. In case that a port had to select a methodology out of the two, group decision making techniques could enhance the validity of the final result and include all parties in the discussion (upper and middle management, other parties involved, etc). However a group decision making problem does not fit in the current research context and is well described in the literature (e.g. Saaty, 2001). In such a case more or other criteria can also be incorporated in the hierarchy depending on the formulation of the problem.

The limitations of the model are not severe at the strategic level and basically are the ones of a multi-criteria problem and of the used evaluation methodology: AHP. First of all a hierarchy imposes a stringent limitation of the given number of criteria and their crisp attributes. Despite the fact that a methodology reflects the understanding of a decision-maker there is an error in the whole approach. By using Saaty's ideas and observing the result table, the degree of confidence is high as in most cases the error accounts less than 5%. Furthermore the CRs of the judgment tables are also considerably lower than 10%. The criteria-sets have not been applied only in that case and have produced results of adequate significance in other cases as well (Schinas et al, 2002).

Examining the hierarchy and the results from a numerical point of view through detailed sensitivity analysis as described in the literature one can easily understand the limitations of a model (even rank reversals) as well as improve his decision-making capabilities. The sensitivity analysis for such a problem has to determine the most critical criterion and the most critical performance measure a_{ij}. By using approaches presented in the MCDM literature such as the most critical criterion, the following results are extracted (Triantaphyllou et al, 1997, p.p.8-10):

- in the neutral condition, the 'validity' set of criteria is a 'robust criterion', i.e. the a_{i2} elements affect the ranking of the alternatives the least.
- by altering the weight of a_{12} by 0.0742 (in absolute terms) or by 26% (in relative terms) then the current ranking of the alternatives is violated.
- the most sensitive criterion in terms of weighting is the one related to *ability*
- regarding the 'validity' scenarios, one can find that in the case of '*moderate validity scenario*' the a_{12} element is the most critical in absolute terms (0.1599) and the a_{13} in relative terms (29%). The '*ability-to-support-decisions*' is the most sensitive criterion in this scenario. In the '*enhanced validity scenario*', the a_{13} element is critical element in absolute and relative terms (0.1437 and 25% respectively) and the ability-related criterion the most sensitive one.

In practice is not easy to determine the most sensitive criterion uniquely, as there are at least four theoretical approaches (see more in Triantaphyllou et al, 1997).

CONCLUDING REMARKS

Ports currently face the need to expand their forecasting capabilities as well as to position themselves actively or more efficiently along the logistics chains. Historically ports have not developed or used advanced forecasting techniques as there was no such need due to operational, institutional or other limitations. Currently as markets and regions integrate, ports have to quantify their potential market. As this quantification task shifts from simple import-export (or production-consumption) statistics and more transportation-related data enter into the calculations, ports have to adopt new techniques. Nevertheless, as in any other business unit, resources are limited and expertise is lacking, so port management has to face and decide on trade-offs between available resources, needs and forecasting characteristics, such as accuracy, endurance, and many other described above.

The scope of this paper was to offer an evaluation tool for the port management adequate to support a decision on the dilemma, which of the three major groups of forecasting methodologies to select: time-series analysis, supply-side calculations based on the generalized cost or advanced simulation techniques. The trigger for such a tool stems out of the framework of European Commission (EC) funded research project, called TRAPIST that aims to provide ports with 'soft' tools to improve their operational capabilities.

As the evaluation of alternatives (forecasting methodologies) is a multi-criteria decision making problem, the well known and user-friendly Analytic Hierarchy Process (AHP) methodology has been selected and applied. The hierarchy of the problem used, i.e. the understanding of the insights, is a refined version of similar hierarchies used in other research and development projects. Nevertheless as there is no relevant work in the literature, the application of the hierarchy is considered as innovative.

The scenario analysis is based on alterations of the weighting elements between the major criteria-sets, namely those of usability, validity and ability to support decisions. For six biased cases (enhanced and moderate levels of bias) and one neutral case (all weights equal to the unity so no bias is expressed) the analysis yields the most preferable forecasting methodology. It was expected that in the neutral, enhanced and moderate usability the analysis yielded the time-series analysis as the most favorable one; this is consistent with the current practice and the needs of ports that focus on short-term planning horizons, regardless of the size. On the other hand ports that envisage a better positioning along the logistics chains –enhanced and moderate ability to support decisions cases- would select the supply-side calculations based on the generalized costs of various options, as more insights of the trade and the service are revealed. Regarding the moderate and enhanced validity scenarios the model does not yield a straightforward result as in the previous cases. The selection will be decided among time-series and supply-side modeling. This is merely expected from relevant experience. Simulation techniques do not attract practitioners' interest as expected.

The importance of the results for ports is significant at strategic level; since there is a structured hierarchy that ensures an adequate degree of soundness, as it reflects current practices, and the management of the port can be guided towards a specific methodology according to local needs and limitations. The port can either reconstruct the model in a spreadsheet or even pick up the suggested methodology that fits better to the seeking attribute. The application of the model is not limited to a geographical context (e.g. Europe) or to the size of the port. From the sensitivity analysis it became also clear that the most sensitive criterion is the one related to the ability to support decision, which is considered also as essential for practical applications.

This evaluation approach can be expanded to tactical decision-making by considering the opinion of many port-officers or other interested parties (say the State and the users) by applying group decision making techniques. AHP can accommodate group decision making in a very robust way, which is a reason for its wide application in marketing and relevant problems. Furthermore the hierarchy can be expanded accordingly in order to focus on more tactical issues, yet a basic understanding of AHP and multi-criteria decision making is imperative.

REFERENCES

ESPO (2001). '*Survey of port trade tonnage forecasts*'. European Seaports Organization, 2001.

Psaraftis H.N, Schinas, O. (2003a) *TRAPIST report 2.3* – restricted by the European Commission

Psaraftis H.N, Schinas, O. (2003b) *TRAPIST report 2.6* – restricted by the European Commission

Saaty, T.L, (2001) *Decision Making for Leaders*, RWS Publications

Saaty, T.L., (1994) *Fundamentals of Decision Making and Priority Theory with the Analytic Hierarchy Process*. First Edition, RWS Publications

Saaty, T.L., Forman, E.H. (2003). *The Hierarchon: A Dictionay of Hierarchies*. Volume V of the AHP Series, 3rd Edition, RWS Publications

Schinas, O., Lyridis, D.V., Psaraftis, H.N. (2002). *"Introducing E-brokerage in European Transport Services; the Case of the PROSIT Project"*, ICECR-5, ORMS, Montreal, Canada, October 28th, 2002

Triantaphyllou, E., Sánchez, A. (1997). *A Sensitivity Analysis Approach for Some Deterministic Multi-Criteria Decision Making Methods*. Decision Sciences, Vol. 28, No 1, pp. 151-194

Transport Science and Technology
K.G. Goulias, editor

CHAPTER 34

ESTABLISHMENT OF AN INNOVATIVE TANKER FREIGHT RATE INDEX

D.V. Lyridis, P.G. Zacharioudakis, and D. Chatzovoulos
School of Naval Architecture & Marine Engineering, National Technical University of Athens, Greece

ABSTRACT: The description of the tanker market is probably one of the most useful tools in the hands of all interested parties, such as ship owners, charterers, shipbrokers, etc. The most useful way to describe this market (and other markets as well) is to introduce a statistical method by which prices can be compared to a former level. This is achieved through a proper index. Currently, the Baltic Exchange describes the tanker market with the publication of the BITR (Baltic International Tanker Routes), which consists of two daily published indices: the BDTI (Baltic Dirty Tanker Index), that accounts for the tanker routes of crude oil and the BCTI (Baltic Clean Tanker Routes), that deals with the tanker routes of oil derivatives (gasoline, benzine, etc.). Both indices consider a limited number of tanker routes (ten routes for the publication of BDTI and four for the BCTI) and are based on the reports of "panelists". BDTI and BCTI are rather "static". It is very difficult to substitute, add, or delete a tanker route, because new routes must meet the demands and criteria that the Baltic Exchange sets. Moreover the weighting of a tanker route, for the publication of the indices, is arbitrarily decided by the Baltic Exchange. The aforementioned disadvantages can be surpassed by establishing a new index (referring only to the Dirty Tanker Market) and based on the same rules and regulations as the Baltic indices (number of properly weighted tanker routes). The new index is defined as the sum of the products of the percentages of ton miles in each route by the Worldscale Index for that particular route. The "dirty spot rates" are taken from the monthly magazine "SSE" of the Drewry Shipping Consultants Ltd. The weighting of each route is based on the percentage of the ton-miles. Ton-miles are the product of the distance between two points of a tanker route (in miles) by the total amount of the crude oil carried (in tones). The percentage of ton-miles is derived from the division of the ton-miles of each route by the sum of the ton-miles for all routes. The new index has two major advantages: (a) The formation of the new index consists of a huge number of tanker routes (at least 37 tanker routes instead of only 10 for the BDTI), which gives a more representative sample of the world's dirty tanker market; (b) The new index is a "dynamic" index, in the sense that it doesn't depend on the changes that are made in the formation of the tanker routes. Any substitutions, additions, or deletions can take place, without having to recalculate the entire history of the index. The new index is published every month with the contribution of those tanker routes that really contribute to that particular moment. Hence, the introduction of the new index describes the dirty tanker market in a better way, compared to the Baltic Exchange's BDTI, since it embraces all those factors that have an influence to the market. Furthermore it uses a much bigger and, consequently, more representative sample of the market, that makes the new index more reliable.

INTRODUCTION

Up to now, the tanker market is being described by two daily-published indices: the BDTI (Baltic Dirty Tanker Index), which describes the tanker routes of crude oil and the BCTI (Baltic Clean Tanker Index) which deals with all the tanker routes that carry derivatives of oil (gasoline, benzine, etc.). The Baltic Exchange is responsible for the publication of both indices.

The definition of these indices is the same. They are defined as the sum of the multiplications of the average rate (AV_i) of each route with the weighting factor (WF_i) of that particular route, as shown beneath:

$$BDTI = \sum_{i}^{N} (AV_i \cdot WF_i) \quad \text{and} \quad BCTI = \sum_{i}^{N} (AV_i \cdot WF_i)$$

The difference between the two indices is the N number of the routes. BDTI consists of only ten (10) tanker routes of crude oil, while BCTI consists of just four (4) tanker routes of derivatives of oil. The criteria for the selection of these tanker routes by the Baltic Exchange deal with worldwide coverage, a representative sample of the market, transparency, commercial balance, etc.

The average rate (AV_i) of each route is calculated by the Baltic Exchange with the cooperation of the *panellists*. The panellists are shipbrokers of big companies, who estimate the height of the rate (in $/ton) of each route. The Baltic Exchange gathers all those estimations (for each route) and calculates the average rate for each route.

The Baltic Exchange decides about the weighting factor (WF_i) of each route. This decision is more based on the experience of the Exchange and some expert's judgment, rather on a mathematical frame.

Both indices (BDTI and BCTI) are rather "static", which means that their formation of tanker routes can't be easily changed. It is very difficult to substitute, or add, or even delete a tanker route, because the new routes must meet the demands and criteria that the Baltic Exchange sets as shown beneath.

To summarize, each route makes a contribution to the *Index*, which is derived by multiplying the rate on that route by its weighting factor. When the Index is revised, it is necessary to derive new weighting factors only for those routes on which the committee has decided that new weightings are to apply. If a new route replaces an old one, the new weighting factor is derived in such a way that the contribution of the new route to the Index is the same as the contribution to the old one. Because of the constraints imposed on the rate of change of the composition of the Index by the other rules, and because there is no change in the level of the Index on the date of revision, the Index should show no noticeable change of behavior immediately following revision. Moreover the weighting of a tanker route, for the publication of the indices, is arbitrary decided by the Baltic Exchange, as it was shown and above.

The routes that are used by the Baltic Exchange (according to the manual of panellists) are:

BDTI Route 1
280,000 mt, Middle East Gulf to US Gulf. Ras Tanura to LOOP with laydays / canceling 20/30 in advance. Maximum age 20 years.

BDTI Route 2
260,000mt, Middle East Gulf to Singapore. Ras Tanura to Singapore with laydays / canceling 20/30 days in advance. Maximum age 20 years.
BDTI Route 3
250,000 mt, Middle East Gulf to Japan. Ras Tanura to Chiba with laydays / canceling 30/40 days in advance. Maximum age 15 years.
BDTI Route 4
260,000 mt, West Africa to US Gulf. Off Shore Bonny to LOOP with laydays / canceling 15/25 days in advance. Maximum age 20 years.
BDTI Route 5
130,000 mt, West Africa to USAC. Off Shore Bonny to Philadelphia with laydays / canceling 15/25 days in advance. Maximum age 20 years.
BDTI Route 6
135,000 mt, Cross Mediterranean. Sidi Kerir to Lavera with laydays / canceling 10/15 days in advance. Maximum age 20 years.
BDTI Route 7
80,000 mt, North Sea to Continent. Sullom Voe to Wilhelmshaven, with laydays / canceling 7/14 days in advance. Maximum age 20 years.
BDTI Route 8
80,000mt, Kuwait to Singapore (Crude and/or DPP Heat 135F). Mena al Ahmadi/Singapore with laydays / canceling 20/25 days in advance. Maximum age 20 years.
BDTI Route 9
70,000 mt, Caribbean to US Gulf. Puerto La Cruz to Corpus Christi with laydays / canceling 7/14 days in advance. Maximum age 20 years. Assessment basis - Oil Pollution Act premium paid.
BDTI Route 10
50,000 mt, Caribbean to USAC Fuel Oil. Aruba to New York with laydays / canceling 7/14 days in advance. Maximum age 20 years.

BCTI Route 1
75,000 mt, Middle East Gulf to Japan (CPP/UNL Naphtha Condensate). Ras Tanura to Yokohama with laydays / canceling 30/35 days in advance. Maximum age 12 years.
BCTI Route 2
33,000mt, CPP/UNL Continent to USAC. Rotterdam to New York with laydays / canceling 10/14 days in advance. Maximum age 15 years.
BCTI Route 3
30,000 mt, CPP/UNL Caribbean to USAC. Aruba to New York with laydays / canceling 6/10 days in advance. Maximum age 20 years. Assessment basis - Oil Pollution Act premium paid.
BCTI Route 4
30,000 mt, CPP/UNL Singapore to Japan. Singapore to Chiba with laydays / canceling 7/14 days in advance. Maximum 15 years.
(Note: All vessels reported to have major oil company approval.)

ESTABLISHMENT OF A NEW INDEX

The aforementioned disadvantages of the Baltic Indices (limited routes, static indices, arbitrary decided weightings) are surpassed by a new index that is introduced in this paper. This new index (which refers only to the Dirty Tanker Market) is based on the same rules and regulations as the Baltic's indices, that is a number of tanker routes each one having a weighting.

The "Dirty Spot Rates" table of the monthly-published magazine "SSE" of the Drewry Shipping Consultants Ltd derives these tanker routes. This magazine publishes information (such as the WS index for each route, the amount of oil that is being carried, etc.) approximately about forty tanker routes of the dirty tanker market, every month (Table 1). In this huge database (1981 – 2003), appear seventy-eight tanker routes of the dirty tanker market, totally.

Table 1 contains the following information: the monthly Worldscale Index for each route, the number of the fixtures that are carried out every month in each route and furthermore the average amount of the crude oil (dirty) of each route, that is being transferred in each route.

Each route is discerned by another, by the type of the vessel that is being used. This means that a tanker route, in which a ship belongs to a category of 40 – 70 thousand tones and carries crude oil from Northwest Europe (NWE) to the US East States (USES), is different from a tanker route, in which a ship belongs to a category of 100 – 160 thousand tones and carries crude oil from Northwest Europe (NWE) to the US East States (USES).

As an example and with the help of Table 1, in December 1990, for the tanker route from Caribbean (CAR) to the US East States (USES) and for a ship that belongs to a category of 40 – 70 thousand tones, the WS Index was 167, and at that particular month 9 fixtures were carried out in that route with an average capacity of 65 thousand tones for each vessel (40<65<70 thousand tones, as it was expected).

The new index is defined as the sum of the multiplications of the percentages of the ton-miles (see also the next page) of each route with the Worldscale Index of that particular route, as shown beneath:

$$New\ Index = \sum_{i}^{N} (Percentage\ \ of\ \ Ton - Miles)_i \cdot WS_i ,$$

Where i = a, b… N are the tanker routes of the dirty tanker market as described by the monthly-published magazine "SSE" of the Drewry Shipping Consultants Ltd, WS_i is the monthly Worldscale index of each route and the percentage of ton-miles is described below.

As it was mentioned and above, there is a connection between the Baltic Indices (and particular the BDTI) and the new index. This new index is based on the same rules and regulations as the Baltic's indices, that is a number of tanker routes each one having a weighting: WF_i for BDTI and percentage of ton-miles for the new index. Both weightings are multiplied with the rate of the tanker route: WS_i for the new index and the rate in \$/ton for the BDTI (AV_i).

But what is the percentage of ton-miles? Each route contributes to the *new index* with its ton-miles. The ton-miles of each route are calculated every month by the multiplication of the number of the fixtures that are carried out each month in that particular route, with the average amount of crude oil of each route that is being transferred in each route every month (the multiplication of these two, gives the total amount of crude oil that is being transferred in that particular tanker route in that particular month), with the destination that a tanker covers in that particular route (see also Figure 1).

Ton-miles = Destination (in miles) ×

No. of fixtures × *Average cargo (in tons)*

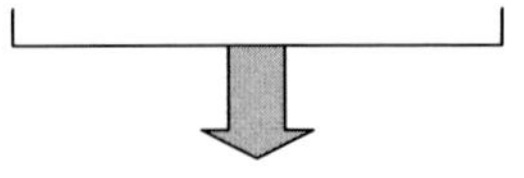

Total cargo that is being transferred (in tons)

Figure 1. Calculation of Ton-Miles

For example and with the help of Table 1, the total ton-miles for the tanker route CAR – USES (40 – 70 thousand tones), for December 1990, are:
9 fixtures × 65000 tones of average crude oil × 1482 nautical miles = 866970000 ton-miles.

The calculation of the destination is based on some characteristic ports of the areas that are described in Table 1. For the above-mentioned example, the destination was calculated between Amuay Bay (CAR) and Morehead City (USES).

The percentage of ton-miles of each route derives from the division of the ton-miles of that particular route (in that particular month), by the total ton-miles of all tanker routes in that particular month.

The WS_i symbol stands for the Worldscale Index of each route (i).

Characteristics of the new index

The new index can be published only in a monthly base, since it uses the "Dirty Spot Rates" table of the monthly-published magazine "SSE" of the Drewry's Shipping Consultants Ltd.

Table 1. Dirty Spot Rates (Drewry's Shipping Consultants Ltd.)

ISSUE JANUARY 1991 – INFORMATION FOR DECEMBER 1990

DIRTY SPOT RATES

(Worldscale prevailing at time of fixing)

LOAD	DISCHARGE	CARGO (M.DWTC)	1989	1990						FIXTURES	
			DEC	JUL	AUG	SEP	OCT	NOV	DEC	NO.	AV. M.DWTC
CAR	USES	30- 40	-	180	220	228	302	267	-	-	-
	CAR	40- 70	-	161	-	-	195	158	280	1	50
	USES	40- 70	206	140	144	148	163	173	167	9	65
	EUR	40- 70	-	85	123	97	115	115	130	2	59
	USES	70-100	190	117	130	115	121	144	150	16	72
	EUR	70-100	112	77	80	89	115	108	116	5	81
ECM	USES	40- 70	175	127	133	154	131	172	209	4	61
	USES	70-100									
	NWE	70-100									
MED	NWE	40- 70	151	115	159	158	185	183	176	3	47
	MED	40- 70	157	147	140	149	188	139	160	1	50
	CAR/USES	40- 70	158	132	136	154	167	162	159	9	56
	NWE	70-100	128	104	98	114	111	126	138	2	80
	MED	70-100	132	108	96	117	117	127	143	21	80
	CAR/USES	70-100	110	116	-	103	106	117	115	3	76
	MED	100-160	112	104	89R	98R	75R	101	98	8	132
	CAR/USES	100-160									
	NWE	100-160	102	93	103R	81R	77R	-	100	2	130
	NWE	200-300									
NWE	NWE	40- 70	123	141	130	139	150	170	146	6	65
	CAR/USES	40- 70	136	137	122	148	163	162	144	3	61
	NWE	70-100	126	112	110	121	126	135	125	11	75
	CAR/USES	70-100	95	120	99	114	124	129	134	4	74
	NWE	100-160	98	120	82	90	104	119	130	4	113
	CAR/USES	100-160	90	102	75	-	88	-	90	1	145
WA	EUR	40- 70	-	-	141	-	-	-	-	-	-
	CAR/USES	40- 70	-	-	-	-	-	-	-	-	-
	EUR	70-100	112	-	93	125	124	150	145	1	70
	CAR/USES	70-100	83	-	105	133	103	-	120	1	80
	EUR	100-160	100	87	84	77	85	99	98	29	124
	CAR/USES	100-160	94	87	79	73	82	97	97	32	127
	FE	200-300									
INDO	FE	40- 70	180	183	195	162	176	133	170	1	58
	USWS	40- 70									
	FE	70-100									
	USWS	70-100	-	-	-	-	-	-	-	-	-
	USES	70-100									
	USWS	100-160									
AG/RS	EUROMED	40- 70									
	FE	70-100									
	CAR/USES	70-100									
	SEA	70-100									
	FE	100-160									
	EUR	100-160									
	CAR/USES	100-160									
	FE	160-200									
	FE	200-300									
	EUR	200-300									
	CAR/USES	200-300									
	EUR	300+									
AG	EUR	40- 70	-	-	-	150	-	-	-	-	-
	EAST	70-100	172	112	129	121	123	112	144	10	83
	CAR/USES	70-100	-	110	110	-	91	-	-	-	-
	EAST	100-160	108	-	99	110	144	86	207	1	100
	EUR	100-160	101	88	100	95	93	94	110	7	127
	CAR/USES	100-160	-	-	-	-	-	-	-	-	-
	EAST	200-300	67	63	56	70	57	68	96	20	237
	WEST	200-300	58	54	51	59	51	58	76	18	256
	EUR	200-300	63	55	55	59	53	67	78	2	263
	JAPAN	200-300									
	S.KOREA	200-300									
	S.AFRICA	200-300									
	CAR/USES	200-300	60	58	57	57	52	65	80	1	260
	RS	200-300									
	WEST	300+									
	USES	300+									
	RS	300+									
	EUR	300+	60	50	-	59	-	-	-	-	-
	CAR/USES	300+	-	53	-	-	-	-	80	2	328
RS	EUR	40- 70	-	195	-	-	-	-	260	1	40
	EUR	70-100	-	104	113	122	116	125	146	2	84

	EUR	100-160	109	103	96	89	79	110	116	6	123
	MED	70-100									
	WEST	100-160									
	MED	100-160									
	EAST	70-100	150	110	-	118	-	122	-	-	-
	EAST	200-300									
KI	EAST	200-300									

However the new index has some major advantages, which the BDTI (of the Baltic Exchange) hasn't.

The new index is a "dynamic" index, in contradiction with the BDTI, which is rather "static" as we described above. By using the term "dynamic", we mean that the new index doesn't depend on the changes that are made in the formation of the tanker routes that are published every month by the Drewry Shipping Consultants Ltd. Any substitutes, or additions, or even deletions can take place, without having to recalculate the index.

To be more specific, if one or more tanker routes are abolished, then they just simply don't contribute to the index. If a tanker route is substituted by another one (it could be any tanker route), then the "old" route doesn't contribute to the index as described above, while the "new" route starts to contribute to the index directly the moment it is imported by the Drewry Shipping Consultants Ltd. Finally, if a tanker route is added, then it starts to contribute to the index directly the moment it is imported, as described above.

Another characteristic of the new index is the number of the tanker routes that are being used for its formation. The Drewry Shipping Consultants Ltd publishes approximately about forty routes every month, which is a representative sample of the global movement of crude oil.

Furthermore, the weighting of each route is based on the percentage of the ton-miles. Ton-miles have the advantage of combining the destination between two areas of a tanker route (calculated in miles) and the total amount of crude oil that is being carried (calculated in tones). The percentage of the ton-miles, derive from the division of the ton-miles of each route to the sum of the ton-miles of all the routes, as described above.

With the use of the ton-miles (and especially the percentage of the ton-miles), the real contribution of each route is reflected to the new index.

A great feature of the new index is the use of the Worldscale Index for the description of the changes to the rates, instead of the use of an absolute measure such as $/ton.

Comparison of the new index with the BDTI

Although both the BDTI and the new index are based on the same concept: that is a number of tanker routes each one having a weighting, the new index seems to describe the dirty tanker market in a better way.

The disadvantages of the BDTI that were mentioned in the beginning of this paper (limited routes, static index, arbitrary decided weightings), are surpassed by the new index that this paper introduces.

First of all the new index uses a much bigger number of tanker routes and so, more representative sample of the market, than the BDTI. To be more specific, the new index uses about forty tanker routes of crude oil every month, which cover the whole world, but also embraces all kinds of tanker ships (from the very small to the VLCC's and ULCC's). This comes in contradiction with the BDTI, which uses only ten routes as they were described previously.

One of the main features of the new index is the formation of the tanker routes. This formation can be easily changed, which means that a tanker route can be easily substituted or added or even deleted, with these changes having an immediate affect on the new index, as it was described above.

These changes don't take place frequently in the BDTI. The formation of the tanker routes of the BDTI is rather "tight", which means that it cannot be easily changed, and when this happens there must always be brought to (the change: substitution or addition or deletion) the first day the BDTI was published. For more information see the examples on index changes in pages 33 – 35 of the manual of panellists of the Baltic Exchange (PDF).

The main difference between the two indices (new index and BDTI) is the weighting factor of each tanker route of crude oil. The BDTI uses a weighting factor (WF_i) that is based more on an expert's judgment rather on a mathematical type, which gives the impression that is arbitrary decided.

In contradiction the new index uses the percentage of ton-miles, in order to describe the contribution of each tanker route to the new index. As it was described and above the percentage of ton-miles of each route is calculated by the following type:

$$Percentage\ of\ ton-miles = \frac{ton-miles\ of\ each\ route}{ton-miles\ of\ all\ routes}$$

One major difference between the two is the kind of rate that they use, in order to calculate the height of their price. The Baltic Index (BDTI) uses the average of the rates for each tanker route of crude oil that the panellists (the shipbrokers that cooperate with the Baltic Exchange) estimate. These rates are expressed in $/ton.

In contradiction the new index is fully independent of any absolute measures such as $/ton that the BDTI uses. As mentioned and above the new index uses the Worldscale Index of each tanker route of crude oil, instead of the $/ton.

A diagrammatic comparison of the two indices is shown in Figure 2.

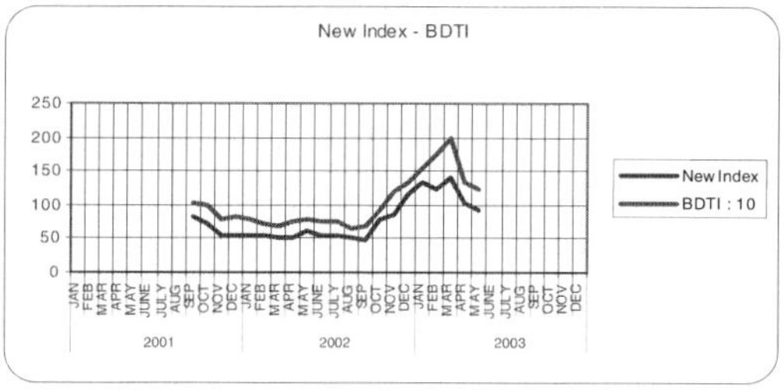

Figure 2. Diagrammatic Comparison of the new index and the BDTI

In Figure 2, we notice that in this small period of time, both indices are moving in almost parallel ways, which means first of all that the new index describes the tanker market of crude oil pretty well.

With a closer look we notice that the new index seems to be more sensitive than the BDTI, despite the fact that the BDTI is moving on higher ranks than the new index (BDTI moves between 750 and 2000 units, while the new index is moving between 50 and 150 units).

In Figure 2, we also notice a couple of inclinations between the two indices: one in the period of October – November 2002 and another on February 2003.

By the early days of September 2002, we notice that there is an intense accession of the market, due to the speculations of an incoming attack to Iraq by the US forces. This of course led to the deposition of enormous quantities of crude oil by all major country-importers of crude oil, and so to the rise of the market.

But in October 2002, something happened that all parties of market were afraid of. Terrorists attacked to the French tanker ship "Limburg" near the coast of Yemen, causing fire on board.

Another incident that occurred in November 2002 was the sinking of the tanker ship "Prestige" near the coast of Spain in the Atlantic Ocean. This incident caused a huge ecological disaster in the area, since thousands of tones of crude oil were spilled in the sea, making the public opinion unrest.

These two incidents seem to have had a temporary affect on the market that particular period, an affect that is reflected to the new index, as shown in Figure 2, but not in the BDTI.

The second inclination is noticed in February 2003. In this particular period the European Union, because of the sinking of the tanker ship "Prestige", took some hard measures that concern the tanker market of crude oil. To be more specific, the European Union decided to set a new deadline for the withdrawal of the single-hull tanker ships by 2010 (the previous deadline was in 2015) and furthermore to forbid temporarily the entrance of those ships (single-hull) to all EU ports.

In the same period there is also intense speculation in press (Reuters) that attacks may occur against tanker ships and crude oil terminals in the Arabian Gulf by members of Al Qaeda in reprisal of an incoming attack to Iraq by the US forces. This news is strengthened by the attack to the French tanker ship "Limburg" near the coast of Yemen in October 2002.

Taking into account all the above-mentioned information, these incidents seem to have had an affect on the market that particular month (February 2003), an affect that is reflected to the new index, as shown in Figure 2, but not in the BDTI.

In this particular period (February 2003) it becomes more obvious the correlation between the new index and the Tanker Earnings (especially with the VLCC / ULCC Earnings), as discussed in the next segment.

Correlation of the new index with the Tanker Earnings

By correlating the new index with the Earnings of the Aframax, Panamax, Suezmax and VLCC / ULCC tankers, we concluded that the degree of correlation is very high. In the following tables and diagrams the above-mentioned correlations are shown.

In these tables and diagrams we can notice that the bigger the capacity of the tanker ship is, the bigger is the degree of the correlation. So the biggest correlation is that between the new index and the VLCC / ULCC tankers Earnings, while the smallest (but still high) is between the new index and the Panamax tankers Earnings.

The BDTI has a high degree of correlation with these Earnings too, but it's not shown in the following tables and diagrams.

In Table 2 and Figure 3, series SERIE 124 and SERIE 125 stand for Aframax Average Earnings Index Built 1990 / 1991 and Aframax Average Earnings Index Modern respectively, while INDEX stand for the new index.

Table 2. Correlation between the new index and the Aframax earnings indices.

		INDEX	SERIE124	SERIE125
INDEX	Pearson Correlation	1.000	.878**	.876**
	Sig. (2-tailed)	.	.000	.000
	N	47	47	47
SERIE124	Pearson Correlation	.878**	1.000	1.000

	Sig. (2-tailed)	.000	.	.000
	N	47	47	47
SERIE125	Pearson Correlation	.876**	1.000	1.000
	Sig. (2-tailed)	.000	.000	.
	N	47	47	47

** Correlation is significant at the 0.01 level (2-tailed).

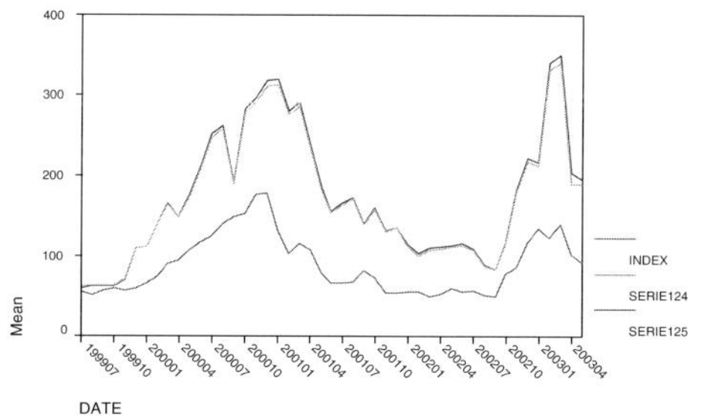

Figure 3. The new index and the Aframax earnings indices.

In Table 3 and Figure 4, series SERIE 103 and SERIE 104 stand for Suezmax Average Earnings Index Built 1990 / 1991 and Suezmax Average Earnings Index Modern respectively, while INDEX stand for the new index.

We notice that the correlation is bigger than the Aframax Earnings (0.878 and 0.876 for the Aframax, while for the Suezmax Earnings are 0.942 and 0.939).

Table 3. Correlation between the new index nd the Suezmax earnings indices.

		INDEX	SERIE103	SERIE104
INDEX	Pearson Correlation	1.000	.942**	.939**
	Sig. (2-tailed)	.	.000	.000
	N	47	47	47
SERIE103	Pearson Correlation	.942**	1.000	.998**
	Sig. (2-tailed)	.000	.	.000
	N	47	47	47
SERIE104	Pearson Correlation	.939**	.998**	1.000
	Sig. (2-tailed)	.000	.000	.
	N	47	47	47

** Correlation is significant at the 0.01 level (2-tailed).

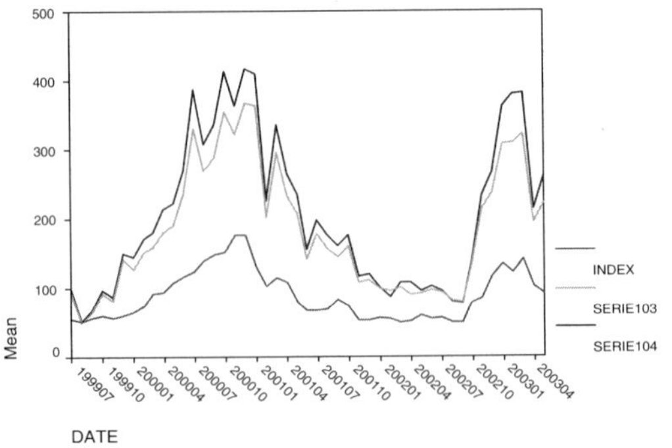

Figure 4. The new index and the Suezmax earnings indices.

In Table 4 and Figure 5, SERIE 59 stand for Dirty Products 50 K Average Earnings Index, while INDEX stand for the new index.

Notice that this is the smallest degree of correlation of all Earnings.

In Table 5 and Figure 6, series SERIE 72, SERIE 93 and SERIE 94 stand for VLCC Average Earnings Index Modern, VLCC Average Earnings Index Built 1970's and VLCC Average Earnings Index Built 1990 / 1991 respectively, while INDEX stand for the new index.

As shown in Figure 6, it is very impressive that the new index is moving almost parallel to the VLCC and ULCC Earnings Indices, especially in the period between September 2002 and March 2003 and taking into account Figure 2, we notice that the new index follows exactly the VLCC and ULCC Earnings Indices, while the BDTI doesn't.

Table 4. Correlation between the new index and the Panamax earnings indices.

		INDEX	SERIE59
INDEX	Pearson Correlation	1.000	.845**
	Sig. (2-tailed)	.	.000
	N	47	47
SERIE59	Pearson Correlation	.845**	1.000
	Sig. (2-tailed)	.000	.
	N	47	47

** Correlation is significant at the 0.01 level (2-tailed).

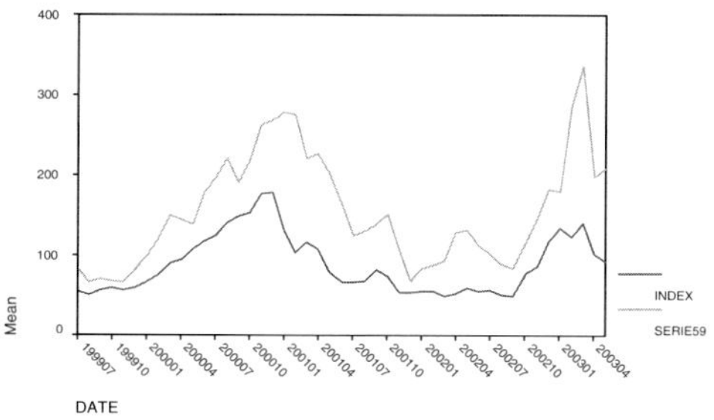

Figure 5. The new index and the Panamax earnings index.

Table 5. Correlation between the new index and the VLCC and ULCC earnings indices

		INDEX	SERIE72	SERIE93	SERIE94
INDEX	Pearson Correlation	1.000	$.980^{**}$	$.975^{**}$	$.980^{**}$
	Sig. (2-tailed)	.	.000	.000	.000
	N	47	47	47	47
SERIE 72	Pearson Correlation	$.980^{**}$	1.000	$.997^{**}$	1.000
	Sig. (2-tailed)	.000	.	.000	.000
	N	47	47	47	47
SERIE 93	Pearson Correlation	$.975^{**}$	$.997^{**}$	1.000	$.997^{**}$
	Sig. (2-tailed)	.000	.000	.	.000
	N	47	47	47	47
SERIE 94	Pearson Correlation	$.980^{**}$	1.000	$.997^{**}$	1.000
	Sig. (2-tailed)	.000	.000	.000	.
	N	47	47	47	47

** Correlation is significant at the 0.01 level (2-tailed).

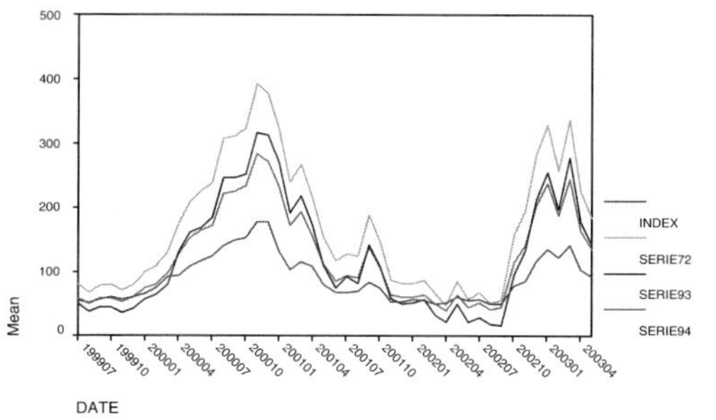

Figure 6. The new index and the VLCC and ULCC earnings indices.

As a conclusion, the tanker market of crude oil seems to be affected by the incidents of that period ("Limburg" terrorist attack, the "Prestige" sinking, EU strong measures, etc) and especially the VLCC's and ULCC's market. But these incidents aren't reflected to the BDTI, as shown in Figure 2, but only in the new index.

CONCLUSIONS AND SUGGESTIONS FOR FURTHER RESEARCH

Conclusions

With the import of the new index the tanker market of crude oil is described in a better way, in comparison with the BDTI of the Baltic Exchange.

The new index embraces all these factors that have an affect on the tanker market in contradiction with the BDTI, even though they are both based on the same principals such as a number of tanker routes each one having a weighting (and a rate) to each index.

But the new index has some special features, which make it unique.

- First of all the new index uses a large number of tanker routes for its formation. The Drewry Shipping Consultants Ltd publishes approximately about forty routes every month, which is a representative sample of the global movement of crude oil, while the BDTI uses only ten.
- Secondly, the definition of the new index itself is based on the ton-miles. The ton-miles and especially the percentage of ton-miles determine the contribution of each tanker route to the Index. This comes to contradiction with the BDTI of the Baltic Exchange, in which the weighting factors of the tanker routes are arbitrary decided, since they are based on an expert's judgment. Ton-miles have the advantage of combining the destination between two areas of a tanker route (calculated in miles) and the total amount of crude oil that is being carried (calculated in tones) reflecting the real contribution of each route to the Index.
- The new index is what we call a "dynamic" index, which means that it doesn't depend on the changes that are made in the formation of the tanker routes. Any substitutes, or additions, or even deletions can take place, without having to recalculate the index. The new index is published every month with the contribution of those tanker routes that can really contribute in this particular moment. This of course doesn't occur in the BDTI, which is rather "static". The formation of the tanker routes of the BDTI can't be easily changed. It is very difficult to substitute, or add, or even delete a tanker route, because the new routes must meet the demands and criteria that the Baltic Exchange sets.
- By comparing the two indices (the new index and the BDTI), we assume that the new index describes the dirty tanker market satisfactory. Furthermore the new index seems to be more sensitive from the BDTI, and this is supported by FIGURE 2, in which incidents such as the "Limburg" terrorist attack, the "Prestige" sinking, EU strong measures, etc. are reflected to the new index, while they have no affect to the BDTI.
- The new index seems to be correlated with the Earnings of the Aframax, Panamax, Suezmax and especially of the VLCC / ULCC tankers. This is probably due to the fact that the larger tanker ships seem to contribute to the new index in a higher degree than the smaller ships (the ton-miles of the VLCC's and ULCC's seem to be bigger than those of the smaller ships). This of course can be noticed also in the formation of the tanker routes that the BDTI uses (most routes refer to VLCC's).

As a conclusion, the introduction of the new index describes the dirty tanker market in a better way, compared to the Baltic's BDTI, since it embraces all those factors that have an influence to the market. Furthermore it uses a much bigger and so, more representative sample of the market, which makes this new index more reliable than the BDTI.

Suggestions for further research

As described previously, the contribution to the new index of each route is based on the ton-miles or to be more specific to the percentage of ton-miles. Ton-miles are calculated by the multiplication of the destination between the two areas that the crude oil is being carried, with the total amount of the cargo. This destination (for each route) was based and decided on an expert's judgment.

Apparently the best way to calculate the new index would be each tanker ship to estimate its own ton-miles. This means that each tanker ship could multiply its cargo (in tones) with the destination between the two ports that the cargo is being transferred. The discretion among those tanker ships could be the same with the one that the Drewry Shipping Consultants Ltd uses, which is by the cargo that a tanker ship is carrying (for example 40 – 70 thousand tones or 100 – 160 thousand tones or 300 thousand tones +, etc. and see also Table 1) and by the areas (for example CAR to USES or ECM to USES or WA to FE, etc.).

By this discretion the new index could be calculated in the same way as it was described in the previous segments of this paper.

But the new index deals only with the tanker market of crude oil (dirty tanker market). What about the clean tanker market, which deals with the derivatives of crude oil such as benzine, etc?

The Drewry Shipping Consultants Ltd publishes each month (along with the "Dirty Spot Rates") the "Clean Spot Rates" table, which is shown in Table 6.

The above table (TABLE 6) is the same as the "Dirty Spot Rates" table, but refers to the derivatives of crude oil. We could easily calculate a new index for this market (clean), but unfortunately the destinations between the areas that are described in TABLE 6 could not be estimated due to lack of information for these tanker routes of the clean derivatives of crude oil.

If those destinations could be estimated, then it would be very easy to calculate a new index for this clean tanker market, following of course the same procedure as in the dirty tanker market. This new index would certainly have the same advantages as the new index for the dirty tanker market, against of course to the Baltic Index (BCTI). Indicatively the new index for the clean tanker market would use approximately about fifteen tanker routes every month (see TABLE 6), while the BCTI uses only four for its publication.

Table 6. Clean spot rates

ISSUE JANUARY 1991 - INFORMATION FOR DECEMBER 1990												
Table 4. CLEAN SPOT RATES (Worldscale prevailing at time of fixing)												
ROUTE NO.	LOAD	DISCHARGE	CARGO (M.DWCT)	1989	1990						FIXTURES	
				DEC	JUL	AUG	SEP	OCT	NOV	DEC	NO.	AV.M. DWCT
	NWE	NWE	17- 30	-	225	190	214	-	265	295	1	28
		USES	25- 35	242	195	175	170	271	209	250	1	28

		MED	17- 30	367	200	210	371	354	306	350	1	20
		MED	25- 35	267	200	168	246	299	270	285	5	27
	MED	NWE	17- 30	276	179	175	254	320	277	290	2	28
		NWE	30- 40									
		USES	25- 35	230	184	170	195	265	229	-	-	-
		USES	17- 30	297	209	249	251	450	438	390	7	28
		EUR	17- 30	-	-	223	387	420	340	315	2	25
		USES	30- 40	304	198	230	308	445	435	342	4	33
	CAR	EUR	30- 40									
		USES	40- 70									
		EUR	40- 70									
	SING	EAST	25- 35	233	228	227	220	292	283	285	3	29
	WCUS	EAST	25- 35									
		FE	17- 30									
		EAST	25- 35	251	-	-	-	-	340	-	-	-
		INDIA	25- 35									
		EAST	30- 50	204	174	169	158	230	225	235	4	43
	AG	EUR	30- 50	220	-	-	-	-	-	-	-	-
		EAST	50- 80									
		INDIA	35- 50									
		JAPAN	50- 60									
		JAPAN	70- 80									
	RS	EAST	25- 35	270	197	-	-	-	-	241	2	30
		EUR	25- 50	220	240	173	-	233	290	-	-	-

REFERENCES

The Drewry monthly: 1981 – 2003 issues

Manual of panellists 2002, Baltic Exchange

Shasi N. Kumar, *Tanker Transportation*, Maine Maritime Academy.

Peyton Z. Peebles, Jr., Probability, *Random Variables and Random Signal Principles*, McGraw-Hill.

Z. Zannetos (1964): The theory of oil tank ships rates, MIT Press, Boston.

Kumar, S. And Hoffman, J. (2002). Globalization: the Maritime Nexus. In *The Handbook of Maritime Economics and Business* (C. Grammenos, ed.), pp 35-62. LLP, London.

Kumar, S. (1995). *Tanker Markets in the 21st Century: Competitive or Oligopolistic?* Paper presented at the 1st IAME Regional Conference held at MIT, Cambridge, MA on Dec. 15, 1995.

Glen, D. and Martin, B. (2002). The Tanker Market: Current Structure and Economic Analysis. In *The Handbook of Maritime Economics and Business* (C. Grammenos, ed.), pp 251-279. LLP, London.

Platou, R.S. (2002). Annual Reports. R.S. Platou, Oslo.
UNCTAD (2001). *Review of Maritime Transport, 2001*. United Nations, New York.

Transport Science and Technology
K.G. Goulias, editor

CHAPTER 35

THE ROLE OF LINER SHIPPING WITHIN AN INTERMODAL SYSTEM - THE PORT COMMUNITY CASE AND THE PORT AUTHORITY INVESTMENT PROBLEM

Maria Boile (corresponding author)
Assistant Professor and Director of Research and Education
Maritime Infrastructure Engineering and Management Program
Department of Civil and Environmental Engineering
Rutgers, The State University of New Jersey
623 Bowser Road, Piscataway, NJ 08854
Tel: (732) 445-7979, Fax: (732) 445-0577
Email: boile@rci.rutgers.edu

Sotiris Theofanis
Visiting Professor and Director of Program Development
Maritime Infrastructure Engineering and Management Program
Department of Civil and Environmental Engineering
Rutgers, The State University of New Jersey
623 Bowser Road, Piscataway, NJ 08854
Tel.: (732) 445-3257, Fax: (732) 445-0577
Email: stheofan@rci.rutgers.edu

and

Lazar Spasovic
Professor and Director, International Intermodal Transportation Center
School of Management
New Jersey Institute of Technology
287 Tiernan Hall, Newark, NJ 07102
Tel.: (973) 596-6420, Fax: (973) 596-6454
Email: spasovic@njit.edu

ABSTRACT

Substantial structural changes have happened in the liner shipping industry in the last decade. These structural changes have influenced significantly also port policy, port development and port operations, as well as the relationships between port players. This paper presents a mathematical programming model to analyze the port authority investment problem. This approach is used to facilitate the Port Authority's investment decisions on critical links, meaning links that are highly utilized and may require future investment in capacity expansion. The paper examines this problem within the context of an intermodal transportation system considering interactions among the players in this system. The paper introduces the criteria for comparing alternative investment strategies, defines the net social benefit and investment cost within the context of the intermodal system and presents the mathematical formulation of the Port Authority's investment problem. The mathematical approach presented in the paper can provide a sound basis to cope the modeling challenges associated with the structural changes of the liner shipping and the associated complexity of the players' interaction.

Key Words: Intermodal Transportation, Port Authority Investment Policy, Freight Network Modeling, Liner Shipping

INTRODUCTION

Economic growth, infrastructure improvements and technological developments over the last few decades resulted in the increasing role of freight transportation in the national and global economy. From the beginning of containerization to today's global markets, trends in the supply chain have shifted from supplier-driven, high-inventory transportation systems to consumer-driven, low-inventory, just-in-time services. Deregulation of the transportation industry has been driven by the desire to encourage greater price and service competition and to increase opportunities to develop multimodal and intermodal relationships among and within the various modes. Deregulation has resulted in better flexibility, efficiency, connectivity and continuity of service between economic markets, and higher quality of service at low rates.

These new developments in the area of freight transportation have intensified competition between freight transportation service providers in terms of lowering service fares and reducing operating costs. Setting competitive while profitable service fare and minimizing operating costs are vitally important issues for freight transportation service providers.

Increased competition is also crucial in ports rethinking how to bolster capacity and improve service quality, to maintain current and attract new business. The increased competition adds to the pressure for infrastructure investment. In response to these opportunities and challenges, most ports have started to or plan to redesign and reorganize their operations and have come up with a long-term investment plan. This is one of the issues giving rise to the port authority investment problem, which is discussed herein. The proposed framework may assist in forecasting the production, consumption, and link flow pattern on a freight network, evaluating the performance of the port terminal operation, and determining the best investment strategies.

BACKGROUNG

Port facilities consist of channels, berths, docks, and land, managed by a Port Authority, which is typically a public or quasi-public agency operating in the public interest. A Port Authority may have several terminals within its port complex and may operate them as is the case of the operating ports, or lease the land and facilities to private operators as is the case of the landlord ports. The terminal operators together with the other public and private transportation carriers that own and operate transportation facilities serving the port constitute a multimodal freight transportation system. Through this system move vehicles and containers carrying commodities from the shippers to the receivers located in spatially separated markets. Due to the existence of different transportation modes and related complicated interactions between components of the freight system within the port terminal, the analysis of port operations must be considered within the framework of a multimodal freight system.

A "traditional" set of players involved in the port operation include the steamship lines, railroads and motor carriers, brokers, shippers, forwarders, port terminal operators, a Port Authority, and the other regulatory agencies. The emergence of substantial structural changes in liner shipping through horizontal and vertical integration processes (formation of shipping alliances, mergers and acquisitions, co-operation agreements between ocean carriers regarding slot exchange, ocean carrier consortia and joint services, expansion into logistic activities etc.), the emergence of Global Port Operators, the various forms of financial and operating co-ordination between players (e.g. port authorities having financial stake in hinterland terminals, operating type port authorities have strong presence in certain countries mainly through companies owing several ports or landlord type port authorities having financial stake in port operation activities) made relationships far more complex. Nevertheless, following the generalized definition of Harker et al. (1986-a,1986-b), these players could be essentially combined into three major groups: the shippers, the carriers, and the regulatory agency.

The way these players interact is influenced by whether the decision they are making is a long-term or a short-term one and whether the market in which they operate can be described as a monopoly, oligopoly or perfect competition. The long-term decision, once committed, is difficult to change in the short term. Hence, the interaction between the long-term decision-maker and the short-term decision-maker has a sequential nature. The market conditions also influence the interaction by bestowing the decision-maker with different levels of market power under different market conditions. For example, the monopoly supplier or the monopoly consumer has strong control over the market price. On the contrary, under the market condition of perfect competition the supplier or the consumer acts as the price taker. Understanding the short-term or long-term nature of each player's decision and the market conditions is very important to the understanding of the interaction between and among the players.

The interaction between the three levels of players can be summarized as follows: Each shipper makes a commodity production and consumption decision based on knowledge of the pattern of the market prices in the spatially separated markets. Routing decisions are made based on knowledge of the pattern of the service charges set by the carriers and the travel time function between O-D pairs on the carriers' network. This is determined by the purely competitive assumption of shippers' market, which indicates that each individual shipper is a price taker. Each carrier makes pricing and routing decisions based on knowledge of the Port Authority's investment decision and forecast of the shippers' and competing carriers' reaction. The sequential nature of the interaction between the Port Authority and the carriers is determined by the fact that the Port Authority's investment decision is a long-term decision while the carriers' pricing and routing decision is a short-term decision. The shippers' and carriers' behaviors in turn influence the Port Authority's decision. How the shippers and the carriers react to the Port Authority's investment strategy determines the effectiveness of this strategy. The Port Authority makes its investment decision based on its forecast of the effects on carriers' pricing and routing decision and the shift of the production, consumption and flow pattern under different investment strategies.

The interactions discussed above can be formulated using a three-level model. The first level describes the behavior of the Port Authority in choosing the best investment strategy to maximize net social benefit. The second level describes the behavior of the carriers in choosing optimal service charge and routing pattern to maximize their profits. The third level formulates the behavior of the shippers, which is to determine the supply and demand of each commodity at each market (or centroid in transportation planning parlance) and the distribution pattern of each commodity on the network. The structure of the model is shown in Figure 1.

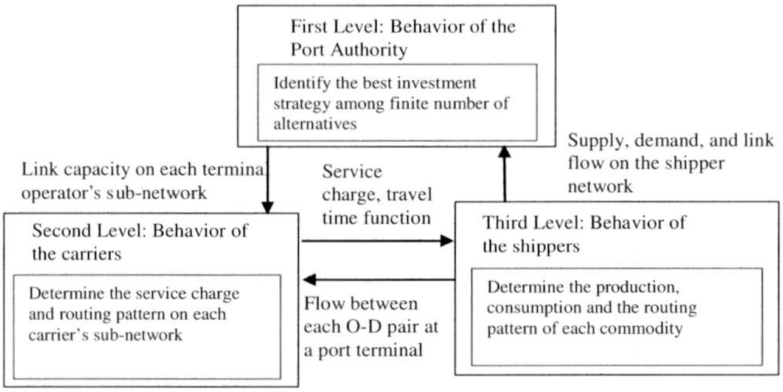

Figure 1 *Interaction between Players in the Intermodal Transportation System*

The behavioral questions raised here include the following:

<u>Behavior of the Port Authority</u>: Which investment strategy out of a finite set of alternatives should the Port Authority implement in order to maximize the net social benefit? What will be the impact of this strategy on the terminal operators and consequently the shippers?

<u>Behavior of the carriers</u>: What is the equilibrium service charge and routing pattern on each carrier's sub-network given the competitive pricing game among the carriers? What is the optimal set of service charge and routing pattern on each terminal sub-network and the resulting profit if the carriers choose to price collusively? What will be the impact of the competitive or collusive pricing on the shippers' decisions?

Behavior of the shippers: What are the optimal locations at which goods are produced and consumed and their optimal quantity and price? What is the equilibrium flow on the shipper network and the resulting cost?

A bi-level programming model to solve the Stackelberg equilibrium between oligopolistic private port terminal operators and shippers along with a solution algorithm for this problem have been presented in Boile and Wang (2002). This bi-level model is very briefly presented and is used in the formulation and solution of a Port Authority's investment problem discussed in here.

THE PORT AUTHORITY'S INVESTMENT PROBLEM

Boile and Wang (2002) provided a bi-level programming approach to solve the Stackelberg equilibrium between carriers and shippers. This approach is used to facilitate the Port Authority's investment decisions as it is demonstrated in the following sections.

Criteria Used in Comparing Alternative Investment Strategies

To facilitate investment decisions and to assist a Port Authority in evaluating alternative strategies, the following criteria have been suggested in the literature: (a) is the current infrastructure sufficient? Investing in the infrastructure of the port whose facilities are underutilized, would be wasteful. However, when the facility operates at or over capacity, investment is warranted; (b) is there an external economy or market inefficiency? In the port operation, in addition to the economic gain occurring directly at the port, there is a substantial spillover of economic benefit to other sectors or industries in the region; (c) is the incremental net social benefit brought by the investment greater than the incremental investment cost? The net social benefit is defined as the sum of the net benefits of all players in the port vicinity affected by the Port Authority's investment in the port infrastructure. The investment cost is the capital expense associated with an investment strategy. For an investment strategy to be feasible, the incremental net social benefit should exceed the incremental investment cost.

Net Social Benefit

The net social benefit (*NSB*) is an important measure of the worthiness of an investment strategy. To make accurate estimation of the net social benefit, the various players impacted by the Port Authority's investment in the port infrastructure need to be identified, and the net benefits of these players need to be estimated. The net social benefit will then be the sum of the net benefits of these players.

The carriers and the shippers are two major types of players impacted by the Port Authority's investment decision. The investment improves the terminal operators' operating cost and the shippers' generalized cost. In response to the improvement in these costs, the terminal operators as well as the other carriers and the shippers will adjust their behavior until a new equilibrium is attained. The bi-level program in Boile and Wang (2002) is used to predict this new Stackelberg equilibrium, based on which the terminal operators' net benefit (*TNB*) and the shippers' net benefit (*SNB*) associated with the Port Authority investment decision are estimated.

Terminal Operators' Net Benefit

Terminal operators are the producers of the port service. For the terminal operators, the monetary value of their net benefits is indicated by the total profits earned from their services. Let (R^u, e^u) denote the terminal operators' decision at the Stackelberg equilibrium under investment strategy u. Then, the terminal operators' net benefit under investment strategy u (TNB^u) is given as Eq. (1).

$$TNB^u = \sum_{t\in TM} Z_t(g_t(R_t^u, R_{-t}^u), R_t^u, e_t^u) = \sum_{t\in TM}\Big(\sum_{v\in V^t}\sum_{c\in C} g_{v,c}(R_t^u, R_{-t}^u) * R_{v,c}^u - \sum_{a\in A^t}\sum_{c\in C} AC_{a,c}(e_{a,c}^u) * e_{a,c}^u\Big) \quad \forall u \in U \qquad (1)$$

The service demand $g_{v,c}(R_t^u, R_{-t}^u)$ and the link flow $e_{a,c}^u$ are in units of flow per hour; TNB^u is in dollars per hour. Assuming that all terminals' profits occur in the port vicinity, 100% of TNB^u are included in the calculation of the net social benefit. As to the carriers other than the terminal operators (who, in this paper, are considered to be a special form of carriers), their profits may or may not occur in the port vicinity. Here, for the sake of simplification, the profits of the carriers other than the port terminal operators are not included in the analysis.

Shippers' Net Benefit

Shippers are the users of the terminal service. According to the economic theory (Wohl, 1984), their net benefit is the monetary value of their total willingness to pay minus the amount they do pay. The shippers can be either the consumers or the producers of the transported commodities. The shippers' net benefit is broken down into two sources: 1. consumer surplus (*CS*) (i.e. the area of triangle AEC at $\rho_{b,c}^u$ in Figure 2) from the consumption of the transported commodities. Consumer surplus is the consumer's total willingness to pay (i.e. the area of trapezoid AEHO in Figure 2) minus what the consumer actually pays for the transported commodities (i.e. the area of rectangle CEHO in Figure 2); and 2. producer surplus (*PS*) (the area of the triangle BCD at $\pi_{b,c}^u$ in Figure 2) from the production of the transported commodities. Producer surplus is the total sales revenue (i.e. the area of rectangle OCDG in Figure 2) minus the total production cost (i.e. the area of trapezoid BDGO in Figure 2).

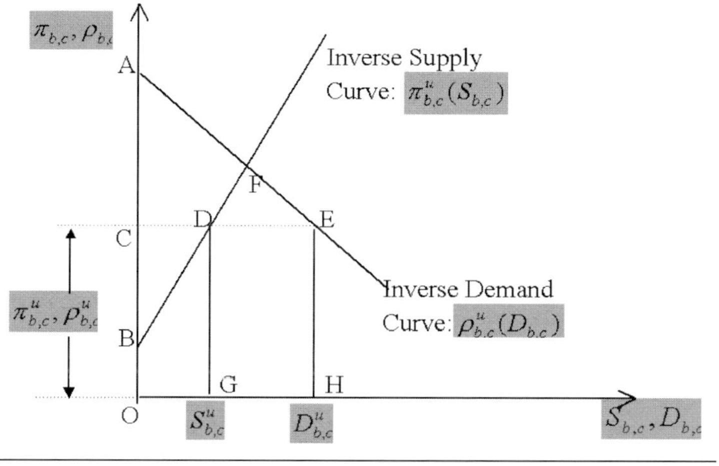

Figure 2 *Shippers' Net Benefit*

Figure 2 illustrates how to estimate various sources of the shippers' net benefit based on the spatial price equilibrium (SPE) solution under investment strategy u (S^u, f^u, D^u, π^u, ρ^u, GC^u). To illustrate this application, the inverse supply and inverse demand functions may be formulated as shown in Eqs. (2) and (3).

$$\pi^u_{b,c}(S_{b,c}) = (\gamma_{b,c} + \sum_{c'\in C, c'\neq c} \lambda_{b,c',c} * S^u_{b,c'}) + \lambda_{b,c,c} * S_{b,c} = \gamma^u_{b,c} + \lambda_{b,c,c} * S_{b,c}$$
$$\text{where } \gamma^u_{b,c} = (\gamma_{b,c} + \sum_{c'\in C, c'\neq c} \lambda_{b,c',c} * S^u_{b,c'}) \qquad \forall b \in CN, c \in C, u \in U \tag{2}$$

$$\rho^u_{b,c}(D_{b,c}) = (\alpha_{b,c} - \sum_{c'\in C, c'\neq c} \beta_{b,c',c} * D^u_{b,c'}) - \beta_{b,c,c} * D_{b,c} = \alpha^u_{b,c} - \beta_{b,c,c} * D_{b,c}$$
$$\text{where } \alpha^u_{b,c} = (\alpha_{b,c} - \sum_{c'\in C, c'\neq c} \beta_{b,c',c} * D^u_{b,c'}) \qquad \forall b \in CN, c \in C, u \in U \tag{3}$$

In Eqs. (2) and (3), $\pi^u_{b,c}(S_{b,c})$ is the inverse supply function of commodity c at centroid b given that the supply vector for the other commodities is $S^u_{b,-c}$. $\rho^u_{b,c}(D_{b,c})$ is the inverse demand function of commodity c at centroid b given that the demand vector for the other commodities is $D^u_{b,-c}$. The curves of $\pi^u_{b,c}(S_{b,c})$ and $\rho^u_{b,c}(D_{b,c})$ are plotted in Figure 2. Using these functions in Eqs. (2) and (3), the consumer surplus and the producer surplus at centroid b for commodity c can be calculated.

The consumer surplus at centroid b from the consumption of commodity c ($CS^u_{b,c}$) is calculated using the formula in Eq. (4).

$$CS^u_{b,c} = \int_0^{D^u_{b,c}} \rho^u_{b,c}(D^u_{b,c})dD^u_{b,c} - \rho^u_{b,c}(D^u_{b,c}) * D^u_{b,c} \quad \forall b \in CN, c \in C, u \in U \qquad (4)$$

The producer surplus at centroid *b* from the production of commodity *c* ($PS^u_{b,c}$) is calculated using the formula in Eq. (5).

$$PS^u_{b,c} = \pi^u_{b,c}(S^u_{b,c}) * S^u_{b,c} - \int_0^{S^u_{b,c}} \pi^u_{b,c}(S^u_{b,c})dS^u_{b,c} \quad \forall b \in CN, c \in C, u \in U \qquad (5)$$

Combining Eqs. (4) and (5), the shippers' net benefit at centroid *b* from the consumption and the production of commodity *c* ($SNB^u_{b,c}$) is obtained as follows:

$$SNB^u_{b,c} = CS^u_{b,c} + PS^u_{b,c} = \int_0^{D^u_{b,c}} \rho^u_{b,c}(D^u_{b,c})dD^u_{b,c} - \rho^u_{b,c}(D^u_{b,c}) * D^u_{b,c} + \pi^u_{b,c}(S^u_{b,c}) * S^u_{b,c} - \int_0^{S^u_{b,c}} \pi^u_{b,c}(S^u_{b,c})dS^u_{b,c}$$

$$\forall b \in CN, c \in C, u \in U \qquad (6)$$

The shippers' net benefit (SNB^u) is calculated as the sum of $SNB^u_{b,c}$ for each centroid and each commodity type as follows:

$$SNB^u = \sum_{b \in CN, c \in C} SNB^u_{b,c} = \sum_{b,c}\left(CS^u_{b,c} + PS^u_{b,c}\right) =$$

$$\sum_{b,c}\left(\int_0^{D^u_{b,c}} \rho^u_{b,c}(D^u_{b,c})dD^u_{b,c} - \rho^u_{b,c}(D^u_{b,c}) * D^u_{b,c} + \pi^u_{b,c}(S^u_{b,c}) * S^u_{b,c} - \int_0^{S^u_{b,c}} \pi^u_{b,c}(S^u_{b,c})dS^u_{b,c}\right)$$

$$= \sum_{b,c}\left(\int_0^{D^u_{b,c}} \rho^u_{b,c}(D^u_{b,c})dD^u_{b,c} - \int_0^{S^u_{b,c}} \pi^u_{b,c}(S^u_{b,c})dS^u_{b,c}\right) - \sum_{b,c}\left(\rho^u_{b,c}(D^u_{b,c}) * D^u_{b,c} - \pi^u_{b,c}(S^u_{b,c}) * S^u_{b,c}\right)$$

$$= \sum_{b,c}\left(\int_0^{D^u_{b,c}} \rho^u_{b,c}(D^u_{b,c})dD^u_{b,c} - \int_0^{S^u_{b,c}} \pi^u_{b,c}(S^u_{b,c})dS^u_{b,c}\right) - \sum_{b1,b2 \in CN,c}\left(GC^*_{b1,b2,c} * Q^u_{b1,b2,c}\right) \quad \forall u \in U \qquad (7)$$

$$= \sum_{b,c}\left(\int_0^{D^u_{b,c}} \rho^u_{b,c}(D^u_{b,c})dD^u_{b,c}\right) - \sum_{b,c}\left(\int_0^{S^u_{b,c}} \pi^u_{b,c}(S^u_{b,c})dS^u_{b,c}\right) - \sum_{l \in L,c}\left(GC^*_{l,c} * f^u_{f,c}\right)$$

In Eq. (7), the first element is the sum of consumers' willingness to pay for each commodity at each market. The second element is the sum of production cost for each commodity at each market. The third element is the total generalized transportation cost. The supply $S^u_{b,c}$, the demand $D^u_{b,c}$, and the link flow $f^u_{l,c}$ are all in units of flow per hour; SNB^u is in dollars per hour.

Adjustments to the Shippers' Net Benefit

It is important to note that two adjustments need to be made before including the shippers' net benefit in the calculation of the net social benefit. First, only part of the commodity transported via the port terminals is produced or consumed in the local region. The rest is just a passing through traffic to the designations outside the region and as such it doesn't contribute to the region's net social benefit. The portion of the shippers' net benefit, which directly contributes to the net social benefit of the local region, is called the local shippers' net benefit. A ratio (υ_c) is used to denote the passing through traffic as a percentage of the total freight of commodity *c*. Then, the local shippers' net benefit as a percentage of the total shippers' net benefit of commodity *c* is given as $1-\upsilon_c$.

Second, besides the shippers' net benefit directly related to the local production and consumption of these traded commodities, other economic sections in the port vicinity are involved and benefited in a meaningful way from the trade in these commodities and all associated manufacturing and services. To account for this external net benefit, a multiplier (ζ) is used to denote the ratio of external benefit to the localized shippers' net benefit.

Taking into account the passing through traffic and the external economy, the adjusted shippers' net benefit under investment strategy *u* ($ASNB^u$) is calculated in Eq. (8).

$$ASNB^u = (1+\zeta)*\sum_{c\in C}\left((1-\upsilon_c)*\sum_{b\in CN} SNB^u_{b,c}\right) = (1+\zeta)*\sum_{c\in C}\left((1-\upsilon_c)*\sum_{b\in CN}(CS^u_{b,c}+PS^u_{b,c})\right)\ \forall u \in U \quad (8)$$

Given the Stackelberg equilibrium (S^u, f^u, D^u, R^u, e^u) (Boile and Wang, 2002), the net social benefit under investment strategy *u* (NSB^u) is calculated as:

$$NSB^u(S^u, f^u, D^u, R^u, e^u) = TNB^u(R^u, e^u) + ASNB^u(S^u, f^u, D^u) \qquad \forall u \in U \quad (9)$$

The above discussion of the various sources of net social benefit is illustrated in Figure 3.

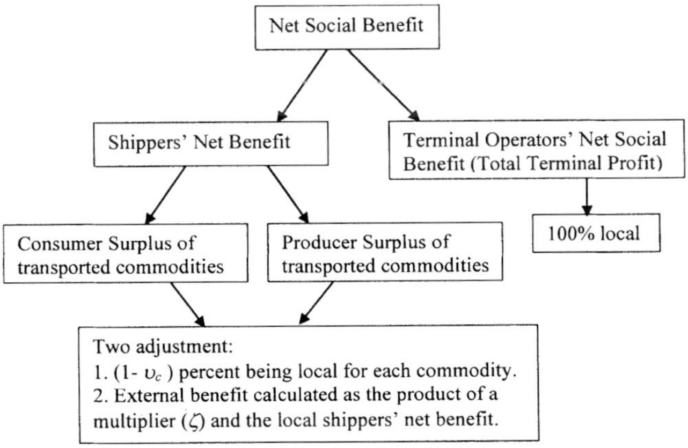

Figure 3 *Net Social Benefit*

Investment Cost

There is a finite number of alternative investment strategies. Associated with each investment strategy ($u \in U$) is a specific vector of capacity improvement pattern $\Delta \overline{E}^{u} = (\cdots, \Delta \overline{E}^{u}_{a}, \cdots)_{|A|}$. Under the do-nothing-strategy, $\Delta \overline{E}^{u} = 0$. The investment cost associated with an investment strategy must be defined and expressed in units that allow for comparison between different investment strategies, which vary in their service lives. For this purpose, investment costs are expressed as hourly costs using a method presented in Boile and Wang (2000).

The hourly investment cost on link a under investment strategy u is a function of the capacity improvement $\Delta \overline{E}^{u}_{a}$, the analysis period designated, the service life of the facility improved, and the discount rate. The total investment cost of the Port Authority is the sum of the investment costs on all improved links.

For the investment cost, a linear function similar to that shown in Yang and Meng (2000) is implemented. The flow dependent investment cost such as the maintenance cost is not considered. The investment cost on link a under investment strategy u (IC^{u}_{a}) is defined as follows:

$$IC^{u}_{a} = p2^{u}_{a} * (p1^{u}_{a} * \Delta \overline{E}^{u}_{a}) \qquad \forall a \in A, u \in U \qquad (10)$$

In Eq. (10), $p1^{u}_{a}$ is a parameter that represents cost of one additional unit of capacity. The value of $p1^{u}_{a}$ is determined by the type of facility represented by link a and the resources such as technology used for investment strategy u. In Eq. (10), $p1^{u}_{a} * \Delta \overline{E}^{u}_{a}$ represents the capital expense for the capacity improvement of $\Delta \overline{E}^{u}_{a}$ on link a. $p2^{u}_{a}$ is a factor that converts the capital expense into hourly investment cost. The value of $p2^{u}_{a}$ depends on the analysis period, the service life of this capital expense, and the discount rate. The method to calculate $p2^{u}_{a}$ is illustrated in Boile and Wang (2000).

Let $p^{u}_{a} = p1^{u}_{a} * p2^{u}_{a}$. Then, Eq. (10) can be restated as: $IC^{u}_{a} = p^{u}_{a} * \Delta \overline{E}^{u}_{a}$. The total hourly investment cost under investment strategy u (IC^{u}) can be calculated as the sum of IC^{u}_{a} over all links. Thereby:

$$IC^{u}(\Delta \overline{E}^{u}) = \sum_{a} IC^{u}_{a}(\Delta \overline{E}^{u}_{a}) = \sum_{a} p^{u}_{a} * \Delta \overline{E}^{u}_{a} \qquad \forall a \in A, u \in U \qquad (11)$$

Mathematical Formulation of the Port Authority's Investment Problem

The Port Authority aims to maximize the ratio between the incremental net social benefit brought about to the region through an investment, and the incremental investment cost. The incremental net social benefit through investment strategy u (ΔNSB^u) is calculated as follows:

$$\Delta NSB^u(S^u,f^u,D^u,R^u,e^u) = NSB^u(S^u,f^u,D^u,R^u,e^u) - NSB^0(S^0,f^0,D^0,R^0,e^0) \quad \forall\, u \in U \tag{12}$$

Where, $NSB^u(S^u,f^u,D^u,R^u,e^u)$ is the net social benefit under investment strategy u. $NSB^0(S^0,f^0,D^0,R^0,e^0)$ is the net social benefit under the do-nothing strategy.

The incremental investment cost is calculated as follows:

$$\Delta IC^u(\Delta\overline{E}^u) = IC^u(\Delta\overline{E}^u) - IC^0(\Delta\overline{E}^0) = IC^u(\Delta\overline{E}^u) \qquad \forall\, u \in U \tag{13}$$

Where, $IC^u(\Delta\overline{E}^u)$ is the hourly investment cost under investment strategy u. $IC^0(\Delta\overline{E}^0)$ is the hourly investment cost under the do-nothing strategy, which equals to zero since $\Delta\overline{E}^0=0$.

Combining Eq. (12) and Eq. (13), the investment problem for the Port Authority is defined and stated in Table 1.

Table 1 Port Authority's Investment Problem

$$\frac{\Delta NSB^{u^*}(S^{u^*},f^{u^*},D^{u^*},R^{u^*},e^{u^*})}{IC^{u^*}(\Delta\overline{E}^{u^*})} = \underset{u\in U}{Max}\left[\frac{\Delta NSB^u(S^u,f^u,D^u,R^u,e^u)}{IC^u(\Delta\overline{E}^u)}\right] \tag{14}$$

s.t. **P1** for competitive game or **P2** for collusive game

Where u^* is the most desirable investment strategy.

P1 competitive game

$$\sum_{t\in T}\left(-\nabla_{R_t}Z_t(g_t(R_V^*),R_t^*,e_t^*)(R_t-R_t^*)-\nabla_{e_t}Z_t(g_t(R_V^*),R_t^*,e_t^*)(e-e^*)\right)\geq 0 \quad \forall (R_V,e)\in KT$$

where $g_{v,c}(R_V)=\sum_l \xi_{l,v} * f_{l,c}(R_L)$

s.t. $\pi(S^*)\bullet(S-S^*)+GC(f^*,R_L^0)\bullet(f-f^*)-\rho(D^*)\bullet(D-D^*)\geq 0 \quad \forall (S,f,D)\in KS$

$R_L^0=\left[\xi_{l,v}\right]_{|L|*|V|} * R_V^*$.

or

P2 collusive game

$$\underset{R_{vl,c},e_{a,c}}{Max}:\sum_{t} Z_t(g_t(R_V),R_t,e_t)$$

s.t. $(R_V, e) \in KT$,

where $g_{v,c}(R_V)=\sum_{l}\xi_{l,v} * f_{l,c}(R_L)$

s.t. $\pi(S^*)\bullet(S-S^*)+GC(f^*,R_L^0)\bullet(f-f^*)-\rho(D^*)\bullet(D-D^*)\geq 0 \;\forall(S,f,D)\in KS$

$$R_L^0=\left[\xi_{l,v}\right]_{|L|*|V|} * R_V^* .$$

Problem 1 (P1) above is the formulation of the bi-level program formulating the Stackelberg game between carriers and shippers for the competitive game. Problem 2 (P2) is the bi-level program for the collusive game. In these formulations the shippers Spatial Price Equilibrium model (SPE) is combined with the carriers' pricing and routing problem. These problems are described in more detail in Boile and Wang (2002). According to Table 1 the port authority investment problem is solved subject to the bi-level shipper / carrier model. According to Figure 1 this is the first level behavioural problem which determines capacity limitations within the marine terminal's sub-network.

An application of the above method is presented in Boile and Wang (2000). This application focuses on solving the Stackelberg equilibrium between two oligopolistic private port terminal operators and several shippers. Various strategies the Port Authority can use to invest in terminals are evaluated. A numerical example demonstrates the capability of the bi-level programming method and the sensitivity analysis method – based heuristic algorithm in solving the Stackelberg equilibrium and the applicability of the model in facilitating the Port Authority's investment decision. Results of the analysis are used to verify the equilibrium conditions for both the shippers' spatial price equilibrium problem and the terminal operators' oligopolistic pricing and routing problem. Four proposed investment strategies are compared, by evaluating criteria from the perspective of various players.

FURTHER RESEARCH CHALLENGES

Since, as mentioned earlier, relationships between players are gradually becoming far more complex, there are several modeling challenges reflecting real life maritime industry relationships that need further attention, modeling effort and elaboration. The modeling approach, presented above, can serve as basis for various cases including, but not limited to the following:

- What is the behavior of an Operating Type Port Authority? (incorporating "authority's" and "carrier – marine terminal" characteristics, according to the modeling approach)

- What is the "spillover effect" for the marine terminal operator net benefit, in case of a Global (or regional) Port Operator, operating other terminals elsewhere?
- What is the problem formulation in case of Port Authorities that are involved in joint investment and operations with one of their terminal operators?
- What is the problem formulation in case of operations where one of the terminal operators is at the same time a shipping line? What is the formulation regarding services to the sister carrier and to third carriers?
- What is the problem formulation if competitive and collusive situations prevail at the same time between carriers or terminal operators?
- What is the problem formulation in case of two competing or collusive operating port authorities, under various circumstances for the other players?

CONCLUSIONS

This paper presents the mathematical formulation of the Port Authority investment problem. The methodology presented herein is unique in that it examines the port within the context of the intermodal transportation system considering the complex interactions among all players in this system and accounts not only for the direct impact of ports, but also for the effect on the inland transportation system. Theories that are well founded in the economic theory are used to formulate this problem. These theories underlie the market behavior and allow transportation planners, managers and decision makers to make predictions of the behavior of the market before implementing any short or long term decisions. Given the complexity of the liner shipping and port industries and their interrelationships, several modeling challenges, based on the approach presented in the paper need further elaboration.

REFERENCES

Boile, M.P. and Wang Y. (2000). A bi-level programming approach for the shipper – carrier network problem. *Working Paper* RU-2000-WP-BW01.

Boilé, M.P. and Wang Y. (2002). Intermodal freight network modeling. In: *The First International Conference on Transport Research in Greece*. Athens, Greece, February 21-22.

Harker, P. T. & Friesz, T. L. (1986a). Prediction of intercity freight flows, I: Theory. *Transportation Research,* **20B,** 139-153.

Harker, P. T. & Friesz, T. L. (1986b). Prediction of intercity freight flows, II: Mathematical formulation. *Transportation Research,* **20B,** 155-174.

Wohl, M. & Hendrickson, C. (1984). *Transportation Investment and Pricing Principles: An Introduction for Engineers, Planners and Economists*. Wiley-Interscience, New York.

Yang, H. & Meng, Q. (2000-a). Highway pricing and capacity choice in a road network under a Build-Operate-Transfer scheme. *Transportation Research,* **34A,** 207-222.

APPENDIX - MATHEMATICAL NOTATION

Parameters and Variables

S: Vector of supplies, $S = \left(\cdots, S_{b\in CN}, \cdots\right)^T_{|CN|}$.

$D_{b,c}$: Demand of commodity c at centroid b.

f: Vector of flows on all shipper links, $f = \left(\cdots; f_{l\in L}, \cdots\right)^T_{|L|}$.

R_l: Vector of commodity specific service charges on link l, $R_l = (\cdots, R_{l,c\in C}, \cdots)^T_{|C|}$.

$e_{a,c}$: Flow of commodity c on the link a.

e_t: Vector of link flows on the carrier t's sub-network, $e_t = \left(\cdots, e_{a\in A^t}, \cdots\right)^T_{|A^t|}$.

R_t: Vector of service charges between O-D pairs on the carrier t's sub-network, $R_t = \left(\cdots, R_{v\in V^t}, \cdots\right)^T_{|V^t|}$.

R_{-t}: Vector of service charges between O-D pairs on all other carriers' sub-networks except the carrier t's,

$\overline{E}_a$: Capacity on link a.

FUNCTIONS

$S_{b,c}(\square_b\square$ Supply function of commodity c at centroid b.

$D_{b,c}(\square_b\square$ Demand function of commodity c at centroid b.

$\square_{b,c}(S_b)$: Inverse supply of commodity c at centroid b.

$\gamma_{b,c}, \lambda_{b,c',c}$:Constants in the inverse supply function.

$\rho_{b,c}(D_b)$:Inverse dem. of commodity c at centroid b.

$\alpha_{b,c}, \beta_{b,c'c}$:Constants in the inverse demand function.

$Z_t(g_t(R_t, R_{-t}), R_t, e_t)$: Profit of carrier t as a fnct of the vector of service charges at this carrier's sub network (R_t) and the vector of service charges at the other carriers' sub-networks (R_{-t}) and the vector of link flows at this carrier's sub-network (e_t)

$g_{v,c}(R_V)$: Dem. Fnct. of commodity c between O-D pair v as a fnct of the vector of service charges, R_V.

$AC_{a,c}(e_{a,c})$: Avg. oper. cost fnct for commodity c on link a

Other Notation

U: Set of investment strategies available to the Port Authority, $u \in U$.

$\overline{E}^u_a$:Capacity on terminal link a under investment strategy u.

$\Delta\overline{E}^u_a$: Capacity impr. on terminal link a under investm. strategy u.

(S^u, f^u, D^u): Spatial price equil. solution under investm. strategy u.

(R^u, e^u): Equil. service charge and link flow under inv. strat. u.

υ_c : % of passing through freight of commodity c for the local region the port authority is located.

ζ : Economic multiplier.

TNB^u:The terminal operators' net benefit under invest. strategy u.

$CS^u_{b,c}$:The consumer surplus at centroid b from the consumption of commodity c under investment strategy u.

$PS^u_{b,c}$:The producer surplus at centroid b from the production of commodity c under investment strategy u.

$SNB^u_{b,c}$:The shippers' net benefit at centroid b from the consum. and the prod. of commodity c under investment strategy u.

SNB^u: The shippers' net benefit under investment strategy u.

$ASNB^u$:The shippers' net benefit adjusted to account for the passing through traffic and the external economy under invest. strat. u.

NSB^u:Net social benefit under investment strategy u,

$$NSB^u = TNB^u + ASNB^u \quad \forall u \in U.$$

I^u_a :Capital expense for the cap. impr. on link a under inv. strat. u.

PIC^u_a :PV of all capital expenses on link a under inv. strat. u

AIC^u_a :Annual Investment cost

IC^u_a : Hourly Investment cost

$$IC^u_a = p2^u_a * (p1^u_a * \Delta\overline{E}^u_a) \qquad \forall a \in A, u \in U.$$

$p1^u_a$, $p2^u_a$:Constants.

IC^u: Total hourly investment cost under investment strategy u,

$$IC^u = \sum_{a\in A} IC^u_a.$$

Transport Science and Technology
K.G. Goulias, editor

CHAPTER 36

INFRASTRUCTURE DEVELOPMENT TO SUPPORT THE FLOATING ACCOMMODATION PROGRAM OF THE ATHENS 2004 OLYMPIC GAMES – PROSPECTS AND CHALLENGES

Sotiris Theofanis (corresponding author)
Visiting Professor and Director of Program Development
CAIT/Maritime Infrastructure Engineering and Management Program
Department of Civil and Environmental Engineering
Rutgers, The State University of New Jersey
100 Brett Road, Piscataway, NJ 08854
Tel.: (732) 445-0357 ext. 110, Fax: (732) 445-0577
Email: stheofan@rci.rutgers.edu

Maria Boile
Assistant Professor and Director of Research and Education
CAIT/Maritime Infrastructure Engineering and Management Program
Department of Civil and Environmental Engineering
Rutgers, The State University of New Jersey
100 Brett Road, Piscataway, NJ 08854
Tel: (732) 445-0357 ext. 129, Fax: (732) 445-0577
Email: boile@rci.rutgers.edu

ABSTRACT

The Athens 2004 Olympic Games floating accommodation, a unique initiative held at the Port of Piraeus is presented. The program resulted in substantial re-organization and infrastructure development in the passenger and cruise sections of the port. Challenges associated with planning, project execution and event operations are analyzed along with a description of the port development projects pertinent to this challenging event.

Key Words: Cruise Terminal, Port Security, Port Infrastructure Development, Floating Accommodation

THE 2004 OLYMPIC GAMES AND THE PORT OF PIRAEUS

Greece is the birthplace of the ancient Olympic Games, which were held in Olympia every four years for almost 12 centuries. Early historic records date the first games to 776 B.C. and indicate that they were abolished after the 393 A.D. games. The games were revived in 1896 in Athens, which was fittingly chosen for the first modern Olympics. They have been held every four years in different cities since then, with an exception during the first and second world wars.

In 1997, Athens was chosen by the International Olympic Committee (IOC) to host the 2004 Games. The 28th Olympiad, a seventeen-day event, opened on August 13th and closed on August 29th with grand celebrations.

A Special Purpose Vehicle named "Athens 2004", having the legal status of a corporate body, was formed to accommodate the role of the Organizing Committee, to pursue all necessary preparatory actions, formulate all necessary structural arrangements and eventually manage the games.

In preparation for the Olympic Games, the Athens 2004 accommodation department was responsible for providing accommodation and related services to members of the IOC, heads of state, prime ministers and sovereigns, national Olympic committees, international federations, media representatives, sponsors and marketing partners. Over 17,000 hotel rooms were secured through the Olympic Hotel Agreement and 6,000 private houses out of a supply of 20,000 available houses were secured through the Private Home Rental Program. In addition, luxury cruise ships were to be berthed at the Port of Piraeus to provide top class accommodation capacity, through an ambitious specifically designed program called "the Floating Accommodation Program".

The city of Piraeus, one of the largest and most important cities in Greece and one of the busiest passenger ports in the world, had a vital role during the 2004 Olympics. In addition to being the home of several Olympic venues, Piraeus was given the responsibility of housing thousands of Olympic dignitaries on board cruise ships docked on the port.

The Athens Organizing Committee for the Olympic Games' (ATHOC's) floating hotels' accommodation program was the biggest ever in the history of the Olympic Games. This floating accommodation program required careful planning, preparatory work, execution and management. Several projects were to be implemented to provide the nucleus of the facilities and infrastructure needed for the Olympic accommodation, minimize the disruptions to the regular functions of the port, and ensure long lasting value to the port and the city in general. Moreover careful planning for the effective execution of the program was considered as a matter of vital importance for the overall success of the event.

PORT OF PIRAEUS – PAST AND PRESENT

The port of Piraeus is located about five miles to the southwest of Athens. The city of Piraeus has been inhabited since about 2,600 B.C. It was incorporated into the city-state of Athens in the ninth century B.C. and reached its golden age in the fifth century B.C. The "Long Walls" between Athens and Piraeus were started in 480 B.C. but they were destroyed during the Roman Empire. Following a very long period of decline, Piraeus started revitalizing with new buildings and modern factories after Athens became the Greek capital in 1834.

Today, Piraeus is Greece's third largest city in terms of population and its biggest and busiest port by far. Piraeus is at the crossroads of three continents, Europe, Asia and Africa, it is the southern gate of European Union in the Mediterranean Sea, and it is located at the point where major maritime routes meet.

The commercial port of Piraeus is one of the most important in the Mediterranean. With a throughput of more than 1.5 million TEUs (twenty-foot equivalent units, measuring container traffic in terms of standard twenty-foot containers) for the year 2004 Piraeus is placed among the top 50 ports in the world in terms of container traffic. A modern container terminal is in operation and is considered to be the biggest one in the Balkan area, serving both imports/exports as well as transshipment traffic.. The container terminal operates within a short distance from the center of Piraeus, but its location is physically separated from the passenger port, within adjacent bays.

The port of Piraeus is servicing 20 million passengers annually including domestic and international ferry and cruise ship traffic, and a commuting service. It is one of the busiest passenger ports in the world, and in terms of combined ferry and commuting traffic the largest one in Europe. It is an important destination for cruise ships in the Mediterranean and can accommodate the simultaneous berthing of 12 vessels, including even the largest cruise ships. The port operates passenger terminals with customs office, tourist police and duty free shops. Parking is available adjacent to the terminal and additional transportation services are provided between individual berthing places and the passenger terminals.

In addition to containerized cargo and passenger facilities, the port operates other dry bulk and general cargo facilities and two car terminals, which cover a significant portion of the country's car imports, as well as transshipment traffic.

The port of Piraeus is part of the Athens greater metropolitan area. The greater area of the city of Piraeus provides important support services for the port and the maritime industry, since several ship repairing, machine shops, shipping and cargo agencies and numerous other maritime related enterprises are located in close proximity to the port and they are continuously interfering with port facilities and port operations.

Because of the multiple important roles that the port plays and due to the functions it serves in relation with regional and state economic prosperity, maintaining smooth operations in the pre-Olympics period as well as during the Games was of utmost importance.

THE FLOATING ACCOMMODATION PROGRAM DEVELOPMENT

A contractual arrangement was signed between "Athens 2004" and Piraeus Port Authority S.A. to cater for the Floating Accommodation Program.. More specifically, PPA S.A. were to develop the berthing capacity needed and make it available for the berthing of cruise vessels during the course of the games; provide the necessary landside area for storage, transit terminal facilities, emergency services, etc.; and provide terminals, ancillary services and administrative building space to accommodate the Olympic cruise vessels and their passengers.

The goal was to provide for safe and convenient accommodation and transportation of the Olympic family and guests of the Athens 2004 Olympic Games, while eliminating any disruptions to the regular services, primarily those served by the central passenger port, and maintaining smooth operation of the port throughout the course of the games. Given that the games were to take place during the busiest period of the year for the central port and the supporting highway port access network, achieving the goals set for the port during the Olympics presented a major challenge.

Although a substantial berthing capacity for the Accommodation Program was provided, thus making berthing capacity for the domestic passenger coastal shipping traffic rather limited, the later was substantially upgraded. Vessels traveling to the eastern Aegean, Crete and Rhodes were moved from a cramped central port section, which was sequenced as part of the 3 km stretch of Olympic berthing space, to a brand new section. Shuttle bus service between sections and gates was provided by the PPA S.A. In addition, regular cruise ship lines that had been using the east side of the port would have to be relocated to nearby Keratsini, were supporting facilities would be provided, according to the Program's master planning.

Figure 1 below shows the Piraeus Central Port, according to the Olympics Accommodation master planning. Highlighted at the bottom of the figure is the east side of the port, the site of the floating accommodation program. The legends indicate the location of specific supporting facilities and services, security checkpoints and building locations.

A closer image of the berthing locations of Olympic accommodation vessels is shown in Figures 2 and 3. Figure 2 shows the existing cruise vessels locations adjacent to the Exhibition Center and the Piraeus Port Authority S.A. administration building. Figure 3 shows the new extension of the cruise terminal with the berthing locations and other new infrastructure developed to facilitate the Floating Accommodation Program.

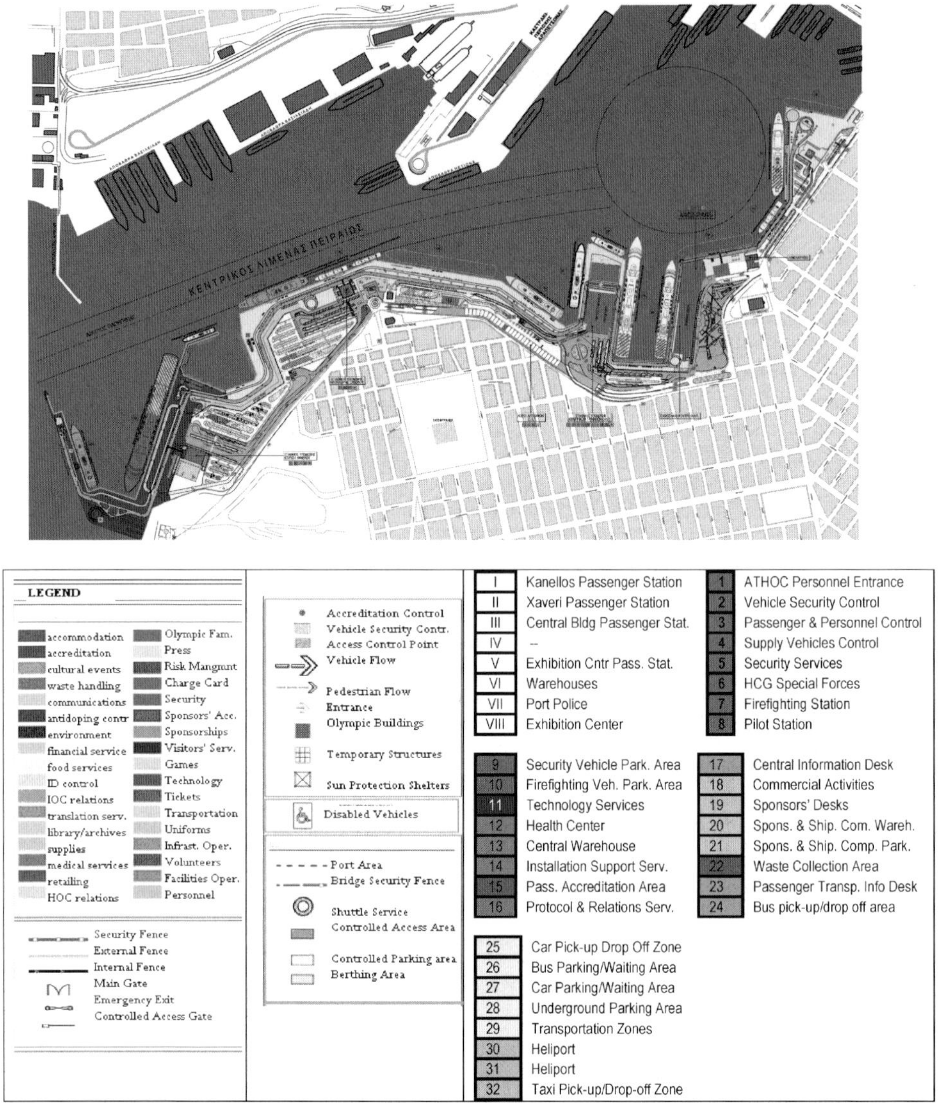

Figure 1 *Piraeus Central Port*

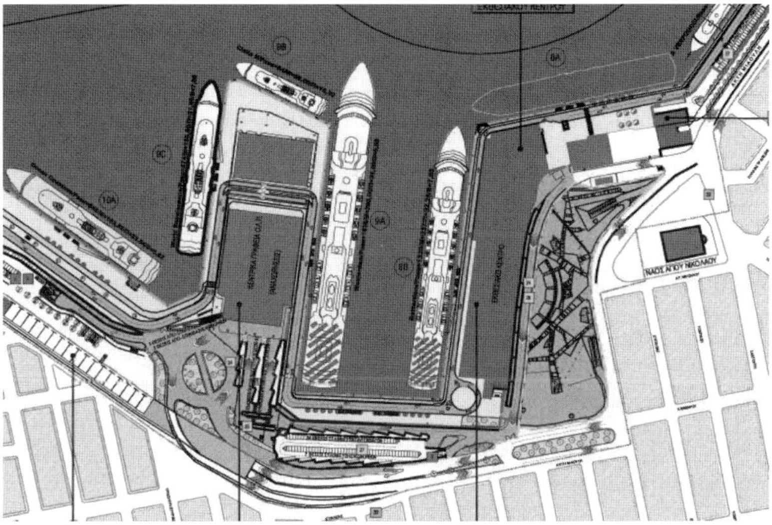

Figure 2 *Existing Cruise Vessel Berthing Locations*

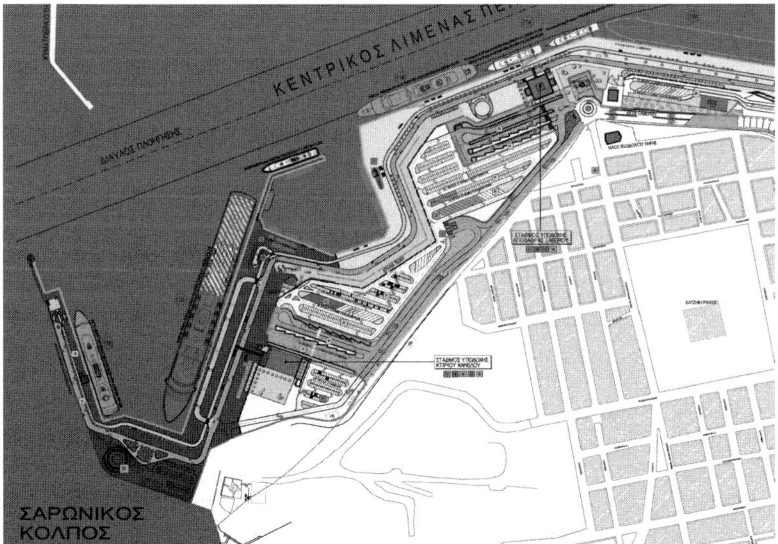

Figure 3 *New Extension of the Cruise Terminal*

Major challenges in the overall project were directly related to site selection criteria. These site-related challenges included the following:

(a) provide extended and not segmented berthing for 10 to 13 large cruise vessels, which translates to about 3 to 4.5 km of berths

(b) provide extended adjacent land side area to cater for parking and internal traffic, to accommodate a total planning number of 13,000 passengers.

(c) provide proximity to city center

(d) provide adequate buildings evenly scattered to cater for guests' services and administrative needs

(e) examine the possibility to enclose existing facilities, such as cruise facilities, to the project area

(f) cater for an effective water supply and sewage collection plant

(g) ensure mobility inside the port area

(h) avoid congestion at the gates, particularly during the peak traffic period of the day

(i) ensure physical site configuration suitable to apply an integrated security plan, which would meet and even exceed those requirements set by the IMO's ISPS Code

(j) provide for stepwise security control at the gates, for passengers, crews and supporting personnel by designing and implementing, both physically and operationally an appropriate zoning process, and

(k) ensure synergies with other port facilities for after event use

The project site included the existing cruise terminal, part of the passenger coastal shipping zone and an old breakbulk terminal, which was totally refurbished. The completion of the overall project doubled the cruise terminal berthing capacity, providing a length of approximately 3 km.

The specific projects that were to be undertaken by the PPA S.A. may be classified in three main categories, which are reviewed next.

Projects relating to Contractual Obligations

The first category includes projects for which PPA S.A. had a contractual obligation to the ATHOC and the Ministry of Mercantile Marine. These projects are the following:

1. Development and implementation of a traffic control system and new parking facilities, and improvement of land adjacent to the Olympic accommodations.

2. Development of an innovative water supply and sewage collection plant, which exceeded MARPOL 73/78 ANNEX IV requirements. The new system has high capacity in terms of water supplied to the vessels. In parallel to the water supply system channels, a network of fiber optics was created. In addition, a model for Mediterranean port standards sewage system which connects port wastewater collection facilities to the Athens off-shore wastewater treatment plant was developed.

3. Renovate, restore, and/or modify existing buildings

    a. The PPA S.A. administration building

    b. The Kanelos passenger terminal, an old warehouse circa-1920 (figure 4) and Palataki pavilion. Special attention had to be given to restore these historic for the area buildings, to maintaining their original character (figure 5)

    c. The Xaveris passenger terminal (figure 6)

4. Development of new port infrastructure for the berthing of cruise ships in the Palataki area. This project included the creation of 900 m of new quay infrastructure, which would be able to accommodate even the largest cruise vessels, with a 12-meter draft. Figure 7 shows part of this new infrastructure, which is the location where the Queen Mary 2 were to berth during the games

5. Additional developments in the port area of Drapetsona – Keratsini for the accommodation of regular cruise traffic during the course of the games.

Figure 4 *Kanelos Passenger Terminal*

Additional Supporting Projects

In addition to the projects under contractual obligation with ATHOC and the Ministry of Mercantile Marine, the PPA S.A. implemented a series of other projects, which would directly assist the floating accommodation program. These projects include:

1. Development of an underground parking garage under Merkouri square, at the PPA S.A. Exhibition Center. The facility provided 750 new parking spaces
2. Development of a new firefighting station in the Palataki area of the central port
3. Development of a new pilot station in the Palataki area (figure 8)
4. Development of a new heliport

Within the scope of the above projects, various older warehouse buildings were demolished and new paving, landscaping, and pedestrian walkway development projects were undertaken. In addition, part of the ancient Long Wall was restored at that site.

Figure 5 *Palataki Historic Building*

Figure 6 *Xaveris Passenger Terminal*

Other Projects

Other projects undertaken by the PPA S.A. and which could support the floating accommodation program include the following:

1. Completion of a ring road and connection of the central port to a national roadway through the Shisto – Skaramaga avenue.

2. Development of new car terminal facilities in Keratsini and transfer of cargo activity from the central port to this area

Security

Security in the Olympic zone of the port was the responsibility of the Olympic Games Security Division, which in collaboration with the Ministry of Public Order and Athens 2004 developed a security plan for the port of Piraeus. A consortium of private companies led by SAIC was involved in the implementation of the security system plan,the development of the necessary infrastructure and the installation of the necessary equipment.

To accommodate the port security plan several projects were undertaken. Motion sensors and cameras were installed along the reinforced steel gates and the high fences of the port area; a closed-circuit television system was installed within the port; coast guard vessels and divers were patrolling the port basin and the adjacent seaside access area (Fig. 7) and underwater sonar equipment and sensors were used to prevent any attack from the sea; elite commandos and soldiers were guarding the area; radiological, chemical and biological materials detectors and X-ray machines were used. Access to the Olympic zone of the port was controlled on a 24 hour basis.

Operating the security infrastructure and combining that to the landside traffic control for such a major importance event presented a great challenge and provided a significant know how to those involved in the operation. Handling such large scale passenger traffic , with daily peaks associated with the Olympics events, in a Cruise port presented a unique experience worldwide.

***Figure** 7 The Floating Accommodation Program in operation*

Figure 8 *The New Pilot Station*

SOME FACTUAL DATA FOR THE PROGRAM

During the 2004 Olympics, eight cruise ships were finally berthed in the port of Piraeus. The total capacity of the ships was 4500 cabins, or 9000 passengers. Among the eight cruise ships were the world's largest and most technologically advanced luxury Queen Mary 2, and the Ocean Countess, World Renaissance, Silver Whisper, Olympia Explorer, Olympia Voyager, Olympia Countess, Oosterdam and Aidaura. Queen Mary 2 has an l.o.a. of 343 m. and accommodates 2,600 people in 1,310 cabins. A total of 16500 passengers and 4400 crew members were accommodated onboard the cruise ships during the Games period

In overall, the Floating Accommodation Program was considered to be highly successful from service, security and operation point of view, despite the fact that berthing and vessel capacity were rather underutilized, since accommodation needs were proved to be overestimated.

CONCLUSIONS

The Floating Accommodation Program of the Athens 2004 Olympic Games, held in Piraeus, represented a unique in scale and features floating accommodation initiative.
The Program led to a substantial reorganization of the passenger port of Piraeus and particularly of the Cruise sector of the port and resulted in major additions in infrastructure and equipment regarding berthing capacity, supporting land area, terminal buildings, supporting facilities, utilities and security infrastructure.

Planning, implementation of infrastructure projects and equipment installation, as well as planning and execution of operations for the event created many challenges for those involved. These challenges included the need to avoid distortion of ordinary port functions and services, planning of new infrastructure and facilities in a way that they can be embedded in the existing port facilities, cater for after event added value and use of the investments realized for the Program, cater for adequate security in a multifunctional environment for a very sensitive and large accommodation population, ensure an effective personnel coordination platform for all the stages of the project, cope with mobility and transportation challenges with sharp daily peaks.